Random Functions and Hydrology

Rafael L. Bras
Massachusetts Institute of Technology

Ignacio Rodríguez-Iturbe
Texas A & M University

Dover Publications, Inc.
New York

Copyright

Copyright © 1985, 1993 by Rafael L. Bras and Ignacio Rodríguez-Iturbe.
All rights reserved under Pan American and International Copyright Conventions.

Published in Canada by General Publishing Company, Ltd., 30 Lesmill Road, Don
Mills, Toronto, Ontario.
Published in the United Kingdom by Constable and Company, Ltd., 3 The Lanchesters,
162–164 Fulham Palace Road, London W6 9ER.

Bibliographical Note

This Dover edition, first published in 1993, is an unabridged and corrected republication
of the work first published by the Addison-Wesley Publishing Company, Reading, Massachusetts, in 1985.

Library of Congress Cataloging-in-Publication Data

Bras, Rafael L.
    Random functions and hydrology / Rafael L. Bras, Ignacio Rodríguez-Iturbe.
        p.   cm.
    Originally published: Reading, Mass. : Addison-Wesley, c1985.
    Includes bibliographical references and index.
    ISBN 0-486-67626-9 (pbk.)
    1. Hydrology—Mathematical models.   2. Hydrological forecasting.
I. Rodríguez-Iturbe, Ignacio.   II. Title.
[GB656.2.M33B73   1993]
551.48'01'5118—dc20   93-29005
                                                                            CIP

Manufactured in the United States of America
Dover Publications, Inc., 31 East 2nd Street, Mineola, N.Y. 11501

Preface

The study of hydrology has developed unusually fast in the last twenty years, particularly in the last decade. What used to be a purely descriptive discipline taught as an appendix to hydraulic courses, with little research of its own, is now, in our opinion, the most exciting field in the water sciences. One of the most significant factors in this emergence is the full-fledged use of probabilistic and random processes techniques in the study of hydrologic problems. In this book, we have embraced this probabilistic point of view, which has proven its usefulness throughout all aspects of hydrology and is the key to the analysis and synthesis of hydrologic processes.

This book is unique in its ambitious coverage of topics including time-series analysis, optimal estimation, optimal interpolation (Kriging), frequency-domain analysis of signals, and linear systems theory. *Random Functions and Hydrology* synthesizes tools originally developed under the auspices of many different disciplines: electrical engineering, communications theory, aeronautics and astronautics, mechanical engineering, civil engineering, hydrology, mining, and geophysics. All tools and topics are related and have in common their applied mathematics origins. Nowhere else are they treated together as they are here. The problems of simulation, forecasting, and monitoring network design are complementary issues in many branches of science that we also discuss together rather than separately.

We hope to reach a wide audience—principally all those interested in random functions (processes and fields), i.e., mathematicians, engineers, and scientists. Mathematicians can benefit from the breadth of subjects and the definition of practical problems. Engineers and scientists can extrapolate to their own interests and achieve working knowledge of a wide variety of subjects.

As hydrologists we trust that we can convince more colleagues, practitioners, and academicians that the theory of random processes and fields is valid, useful in our profession, and here to stay. All examples, problems, and illustrations in the book are flavored by hydrology.

Criticisms of our efforts can be many, all probably reasonable, but we hope none serious. We have carefully selected the topics, taking into account the space and time constraints and our own biases and expertise. We believe the final work is the best possible for our goals of bringing together so wide a spectrum of subjects for such a diverse audience. We hope the majority of readers agree and that the minority excuse the limitations.

This book is intended to be a reference book and a textbook. It assumes familiarity with probability and statistics and a fair knowledge of random processes. The hydrologists among the readers will probably identify best with the examples chosen. As a textbook, it is intended for graduate students with the necessary background.

All chapters center on three basic hydrologic problems: sampling, simulation, and forecasting. Chapter 1 provides a refresher on concepts of random processes and defines hydrologic simulation, hydrologic forecasting, and network design under a very general but quantitative framework. Chapters 2 and 3 discuss hydrologic processes in the time domain; they deal with univariate and multivariate hydrologic time series. One aim in these chapters is to identify the mechanism that gives origin to the values of the series in time. Coupled with the model identification problem is the synthetic generation of hydrologic series, which has become a standard tool in the design of water projects. Also treated is the use of stochastic time-series models for hydrologic forecasting.

Chapter 4 is intended to familiarize the reader with the concepts of frequency-domain analyses and their use in hydrologic analysis and synthesis. Emphasis is given to the insights into the behavior of hydrologic processes that may be obtained through frequency-domain techniques. Long-term persistence, a problem of singular importance in hydrology, is the topic of Chapter 5; the techniques of the previous chapters are combined to analyze the Hurst effect and its influence on the design of reservoirs. Chapter 5 treats simulation models that incorporate characteristics of long-term persistence. Appendix A presents the details of one of these models, operational both at the univariate and multivariate levels.

The first five chapters deal with discrete processes in time and space; Chapter 6 addresses the analysis and synthesis of multidimensional hydrologic processes, varying in a continuous way over space and time. The problem of rainfall network design is studied in this chapter as the sampling of a random field.

Chapters 7 and 8 discuss the estimation of static and dynamic linear hydrologic systems. Among the static estimation tools, Chapter 7 covers Kriging techniques and their use in the estimation of stationary and nonstationary random fields. This tool has been heavily used in mining geostatistics, meteorology, and hydrology. Chapter 8 gives special emphasis to the problem of hydrologic forecasting using the Kalman filter to combine mathematical models with uncertain observations. This chapter assumes linearity and complete knowledge of model and filter parameters. Chapter 9 extends the techniques of Chapter 8 to the nonlinear case and also discusses the adaptive estimation of model dynamics and filter parameters.

According to the outline of the course being taught and the level of the students, one may vary the topics to be covered. The main aspects of all chapters have been the basis of graduate courses at the Massachusetts Institute of Technology and the Universidad Simón Bolivar although it is probably difficult to teach all the material in the book in a one-semester course. Chapters 1, 2, 3, 7.1–7.6, 8, and 9 may form a valid sequence with Chapters 4, 5, 6, and 7.6–7.11 as complementary material or another course. In some topics we have gone beyond what is normally covered in courses in the hope the book will also be useful to research hydrologists and engineers designing water resources projects.

We are indebted to many persons and institutions. The Massachusetts Institute of Technology, Ralph M. Parsons Laboratory, Department of Civil Engineering, and Universidad Simón Bolivar have provided a very stimulating environment. Particular acknowledgement goes to Drs. D. R. F. Harleman, P. S. Eagleson, D. H. Marks, J. C. Schaake, F. E. Perkins, J. R. Cordova, J. B. Valdes, M. Gonzalez, J. Mejía, and L. Perichi for being the best of colleagues and providing ideas, help and encouragement. We also want to thank Professors S. J. Burges, W. Yeh, P. Kitanidis, J. Salas, Dr. R. Budzianowski, and Dr. E. Johnson who reviewed and commented on the last versions of this book. It has been our privilege to work with outstanding graduate students from whom we have learned much; their contributions cannot be ever properly acknowledged. Particular thanks go to Mr. R. Colón, Mr. Siong Chua, Dr. Carlos Puente, Mr. Kevin Curry, Mr. Alonso Rhenals, Dr. Konstantine Georgakakos, Dr. Peter Kitanidis, Dr. Aristotelis Mantoglou, and Dr. Pedro Restrepo, all former students and collaborators on whose work we rely in many sections of the book.

We also wish to thank the Guggenheim Foundation, the Venezuelan Council for Scientific and Technological Research (CONICIT), and the Instituto Internacional de Estudios Avanzados for their support during the sabbatical leave which one of us (R.L.B.) spent at the Universidad Simón Bolivar. Many have been our research sponsors. We thank them all and in particular acknowledge the continuous support and confidence given to us by the U.S. National Weather Service (Hydrologic Research Laboratory), the National Science Foundation, the U.S. Agency of International Development (through M.I.T.'s Technology Adaptation Program), the National Science Foundation (Water Resources and Environmental Engineering), and the Venezuelan research agency CONICIT. Special thanks are due to Ms. Elaine Healy for the preparation of the many drafts of the manuscript. We thank Dr. Ying Fan for her diligent and conscientious help in preparing the book for its second printing.

Our parents, our children, and especially our wives have made possible this book and all other academic endeavors that we have pursued in our lives. The dedication of this book to them does not begin to pay our debt to them.

May 1984
Cambridge, Massachusetts   **R.L.B.**
Caracas, Venezuela   **I.R.-I.**

Contents

CHAPTER 1

Introduction

1.1 INTRODUCTION

This book will develop the concepts of time-series analysis and synthesis within the framework of three important hydrologic problems: data collection, simulation, and forecasting. These three problems will be studied with the techniques of: (1) random-process theory, (2) time-series analysis in the frequency and time domains, and (3) optimal filtering and forecasting in the time domain.

Obviously, this is an ambitious task; many books have been and will be written on these topics and their subtopics. Nevertheless, the techniques and examples herein should illustrate the latest advances in hydrologic-signals analysis and point out the many possible uses of the discussed topics.

We chose the three problems because they dominate the hydrologic literature that acknowledges natural processes as essentially random phenomena. This is the basic hypothesis of this work. All hydrologic processes (e.g., rainfall, runoff, evaporation, infiltration) are random. Randomness and the applicability of random-process theory may be inherent in the structure of the process or may result from lack of knowledge or from the scale of observation.

Many arguments, mainly philosophical, exist to refute or justify the above statement. The techniques and philosophy in this book have proved their usefulness to us. The nonbeliever hopefully will be impressed by the power of the various techniques and therefore accept them.

1.2 REFRESHER ON CONCEPTS OF RANDOM PROCESSES

Simple random variables are defined by probability density functions (pdf). Given the pdf, $f(x)$, the probability of the random variable X taking a value between x_1 and x_2 is given by

$$P[x_1 < X < x_2] = \int_{x_1}^{x_2} f(x)\,dx. \tag{1.1}$$

Similarly, the cumulative density function (cdf) is defined by

$$P[X \le x_1] = F(x_1) = \int_{-\infty}^{x_1} f(x)\,dx. \tag{1.2}$$

Given the cdf, a random variable with the correct distribution can be obtained by solving the equation

$$x = F^{-1}(p), \tag{1.3}$$

where p is a uniformly distributed number between 0 and 1 corresponding to the probability of not exceeding x. Equation (1.3) is the basis for Monte Carlo simulation of independent random variables.

A random variable whose properties depend on an argument can be interpreted as a random process. The hydrologist usually deals with continuous processes with continuous or discrete arguments. The best example is streamflow at a point. Figure 1.1 represents a possible time series of river discharges. The discharge is a function of the argument time $x(t)$. A time series constitutes a possible realization of the random process. The infinite set of possible realizations is called the ensemble of the process. The probabilistic definition of the ensemble defines the random process. Figure 1.2 illustrates an ensemble of the random process $x(t)$.

A random process is fully described by defining the joint probability density function of the random process at all possible values of the argument. At any time t_1 (see Figs. 1.1 and 1.2) the random process takes a value $x(t_1)$, which is a random variable. As a random variable, $x(t_1)$ is described by the first-order pdf:

$$f(x(t_1)). \tag{1.4}$$

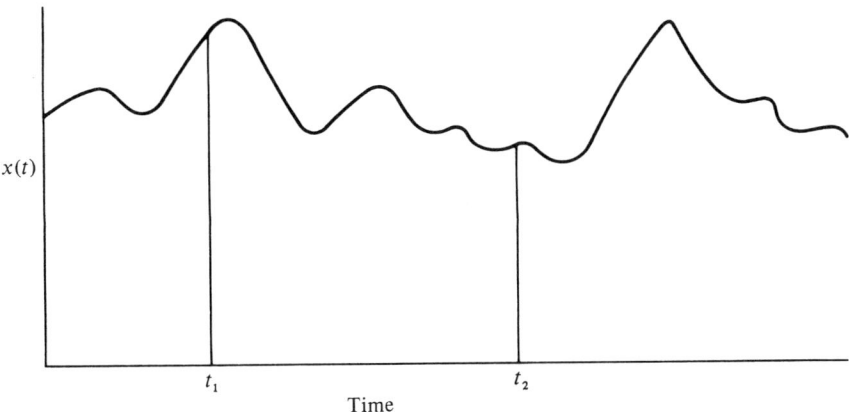

Figure 1.1 Example of a time series of river discharges.

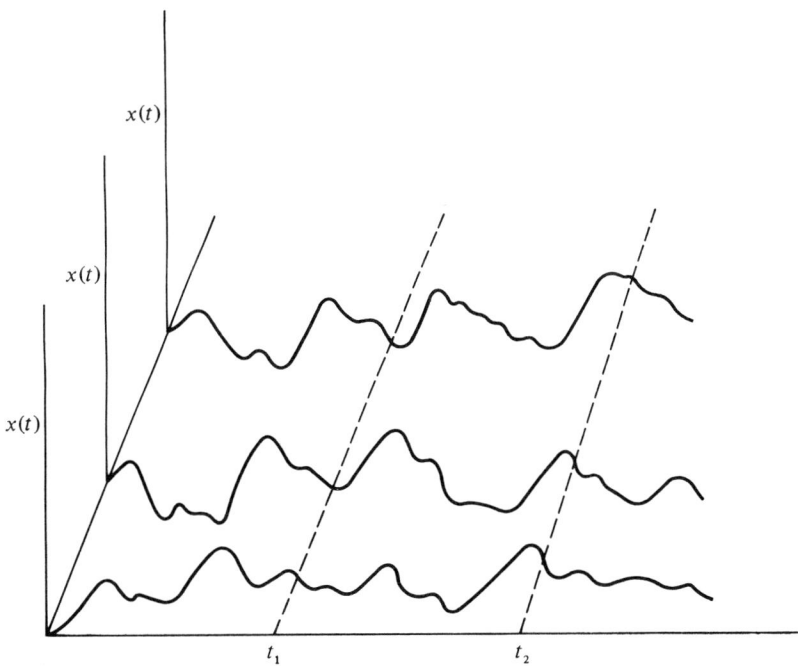

Figure 1.2 Concept of the ensemble of a random process.

Note that the pdf is a function of the value of the argument t_1. At any other time t_2 the pdf of $x(t_2)$ is generally defined as:

$$f(x(t_2)). \tag{1.5}$$

Given the first-order pdf, it is possible to obtain the mean of the random function at any time:

$$m(t) = \int_{-\infty}^{\infty} x(t)f(x(t))\,dx(t). \tag{1.6}$$

The variance of the process at any time is similarly obtained as a function of time:

$$\sigma^2(t) = \int_{-\infty}^{\infty} [x(t) - m(t)]^2 f(x(t))\,dx(t). \tag{1.7}$$

Note that the first two moments are defined in terms of a pdf that holds over the ensemble of possible realizations of the process.

It is possible to define higher-order probability density functions. For example, a second-order description of the process is provided by the joint distribution of the random variables $x(t_1)$ and $x(t_2)$ (the process evaluated at two different times). This joint distribution is represented by

$$f(x(t_1), x(t_2)). \tag{1.8}$$

The joint distribution is a function of the absolute values of the arguments t_1 and t_2. The second-order distribution can be used to define the covariance between $x(t_1)$ and $x(t_2)$:

$$\text{cov}(t_1, t_2) = \int_{-\infty}^{\infty} \int_{-\infty}^{\infty} [x(t_1) - m(t_1)][x(t_2) - m(t_2)]$$
$$\times f(x(t_1), x(t_2))\,dx(t_1)\,dx(t_2). \tag{1.9}$$

Therefore the correlation between $x(t_1)$ and $x(t_2)$ is

$$\rho(t_1, t_2) = \frac{\text{cov}(t_1, t_2)}{\sigma(t_1)\sigma(t_2)}. \tag{1.10}$$

Higher-order density functions can be defined similarly. In the limit, the highest-order distribution possible defines the random process. Only Gaussian and Markov processes are fully defined with second-order distributions.

In hydrology and other geophysical sciences it is hard, and many times impossible, to deal with the general definition of a random process given above. Several assumptions are made to ease the use of probability theory. The first and most common assumption is stationarity. This implies that the distributions of all orders are not a function of absolute argument (i.e., time) values.

In first-order distributions, this implies that

$$f(x(t_1)) = f(x(t_2)) = \cdots = f(x(t_n)) = f(x),\qquad(1.11)$$

which leads to the well-known result

$$m = \int_{-\infty}^{\infty} xf(x)\,dx,\qquad(1.12)$$

which shows that the mean is not a function of time. Similarly, the variance will also be independent of the process's argument.

In the second-order sense, stationarity implies that the joint distribution of two random variables defined by $x(t_1)$ and $x(t_2)$ is not a function of the absolute values of t_1 and t_2 but only of the lag, $\tau = t_2 - t_1$.

Defining t_1', t_2' such that $\tau = t_2' - t_1' = t_2 - t_1$, stationarity in the second-order pdf implies that

$$f(x(t_1'), x(t_2')) = f(x(t_1), x(t_2)).\qquad(1.13)$$

The above is true only if the joint pdf is a function of the equal argument τ. Therefore

$$f(x(t_1), x(t_2)) = f(x(t_1'), x(t_2')) = f(x_1, x_2, \tau).\qquad(1.14)$$

Stationarity of second order implies stationarity of the first order because

$$f(x(t_1)) = \int_{-\infty}^{\infty} f(x(t_1), x(t_2))\,dx(t_2)$$
$$= \int_{-\infty}^{\infty} f(x(t_1'), x(t_2'))\,dx(t_2') = f(x(t_1')).\qquad(1.15)$$

The above certainly is true if Eq. (1.14) is satisfied; the inverse is not true. Lower-order stationarity does not imply stationarity of the higher orders, but higher orders always dictate lower-order properties.

Because of the difficulties of determining and obtaining higher-order pdf's, hydrologists usually deal, or like to deal, with stationary processes of the first and second order. Full stationarity (in the strict sense) is hard to prove. Only in Gaussian and Markov processes does limited second-order stationarity lead to the full assumption.

Remember that the above assumptions lead to:

$$m(t) = m$$
$$\sigma^2(t) = \sigma^2$$
$$\operatorname{cov}(t_1, t_2) = \operatorname{cov}(\tau)\qquad(1.16)$$
$$\rho(t_1, t_2) = \rho(\tau)$$

The previously discussed probability density functions and their respective moments are defined over the ensemble of the random process, the collection of all possible realizations. Hydrologists and other scientists studying natural processes usually are able to observe only one realization of the random processes of interest. How then can the distributions, or for that matter, the moments of the random process be estimated from one realization? How do hydrologists bypass ensemble averaging? The answer to the above questions is the assumption of ergodicity. Ergodicity states that averaging over the ensemble is equivalent to averaging over a realization

$$m = \int_{-\infty}^{\infty} xf(x)\,dx = \lim_{T \to \infty} \frac{1}{T} \int_0^T x(t)\,dt \tag{1.17}$$

$$\sigma^2 = \int_{-\infty}^{\infty} (x - m)^2 f(x)\,dx = \lim_{T \to \infty} \frac{1}{T} \int_0^T [x(t) - m]^2 dt \tag{1.18}$$

$$\mathrm{cov}(\tau) = \int_{-\infty}^{\infty} \int_{-\infty}^{\infty} (x(t_1) - m)(x(t_2) - m) f(x_1, x_2, \tau)\,dx_1 dx_2$$

$$= \lim_{T \to \infty} \frac{1}{T} \int_0^T [x(t + \tau) - m][x(t) - m]\,dt. \tag{1.19}$$

Ergodicity is theoretically defined, but practically impossible to prove for any natural series. Hydrologists usually use estimates (finite summations) of Eqs. (1.17), (1.18), and (1.19) in obtaining sample values of the mean, variance, and covariance of a random process.

1.3 SIMULATION OF HYDROLOGIC PROCESSES

In the design and operation of water-resource systems, engineers have always recognized the variability and uncertainty of the hydrologic inputs. Rainfall, streamflow, evapotranspiration, and groundwater flow are all more or less unpredictable processes. A sequence of hydrologic events will rarely repeat itself. Faced with the unavoidable decision of designing reliable water-management structures, the traditional engineer relies on so-called frequency analysis. Assuming independence of events and using various parametric or nonparametric techniques, the probability of occurrence of critical events can be obtained. This probability, or mean recurrence period, helps select the events for which engineering works are designed.

The water-resource–system analyst extends the above concepts into mathematical programming formulations to handle uncertain inputs. Chance-constraint linear programming is used extensively in screening models. These models are devised to determine optimal configurations for basinwide waterworks developments and operations. They rely on the same concepts of recurrence and reliability as does traditional frequency analysis. Uncertainty about hydrologic inputs such as streamflow is handled by converting an essentially stochastic problem into a deterministic linear programming formulation, which is to be satisfied with some predetermined reliability or frequency of occurrence.

Stochastic linear programming and stochastic dynamic programming handle the variable inputs by optimizing an objective function. Finding the expected value of possible outcomes is the procedure by which we obtain a single answer to what is essentially a problem with an infinite number of solutions.

All the above screening procedures are useful in obtaining good approximations of water-resource–system behavior, but they may fail in several ways:

1. Stochastic mathematical programming solutions may correspond to either one critical hydrologic scenario or usually result in the assessment of some sort of mean behavior of the system.

2. In relation to the above, water-resource systems are studied under a limited set of hydrologic conditions.

3. Mathematical programming formulations often require considerable simplifications in system representation so that they can be tractable.

4. Solutions are static. Usually, mathematical programming handles a fixed configuration for a given system. Therefore during river-basin planning, reservoirs, diversions, operating rules, and irrigation sites must be studied in unchanging configurations. At best different configurations may be sequentially studied.

The above limitations can be bypassed by doing simulation exercises or by appropriately combining simulation and optimization. A river basin or any water-resource system can be simulated in considerable detail. Given a series of hydrologic inputs, the response of the system can be studied. Different development configurations may be simulated. Furthermore, the dynamic nature of the basin developments can be reproduced. Unfortunately, historical hydrologic inputs are usually limited, leading to sparse documentation of system response in different hydrologic scenarios.

Operational, or synthetic, hydrology is used to solve the limitation of historical hydrologic inputs. Recognizing that streamflows and other hydrologic time series are random processes, operational hydrology attempts to generate random sequences with the correct probabilistic behavior. These sequences can then be used in a series of Monte Carlo experiments directed toward defining the probabilistic behavior of the output. Monte Carlo experiments are essentially a numerical method for solving a derived distribution problem. This is illustrated in Fig. 1.3. Inputs (streamflow) with given probabilistic behavior represented by known probability laws (probability density functions) are used with a given deterministic model to produce output sequences. The output sequences can then be analyzed to determine their particular probability laws.

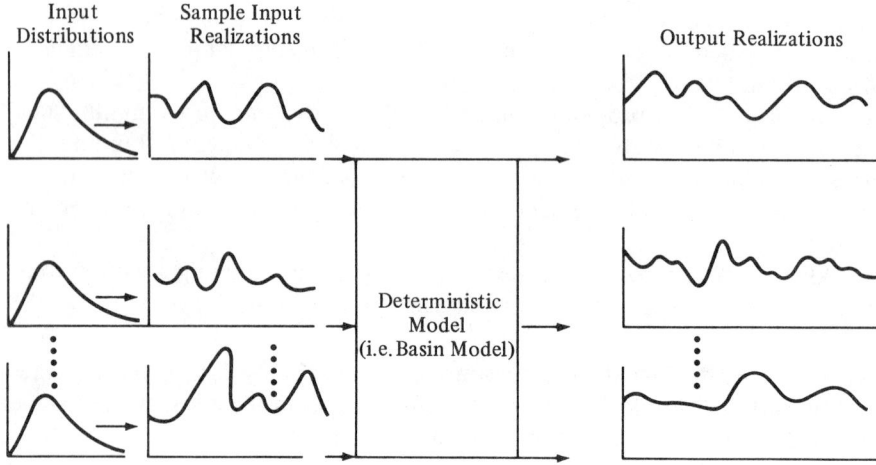

Figure 1.3   Concept of Monte Carlo experiments.

Given the input probability laws, Monte Carlo simulations can solve the derived distribution problem, which would otherwise be intractable. Nevertheless, it must be realized that the accuracy of the inference of the output behavior can never exceed the accuracy of the inference of the input probabilistic model. On the other hand, if you are willing to accept the input model you can observe outputs that would otherwise be unattainable with the available historical input data.

Observed realizations of a random process are usually called time series. Time-series analysis is the exercise of estimating the properties of the underlying processes that lead to an observed series. Operational, or synthetic, hydrology utilizes the results of time-series analysis to hypothesize mathematical models capable of producing realizations that would be statistically indistinguishable from the observed hydrologic series. Ideally, if the underlying model could be identified, the hydrologist could conduct an infinite number of experiments "repeating history" to any extent desired.

The mathematical equivalent of Eq. (1.3) for the generation of a stationary random process is to solve for $x(t_n)$ in the stochastic equation

$$p = F\big(x(t_n)|x(t_{n-1}), x(t_{n-2})...\big), \qquad (1.20)$$

where p is a uniformly distributed number (representing probability) between 0 and 1. $F(x(t_n)|x(t_{n-1}), x(t_{n-2})...)$ is the conditional distribution of the random variable of interest at time t_n, $x(t_n)$, conditional on the past values of the random variable. Assuming, for notational simplicity, a stationary Markovian behavior, the conditional probability density function is given by

$$f\big(x(t_2)|x(t_1)\big) = \frac{f\big(x(t_1), x(t_2)\big)}{f\big(x(t_1)\big)}. \qquad (1.21)$$

The conditional cumulative distribution is defined as

$$F\big(x(t_2)|x(t_1)\big) = \int_{-\infty}^{x} f\big(x(t_2)|x(t_1)\big)\,dx(t_2). \tag{1.22}$$

Operationally, hydrologists usually solve Eq. (1.20) only partially, by generating a random process preserving limited moments of the conditional distribution. The preservation of the complete distribution is usually impossible in the general case.

Many different models for the generation\of random hydrologic processes exist. Chapters 2, 3, and 5 will cover some of the basic formulations. Chapter 6 will extend the concept to generation of random fields (multidimensional processes dependent on vector arguments; e.g., rainfall over a region, dependent on time and spatial coordinates).

1.4   HYDROLOGIC FORECASTING

Equation (1.21) (still using the Markovian assumption for simplicity) also represents a generalized forecasting algorithm where the past state of the system, $x(t_1)$, is perfectly observed and where the system probability law, $f(x(t_1), x(t_2))$, is known at all times. The forecast is in the form of the full conditional density function of the random variable at time t_2. If the past contains considerable information of the future, it is expected that the conditional distribution of the random variable will exhibit less dispersion than the marginal density $f(x)$. This is illustrated in Fig. 1.4. The difference between simulation and forecasting, at this stage, is that the former involves sampling a random variable from the conditional distribution to generate a time series.

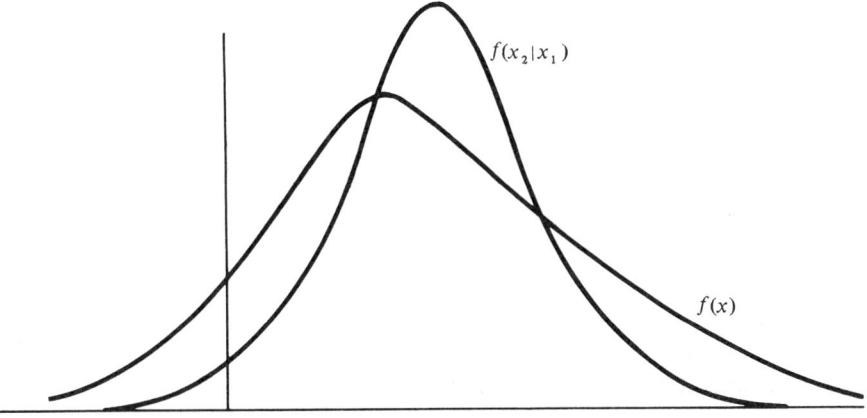

Figure 1.4   General forecasting problem, the conditional distribution.

Operationally the state of the system is rarely perfectly observed. The measuring device or methodology is usually modeled with a particular signal distribution, conditional on the true value of the state or natural variable of interest,

$$f(\mathbf{z}(t + \Delta t)|\mathbf{x}(t + \Delta t)), \qquad (1.23)$$

where now, for generality, $\mathbf{z}(t + \Delta t)$ is a random vector of observations at time $t + \Delta t$ and $\mathbf{x}(t + \Delta t)$ is the random state vector at time $t + \Delta t$; Δt is a time increment.

The forecast is now given by the conditional distribution of the state at time $t + \Delta t$, $\mathbf{x}(t + \Delta t)$, given an observation, $\mathbf{z}(t)$, which is uncertain. This conditional distribution is

$$f(\mathbf{x}(t + \Delta t)|\mathbf{z}(t)). \qquad (1.24)$$

Once a new observation at time $t + \Delta t$ is available, the marginal distribution of $\mathbf{x}(t + \Delta t)$ should be updated to incorporate the observation and acknowledge the lack of perfect knowledge of what actually occurred. Had the observation been perfect, the distribution of $\mathbf{x}(t + \Delta t)$ would become a spike at the observed value.

This updating or filtering step is accomplished using the well-known Bayes theorem, which yields the distribution of $\mathbf{x}(t + \Delta t)$ given the observation $\mathbf{z}(t + \Delta t)$,

$$f(\mathbf{x}(t + \Delta t)|\mathbf{z}(t + \Delta t))$$

$$= \frac{f(\mathbf{z}(t + \Delta t)|\mathbf{x}(t + \Delta t))f(\mathbf{x}(t + \Delta t)|\mathbf{z}(t))}{\int_{-\infty}^{\infty} f(\mathbf{z}(t + \Delta t)|\mathbf{x}(t + \Delta t))f(\mathbf{x}(t + \Delta t)|\mathbf{z}(t))\,d\mathbf{x}(t + \Delta t)},$$

$$(1.25)$$

where

$$f(\mathbf{x}(t + \Delta t)|\mathbf{z}(t)) = \int_{-\infty}^{\infty} f(\mathbf{x}(t + \Delta t)|\mathbf{x}(t))f(\mathbf{x}(t)|\mathbf{z}(t))\,d\mathbf{x}(t). \quad (1.26)$$

As will be seen in later chapters, the conditional distribution, Eq. (1.24), and the measuring system model, Eq. (1.23), usually result from a description of the hydrologic and measuring systems, respectively.

Forecasting in this book will then be interpreted as a prediction-filtering cycle. The prediction results from a known conditional distribution of the state given a past observation. The filtering updates and readies, for the next prediction, the distribution of the state given a concurrent observation.

The general forecasting algorithm previously outlined is rarely feasible. The full definition and integration of Eqs. (1.23) through (1.26) are usually impossible. Therefore it is common practice to forecast and update particular moments of the state probabilistic distribution. For example, one may choose to represent the distributions of interest by their means and variances. As will be seen later, the analytical sequential propagation and updating of means and variances is feasible when dealing with linear additive models. The algorithm for filtering and predicting first- and second-order statistics is usually called the Kalman filter. For normally distributed variables, the above formulation amounts to the exact solution to the filtering–forecasting cycle. This is due to the regenerative property of the normal distribution under additive linear models as well as to its full definition with first- and second-order statistics. The problem of estimation and forecasting using the Kalman filter will be presented in Chapters 8 and 9.

1.5 DATA-COLLECTION–NETWORK DESIGN

Hydrologic experience and research continuously point out that one of the major reasons for inaccurate hydrologic forecasting is the inaccuracy of the data base. Although errors in data may be due to many factors (i.e., instrument deficiencies, human failure, and spatial and temporal sampling frequency), conscious scientific design of measuring networks could greatly improve the quality of data.

The objectives of the network-design exercise are neither obvious nor easy. Clearly, the value of data lies in its ultimate uses. As pointed out by Moss and Tasker (1979):

> The use of the data often is comprised of two steps, each of which affects the optimality of the network. The first step is the computation of a set of parameters that describe the hydrologic phenomenon of interest. The raw data are used in this step. The parameters may be either statistical or deterministic. Secondly, the parameters are used in a planning or design scheme to make a water-resources-related decision. Although the data are only indirectly involved in this second step, the planning or design scheme exerts a major control on the effectiveness of the data network. For example, a poor design scheme cannot avail itself of the information that is contained in a data set, and thus the utility of collecting the data has been destroyed partially by the means in which they were used. Nevertheless, most network design studies to date have dealt only with the maximization of the efficiency of the network in defining specific hydrologic parameters.

We will not attempt to cover all aspects of network design, particularly the handling of multiple utilities of a set of collected data. Emphasis will be given to methods of quantifying the efficiency of a network in collecting data to yield accurate parameter estimates. Particular attention will be given to methodologies using optimal-estimation–theory concepts.

There are usually five variables to determine in the design of hydrologic data-collection networks:

1. the number of sampling stations,
2. the location of the stations,
3. the frequency, in time, of sampling,
4. the parameters and variables to sample,
5. the duration of the sampling program.

Decisions on the above will dictate the accuracy, measured by some prespecified criteria, of the sampling network.

In this book, the above problem is interpreted as one of estimation and optimal interpolation of a random field in time and space based on incomplete and usually noisy information. Clearly, if a random field could be perfectly sampled everywhere, there would be no data-related operational errors. Estimation of random processes and fields with emphasis on the monitoring network design is discussed in Chapters 4, 6, and 7.

A very simple example, familiar to all those who have studied statistics, will serve to illustrate the data-collection problem. Assume you have a population of independent Gaussian random variables with mean μ and variance σ^2 ($X \sim N(\mu, \sigma^2)$). The objective is to sample the population such that you can obtain the sample mean $\bar{X}$ with some degree of accuracy, quantified by confidence limits around the mean. Given that you obtain N samples and use the common unbiased estimator of the mean,

$$\bar{X} = \frac{1}{N} \sum_{i=1}^{N} x_i, \tag{1.27}$$

it is common knowledge that $\bar{X}$ is also a random variable normally distributed, with mean μ and variance

$$\sigma_{\bar{X}}^2 = \frac{\sigma^2}{N}. \tag{1.28}$$

Given the distribution (normal) and its parameters, the design problem amounts to varying the number of samples, N, until a desired level of accuracy (confidence interval) is achieved.

The problem increases in difficulty if the variance σ^2 is not known. Therefore an intervening estimation problem, obtaining the sample variance S^2, is needed. In such a case, it is again well known that $\bar{X}$ is a random variable with mean μ and variance S^2/N obeying a Student t distribution with $N-1$ degrees of freedom. The design problem remains the same, but as would be expected more samples (larger N) will be needed to achieve a given level of accuracy.

If the population of random variables was serially positively correlated with correlation function $\rho(\tau)$ in the limit (for large N), the problem would be similar to the first one seen (if σ^2 is known) but with

$$\sigma_{\bar{X}}^2 = \frac{\sigma^2}{N}\left[1 + 2\sum_{\tau=1}^{N-1}(1 - \tau/N)\rho(\tau)\right] \qquad (1.29)$$

(Vicens and Schaake, 1972).

A study of Eq. (1.29) will confirm that the correlation implies that a larger number of samples will be needed to achieve a given accuracy in the computation of the sample mean. Correlation implies that the marginal information content of each new sample is less, relative to independent processes.

To illustrate the influence of the objective in the design of the experiments imagine that the goal is now to estimate a realization of a random process. Say that $X_1 \dots X_N$ is a sequence of independent random variables. Observing $m(< N)$ of these variables is no help if the objective is to deduce the value of the $N - m$ unobserved values. With that objective and model, there is no choice but to make N observations. On the other hand, if the time series is correlated, estimates of the unobserved values are possible as well as statements of the accuracy of estimation. We will devote a considerable amount of time to designing experiments of the latter type.

For the sake of simplicity, this introduction has dealt with simple univariate examples. The main text will concentrate on processes that present considerably more difficulties and generally must be considered multidimensional random processes or fields.

CHAPTER 2

Generalized Univariate Time-Series Analysis in Hydrology

2.1 INTRODUCTION AND NOTATION

The records of the measure of some hydrologic process in time constitute a time series. One of the aims of time-series analysis is to identify the mechanism which gives origin to the different values in time. Identification is never perfect, however, because of the limited duration of the phenomenon. Therefore in constructing a model for a specific purpose, the time-series analyst attempts to preserve the properties relevant to the problems with which he or she is dealing. In order to do this, he must identify the parameters leading to such properties and then obtain the best possible estimators of these parameters, based on the history of the phenomenon.

Models of hydrologic time series, such as streamflow and rainfall at one point, have played a major role in the water-resources literature over the past 20 years. These models are constructed to facilitate the analysis and planning of various water uses through Monte Carlo simulation or to help in the real-time operation of water works by forecasting hydrologic events.

The synthetic generation of hydrologic series (Monte Carlo simulation) was popularized in the early 1960s by the work of the Harvard Water Program

(Maass et al., 1962). Synthetic streamflows are used in problems of reservoir design and of operation of river-basin water-resource systems. The aim of synthetic hydrologic simulation is to produce a set of equally likely traces (as long as needed) that are statistically indistinguishable from the historical data. These traces show many possible hydrologic conditions that do not explicitly appear in the historical record. Different designs and operational schemes can then be tested under the many different conditions contained in the synthetic record. The study of the Delaware River by Hufschmidt and Fiering (1966) has become the classic example of operational, synthetic, or stochastic hydrology. In 1914, Hazen suggested combining data from several recording stations in order to augment available information and improve the design of municipal water-supply reservoirs. Sudler (1927) used cards to generate independent sequences of streamflow to use in streamflow-regulating reservoirs. Hurst (1951) interpreted annual streamflows as a random sequence and studied the implication of that stochasticity in storage needs to control the Nile River in Egypt.

The use of stochastic time-series models for hydrologic forecasting has come about more recently. A good review of the literature can be found in the papers presented at the Workshop on Recent Developments in Real Time Forecasting/Control of Water Resource Systems (Wood, 1980). Of unique historical importance is the work of Jamieson et al. (1972a, 1972b, 1976); this is the first clear attempt to control a multipurpose river reservoir system using stochastic models to forecast hydrologic events.

One of the most complete popular textbooks on time-series analysis is that by G. E. Box and G. M. Jenkins (1976), where the general class of autoregressive integrated moving average models (ARIMA) is studied in detail. This chapter will be based mostly on their work. We will not attempt to cover every aspect of the theory; more than one book exists on the topic. We wish to introduce the reader to the concepts of generalized time-series–analysis theory and to provide an operational knowledge of the available tools. This should be enough for most practicing hydrologists and will be a good start for those interested in pursuing the topic further. Examples of the applications of ARIMA models in hydrology will be given.

Definitions

Understanding generalized time-series analysis (in the time domain) requires knowing some basic definitions.

A linear filter is an operator (linear) that converts a sequence of uncorrelated (usually normal) random variables a_t to a sequence of correlated variables Z_t. This is illustrated in Fig. 2.1. The linear operator representing the filter will be called (following Box and Jenkins' notation) $\psi(B)$. The sequence of uncorrelated discrete random variables a_t will be referred to as white noise.

The linear operator $\psi(B)$ is a polynomial in the backward shift operator B. If B operates on the random variable Z_t it yields the random variable Z_{t-1},

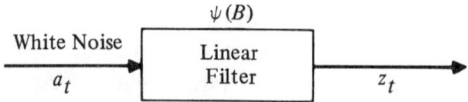

Figure 2.1 A linear filter acting on uncorrelated noise to create a time series (from Box and Jenkins, 1976).

essentially the process delayed by one discrete time unit. Mathematically,

$$BZ_t = Z_{t-1}.$$

Powers of B are logically defined,

$$B^2Z_t = BBZ_t = BZ_{t-1} = Z_{t-2}.$$

In general,

$$B^mZ_t = Z_{t-m}. \tag{2.1}$$

A polynomial in B operates in a similar manner; for example,

$$\left(a_0 + a_1B + a_2B^2\right)Z_t = a_0Z_t + a_1Z_{t-1} + a_2Z_{t-2}. \tag{2.2}$$

A forward shift operator F shifts the random process one time unit into the future,

$$FZ_t = Z_{t+1} \tag{2.3}$$

or

$$F^mZ_t = Z_{t+m}. \tag{2.4}$$

Premultiplying Eq. (2.3) by B,

$$BFZ_t = BZ_{t+1} = Z_t,$$

which implies $F = B^{-1}$.

The backwards difference operator ∇ is defined as

$$\nabla Z_t = Z_t - Z_{t-1}$$
$$= (1 - B)Z_t. \tag{2.5}$$

Powers of ∇ are also well defined,

$$
\begin{aligned}
\nabla^2 Z_t = \nabla \nabla Z_t &= \nabla(Z_t - Z_{t-1}) \\
&= \nabla Z_t - \nabla Z_{t-1} \\
&= Z_t - Z_{t-1} - Z_{t-1} + Z_{t-2} \\
&= Z_t - 2Z_{t-1} + Z_{t-2} \\
&= (1 - B)^2 Z_t.
\end{aligned}
\tag{2.6}
$$

The inverse of the backward difference operator ∇^{-1} is the backward summation operator S. S is defined as

$$
\begin{aligned}
SZ_t &= \sum_{j=0}^{\infty} Z_{t-j} \\
&= (1 + B + B^2 + \cdots) Z_t.
\end{aligned}
\tag{2.7}
$$

2.1.1 The General Stationary Linear Filter (Model)

Figure 2.1 represented the general linear model with operand $\psi(B)$. Assume from now on that the output Z_t is a stationary process that has been normalized by subtracting its constant mean from all its discrete values; Z_t is a zero-mean process. Therefore Z_t is related to the white noise a_t by a discrete convolution.

$$
\begin{aligned}
Z_t = a_t + \psi_1 a_{t-1} &+ \psi_2 a_{t-2} + \psi_3 a_{t-3} + \cdots \\
&= a_t + \sum_{j=1}^{\infty} \psi_j a_{t-j} \\
&= \sum_{j=0}^{\infty} \psi_j a_{t-j} \\
&= \psi(B) a_t,
\end{aligned}
\tag{2.8}
$$

where $\psi_0 = 1$ and

$$
\psi(B) = 1 + \psi_1 B + \psi_2 B^2 + \psi_3 B^3 + \cdots.
$$

The sequence of white noise a_t has zero mean and variance, σ_a^2; the autocovariance function of a_t is, then,

$$
\gamma_k = E[a_t a_{t+k}] = \begin{cases} \sigma_a^2 & k = 0 \\ 0 & k \neq 0 \end{cases},
\tag{2.9}
$$

which implies that the autocorrelation function ρ_k, takes a value of 1 at $k = 0$ and is zero elsewhere.

Using Eq. (2.8) and the properties of a_t, the covariance function of Z_t becomes,

$$\gamma_k = E[Z_t Z_{t+k}] = E\left[\sum_{j=0}^{\infty} \sum_{h=0}^{\infty} \psi_j \psi_h a_{t-j} a_{t+k-h}\right]$$

$$= \sigma_a^2 \sum_{j=0}^{\infty} \psi_j \psi_{j+k}.$$

(2.10)

For $k = 0$, the above equation yields the variance of the process

$$\gamma_0 = \sigma_z^2 = \sigma_a^2 \sum_{j=0}^{\infty} \psi_j^2.$$

(2.11)

Unless the infinite summation of Eq. (2.11) converges, the variance of the generalized linear filter would be infinite. Finite variance and stationarity will require convergence of the polynomial $\psi(B)$ for $|B| \leq 1$ (see Box and Jenkins, 1976). In the above statement, the operator B in Eq. (2.8) is interpreted as a dummy variable, which may take complex values. The stationarity condition then states that the infinite polynomial $\psi(B)$ must converge for all values of B within or on the unit circle, $|B| \leq 1$.

The linear stationary filter given in Eq. (2.8) is a moving average of infinite past white-noise sequences. It is possible to express the same process as an infinite autoregressive sequence,

$$Z_t = \pi_1 Z_{t-1} + \pi_2 Z_{t-2} + \cdots + a_t = \sum_{j=1}^{\infty} \pi_j Z_{t-j} + a_t$$

$$\text{or,} \quad \pi(B)Z_t = a_t$$

(2.12)

where $\pi(B) = 1 - \pi_1 B - \pi_2 B^2 - \cdots$.

To illustrate the equivalence between the moving average and autoregressive model, assume the following model,

$$Z_t = (1 - \theta B)a_t$$

$$= \psi(B)a_t,$$

(2.13)

which is stationary for any value of θ.

By iteratively substituting for the a_t terms, Eq. (2.13) can be expressed as

$$a_t = Z_t + \theta Z_{t-1} + \theta^2 Z_{t-2} + \cdots + \theta^n a_{t-n}.$$

(2.14)

If $|\theta| > 1$, Z_t depends on the past values with weights that increase with distance into the past. To avoid this situation, the invertibility condition, $|\theta| < 1$, is imposed. This allows the association of the present with the past in a more sensible manner. Such a condition is equivalent to stating that $(1 - \theta B)^{-1}$ can be expressed as a convergent infinite geometric series, $(1 + \theta B + \theta^2 B^2 + \theta^3 B^3 + \cdots)$.

By analogy to Eq. (2.12) the autoregressive version of the model in Eq. (2.13) has coefficients given by

$$\pi_j = -\theta^j \qquad \text{for } j = 1, \ldots, \infty. \tag{2.15}$$

Generalizing, a model is invertible if the polynomial in B,

$$\pi(B) = \psi^{-1}(B), \tag{2.16}$$

converges for all $|B| \leq 1$ (within or on the unit circle).

2.2 AUTOREGRESSIVE MODELS OF ORDER p, AR(p)

Autoregressive models of infinite order, such as that given in Eq. (2.12), are of little use in practice. A finite-order model is more useful,

$$\left(1 - \phi_1 B - \cdots - \phi_p B^p\right) Z_t = a_t$$
$$\phi(B) Z_t = a_t. \tag{2.17}$$

An example would be a lag-one model of the form

$$(1 - \phi_1 B) Z_t = a_t. \tag{2.18}$$

Because the polynomial $\phi(B)$ is finite, Eq. 2.18 is unconditionally invertible, a property held by all AR models of finite order p.

Stationarity can be investigated by reformulating the model in the form of Eq. (2.8),

$$Z_t = (1 - \phi_1 B)^{-1} a_t = \sum_{j=0}^{\infty} \phi_1^j a_{t-j}$$
$$= \left(1 + \phi_1 B + \phi_1^2 B^2 + \cdots\right) a_t = \psi(B) a_t, \tag{2.19}$$

which shows that a finite AR model is equivalent to an infinite moving average. We have seen that for the infinite-moving-average model to be stationary, $\psi(B)$ must converge for $|B| \leq 1$. For the AR(1) model given above, this implies that $|\phi_1| < 1$ is required to ensure stationarity. Note that the root of $1 - \phi_1 B = 0$ is $B = \phi_1^{-1}$. Since $|\phi_1| < 1$, then the stationarity condition is that the root of the polynomial in B, $\phi(B)$, must lie outside the unit circle. The above conclusion is valid for autoregressive models of any order.

The autocovariance function of $AR(p)$ is obtained by multiplying $Z_t = \phi_1 Z_{t-1} + \phi_2 Z_{t-2} + \cdots + \phi_p Z_{t-p} + a_t$ by Z_{t-k} and taking expected values:

$$
\begin{aligned}
E[Z_{t-k}Z_t] &= E\big[Z_{t-k}(\phi_1 Z_{t-1} + \phi_2 Z_{t-2} + \cdots + \phi_p Z_{t-p} + a_t)\big] \\
&= E\big[\phi_1 Z_{t-k}Z_{t-1} + \phi_2 Z_{t-k}Z_{t-2} + \cdots + \phi_p Z_{t-k}Z_{t-p} + Z_{t-k}a_t\big].
\end{aligned}
\tag{2.20}
$$

Using the lack of correlation between Z_{t-k} and a_t and invoking stationarity, the autocovariance at lag k, γ_k, becomes

$$
\gamma_k = \phi_1 \gamma_{k-1} + \phi_2 \gamma_{k-2} + \cdots + \phi_p \gamma_{k-p},
\tag{2.21}
$$

where $\gamma_{k-i} = E[Z_{t-k}Z_{t-q}]$.

Dividing by the stationary process variance, γ_0, yields

$$
\rho_k = \phi_1 \rho_{k-1} + \phi_2 \rho_{k-2} + \cdots + \phi_p \rho_{k-p}.
\tag{2.22}
$$

Therefore, the autocorrelation of an $AR(p)$ obeys the equation

$$
\phi(B)\rho_k = 0.
$$

Writing Eq. (2.22) for $k = 1, 2, \ldots, p$ yields the Yule–Walker system of equations

$$
\begin{aligned}
\rho_1 &= \phi_1 &&+ \phi_2\rho_1 &&+ \cdots + \phi_p\rho_{p-1} \\
\rho_2 &= \phi_1\rho_1 &&+ \phi_2 &&+ \cdots + \phi_p\rho_{p-2} \\
&\ \ \vdots &&\ \ \vdots &&\qquad\ \ \vdots \\
\rho_p &= \phi_1\rho_{p-1} &&+ \phi_2\rho_{p-2} &&+ \cdots + \phi_p
\end{aligned}
\tag{2.23}
$$

Note that in formulating the above system, the identities $\rho_0 = 1$ and $\rho_k = \rho_{-k}$ were used. The Yule–Walker equations can be used to estimate parameters of $AR(p)$ by substituting sample correlations, r_k, in Eq. (2.23) and solving for the ϕs. Given that sample correlations are not very stable estimates, particularly for high lags, care must be taken in using this parameter-estimation procedure. Note that in matrix form

$$
\boldsymbol{\Phi} = \boldsymbol{\Sigma}^{-1}\mathbf{R},
\tag{2.24}
$$

where

$$
\boldsymbol{\Phi} = \begin{bmatrix} \phi_1 \\ \vdots \\ \phi_p \end{bmatrix}, \qquad
\mathbf{R} = \begin{bmatrix} \rho_1 \\ \vdots \\ \rho_p \end{bmatrix}, \qquad
\boldsymbol{\Sigma} = \begin{bmatrix}
1 & \rho_1 & \rho_2 & \cdots & \rho_{p-1} \\
\rho_1 & 1 & \rho_1 & \cdots & \rho_{p-2} \\
\vdots & \vdots & \vdots & & \vdots \\
\rho_{p-1} & \rho_{p-2} & \rho_{p-3} & \cdots & 1
\end{bmatrix}.
$$

The variance of the AR(p) process is obtained from Eq. (2.20) for $k = 0$. Now, the expectation $E[Z_t a_t]$ will yield σ_a^2 and the variance is given by

$$\gamma_0 = \phi_1 \gamma_1 + \phi_2 \gamma_2 + \cdots + \phi_p \gamma_p + \sigma_a^2. \tag{2.25}$$

Dividing through by $\gamma_0 = \sigma_z^2$ and solving for σ_z^2 results in

$$\sigma_z^2 = \frac{\sigma_a^2}{1 - \phi_1 \rho_1 - \phi_2 \rho_2 - \cdots - \phi_p \rho_p}. \tag{2.26}$$

It is interesting to study the form of the correlation function ρ_k, which was seen to satisfy

$$\phi(B)\rho_k = 0. \tag{2.27}$$

The polynomial operator $\phi(B)$ can be expanded as

$$\phi(B) = \prod_{i=1}^{p} (1 - G_i B) \tag{2.28}$$

(Box and Jenkins, 1976).

Given Eq. (2.28), the general solution of ρ_k is

$$\rho_k = A_1 G_1^k + A_2 G_2^k + \cdots + A_p G_p^k, \tag{2.29}$$

where $G_1^{-1} \cdots G_p^{-1}$ are the roots of the polynomial (characteristic equation) $\phi(B)$. For stationarity, it was previously seen that $|G_i| < 1$. Following Box and Jenkins, if G_i is real and distinct, then Eq. (2.29) geometrically decays to zero as k increases. If a pair of roots G_i, G_j is complex, they contribute a term

$$d^k \sin(2\pi f k + F)$$

to ρ_k, which is a damped sine wave as k increases.

Before moving ahead, it is important to point out that the AR(p) model has $p + 2$ parameters to be estimated $\{\phi_1, \ldots, \phi_p, \mu, \text{ and } \sigma_a^2\}$.

2.2.1   First-Order Autoregressive Model, AR(1)

The lag-one autoregressive model, or Markov model, is

$$Z_t = \phi_1 Z_{t-1} + a_t$$

or

$$\begin{aligned} (1 - \phi_1 B) Z_t &= a_t \\ \phi(B) Z_t &= a_t. \end{aligned} \tag{2.30}$$

From the Yule–Walker equations,

$$\rho_1 = \phi_1 \rho_0$$
$$= \phi_1$$
$$\rho_2 = \phi_1 \rho_1 \tag{2.31}$$
$$= \rho_1^2$$

or in general

$$\rho_k = \rho_1^k = \phi_1^k. \tag{2.32}$$

From Eq. (2.26), the variance is

$$\sigma_z^2 = \frac{\sigma_a^2}{1 - \phi_1 \rho_1} = \frac{\sigma_a^2}{1 - \rho_1^2}. \tag{2.33}$$

Stationarity conditions require $|\phi_1| < 1$, which is always the case since $\phi_1 = \rho_1$.

For positive ϕ_1, it is clear that the autocorrelation of AR(1) is of the decaying exponential type. For negative ϕ_1, the autocorrelation will alternate in sign and decay as k increases. This behavior is illustrated in Figs. 2.2a and

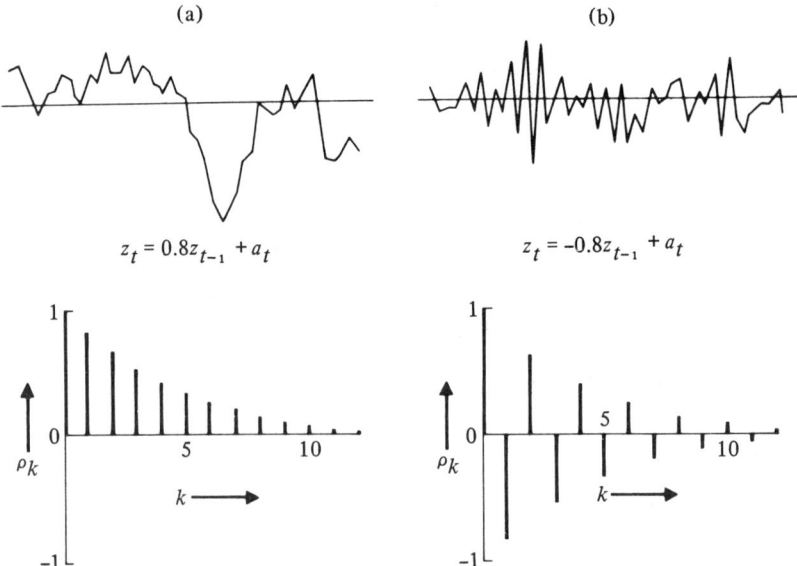

(a)

$$z_t = 0.8 z_{t-1} + a_t$$

(b)

$$z_t = -0.8 z_{t-1} + a_t$$

Theoretical Autocorrelation Functions

Figure 2.2   Autocorrelation functions of an AR(1) (from Box and Jenkins, 1976).

2.2b. As will be corroborated in Chapter 4, the AR(1) with positive parameters is dominated by low-frequency fluctuations while that with negative parameters exhibits dominating high frequencies.

The AR(1) model is the most popular model of time-series simulation and forecasting in hydrology and other fields. In hydrology, it is commonly called the Thomas–Fiering model (Maass et al., 1962). It is usually expressed in terms of the moments of the random process X_t,

$$(X_t - \mu) = \rho_1(X_{t-1} - \mu) + \sigma_x(1 - \rho_1^2)^{1/2}W_t, \tag{2.34}$$

where $X_t - \mu$ is Z_t in the previous paragraphs.

In the above expression, μ is the stationary mean of the process X_t. The random term a_t is substituted by the expression $\sigma_x(1 - \rho_1^2)^{1/2}W_t$, where W_t is a zero mean, variance 1, normally distributed random variable. The expression $\sigma_x(1 - \rho_1^2)^{1/2}$ is, according to Eq. (2.33), the standard deviation of a_t.

The AR(1) model is frequently used with the sole objective of preserving first- and second-order moments of time series. Such an objective is sometimes sufficient for simulation purposes and generally adequate for short-term forecasting. Parameter estimation using the method of moments simply requires substitution of sample moments $\bar{X}$, r_1, and S_x^2 for population moments, μ, ρ_1, and σ_x^2, respectively, in Eq. (2.34). Common equations for these statistics are

$$\bar{X} = \frac{1}{N}\sum_{i=1}^{N} x_i \tag{2.35}$$

$$S_x^2 = \frac{1}{N}\sum_{i=1}^{N}(x_i - \bar{X})^2$$
$$= \frac{1}{N}\sum_{i=1}^{N} x_i^2 - \bar{X}^2 \tag{2.36}$$

$$r_1 = \frac{\frac{1}{N}\sum_{i=1}^{N-1}(x_i - \bar{X})(x_{i+1} - \bar{X})}{S_x^2} \tag{2.37}$$

where the lower case variable x_i denotes sample values of the random variable X.

Keep in mind that the above statistics are random variables and can be expected to vary (some considerably) when estimated from different samples. The properties of the above and other estimators are discussed in the statistics literature. Users of synthetic hydrology should be aware of these properties. This opens for discussion the wisdom of fitting models to highly variable sample statistics. Because of sampling uncertainty, there is no assurance that the moments of the underlying populations are preserved. Synthetic hydrology and simulation cannot improve basic statistical reliability over historically available information. It only offers the opportunity to experiment with statistically similar realizations. Methods that take into account the stochasticity of the parameters are discussed by Vicens et al. (1975), Valdes et al. (1977), and McLeod and Hipel (1978).

The random term of Eq. (2.34), W_t, has been assumed throughout the previous discussions to be normally distributed. Since the addition of normal variates generates other Gaussian variables, Eq. (2.34) represents a Gaussian process. As such, the process is fully defined with the first- and second-order distributions, both Gaussian. Given a value of X_{t-1}, it is clear that the distribution of X_t, conditional on x_{t-1}, is also normal with mean

$$E[X_t|x_{t-1}] = \mu + \rho_1(x_{t-1} - \mu) \tag{2.38}$$

and variance

$$\sigma^2_{X_t|x_{t-1}} = \sigma_x^2[1 - \rho_1^2]. \tag{2.39}$$

Note that the variance of the process is reduced by $1 - \rho_1^2$ when a previous value is available.

It can be shown that Eqs. (2.38) and (2.39) result from obtaining the conditional distribution using Eq. (1.21) where the joint distribution of consecutive flows is normal and so is the marginal distribution. Since the conditional distribution is also normal, the propagation of means and variances given in the above equations represents an exact solution to the simulation problem as defined in Section 1.3. Sampling from the conditional distribution is automatically achieved by generating the standard normal deviate W_t.

Equations (2.38) and (2.39) also form a prediction model, where the prediction of the state X_t is the conditional mean value, which, for the Gaussian model, happens to be a linear function of previously observed quantities. Theoretically then, a model of the type of Eq. (2.34) should not be used unless there is reason to suspect that the underlying process of interest has approximately linear expressions for the conditional mean. Nevertheless, as mentioned in Chapter 1, many times the interest is in simulation and preservation of finite numbers of moments of the historical observations. If the user is interested in mean, variance, and lag-one correlation, and the process is not highly non-Gaussian, then use of a model of the type given in Eq. (2.34) is adequate.

Hydrologic simulation sometimes requires that third moments or skewness of historical time series be preserved. Statistically, the skewness coefficient is estimated as:

$$\gamma = \frac{\frac{1}{N}\sum_{i=1}^{N}(x_i - \overline{X})^3}{S_x^3}. \tag{2.40}$$

This is known to be a bounded and highly variable estimator (Kirby, 1974). It is therefore questionable whether sample skewness should be preserved. Nevertheless, if skewness is desired, the present formulation of Eq. (2.34) is inadequate.

Several schemes have been proposed to force skewness on the results of autoregressive models. One procedure is to modify the model:

$$X_t = \mu + \rho_1(X_{t-1} - \mu) + \sigma_x(1 - \rho_1^2)^{1/2}\varepsilon_t, \tag{2.41}$$

where ε_t is an approximately gamma-distributed variate with zero mean, variance 1, and skewness coefficient γ_ε. Such a model would resemble a process X_t with mean μ, variance σ_x^2, lag-one correlation ρ_1, and skewness γ_x. The skewness coefficient of the process is related to the skewness coefficient of the generating deviate ε_t by

$$\gamma_\varepsilon = \frac{1 - \rho_1^3}{(1 - \rho_1^2)^{1.5}}\gamma_x.$$

The εs can be generated with the skewness given by the above equation using an approximate expression

$$\varepsilon_t = \frac{2}{\gamma_\varepsilon}\left[\left(1 + \frac{\gamma_\varepsilon W_t}{6} + \frac{\gamma_\varepsilon^2}{36}\right)\right]^3 - \frac{2}{\gamma_\varepsilon} \tag{2.42}$$

(Wilson and Hillerty, 1931), where W_t is a normally distributed variable with zero mean and variance 1.

Equation (2.42) is in fact an approximation of a chi-square–distributed variable, which is theoretically valid for degrees of freedom ν greater than 30. This implies a very small possible skewness coefficient, since $\gamma = \sqrt{8/\nu}$. Nevertheless, some investigators (e.g., Mejia, 1971) claim that the bias introduced by using Eq. (2.42) in generating random variables with skewness greater than $\sqrt{8/30} \approx (0.52)$ is acceptable within the range of commonly observed streamflow statistics.

Another method suggested by Yevjevich (1966) and discussed by Matalas (1967) involves a three-step procedure to form approximately gamma-distributed variables. First, a zero mean, unit variance, lag-one correlation ρ_y process is formed using an autoregressive model:

$$Y_t = \rho_y Y_{t-1} + (1 - \rho_y^2)^{1/2}W_t. \tag{2.43}$$

A number m of variates Y_t is selected and used to form a new process Z:

$$Z_i = \sum_{j=(i-1)m+1}^{im} Y_j^2. \tag{2.44}$$

Note that Z_i consists of groupings of squares of the original variate Y_j. So, for example, 100 Ys and $m = 10$ lead to only 10 values of Z. The variate Z is approximately gamma with mean m, variance $2m$, skewness coefficient $2\sqrt{2/m}$, and lag-one correlation ρ_y^2. By renormalizing the time series of Zs, it is possible to obtain a new series X_t with any desired mean and variance:

$$X_t = \overline{X} + S_x \left[(2m)^{-0.5} Z_t - (m/2)^{0.5} \right]. \tag{2.45}$$

The resulting series of X_ts will have mean $\overline{X}$, variance S_x^2, lag-one correlation ρ_y^2, and skewness coefficient $\gamma_x = 2\sqrt{2/m}$. Clearly, m is a parameter to be determined as a function of the desired skewness. The highest skewness the model can preserve is $2\sqrt{2} \approx 2.8$.

Possibly the most common attempt to preserve skewness is the generation of log-normally distributed variables. Assume that X_t is log-normal, such that

$$Y_t = \ln(X_t - a) \tag{2.46}$$

is a normally distributed variable. The statistics of Y_t and X_t are related through the well-known equations (Matalas, 1967)

$$\mu_x = a + \exp\left[\sigma_y^2/2 + \mu_y \right] \tag{2.47}$$

$$\sigma_x^2 = \exp\left[2\left(\sigma_y^2 + \mu_y \right) \right] - \exp\left[\sigma_y^2 + 2\mu_y \right] \tag{2.48}$$

$$\gamma_x = \frac{\exp\left(3\sigma_y^2 \right) - 3\exp\left(\sigma_y^2 \right) + 2}{\left[\exp\left(\sigma_y^2 \right) - 1 \right]^{3/2}} \tag{2.49}$$

$$\rho_x = \frac{\exp\left[\sigma_y^2 \rho_y \right] - 1}{\exp\left(\sigma_y^2 \right) - 1}. \tag{2.50}$$

The statistics μ_y, σ_y, and ρ_y, directly computed from the transformed historical sequences, could be used in an autoregressive formulation leading to the generation of normal variates, which, upon exponentiation would yield a log-normal process. Nevertheless, such a procedure will preserve the sample statistics of the transformed historical sequences but produce a biased estimate of the sample statistics of the original variables. Furthermore, the generation of transformed variables will not necessarily preserve the historical skewness of the process. The procedure is so simple, though, that the practitioner many times settles on slightly biased results.

If the unbiased preservation of the sample statistics of the original variables is the goal, then Eqs. (2.47) through (2.50) must be used. The idea is to estimate the parameters μ_x, σ_x, γ_x, and ρ_x of the original variables and simultaneously solve Eqs. (2.47) through (2.50) for the corresponding statistics a, σ_y, μ_y, and ρ_y. The latter statistics are used in developing an autoregressive model, which, in turn, is used to generate a normally distributed series, Y_i, $i = 1, \ldots, n$. Transforming Y_i

$$X_i = \exp(Y_i) + a \qquad (2.51)$$

results in a sequence X_i with the historical sample statistics μ_x, σ_x, γ_x, and ρ_x.

Authors such as Burges et al. (1975), Stedinger (1980), and Bobée and Robitaille (1975) have extensively discussed the value of the log-normal transformation in hydrology. Their discussions of estimation bias and model behavior with small data samples may be of interest to the reader.

EXAMPLE 2.1   (adapted from notes by John C. Schaake, Jr.)

Following are 10 years of observations of annual streamflows in millions of cubic meters:

Year	Discharge	Year	Discharge
1	145.78	6	175.02
2	95.43	7	101.98
3	116.66	8	146.14
4	96.12	9	126.01
5	122.09	10	132.73

Use the above data to estimate (by the method of moments) the parameters of an AR(1) model that preserves first, second, and third moments. Generate 40 years of synthetic annual streamflows from the resulting model.

Relevant sample moments are obtained from the given data as

$$\overline{Q} = \frac{1}{10} \sum_{i=1}^{10} Q_i = 125.80$$

$$S = \left(\frac{1}{10} \sum_{i=1}^{10} Q_i^2 - \overline{Q}^2 \right)^{1/2} = 25.3$$

$$r_1 = \frac{\frac{1}{10} \sum_{i=1}^{10} Q_i Q_{i+1} - \overline{Q}^2}{S^2} = -0.28.$$

(*Note:* in the above, $Q_{11} = Q_1$; this is a circular definition of correlation. Also an unusual negative correlation, probably due to the small sample used in the estimation, is observed.)

$$\gamma = \frac{\frac{1}{10} \sum_{i=1}^{10} (Q_i - \overline{Q})^3}{S^3} = 0.42.$$

Following are three sets of 40 years of simulated streamflows and corresponding summary statistics. The first column corresponds to streamflows generated by taking the logarithms of the original data, generating normally distributed variables, and then exponentiating the results. Column 2 uses the gamma approximation given in Eq. (2.41). Column 3 used the full log transformation as described in Eqs. (2.46) through (2.51).

Run 1 $(\times 10^2)$	Run 2 $(\times 10^2)$	Run 3 $(\times 10^2)$
1.25796	1.25796	1.25796
1.04610	1.14337	1.23635
1.45375	1.53361	1.61603
1.09044	1.14088	1.72124
1.10708	1.06840	1.06668
1.37060	1.27093	1.33715
1.18291	1.38975	1.44948
1.08402	1.36419	1.31257
1.38411	0.81018	1.23944
1.03360	0.98727	1.30080
1.57609	1.81578	1.31322
0.81442	1.34882	0.77345
1.90717	1.09977	1.94454
0.97410	1.65840	1.01594
1.33417	1.09074	1.04225
0.81967	1.30820	1.27225
1.22386	1.63686	1.10791
1.21820	0.93435	1.30460
0.65722	1.44445	1.33684
1.30744	0.84896	1.24458
1.48888	1.32676	1.08977
1.39215	1.64438	1.45393
1.31445	1.03494	1.23296
1.31683	1.45548	1.38069
1.16038	0.89716	1.32512
1.37676	1.71125	1.14234
1.36500	0.83465	1.33604
1.53774	1.34727	1.25665
1.41753	1.54085	1.12521
1.52989	1.41571	1.05186
1.38218	1.21430	1.28745
1.42435	1.78309	1.18783
1.41133	0.87965	1.42316
1.29935	1.84489	0.82781
1.10013	1.01661	1.41398
1.40968	1.17430	1.07725
1.20144	1.75676	1.36271
1.25377	0.89639	0.76215

(cont'd)

	Run 1 $(\times 10^2)$	Run 2 $(\times 10^2)$	Run 3 $(\times 10^2)$
	1.03618	1.06355	1.36016
	0.49733	1.86203	1.10057
Mean	124.4	129.6	125.2
Standard deviation	26.0	31.4	22.7
Lag-one correlation	-0.17	-0.47	-0.36
Skewness coefficient	-0.56	0.22	0.33

2.2.2   Second-Order Autoregressive Model

Second order is usually the highest lag necessary in representing hydrologic time series. The model takes the form

$$\phi(B)Z_t = a_t$$
$$\left(1 - \phi_1 B - \phi_2 B^2\right) Z_t = a_t. \tag{2.52}$$

The more familiar representation is

$$Z_t = \phi_1 Z_{t-1} + \phi_2 Z_{t-2} + a_t. \tag{2.53}$$

For stationarity, the roots (solutions of B) of the quadratic polynomial $\phi(B)$ must lie outside the unit circle. Using the well-known solution for quadratic equations, it can be shown that the implied conditions on parameters ϕ_1 and ϕ_2 are

$$\phi_2 + \phi_1 < 1$$
$$\phi_2 - \phi_1 < 1 \tag{2.54}$$
$$-1 < \phi_2 < 1.$$

Figure 2.3 shows the triangular parameter space defined by Eq. (2.54). The Yule–Walker equations for the AR(2) model are

$$\rho_1 = \phi_1 + \phi_2 \rho_1$$
$$\rho_2 = \phi_1 \rho_1 + \phi_2, \tag{2.55}$$

which when solved simultaneously yield

$$\phi_1 = \rho_1 (1 - \rho_2)/(1 - \rho_1^2)$$
$$\phi_2 = \frac{\rho_2 - \rho_1^2}{1 - \rho_1^2}, \tag{2.56}$$

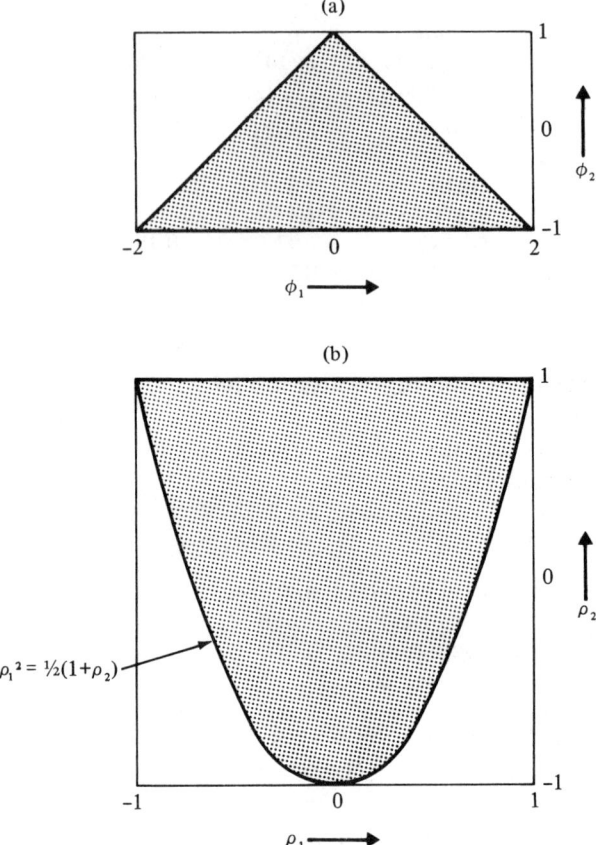

Figure 2.3 Valid regions of the parameters and correlations of a stationary AR(2) process (from Box and Jenkins, 1976).

or inversely,

$$\rho_1 = \frac{\phi_1}{1 - \phi_2}$$

$$\rho_2 = \phi_2 + \frac{\phi_1^2}{1 - \phi_2}.$$

(2.57)

Figure 2.4 gives the solution of Eq. (2.56) for various values of correlation coefficients ρ_1 and ρ_2. In practice, sample estimates could be used for the correlations in order to obtain parameter values.

The parameter limits given in Eq. (2.54) and the relations between correlations and parameters given in Eq. (2.57) define a region of valid correlation

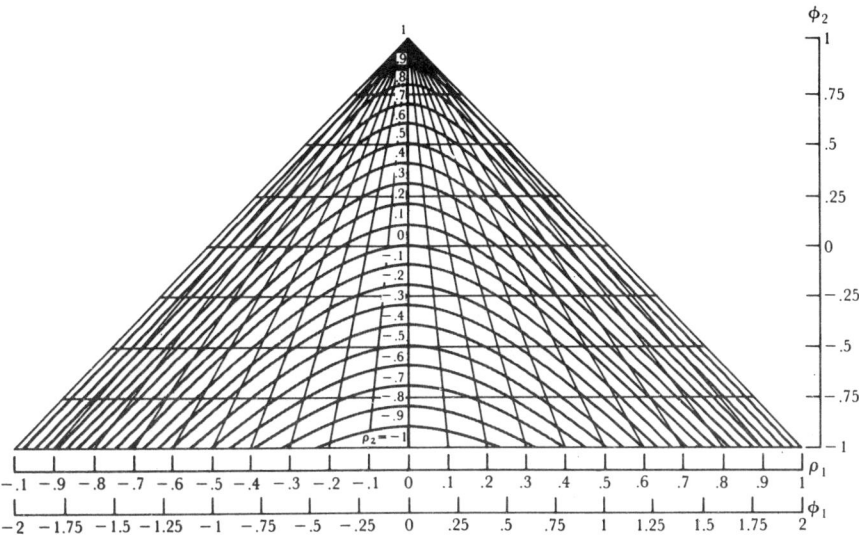

Figure 2.4   Relation between correlations and parameters of an AR(2) model. Diagram may be used for parameter estimation using the method of moments. (From Box and Jenkins, 1976.)

coefficients for the lag-two (second-order) autoregressive model,

$$-1 < \rho_1 < 1$$
$$-1 < \rho_2 < 1 \tag{2.58}$$
$$\rho_1^2 < \tfrac{1}{2}(\rho_2 + 1).$$

The admissible regions of parameters and correlations are shown in Fig. 2.3. Notice that this figure could be used as a first-cut criteria for the possible use of an AR(2) with a given set of data.

The correlation function of the AR(2) model is easily obtained by recursively solving

$$\rho_k = \phi_1 \rho_{k-1} + \phi_2 \rho_{k-2} \tag{2.59}$$

with initial conditions,

$$\rho_0 = 1$$
$$\rho_1 = \frac{\phi_1}{1 - \phi_2}.$$

Box and Jenkins (1976) give the general solution to Eq. (2.59) as

$$\rho_k = \frac{G_1(1 - G_2^2)G_1^k - G_2(1 - G_1^2)G_2^k}{(G_1 - G_2)(1 + G_1 G_2)},$$

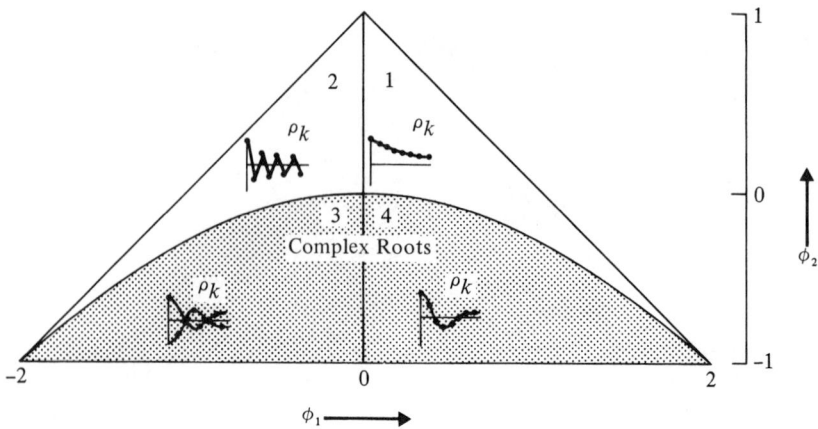

Figure 2.5 Autocorrelation functions of various stationary AR(2) models (from Box and Jenkins, 1976).

where G_1^{-1} and G_2^{-1} are the roots of $\phi(B)$. Real roots result when $\phi_1^2 + 4\phi_2 \geq 0$. Complex roots are obtained when the discriminant is less than 0, $\phi_1^2 + 4\phi_2 < 0$. Regions 1 and 2 of Fig. 2.5 illustrate the behavior of the correlation for real roots—Region 1 for a dominant positive root and Region 2 for a negative dominant root. The correlation decays geometrically either monotonically (Region 1) or alternating signs (Region 2). Conjugate complex roots result in a damped sinusoid with a phase angle between 90 degrees and 180 degrees in Region 4.

Using Eq. (2.26), the variance of the AR(2) model is

$$\sigma_z^2 = \frac{\sigma_a^2}{1 - \rho_1\phi_1 - \rho_2\phi_2}$$

$$= \left(\frac{1 - \phi_2}{1 + \phi_2}\right) \frac{\sigma_a^2}{\left[(1 - \phi_2)^2 - \phi_1^2\right]}. \tag{2.60}$$

EXAMPLE 2.2

The autocorrelation of the St. Lawrence River at Ogdensburg, New York, as given by Yevjevich (1972) and McLeod et al. (1977), is shown in Fig. 2.6.

The lag-one autocorrelation is observed at around 0.7 and the lag-two value is about 0.5. The mean discharge is 6825 m^3/sec with standard deviation of 544 m^3/sec and skewness coefficient of -0.286 (Yevjevich, 1963).

A lag-one autoregressive model of the St. Lawrence fitted by the method of moments would be

$$Z_t = 6825 + 0.7(Z_{t-1} - 6825) + 544(1 - 0.7^2)^{1/2}W_t$$
$$= 6825 + 0.7(Z_{t-1} - 6825) + 388.49W_t.$$

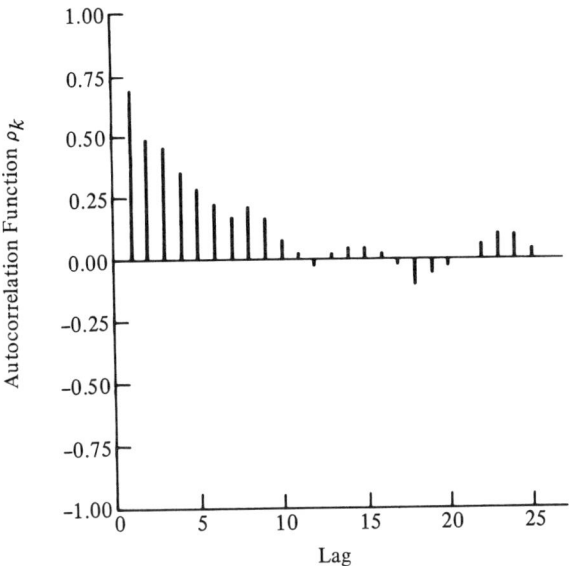

Figure 2.6 Autocorrelation function for the Saint Lawrence River (from McLeod, et al. in *Water Resources Research* 13(3):577-86, 1977).

The lag-one and lag-two correlations are within the admissible regions of Fig. 2.3. Using Eq. (2.56) (or Fig. 2.4) the method of moments fit of an AR(2) model would yield the following parameters

$$\phi_1 = r_1(1 - r_2)/(1 - r_1^2) = 0.69$$

$$\phi_2 = \frac{r_2 - r_1^2}{1 - r_1^2} = 0.02.$$

Equation (2.60) yields

$$S_a^2 = S^2(1 - \rho_1\phi_1 - \rho_2\phi_2)$$

$$= 295,936[1 - 0.7(0.69) - 0.5(0.02)] = 150,040$$

or

$$S_a = 387.$$

Therefore an estimated AR(2) would be

$$Z_t = 6825 + 0.69(Z_{t-1} - 6825) + 0.02(Z_{t-2} - 6825) + 387W_t.$$

Given the inherent variability in moment estimation the reader should suspect the statistical significance of ϕ_2.

2.3 MOVING-AVERAGE MODELS OF ORDER $q, \text{MA}(q)$

As in the autoregressive models, moving-average formulations of infinite order have no practical use. A finite-order model is represented by

$$
\begin{aligned}
Z_t &= a_t - \theta_1 a_{t-1} - \cdots - \theta_q a_{t-q} \\
&= \left(1 - \theta_1 B - \cdots - \theta_q B^q\right) a_t \\
&= \theta(B) a_t.
\end{aligned}
\tag{2.61}
$$

Since the polynomial $\theta(B)$ is finite, it always converges, implying unconditional stationarity, according to the criteria discussed in Section 2.1.1. The invertibility of Eq. (2.61) requires $\theta^{-1}(B)$ to converge for $|B| \leq 1$. Box and Jenkins (1976) show that this is equivalent to stating that the roots of the characteristic equation $\theta(B)$ must lie outside the unit circle. It is important to note that a finite moving-average model is equivalent to an infinite autoregressive model.

The autocovariance function of the MA(q) process is obtained by performing the following expectation:

$$
\gamma_k = E\left[\left(a_t - \theta_1 a_{t-1} - \cdots - \theta_q a_{t-q}\right)\left(a_{t-k} - \theta_1 a_{t-k-1} - \cdots - \theta_q a_{t-k-q}\right)\right].
$$

For $k = 0$ and using the "whiteness" properties of the series a_k, the process variance becomes

$$
\gamma_0 = \left(1 + \theta_1^2 + \theta_2^2 + \cdots + \theta_q^2\right)\sigma_a^2.
\tag{2.62}
$$

For $k \neq$ zero, it is clear that the autocovariance is

$$
\gamma_k =
\begin{cases}
\left(-\theta_k + \theta_1 \theta_{k+1} + \cdots + \theta_{q-k}\theta_q\right)\sigma_a^2 & k = 1, 2, \ldots, q \\
0 & k > q.
\end{cases}
$$

Upon normalizing the autocovariance by γ_0 the autocorrelation function becomes

$$\rho_k = \begin{cases} \dfrac{-\theta_k + \theta_1\theta_{k+1} + \cdots + \theta_{q-k}\theta_q}{1 + \theta_1^2 + \cdots + \theta_q^2} & k = 1, 2, \ldots, q \\ 0 & k > q. \end{cases} \tag{2.63}$$

Therefore the autocorrelation of a moving-average model is zero beyond the order of the model. This is in contrast to the infinite extent of ρ_k in the AR(p) model.

The nonlinearity of the autocorrelation expressions makes parameter estimation for the MA(q) model using the method of moments highly unstable. Iterative least-squares procedures are usually required (Box and Jenkins, 1976).

2.3.1 First-Order Moving-Average Model, MA(1)

The form of the MA(1) is

$$\begin{aligned} Z_t &= a_t - \theta_1 a_{t-1} \\ &= (1 - \theta_1 B) a_t, \end{aligned} \tag{2.64}$$

which is stationary for all values of θ_1 but invertible only for $|B| > 1$, which implies (since $B = \theta^{-1}$) that $|\theta_1| < 1$.

The variance of the process is

$$\gamma_0 = \left(1 + \theta_1^2\right)\sigma_a^2 \tag{2.65}$$

and its autocorrelation function

$$\rho_k = \begin{cases} \dfrac{-\theta_1}{1 + \theta_1^2} & k = 1 \\ 0 & k \geq 2, \end{cases} \tag{2.66}$$

which (for $k = 1$) leads to the following relation between θ_1 and ρ_1

$$\theta_1^2 + \frac{\theta_1}{\rho_1} + 1 = 0. \tag{2.67}$$

Note that the autocorrelation disappears after lag one. The two roots of Eq. (2.67) must then be θ_1 and θ_1^{-1}. The invertibility conditions will be satisfied if the root $|\theta_1| < 1$ is used.

Table 2.1
**Table relating ρ_1 to θ for a first-order
moving-average process**

θ	ρ_1	θ	ρ_1
0.00	0.000	0.00	0.000
0.05	-0.050	-0.05	0.050
0.10	-0.099	-0.10	0.099
0.15	-0.147	-0.15	0.147
0.20	-0.192	-0.20	0.192
0.25	-0.235	-0.25	0.235
0.30	-0.275	-0.30	0.275
0.35	-0.315	-0.35	0.315
0.40	-0.349	-0.40	0.349
0.45	-0.374	-0.45	0.374
0.50	-0.400	-0.50	0.400
0.55	-0.422	-0.55	0.422
0.60	-0.441	-0.60	0.441
0.65	-0.457	-0.65	0.457
0.70	-0.468	-0.70	0.468
0.75	-0.480	-0.75	0.480
0.80	-0.488	-0.80	0.488
0.85	-0.493	-0.85	0.493
0.90	-0.497	-0.90	0.497
0.95	-0.499	-0.95	0.499
1.00	-0.500	-1.00	0.500

Source: Box and Jenkins (1976).

Table 2.1 gives the invertible solution for Eq. (2.67) for lag-1 correlation between -0.5 and 0.5 (Box and Jenkins, 1976).

2.3.2 Second-Order Moving-Average Model, MA(2)

The MA(2) model is given by

$$Z_t = a_t - \theta_1 a_{t-1} - \theta_2 a_{t-2}$$
$$= \left(1 - \theta_1 B - \theta_2 B^2\right) a_t \qquad (2.68)$$
$$= \theta(B) a_t.$$

Since the polynomial $\theta(B)$ converges for $|B| \leq 1$, the model is unconditionally stationary. Invertibility requires that the roots of $\theta(B)$ lie outside the unit circle. This is analogous to the stationarity conditions of the AR(2) model given by Eq. (2.54). The implied parameter space is then the same:

$$\theta_2 + \theta_1 < 1$$

$$\theta_2 - \theta_1 < 1 \tag{2.69}$$

$$-1 < \theta_2 < 1.$$

From Eq. (2.62), the variance is

$$\gamma_0 = \sigma_a^2 \left(1 + \theta_1^2 + \theta_2^2 \right). \tag{2.70}$$

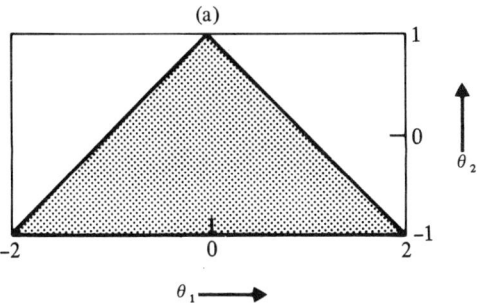

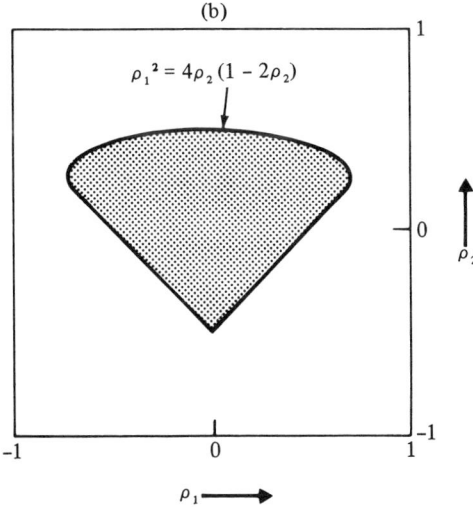

Figure 2.7 Valid regions of the parameters and the correlations of an invertible MA(2) process (from Box and Jenkins, 1976).

The autocorrelation function, Eq. (2.63), is

$$\rho_1 = \frac{-\theta_1(1-\theta_2)}{1+\theta_1^2+\theta_2^2} \qquad \rho_2 = \frac{-\theta_2}{1+\theta_1^2+\theta_2^2} \qquad (2.71)$$

$$\rho_k = 0 \qquad k \geq 3.$$

Invertibility conditions on parameters and their explicit relationship with ρ_1 and ρ_2 force the following limits on correlations of MA(2) models

$$\rho_2 + \rho_1 = -0.5$$

$$\rho_2 - \rho_1 = -0.5 \qquad (2.72)$$

$$\rho_1^2 = 4\rho_2(1-2\rho_2).$$

Figure 2.7 gives the invertible-parameter region and the limits on correlations imposed by the second-order moving-average model. Figure 2.8 is the solution to the nonlinear Eq. (2.71), giving θ_1 and θ_2 as a function of ρ_1 and ρ_2. Sample estimates of the correlations can then be used to obtain initial estimates of the parameter values. Again, the variability of sample correlation estimates and the nonlinearity of the relationship between moments and parameters may lead to serious problems in parameter estimation.

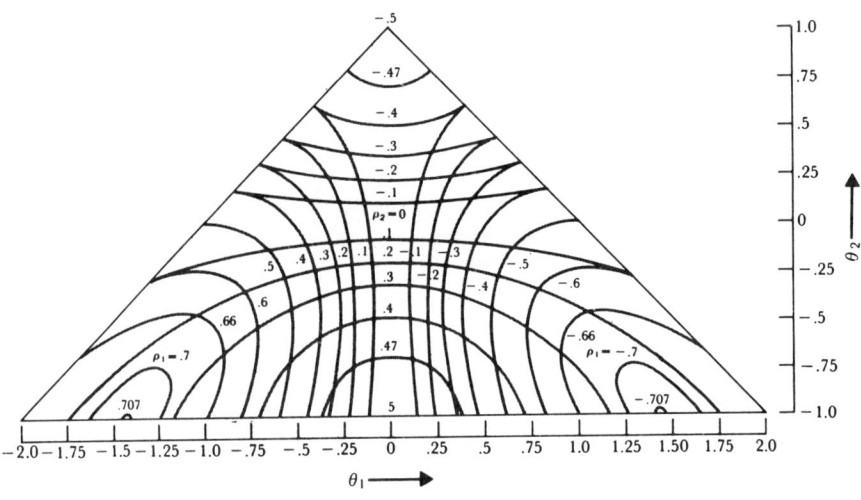

Figure 2.8 Relation between correlations and parameters for a second-order moving-average model. Diagram may be used for parameter estimation using the method of moments. (From Box and Jenkins, 1976.)

2.4 AUTOREGRESSIVE–MOVING-AVERAGE MODELS, ARMA(p, q)

Autoregressive and moving-average models can be combined to model processes that otherwise would be operationally impossible to represent with single finite AR or MA models. An ARMA(p, q) model takes the form

$$\left(1 - \phi_1 B - \phi_2 B^2 - \cdots - \phi_p B^p\right) Z_t = \left(1 - \theta_1 B - \theta_2 B^2 - \cdots - \theta_q B^q\right) a_t$$
$$\phi(B) Z_t = \theta(B) a_t. \tag{2.73}$$

The stationarity and invertibility conditions of the ARMA(p, q) model correspond to those of the component MA and AR models. For stationarity and invertibility, the roots of $\phi(B)$ and of $\theta(B)$ must lie outside the unit circle.

The autocovariance function is found by multiplying $Z_t = \phi_1 Z_{t-1} + \cdots + \phi_p Z_{t-p} + a_t - \theta_1 a_{t-1} - \cdots - \theta_q a_{t-q}$ by Z_{t-k} and finding expected values,

$$\gamma_k = E[Z_t Z_{t-k}] = \phi_1 \gamma_{k-1} + \cdots + \phi_p \gamma_{k-p} + \gamma_{za}(k)$$
$$- \theta_1 \gamma_{za}(k-1) - \cdots - \theta_q \gamma_{za}(k-q), \tag{2.74}$$

where

$$\gamma_{za}(k) = E[Z_{t-k} a_t]$$
$$\gamma_{za}(k-1) = E[Z_{t-k} a_{t-1}]. \tag{2.75}$$

The value for $\gamma_{za}(k)$ will be zero as long as $k > 0$, since no correlation exists between a_t and values of Z before t. The value for $\gamma_{za}(k)$ will not be zero for $k \leq 0$.

With the above in mind, it should be clear that for $k > q$, the autocovariance (and autocorrelation) in Eq. (2.74) reduces to that of an AR(p) model:

$$\gamma_k = \phi_1 \gamma_{k-1} + \cdots + \phi_p \gamma_{k-p} \qquad \text{for } k > q$$
$$\rho_k = \phi_1 \rho_{k-1} + \cdots + \phi_p \rho_{k-p} \qquad \text{for } k > q.$$

For values of k less than or equal to q, the autocovariance will be a function of the moving-average terms and will depend on all coefficients $\phi_1, \ldots, \phi_p, \theta_1, \ldots, \theta_q$, and the variance σ_a^2. The ARMA(p, q) model then has the convenient property that its first q autocorrelations depend on moving-average terms as well as autoregressive terms. After q lags, autoregressive behavior takes over from the last correlation value.

The variance of the process is given by Eq. (2.74) for $k = 0$. Evaluation of the variance requires the solution of $\gamma_1, \ldots, \gamma_p$.

2.4.1 ARMA(1,1)

A popular, and useful, model in hydrology is

$$Z_t - \phi_1 Z_{t-1} = a_t - \theta_1 a_{t-1}$$
$$(1 - \phi_1 B) Z_t = (1 - \theta_1 B) a_t. \tag{2.76}$$

Stationarity and invertibility conditions correspond to the individual AR(1) and MA(1) models and so imply that the parameter region is

$$-1 < \phi_1 < 1$$
$$-1 < \theta_1 < 1.$$

Figure 2.9 shows this admissible parameter space.

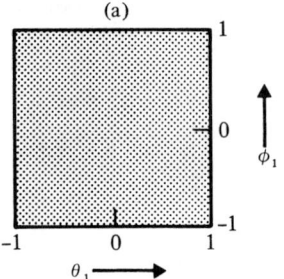

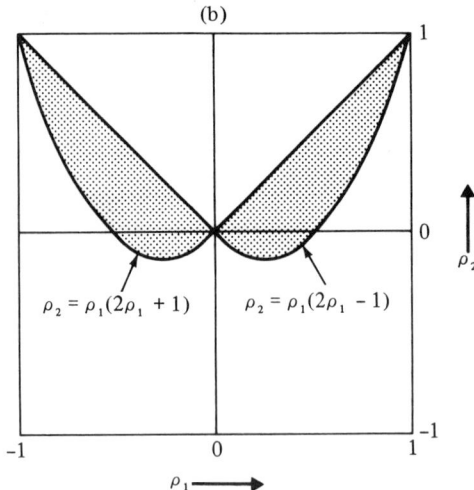

Figure 2.9 Valid regions for the parameters and correlations of a stationary and invertible ARMA(1,1) process (from Box and Jenkins, 1976).

Using Eq. (2.74), the autocovariance function is

$$\gamma_0 = \phi_1 \gamma_1 + \sigma_a^2 - \theta_1 \gamma_{za}(-1)$$
$$\gamma_1 = \phi_1 \gamma_0 - \theta_1 \sigma_a^2 \tag{2.77}$$
$$\gamma_k = \phi_1 \gamma_{k-1} \qquad k \geq 2.$$

To obtain $\gamma_{za}(-1)$ Eq. (2.76) is multiplied by a_{t-1} and expectations are taken:

$$\gamma_{za}(-1) = E[Z_t a_{t-1}] = (\phi_1 - \theta_1) \sigma_a^2. \tag{2.78}$$

Using Eq. (2.78) in Eq. (2.77) the autocovariance function of the process is obtained as

$$\gamma_0 = \frac{1 + \theta_1^2 - 2\phi_1\theta_1}{1 - \phi_1^2} \sigma_a^2$$

$$\gamma_1 = \frac{(1 - \phi_1\theta_1)(\phi_1 - \theta_1)}{1 - \phi_1^2} \sigma_a^2 \tag{2.79}$$

$$\gamma_k = \phi_1 \gamma_{k-1} \qquad k \geq 2.$$

Note that the autocovariance will decay exponentially from a starting value γ_1, which is dependent on θ_1. The sign of γ_1 (and ρ_1) is defined by $\phi_1 - \theta_1$. The sign of ϕ_1 determines if the correlation decay is smooth or alternates in sign.

The correlation function is given by

$$\rho_1 = \frac{(1 - \phi_1\theta_1)(\phi_1 - \theta_1)}{1 + \theta_1^2 - 2\phi_1\theta_1}$$

$$\rho_k = \phi_1 \rho_{k-1} \qquad k \geq 2. \tag{2.80}$$

The relationships shown in Eq. (2.80) and the invertibility–stationarity parameter space define an admissible region for the first two correlations

$$|\rho_2| < |\rho_1|$$
$$\rho_2 > \rho_1(2\rho_1 + 1) \qquad \rho_1 < 0 \tag{2.81}$$
$$\rho_2 > \rho_1(2\rho_1 - 1) \qquad \rho_1 > 0.$$

Figure 2.9 illustrates the above region; correlations outside that space indicate that the ARMA(1, 1) is not a good model. Figure 2.10 diagrams the solution of parameter ϕ_1 and θ_1 in terms of ρ_1 and ρ_2 as given by Eq. (2.80). Figure 2.11 gives typical forms of the autocorrelation expected for various regions of the parameter space.

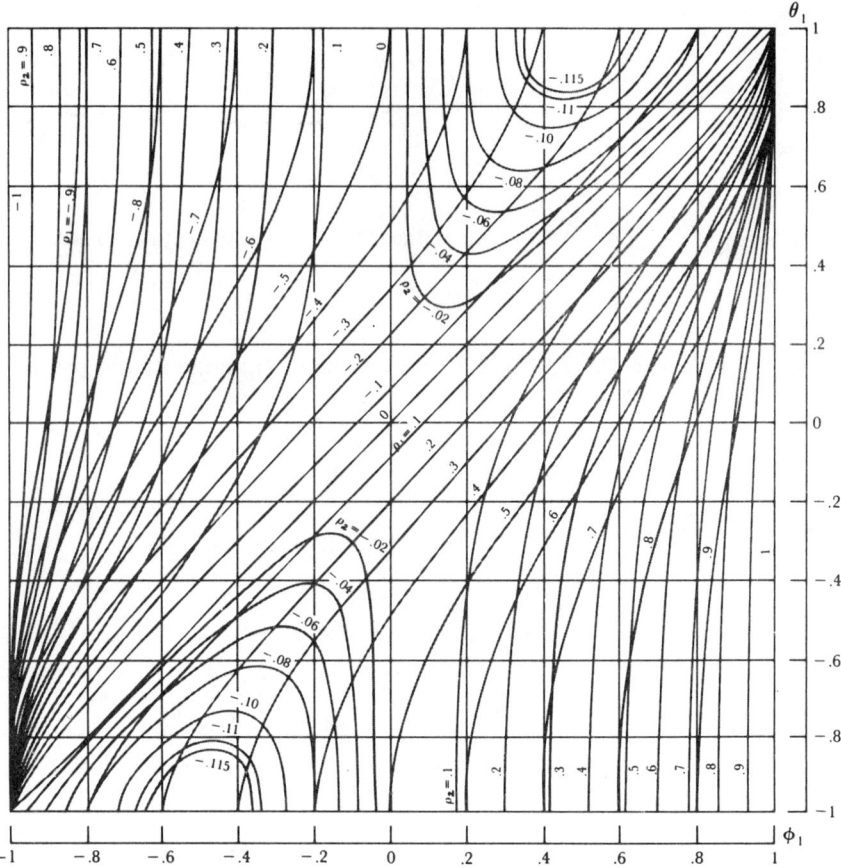

Figure 2.10 Relation between correlations and parameters for a stationary and invertible ARMA(1,1). Diagram may be used for parameter estimation using the method of moments. (From Box and Jenkins, 1976).

EXAMPLE 2.3

Figure 2.12 gives the sample autocorrelation of the Niger River annual streamflows, which are shown in Fig. 2.13. The observed lag-one autocorrelation r_1 is 0.554 and the lag-two autocorrelation r_2 is around 0.45. Clearly, the rate of correlation decay is much slower than that of an AR(1) or MA(1) model. The autocorrelation does fall relatively quickly for higher lags. Carlson et al. (1970), following identification and estimation procedures discussed in Section 2.6, concluded that an ARMA(1,1) model provided the best fit to the data. They suggested that

$$Z_t = 0.82 Z_{t-1} + a_t - 0.4 a_{t-1}.$$

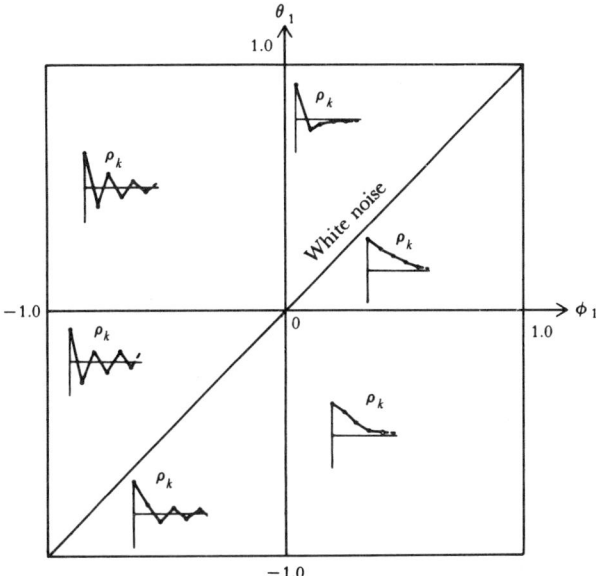

Figure 2.11   Autocorrelation functions for various ARMA(1,1) models (from Box and Jenkins, 1976).

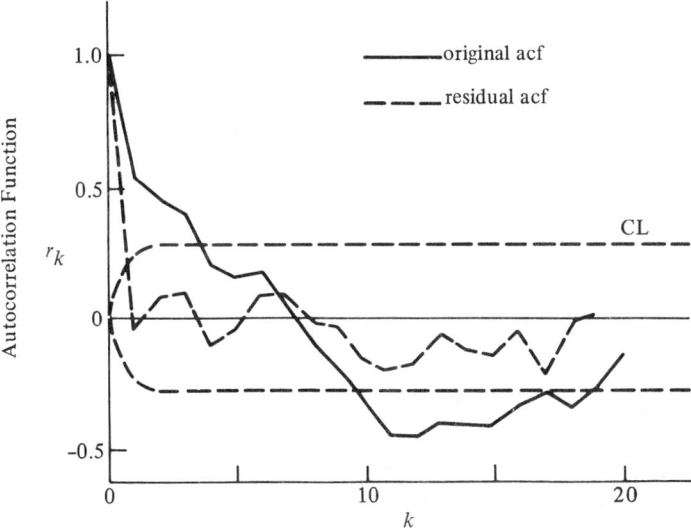

Figure 2.12   Sample autocorrelation function (acf) of Niger data series and the autocorrelation function of the residual series from best fit model for Niger series. The approximate 95% confidence limits (CL) of the autocorrelation function of the residual series are indicated by dashed lines. (From Carlson et al. in *Water Resources Research* 6(4):1070-8, 1970.)

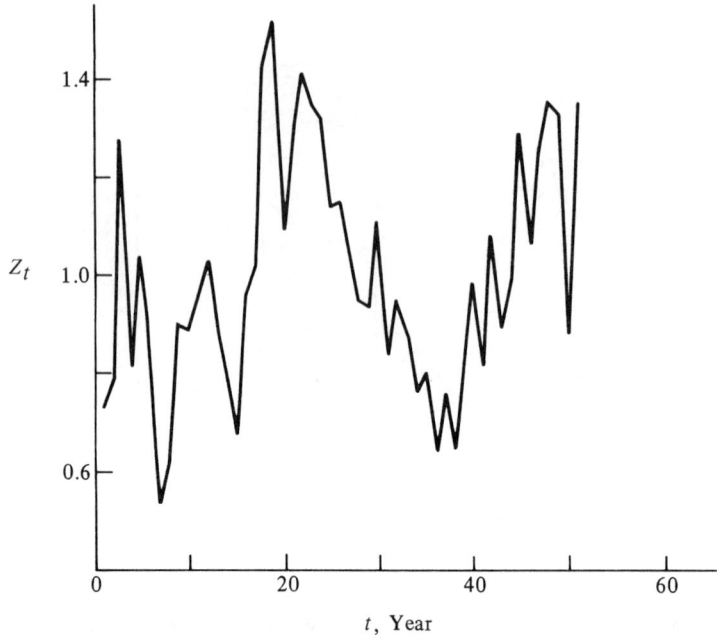

Figure 2.13 Flow of Niger River at Koulicoro, Africa, 1906–1957. Z_t is given as a modular coefficient, and t is time from beginning of series. (From Carlson et al. in *Water Resources Research* 6(4):1070-8, 1970).

The coefficients were obtained using the least-squares procedures discussed in Section 2.6. Using the method of moments and Fig. 2.10 yield

$$Z_t = 0.8 Z_{t-1} + a_t - 0.4 a_{t-1}.$$

2.5 NONSTATIONARY MODELS: THE AUTOREGRESSIVE INTEGRATED MOVING-AVERAGE MODEL (ARIMA)

2.5.1 Local Nonstationarities

A relatively simple extension of the theory for ARMA processes permits the handling of a limited type of nonstationary model, essentially nonstationarities in the mean of the process.

In previous sections, it was seen that the model

$$\psi(B) Z_t = \theta(B) a_t \tag{2.82}$$

is stationary if the roots of $\psi(B)$ are outside the unit circle. Assume now that some of the roots of $\psi(B)$ lie on the unit circle. The model can then be expressed as

$$\psi(B)Z_t = \phi(B)(1-B)^d Z_t = \theta(B)a_t \qquad (2.83)$$

or

$$\phi(B)\nabla^d Z_t = \theta(B)a_t,$$

where the roots of $\phi(B)$ are outside the unit circle. Because of the differencing operation, ∇^d, $(d \geq 1)$, the model can be reformulated in terms of the nonzero mean process X_t, since $\nabla^d Z_t = \nabla^d X_t$. So the above is equivalent to $\phi(B)\nabla^d X_t = \theta(B)a_t$, where the roots of $\phi(B)$ are outside the unit circle. Defining

$$Y_t = \nabla^d X_t, \qquad (2.84)$$

Eq. (2.83) represents a stationary–invertible ARMA process on the variable Y_t. Given Y_t, X_t can be obtained by performing the summation operation,

$$X_t = S^d Y_t, \qquad (2.85)$$

where

$$SY_t = \sum_{h=-\infty}^{t} Y_h$$

$$S^2 Y_t = \sum_{i=-\infty}^{t} \sum_{h=-\infty}^{i} Y_h$$

and so on.

Since X_t is then essentially the integration of an ARMA model, this particular formulation is called the autoregressive integrated moving average of orders p, d, and q [ARIMA(p,d,q)], where d is the order of differentiation of the original data necessary to obtain a stationary process. Note that,

$$\text{ARMA}(p,q) = \text{ARIMA}(p,0,q)$$
$$\text{AR}(p) = \text{ARIMA}(p,0,0)$$
$$\text{MA}(q) = \text{ARIMA}(0,0,q).$$

It should be clear that ARIMA models handle particular types of non-stationarities. By taking first-order differences, it is possible to eliminate unknown stochastic biases in the data. The differenced data should be stationary if the behavior was otherwise homogeneous. If a process exhibits random changes in slope and level but is otherwise homogeneous, then differencing it twice will yield a stationary process.

Deterministic linear trends or higher-order polynomial trends can be incorporated by stating the ARIMA model as

$$\phi(B)\nabla^d X_t = \theta_0 + \theta(B)a_t. \tag{2.86}$$

ARIMA models can be expressed in three different forms. The first is a difference equation by expanding the polynomial

$$\psi(B) = \phi(B)(1-B)^d = 1 - \psi_1 B - \psi_2 B^2 - \cdots - \psi_{p+d}B^{p+d}$$

so that

$$X_t = \psi_1 X_{t-1} + \cdots + \psi_{p+d}X_{t-p-d} - \theta_1 a_{t-1} - \cdots - \theta_q a_{t-q} + a_t. \tag{2.87}$$

For example, ARIMA(1,1,1),

$$(1-\phi B)(1-B)X_t = (1-\theta B)a_t,$$

is equivalent to

$$\left\{1-(1+\phi)B + \phi B^2\right\}X_t = (1-\theta B)a_t.$$

The model can be expressed as a function of past random shocks a_t,

$$X_t = \Omega(B)a_t. \tag{2.88}$$

Operating with $\psi(B)$ on both sides of the above equation,

$$\psi(B)X_t = \psi(B)\Omega(B)a_t = \theta(B)a_t, \tag{2.89}$$

which indicates that

$$\psi(B)\Omega(B) = \theta(B). \tag{2.90}$$

The weights Ω_j are then obtained by equating coefficients of B in the two polynomials of Eq. (2.90):

$$\left(1-\psi_1 B - \cdots - \psi_{p+d}B^{p+d}\right)\left(1+\Omega_1 B + \Omega_2 B^2 + \cdots\right)$$
$$= \left(1 - \theta_1 B - \theta_2 B^2 - \cdots - \theta_q B^q\right). \tag{2.91}$$

For example, in the previously seen ARIMA(1,1,1)

$$\psi(B) = 1 - (1+\phi)B + \phi B^2$$
$$\theta(B) = 1 - \theta B.$$

Substituting in Eq. (2.90) and equating coefficients of B yields

$$\Omega_0 = A_0 + A_1 = 1$$
$$\Omega_1 = A_0 + A_1\phi$$
$$\Omega_2 = A_0 + A_1\phi^2$$
$$\vdots$$
$$\Omega_j = A_0 + A_1\phi^j,$$

where

$$A_0 = \frac{1-\theta}{1-\phi}$$

$$A_1 = \frac{\theta-\phi}{1-\phi}.$$

So the ARIMA$(1,1,1)$ model is equivalent:

$$X_t = \sum_{j=0}^{\infty} \left(A_0 + A_1\phi^j\right) a_{t-j}. \tag{2.92}$$

Finally, the ARIMA model can be expressed in terms of previous Xs and the current shock a_t. Starting with the general model

$$\psi(B)X_t = \theta(B)a_t \tag{2.93}$$

the goal is to obtain,

$$\pi(B)X_t = a_t, \tag{2.94}$$

where

$$\pi(B) = \left(1 - \sum_{j=1}^{\infty} \pi_j B^j\right).$$

Using Eq. (2.93) in Eq. (2.94) yields

$$\psi(B)X_t = \theta(B)\pi(B)X_t \tag{2.95}$$

or

$$\left(1 - \psi_1 B - \cdots - \psi_{p+d}B^{p+d}\right)$$
$$= \left(1 - \theta_1 B - \cdots - \theta_q B^q\right)\left(1 - \pi_1 B - \pi_2 B^2 - \cdots\right) \tag{2.96}$$

The coefficients π_j are obtained by equating coefficients of B^j. It is easy to corroborate that the coefficients π_j must add up to 1 for $d \geq 1$.

Again, using the ARIMA(1,1,1) as an example, Eq. (2.96) takes the form

$$\pi(B) = \psi(B)\theta^{-1}(B)$$
$$= \left[1 - (1+\phi)B + \phi B^2\right](1 + \theta B + \theta^2 B^2 + \cdots)$$

resulting in

$$\pi_1 = \phi + (1-\theta),$$
$$\pi_2 = (\theta - \phi)(1-\theta), \tag{2.97}$$
$$\vdots \qquad \vdots$$
$$\pi_j = (\theta - \phi)(1-\theta)\theta^{j-2} \qquad j \geq 3.$$

2.5.2 Seasonal Nonstationarities

Hydrologic time series of time scales less than a year usually exhibit strong seasonal variability or nonstationarity. The nonstationarity of monthly

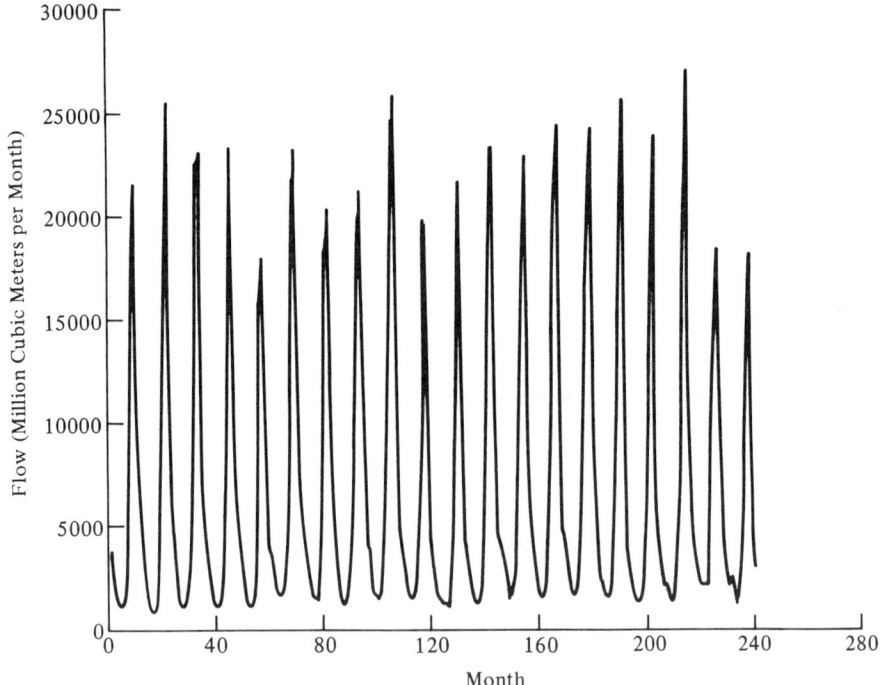

Figure 2.14 Wadi Halfa Streamflow (1921–1940).

streamflows and rainfall volumes in the Nile basin are evident in Figs. 2.14 and
2.15. Figure 2.16 gives the correlation of monthly streamflows of the South
Saskatchewan River (computed around the annual mean). The reader will
recognize the nondecaying nature of the correlation function, with similar high
correlations every 12 months. Such behavior represents a particular type of
nonstationarity, where

$$X_t = X_{t-s} + \varepsilon_t, \tag{2.98}$$

and s is the basic period of the nonstationarity.

Equation (2.98) is of the form,

$$(1 - B^s) X_t = \varepsilon_t, \tag{2.99}$$

where the operator B^s implies

$$B^s X_t = X_{t-s},$$

and ε_t is a residual with undefined structure. In fact, ε_t can be modeled by a

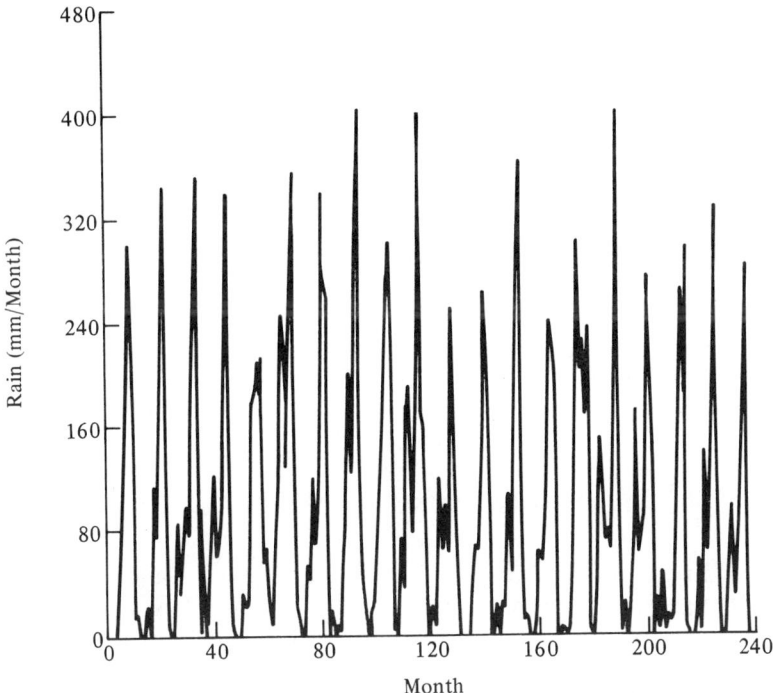

Figure 2.15 Addis Ababa Rainfall (1921–1940).

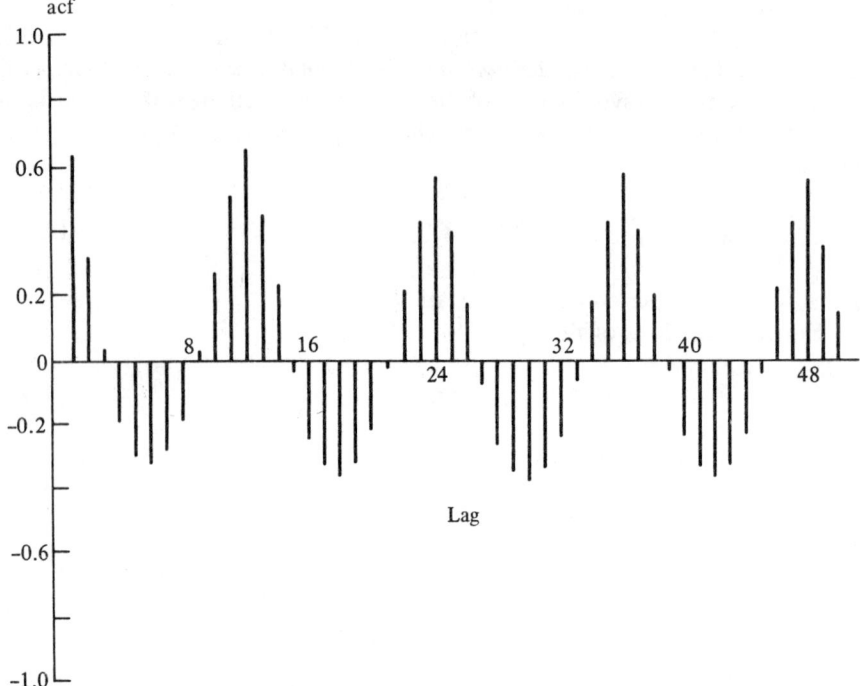

Figure 2.16 Autocorrelation function for data given for the South Saskatchewan River (from Hipel, 1975).

general ARIMA(p, d, q) model. The concept of first-order seasonal differences can be extended to higher-order, D, differences. It is further possible to suggest a seasonal ARMA structure for the ε_t on top of the nonseasonal ARIMA(p, d, q) structure. In summary the general seasonal multiplicative ARIMA model becomes

$$\phi(B)\phi(B^s)(1-B)^d(1-B^s)^D X_t = \theta(B)\theta(B^s)a_t,$$

or

$$\phi(B)\phi(B^s)\nabla^d \nabla_s^D X_t = \theta(B)\theta(B^s)a_t, \qquad (2.100)$$

where $\phi(B)$ are $\theta(B)$ are polynomials of B of degrees p and q, respectively, and $\phi(B^s)$ and $\theta(B^s)$ are polynomials of B^s of degrees P and Q, respectively. The basic seasonal period is s, d is the order of local differentiation, and D is the order of seasonal differentiation. The model in Eq. (2.100) is called an ARIMA(p, d, q)×(P, D, Q) model. Note that Eq. (2.100) represents an ARMA model (with seasonal components) operating on a stationary series, resulting from differences of the original X_t sequence,

$$Z(t) = \nabla^d \nabla_s^D X_t. \tag{2.101}$$

Traditionally, hydrologists have handled seasonal nonstationarities in the mean as well as variance by externally estimating the first two moments and then normalizing to a zero mean, unit-variance process. Ideally, given the continuous process $X(t)$, normalization yields,

$$Z(t) = \frac{X(t) - \mu(t)}{\sigma(t)}. \tag{2.102}$$

If nonstationarity is limited to the mean and variance, then $Z(t)$ can be modeled and studied as a stationary process. Generally continuous estimates of $\mu(t)$ and $\sigma(t)$ result from deterministic Fourier-series analysis on the original series and on discrete estimates of the variance, respectively. Note that for Fourier-series analysis it is inherently assumed that the signal is periodic. Furthermore, discrete data must be given at equal intervals. Chapter 4 details the concept of Fourier series. The purpose of Fourier-series analysis in the present context is to identify the dominant harmonics of the process, which can then be interpreted as the seasonally varying elements to be taken from the original series. For example, monthly data should have a fundamental period of 12 months with the additional possibility of biannual and quarterly components. Many investigators (Roesner and Yevjevich, 1966; Rodríguez-Iturbe, 1968; Quimpo, 1968; and Rodríguez-Iturbe and Nordin, 1968) have used the above procedure to handle seasonal nonstationarity.

Another approach, particularly useful when dealing with monthly data (or data at larger time intervals) is simply to compute seasonal (e.g., monthly, quarterly) means and variances to be used in the standardizing equation (Eq. 2.102). This and the previous method requires a considerable number of years of data. For example, the mean and variance of the month of January may be considered constant for all months of January and must be computed from several years of observations on that month.

The time series Z_t resulting from the standardization of Eq. (2.102) may still result in a series with correlations between seasons of nonstationary nature. That is, the correlation is not only a function of lag but may depend on the absolute time or season. Thomas and Fiering (in Maass et al., 1962; and in Fiering and Jackson, 1971) suggest an AR(1), with time-varying coefficients, that handles seasonally varying lag-one correlations. It is commonly called the univariate seasonal, lag-one, autoregressive model. It takes the following form,

$$\left(X_{t,j} - m_j\right) = \rho_j \frac{\sigma_j}{\sigma_{j-1}} \left(X_{t,j-1} - m_{j-1}\right) + \sigma_j \left(1 - \rho_j^2\right)^{1/2} W_{t,j}, \tag{2.103}$$

where $X_{t,j}$ is the value of random process (i.e., streamflow) at year t and season j,

ρ_j is the lag-one correlation coefficient between seasons j and $j-1$,

m_j is the mean value of season j,

σ_j is the standard deviation at season j, and

$W_{t,j}$ is an independent normal random variable with mean 0 and variance 1.

The above model will preserve not only the seasonal means, but also the seasonal variances and correlations of the process. To complete the definition, it must be stated that

$$X_{t,j*+1} = X_{t+1,1},\qquad(2.104)$$

where $j*$ is the last season of year t.

By taking expected values of Eq. (2.103), it is obvious that a zero-mean process is defined, implying that the correct mean values are preserved. Redefining this zero-mean process as

$$Z_{t,j} = X_{t,j} - m_j,$$

the variance is

$$E\left[Z_{t,j}^2\right] = \rho_j^2 \frac{\sigma_j^2}{\sigma_{j-1}^2} E\left[Z_{t,j-1}^2\right] + \sigma_j^2\left(1-\rho_j^2\right) E\left[W_{t,j}^2\right]$$

$$= \rho_j^2 \frac{\sigma_j^2}{\sigma_{j-1}^2} \sigma_{j-1}^2 + \sigma_j^2\left(1-\rho_j^2\right)$$

$$= \sigma_j^2,$$

which proves preservation of each σ_j^2 assuming that σ_{j-1}^2 is preserved.

The above uses a circular argument. Assuming a two-season model, the proof of the preservation of the variance is

$$E\left[Z_{t,2}^2\right] = \rho_2^2 \frac{\sigma_2^2}{\sigma_1^2} E\left[Z_{t,1}^2\right] + \sigma_2^2\left(1-\rho_2^2\right)$$

$$= \rho_2^2 \frac{\sigma_2^2}{\sigma_1^2}\left[\rho_1^2 \frac{\sigma_1^2}{\sigma_2^2} E\left[Z_{t-1,2}^2\right] + \sigma_1^2\left(1-\rho_1^2\right)\right] + \sigma_2^2\left(1-\rho_2^2\right).$$

Using the stationarity assumption,

$$E\left[Z_{t,2}^2\right] = \rho_2^2\rho_1^2 \frac{\sigma_2^2}{\sigma_1^2} \frac{\sigma_1^2}{\sigma_2^2} E\left[Z_{t,2}^2\right] + \rho_2^2 \frac{\sigma_2^2}{\sigma_1^2}\sigma_1^2\left(1-\rho_1^2\right) + \sigma_2^2\left(1-\rho_2^2\right)$$

$$= \rho_2^2\rho_1^2 E\left[Z_{t,2}^2\right] + \rho_2^2\sigma_2^2\left(1-\rho_1^2\right) + \sigma_2^2\left(1-\rho_2^2\right)$$

$$= \rho_2^2\rho_1^2 E\left[Z_{t,2}^2\right] + \sigma_2^2\left(1-\rho_2^2\rho_1^2\right)$$

$$E\left[Z_{t,2}^2\right] = \frac{\sigma_2^2\left(1-\rho_2^2\rho_1^2\right)}{1-\rho_2^2\rho_1^2} = \sigma_2^2.$$

The lag-one covariance of the seasonal model is

$$E\left[Z_{t,j}Z_{t,j-1}\right] = \rho_j\frac{\sigma_j}{\sigma_{j-1}}E\left[Z_{t,j-1}\right]^2 + \sigma_j\left(1-\rho_j^2\right)^{1/2}E\left[W_{t,j}Z_{t,j-1}\right]$$

$$= \rho_j\frac{\sigma_j}{\sigma_{j-1}}\sigma_{j-1}^2 = \rho_j\sigma_j\sigma_{j-1}.$$

The above in fact is the covariance between seasons j and $j-1$. By definition, this leads to

$$\rho_j = \frac{E\left[X_{t,j}X_{t,j-1}\right]}{\sigma_j\sigma_{j-1}},$$

which shows that the lag-one correlation is preserved.

Attempts to preserve skewness with the seasonal autoregressive model follow the same ideas as presented for the single season case (Eq. 2.41). The use of log-normal transformations will be illustrated in Chapter 3 on seasonal multivariate models.

Following Fiering and Jackson (1971), if γ_j is the skewness of season j, it is also possible to generate a process approximately gamma-distributed with the correct skewness by using an equation of the form

$$(X_{t,j} - m_j) = \rho_j\frac{\sigma_j}{\sigma_{j-1}}\left(X_{t,j-1} - m_{j-1}\right) + \sigma_j\left(1-\rho_j^2\right)^{1/2}\varepsilon_{t,j} \qquad (2.105)$$

where $\varepsilon_{t,j}$ is an approximately gamma-distributed variate with skewness given by

$$\gamma_{\varepsilon j} = \frac{\left[\gamma_j - \rho_{j-1}^3\gamma_{j-1}\right]}{\left(1-\rho_j^2\right)^{1.5}}. \qquad (2.106)$$

The variates $\varepsilon_{t,j}$ are generated using the familiar expression

$$\varepsilon_{t,j} = \frac{2}{\gamma_{\varepsilon j}}\left(1 + \frac{\gamma_{\varepsilon j}W_{t,j}}{6} - \frac{\gamma_{\varepsilon j}^2}{36}\right)^3 - \frac{2}{\gamma_{\varepsilon j}}. \qquad (2.107)$$

Salas and Yevjevich (1972) have extended the concept of seasonally varying coefficients to the AR(2) and AR(3) models while Tao and Delleur (1976) and Delleur and Kavvas (1978) have done the same for the general ARMA(p,q) model, which now takes the form

$$Z_{t,j} = \sum_{k=1}^{p}\phi_{k,j}Z_{t,j-k} - \sum_{k=1}^{q}\theta_{k,j}a_{t,j-k} + a_{t,j}, \qquad (2.108)$$

where $Z_{t,j}$ is a standardized series resulting from Eq. (2.102). The reader is referred to their work for details of model identification, parameter estimation, and model testing. See also Salas et al. (1982) for parameter estimation of such models.

Rao and Kashyap (1973, 1974) and Kashyap and Rao (1976) suggested another stochastic model for monthly streamflow that is nonstationary in mean and variance. The model structure is

$$X(k,i) = \sum_{\ell=1}^{n_1} \alpha_\ell X(k, i-\ell) + U(i) + V(k,i) + \sum_{\ell=1}^{n_2} \alpha_{n_1+\ell} V(k, i-\ell)$$

$$(i = 0,1,\ldots,11; \ k = 0,1,2,\ldots),$$ (2.109)

where $X(k,i)$ is discharge during the ith month of the kth year.

$$\left.\begin{array}{l} X(k, i-\ell) = X(k-1, 12+i-\ell) \\ V(k, i-\ell) = V(k-1, 12+i-\ell) \end{array}\right\} \quad \text{if } i - \ell = 0.$$

$$U(i) = \alpha_0 + \sum_{j=1}^{n_3} \left(\alpha_{n_1+n_2+2j-1}\cos w_j i + \alpha_{n_1+n_2+2j}\sin w_j i \right)$$

$$w_j = 2\pi j/12$$

$$V(k,i) = \psi(i) W(k,i)$$

$$\psi(i) = \beta_0 + \sum_{j=1}^{n_4} \left(\beta_{2j-1}\cos w_j i + \beta_{2j}\sin w_j i \right)$$

$$\mathbf{N} = \{ n_1, n_2, n_3, n_4 \} \text{ are structural parameters}$$

$$\boldsymbol{\alpha} = \{ \alpha_0, \alpha_1, \ldots, \alpha_{n_1+n_2+2n_3} \} \text{ are model coefficients}$$

$$\boldsymbol{\beta} = \{ \beta_0, \beta_1, \ldots, \beta_{2n_4} \} \text{ are model coefficients.}$$

The random sequence $W(k,i)$ is assumed to satisfy

$$E[W(k,i)] = 0 \quad \forall k, i$$

$$E[W(k,i)W(k',i')] = \delta(k-k')\delta(i-i')\sigma_w^2 \quad \forall k, k', i, i'$$

$$E[W(k,i)X(k, i-j)] = 0 \quad \forall k, i; \qquad j > 0$$

$\delta(\cdot)$ is the Kronecker delta function.

Peculiarities of the above model are the periodic deterministic function $U(i)$, which reflects the annual variation in the mean flow and the random noise of variable amplitude and variance introduced by the $V(k,i)$ term. For monthly flows the period is 12.

Identification of the order and the parameters of the above model requires considerable effort and is based on iterative procedures. The model can be used for simulation and forecasting. Chapter 3 will discuss in more detail the multivariate version of this model developed by Curry and Bras (1980).

The main difficulty of all the seasonal models previously discussed, and particularly those with time-varying coefficients, is their lack of parsimony. The number of coefficients involved can be very large and, with limited data, are poorly estimated. As an illustration of this problem, the simple univariate seasonal autoregressive model has 36 parameters when used with a monthly time series.

2.6 MODEL IDENTIFICATION, ESTIMATION, AND VERIFICATION

Box and Jenkins (1976) proposed an iterative algorithm for model identification, parameter estimation, and model verification. Figure 2.17 illustrates a version of their approach.

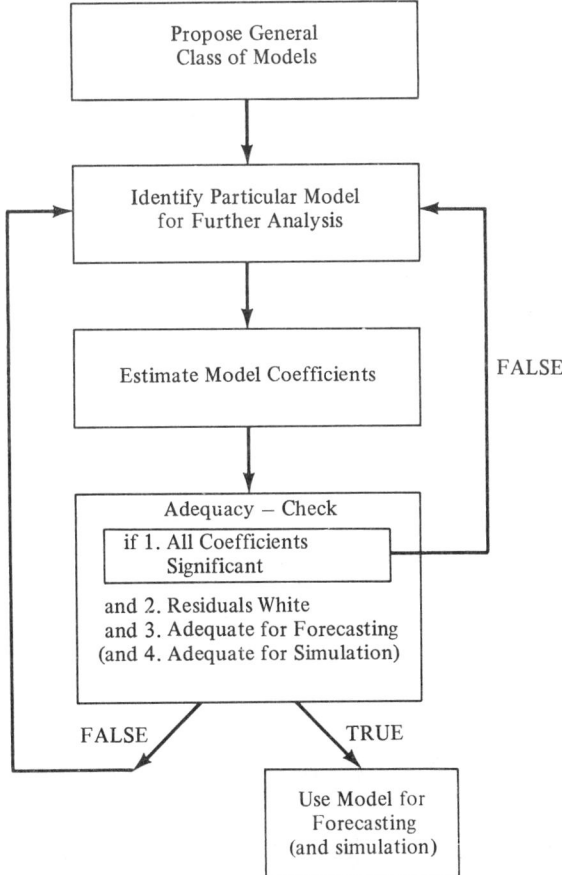

Figure 2.17   Iterative approach to model building (from Curry and Bras, 1980).

The identification phase selects a series or class of models to be studied. The decision is based mostly on the analysis of the statistical properties of the original data and their comparison to the theoretical behavior of possible models. During the second phase the parameters of the set of possible models are identified, mainly by iterative refinement of reasonable first estimates. The third phase checks the reasonableness of the various estimated models and either selects one or eliminates all, and goes back to the identification of other model alternatives. This third phase of verification or diagnosis operates on the expected theoretical behavior of model residuals, the estimates of the white-noise sequence, a_t. Although the iterative approach follows a sequential scheme, the reader should realize that the various steps are not completely independent. This will be clearer in the following detailed discussion of procedures.

2.6.1 Model Identification

The following ideas follow closely the general concepts and sequence of Hipel et al. (1977).

The first step in model identification is visually inspecting a time plot of the original series. Strong seasonalities, nonstationary trends, outliers, and extreme values should be evident from this plot. Less obvious may be the nature of short- and long-term persistence. It may be feasible to decide between seasonal and nonseasonal models, to decide if local differentiation may be required to eliminate trends, and to get a feeling of the order of models (based on persistence) that may be investigated.

Estimates of the autocorrelation function, ρ_k, can be obtained from

$$r_k = \frac{\dfrac{1}{N} \sum_{t=1}^{N-k} (X_t - \overline{X})(X_{t+k} - \overline{X})}{S^2}, \tag{2.110}$$

where $\overline{X}$ and S^2 are given by Eqs. (2.35) and (2.36), and N is the number of available data points. The autocorrelation should be estimated up to lag k on the order of $N/10$ to $N/4$. This limitation is due to increased variability of the above estimator with increasing lag. Bartlett (1946) and Box and Jenkins (1976) give the following approximate expression for the variance of the estimated autocorrelation coefficient of a stationary process,

$$\text{var}[r_k] \approx \frac{1}{N} \sum_{\nu = -\infty}^{\infty} \{ \rho_\nu^2 + \rho_{\nu+k}\rho_{\nu-k} - 4\rho_k\rho_\nu\rho_{\nu-k} + 2\rho_\nu^2\rho_k^2 \}. \tag{2.111}$$

Note that the variance of r_k depends on the true values of ρ_k, which are essentially unknown. In the moving-average model, the correlation is expected to die out for $k > q$; then Eq. (2.111) yields

$$\text{var}[r_k] \approx \frac{1}{N}\left\{1 + 2\sum_{\nu=1}^{q} \rho_\nu^2\right\}; \qquad k > q. \qquad (2.112)$$

In practice, Eqs. (2.111) and, in particular, (2.112), require the use of estimators of the correlations instead of the unknown true ρ_k. In doing so, the error is somewhat underestimated by the given expressions.

The covariance between estimated correlations r_k, r_{k+s} is given by Bartlett (1946) as

$$\text{cov}(r_k, r_{k+s}) \approx \frac{1}{N}\sum_{\nu=-\infty}^{\infty} \rho_\nu \rho_{\nu+s}, \qquad (2.113)$$

implying that large covariances may exist between neighboring values, and care must be exercised in the interpretation of individual and sample autocorrelations.

From the plotted autocorrelation function r_k it is possible to detect wave patterns with fundamental period s. This will clearly indicate seasonality in the series. If the correlation of lag s or multiples of s remains significant and does not seem to decay, then a nonstationary seasonal model is probable. Such behavior is represented by the South Saskatchewan River autocorrelation of monthly flow (measured at Saskatoon, Saskatchewan, Canada) illustrated in Fig. 2.16.

The original time series can then be differenced to eliminate seasonal nonstationarity. The differenced Y_t series exhibits the autocorrelation shown in Fig. 2.18. The properly differenced series exhibits decaying correlations at all lags, although seasonality is still present.

Equation (2.112) can be used to check the significance of decaying correlations. The hypothesis that the differenced sequence Y_t corresponds to a pure MA model of order q can be tested by computing 95% confidence limits approximately as $2\sqrt{\text{var}\,r_k}$. Remember that in a pure MA(q) model the correlation dies after lag q. In a pure ARMA$(0,0,q)\times(0,0,Q)$ the correlation dies after lag $q + sQ$. For example, a white-noise hypothesis implies $\text{var}\,r_k = 1/N$. For the South Saskatchewan example with 626 months of data the 95% confidence limit is $0.08(2\sqrt{1/626}\,)$, which clearly indicates nonwhite structure in the correlation of Fig. 2.18 since the confidence limit is clearly exceeded several times. A similar test for $q = 1$ and $q = 2$ would indicate that a pure MA(1) or a pure MA(2) is not a reasonable model. In fact, the decay of the first few lags seems gradual and autoregressive in nature.

The possibility of a pure seasonal moving-average component of order Q can be tested by computing,

$$\text{var}\,r_k \approx \frac{1}{N}\left(1 + 2\sum_{i=1}^{q+sQ} r_i^2\right) \qquad k > q + sQ.$$

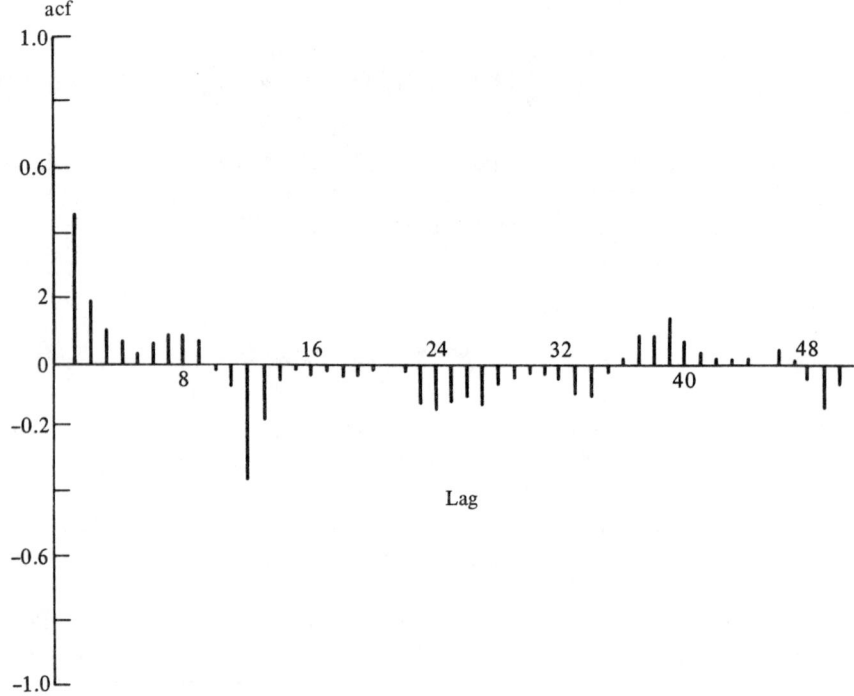

Figure 2.18 Autocorrelation function for the seasonally differenced South Saskatchewan River data (from Hipel, 1975).

With $s = 12$, $q = 0$, and $Q = 1$

$$\text{var } r_{12} \approx \frac{1}{626}\left(1 + 2\sum_{i=1}^{12} r_i^2\right) = 0.0021.$$

From the above, the 95% confidence limit is about 0.1, which makes the lag-12 correlation significant and the other correlations marginal at best. So a seasonal moving-average of first order seems to be a candidate for a component of a good model.

The partial autocorrelation function (pacf) plays the same role in identifying pure autoregressive models that the autocorrelation plays in discovering pure MA models. The pacf dies abruptly for an AR model after the order of the model. In contrast, the pacf of a MA model dies gradually (as the acf does for the AR model). Assume that a time series of unknown origin has correlation function $\rho_k (k = 1, 2, \ldots)$. The idea is to fit the observed correlations to an autoregressive model of order p using the Yule–Walker equations. This allows the reproduction of $\rho_1, \ldots, \rho_p$ by the selection of p parameters ϕ_{pi} $(i = 1, \ldots, p)$. For different values of p, the sequence of coefficients ϕ_{pp} forms the partial autocorrelation function. Different autocorrelations ρ_k will produce different partial autocorrelations ϕ_{pp}. If in fact the infinite-value function ρ_k $(k=1,2,\ldots)$ belongs to an AR($\tilde{p}$) then the partial autocorrelation will be zero for values of $p > \tilde{p}$. This should be obvious given the definition of an AR($\tilde{p}$) model. Note that while the correlation function ρ_k of an autoregressive model exists for values of k going to infinity, its partial autocorrelation is finite, having as many nonzero elements as the order of the model.

The partial autocorrelation is estimated using sample correlation functions in the Yule–Walker equation or by any other method (i.e., least squares) of successively fitting autoregressive models of increasing order.

Box and Jenkins (1976) give a recursive formula, attributable to Durbin (1960), for computing partial autocorrelations:

$$\hat{\phi}_{11} = r_1$$

$$\hat{\phi}_{p+1,p+1} = \frac{r_{p+1} - \sum_{j=1}^{p} \hat{\phi}_{p,j} r_{p+1-j}}{1 - \sum_{j=1}^{p} \hat{\phi}_{pj} r_j} \tag{2.114}$$

where

$$\hat{\phi}_{p+1,j} = \hat{\phi}_{pj} - \hat{\phi}_{p+1,p+1}\hat{\phi}_{p,p-j+1} \qquad j=1,2,\ldots,p$$

(r_i is the ith sample autocorrelation).

Under the hypothesis that the model is AR(p), estimates of the partial autocorrelation $\phi_{\ell\ell}$, $\ell > p$ are independently distributed and approximately normal (Box and Jenkins, 1976). The variance of the estimate is only a function of the number of observations used in obtaining the parameters,

$$\mathrm{var}\left[\hat{\phi}_{\ell\ell}\right] \approx \frac{1}{N} \qquad \ell > p. \tag{2.115}$$

Figure 2.19 shows the partial autocorrelation of the differenced South Saskatchewan River data, including the 95% confidence band ($2\sqrt{1/626}$). It is apparent that an AR(1) component is a strong possibility since correlations at lags >1 and <12 are negligible. The fact that the pacf dies off slowly but is not negligible at lag multiples of 12 (or nearly so) corroborates the previous suspicion of the presence of a lag-one seasonal moving average.

If both the autocorrelation and the partial autocorrelation fail to abruptly cut-off, it will be an indication of an ARMA model, a seasonal one if periodicities are present. Table 2.2 summarizes the expected behavior of AR, MA, and ARMA models.

Hipel and his co-workers (1975, 1977) recommended the use of the inverse partial autocorrelation and the inverse autocorrelation to refine further the

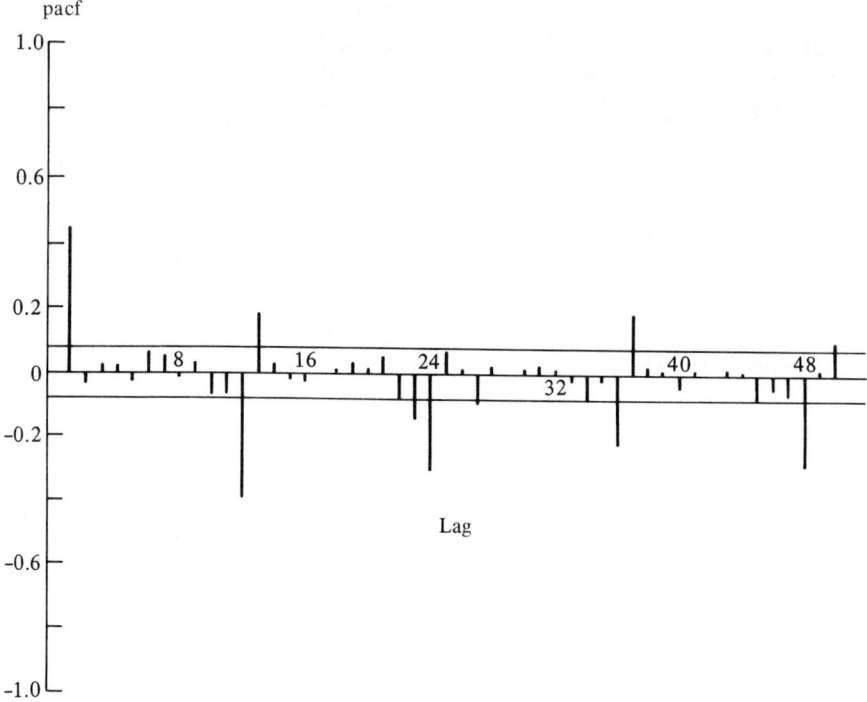

Figure 2.19 Partial autocorrelation function for the seasonally differenced South Saskatchewan River data (from Hipel, 1975).

model identification. Here, the inverse partial autocorrelation plays the same role as does the autocorrelation for pure MA models. The inverse autocorrelation behaves as the partial autocorrelation for AR models. Details of their estimation can be found in the above references.

2.6.2 Parameter Estimation

Given a set of possible models, parameters are first roughly estimated and then refined in several iterative procedures. The following discussion follows the ideas of Delleur (1978) and Salas et al. (1980).

A rough initial parameter estimate can be achieved using the method of moments. For a pure autoregressive model the Yule–Walker equations usually provide an excellent first estimate as long as the sample correlations are well within the valid region of the selected model. Figure 2.4 provides this solution for the common AR(2) model. Example 2.3 illustrated the quality of the estimate.

Table 2.2
Properties of AR, MA, and ARMA processes

Process	Autocorrelation	Partial Autocorrelation
AR(p)	$\rho_k = \sum\limits_{j=1}^{p} \phi_j \rho_{k-j}$	Peaks at lags 1 through p then cuts off; zero for lags greater than p
MA(q)	Zero for lags greater than q	Gradually decays
ARMA(p, q)	Irregular for lags 1 through q, then obeys $\rho_k = \sum\limits_{i=1}^{p} \phi_i \rho_{k-i}$ for $k \geq q+1$	Decays

Initial parameter estimates for the pure moving average are available after iteratively solving the nonlinear Eq. (2.63) [Fig. 2.8 for MA(2)]. The nonlinearity, though, seriously hampers the robustness and accuracy of the initial estimates relative to the AR(p) results. Delleur (1978) and Salas et al. (1980) suggest the following procedure. Equate the theoretical covariance to the sample estimates

$$\hat{\gamma}_0 = \sigma_a^2 \sum_{j=0}^{q} \theta_j^2 \tag{2.116}$$

and

$$\hat{\gamma}_k = \sigma_a^2 \sum_{j=0}^{q-k} \theta_j \theta_{j+k}, \qquad k \leq q, \tag{2.117}$$

from which

$$\sigma_a^2 = \frac{\hat{\gamma}_0}{\sum\limits_{j=0}^{q} \theta_j^2} \tag{2.118}$$

$$\theta_j = -\frac{\hat{\gamma}_j}{\sigma_a^2} + \sum_{k=1}^{q-j} \theta_k \theta_{k+j}. \tag{2.119}$$

First, set $\theta_1, \theta_2, \ldots, \theta_q$ to zero. Using $\theta_0 = 1$ obtain σ_a^2. With the estimated σ_a^2 and Eq. (2.119), obtain estimates of θ_j, $j = 1 \ldots q$. Iterate on Eqs. (2.118) and (2.119) until stable parameter values are obtained. The above is essentially a trial-and-error solution to Eq. (2.63).

Box and Jenkins (1976) give a quadratically (as opposed to linearly in the previous technique) convergent solution to the estimation of $MA(q)$ parameters. The solution is attributed to Wilson (1969). Following Box and Jenkins (1976) define

$$\tau^T = (\tau_0, \tau_1, \ldots, \tau_q),$$

where

$$\tau_0^2 = \sigma_a^2$$

and

$$\theta_j = -\tau_j/\tau_0 \qquad j = 1, 2, \ldots, q. \tag{2.120}$$

τ^T is the transpose of an auxiliary vector related to the parameters as given by Eq. (2.120). The vector τ is iteratively updated.

If τ^i is the ith iteration, then

$$\tau^{i+1} = \tau^i - (\mathbf{K}^i)^{-1}\mathbf{f}_i, \tag{2.121}$$

where

$$\mathbf{f}^T = (f_0, f_1, \ldots, f_q); \qquad f_j = \sum_{i=0}^{q-j} \tau_i \tau_{i+j} - \hat{\gamma}_j \tag{2.122a}$$

and

$$\mathbf{K} = \begin{bmatrix} \tau_0 & \tau_1 & \cdots & \tau_{q-2} & \tau_{q-1} & \tau_q \\ \tau_1 & \tau_2 & \cdots & \tau_{q-1} & \tau_q & 0 \\ \tau_2 & \tau_3 & \cdots & \tau_q & 0 & 0 \\ \cdot & \cdot & & \cdot & \cdot & \cdot \\ \cdot & \cdot & & \cdot & \cdot & \cdot \\ \cdot & \cdot & & \cdot & \cdot & \cdot \\ \tau_q & 0 & & 0 & 0 & 0 \end{bmatrix} + \begin{bmatrix} \tau_0 & \tau_1 & \tau_2 & \cdots & \tau_q \\ 0 & \tau_0 & \tau_1 & \cdots & \tau_{q-1} \\ 0 & 0 & \tau_0 & & \tau_{q-2} \\ \cdot & \cdot & \cdot & & \cdot \\ \cdot & \cdot & \cdot & & \cdot \\ \cdot & \cdot & \cdot & & \cdot \\ 0 & 0 & 0 & & \tau_0 \end{bmatrix}. \tag{2.122b}$$

If the autoregressive order p of the $ARMA(p, q)$ model is known, parameter estimation exploits the mixed behavior of the theoretical ARMA correlation. For $k \geq q+1$ the $ARMA(p, q)$ obeys the Yule–Walker equations. Therefore the covariance estimates can be substituted in the Yule-Walker equations to obtain

$$\hat{\gamma}_{q+1} = \phi_1\hat{\gamma}_q + \phi_2\hat{\gamma}_{q-1} + \phi_3\hat{\gamma}_{q-2} + \cdots + \phi_p\hat{\gamma}_{q-p+1}$$

$$\hat{\gamma}_{q+2} = \phi_1\hat{\gamma}_{q+1} + \phi_2\hat{\gamma}_q + \phi_3\hat{\gamma}_{q-1} + \cdots + \phi_p\hat{\gamma}_{q-p+2} \tag{2.123}$$

$$\vdots \qquad\qquad \vdots$$

$$\hat{\gamma}_{q+p} = \phi_1\hat{\gamma}_{q+p-1} + \phi_2\hat{\gamma}_{q+p-2} + \cdots + \phi_p\hat{\gamma}_q.$$

The above system of equations can be sequentially solved for $\hat{\phi}_1, \ldots, \hat{\phi}_p$. With these initial estimates of the autoregressive terms, a new series is constructed,

$$Z_t^1 = Z_t - \hat{\phi}_1 Z_{t-1} - \cdots - \hat{\phi}_p Z_{t-p}, \tag{2.124}$$

which presumably exhibits pure MA behavior. Once the autocovariance of the new series is computed, the parameters $\theta_1, \ldots, \theta_q$ are estimated from the iterative procedure corresponding to pure MA models.

The reader is reminded that a reasonable method of moments solution is available from Fig. 2.10 for the popular ARMA(1,1) model.

Refinements of parameter estimates is as much a theory as an art. Innumerable procedures exist, in varying degrees of complexity and theoretical purity. Box and Jenkins (1976) remains the most exhaustive and critical study of available approaches. We do not intend to duplicate that work, and we urge the reader to use that reference for any extensive work on time-series modeling. Nevertheless, the main points of refined parameter estimation are clear in an article by Box, Jenkins, and Bacon (1968). The concepts in the following discussion follow that reference.

Most available techniques work on, or are based on, the evaluation of the sum of squares of residuals.

In the general model

$$\phi(B)\nabla^d X_t = \theta(B)a_t \tag{2.125}$$

the goal is to find parameters $\phi^T = \{\phi_1, \ldots, \phi_p\}$ and $\theta^T = \{\theta_1, \ldots, \theta_q\}$ to minimize

$$S(\phi, \theta) = \sum_{t=1}^{N} a_t^2. \tag{2.126}$$

In fact, the log-likelihood function of ϕ and θ is a linear function of $S(\phi, \theta)$.

The residuals a_t must be estimated and clearly are conditional on some parameter values. The basic model can be inverted to yield,

$$a_t = \theta^{-1}(B)\phi(B)\nabla^d X_t,$$

which when expanded in polynomial (of B) form will express a_t as a function of past a_t values and present and past X_t values. For example, an ARIMA $(1,1,1)$ reduces to

$$a_t = \theta a_{t-1} + X_t - (1+\phi)X_{t-1} + \phi X_{t-2}. \tag{2.127}$$

Given a reasonably long time series, it would be possible to iteratively solve for a_t values conditional on assumed parameters ϕ and θ. In the above, the iteration could start with a_3, setting a_2 to its expected value 0, and using observations x_3, x_2, and x_1. This procedure introduces an initial bias and induces a transient effect that will hopefully disappear for long series. Box and Jenkins recommend, particularly for short series, to start the process using maximum likelihood estimates of past values. This is the concept of past forecasting. Essentially, a process obeying Eq. (2.125) also satisfies the reversed process

$$\phi(F)(1-F)^d X_t = \theta(F)a_t, \qquad (2.128)$$

where F is the forward shift operator defined at the beginning of this chapter. In Eq. (2.127), a_1 depends on a_0, X_1, X_0, and X_{-1}. Their maximum likelihood estimates are the predictions using the reversed process, Eq. (2.128), which depends on existing quantities and on the assumed parameters ϕ and θ. The predicted a_0 is its expected value, 0. Note that the backwards prediction starts in the future, that is, the end of the observed series x_t. At that point the same starting difficulties will exist but hopefully the transient will disappear by the time the beginning of the series is reached.

Sum of squares, $S(\phi, \theta)$, can be obtained conditional on several parameter sets, ϕ and θ, the range of which is controlled by the parameter region of the model structure being investigated and the initial parameter estimates resulting from the previously discussed method of moments. It is then possible to evaluate $S(\phi, \theta)$ over a limited space of parameters, draw contours of equal squared errors, and locate the parameter set that minimizes the square of residuals.

It is possible to estimate $1 - \alpha$ confidence intervals on the minimum sum of square residuals resulting from parameter sets ϕ and θ. The $1 - \alpha$ confidence region is included within the contour.

$$S_{1-\alpha}(\phi, \theta) = S(\phi, \theta)\left\{1 + \frac{\chi_\alpha^2(p+q)}{N}\right\}, \qquad (2.129)$$

where $\chi_\alpha^2(\nu)$ is the value of a chi-squared distribution with ν degrees of freedom that may be exceeded with probability α.

Visualizing the sum of squared-residuals contours is difficult when the models have more than two parameters. In such cases it is possible to compute a conditional sum of squares. An example given by Box et al. (1968) is based on

$$(1-\phi B)\nabla X_t = (1 - \theta_1 B - \theta_2 B^2)a_t, \qquad (2.130)$$

which can be decomposed into

$$(1-\phi B)e_t = a_t \qquad (2.131)$$

$$(1 - \theta_1 B - \theta_2 B^2)e_t = \nabla X_t. \qquad (2.132)$$

Given a pair of values θ_1 and θ_2, Eq. (2.132) is used to compute a sequence e_t, which can then be used in the linear Eq. (2.131) to obtain an optimum (in the least-square sense) ϕ. Figure 2.20 gives the results of such a search.

For a pure AR(p) model, the minimizing parameters with a square-error criterion can be analytically obtained. Essentially the linearity of the equation reduces the parameter-estimation problem to a multivariate generalized least-squares problem. From a probabilistic interpretation the problem becomes

$$\min_{\phi}\left\{ S(\phi) = \sum_{t=p}^{n} \left(Z_t - \phi_1 Z_{t-1} - \phi_2 Z_{t-2} - \cdots - \phi_p Z_{t-p} \right)^2 \right\}. \qquad (2.133)$$

Differentiation with respect to the parameters of Eq. (2.133) and equating the result to zero will yield the normal equations, which must then be solved simultaneously.

Box et al. (1968) also give a Bayesian interpretation to the parameter-estimation problem. Given a locally uniform prior distribution, the posterior density function on the parameters is proportional to

$$S^{-N/2}(\phi, \theta).$$

The iso–square-error contours are then posterior probability contours and the $1 - \alpha$ confidence region is the space centered at the posterior mean (the minimum squared error) and over which the posterior density function integrates to $1 - \alpha$.

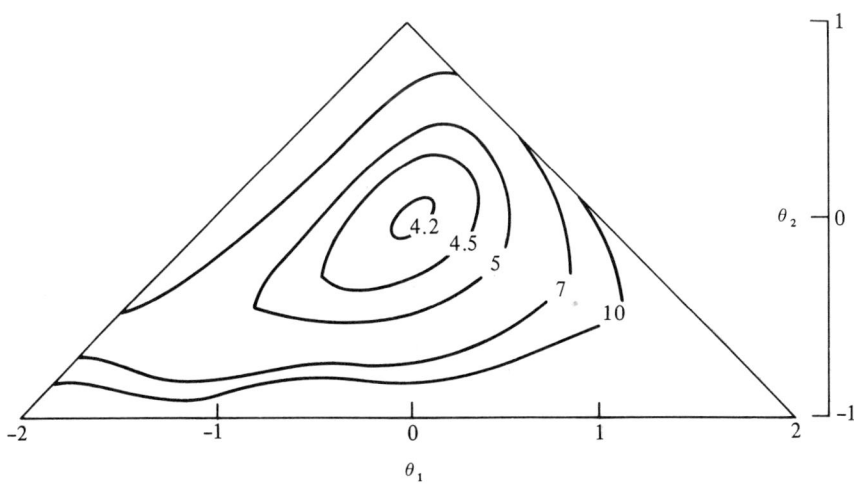

Figure 2.20 Contours of $S(\phi|\theta_1, \theta_2)$ plotted over the admissible parameter space for the θs (from Box and Jenkins, 1976).

Refined estimation is also possible using an iterative nonlinear (but linearized) least squares, assuming the initial parameter values are good. Box and Jenkins (1976), Delleur (1978) and Salas et al. (1980) discuss this procedure. Box et al. (1968) summarize it in the following manner.

The inverted form of Eq. (2.125) is

$$a_t = \phi(B)\theta^{-1}(B)Z_t, \tag{2.134}$$

where

$$Z_t = \nabla^d X_t.$$

The above is a nonlinear equation on the parameters ϕ and θ. Expanding in a Taylor series about present parameter estimates

$$\phi_1^{n-1}, \phi_2^{n-1}, \ldots, \phi_p^{n-1} \text{ and } \theta_1^{n-1}, \theta_2^{n-1}, \ldots, \theta_q^{n-1},$$

and after rearrangement yields approximately

$$a_t^{n-1} = \sum_{i=1}^{p} \left\{ \phi_i - \phi_i^{n-1} \right\} U_{t-i}^{n-1} - \sum_{j=1}^{q} \left\{ \theta_j - \theta_j^{n-1} \right\} V_{t-j}^{n-1}. \tag{2.135}$$

Subscripts and superscripts $n-1$ indicate evaluation or dependence on the $(n-1)$th parameter estimate. Equation (2.135) is then interpreted as a linear least-squares problem where a_t^{n-1} is the dependent variable and U_ts and V_ts are the independent variables. The unknown regression coefficients are ϕ_i $(i = 1, \ldots, p)$ and θ_i $(i = 1, \ldots, q)$, which when estimated become the nth parameter estimates in the iteration. The value of a_t^{n-1} is computed from Eq. (2.134) using the latest estimates of the parameter polynomials $\theta^{n-1}(B)$ and $\phi^{n-1}(B)$. The variables U_t and V_t are

$$U_t = \phi_{n-1}^{-1}(B)a_t^{n-1} = \phi_1^{n-1}U_{t-1} + \cdots + \phi_p^{n-1}U_{t-p} + a_t^{n-1}$$

$$V_t = \theta_{n-1}^{-1}(B)a_t^{n-1} = \theta_1^{n-1}V_{t-1} + \cdots + \theta_q^{n-1}V_{t-q} + a_t^{n-1}.$$

The iteration with Eq. (2.135) may start with estimates of zero for the as, Us, and Vs.

Parameter estimation for seasonal models, particularly multiplicative ones, proceeds in the same manner as outlined in the previous paragraphs. The difficulty lies in computing first estimates of parameters. To aid in this effort, Box and Jenkins (1976), pp. 329-33 give the theoretical autocovariances of the most popular seasonal models (Table 2.3). They may be used to obtain initial parameter estimates using the method of moments. In hydrology the most common seasonal model deals with monthly flows with a 12-month period. Seasonal lags higher than one are rarely required.

Table 2.3
Autocovariances for some seasonal models

Model	(Autocovariances of w_t)/σ_a^2	Special characteristics
(1) $w_t = (1 - \theta B)(1 - \Theta B^s)a_t$ $w_t = a_t - \theta a_{t-1} - \Theta a_{t-s} + \theta\Theta a_{t-s-1}$ $s \geq 3$	$\gamma_0 = (1 + \theta^2)(1 + \Theta^2)$ $\gamma_1 = -\theta(1 + \Theta^2)$ $\gamma_{s-1} = \theta\Theta$ $\gamma_s = -\Theta(1 + \theta^2)$ $\gamma_{s+1} = \gamma_{s-1}$ All other autocovariances are zero	(a) $\gamma_{s-1} = \gamma_{s+1}$ (b) $\rho_{s-1} = \rho_{s+1} = \rho_1\rho_s$
(2) $(1 - \Phi B^s)w_t = (1 - \theta B)(1 - \Theta B^s)a_t$ $w_t - \Phi w_{t-s} = a_t - \theta a_{t-1} - \Theta a_{t-s} + \theta\Theta a_{t-s-1}$ $s \geq 3$	$\gamma_0 = (1 + \theta^2)\left[1 + \dfrac{(\Theta - \Phi)^2}{1 - \Phi^2}\right]$ $\gamma_1 = -\theta\left[1 + \dfrac{(\Theta - \Phi)^2}{1 - \Phi^2}\right]$ $\gamma_{s-1} = \theta\left[\Theta - \Phi - \dfrac{\Phi(\Theta - \Phi)^2}{1 - \Phi^2}\right]$ $\gamma_s = -(1 + \theta^2)\left[\Theta - \Phi - \dfrac{\Phi(\Theta - \Phi)^2}{1 - \Phi^2}\right]$ $\gamma_{s+1} = \gamma_{s-1}$ $\gamma_j = \Phi\gamma_{j-s} \qquad j \geq s + 2$ For $s \geq 4$, $\gamma_2, \gamma_3, \ldots, \gamma_{s-2}$ are all zero	(a) $\gamma_{s-1} = \gamma_{s+1}$ (b) $\gamma_j = \Phi\gamma_{j-s} \qquad j \geq s + 2$

Table 2.3 (Cont'd)

Model	(Autocovariances of w_t)$/\sigma_a^2$	Special characteristics
$(3)\ w_t = (1 - \theta_1 B - \theta_2 B^2)$ $\times(1 - \Theta_1 B^s - \Theta_2 B^{2s})a_t$ $w_t = a_t - \theta_1 a_{t-1} - \theta_2 a_{t-2} - \Theta_1 a_{t-s}$ $+ \theta_1 \Theta_1 a_{t-s-1} + \theta_2 \Theta_1 a_{t-s-2}$ $- \Theta_2 a_{t-2s} + \theta_1 \Theta_2 a_{t-2s-1}$ $+ \theta_2 \Theta_2 a_{t-2s-2}$ $s \geq 5$	$\gamma_0 = (1 + \theta_1^2 + \theta_2^2)(1 + \Theta_1^2 + \Theta_2^2)$ $\gamma_1 = -\theta_1(1 - \theta_2)(1 + \Theta_1^2 + \Theta_2^2)$ $\gamma_2 = -\theta_2(1 + \Theta_1^2 + \Theta_2^2)$ $\gamma_{s-2} = \theta_2 \Theta_1(1 - \Theta_2)$ $\gamma_{s-1} = \theta_1 \Theta_1(1 - \theta_2)(1 - \Theta_2)$ $\gamma_s = -\Theta_1(1 + \theta_1^2 + \theta_2^2)(1 - \Theta_2)$ $\gamma_{s+1} = \gamma_{s-1}$ $\gamma_{s+2} = \gamma_{s-2}$ $\gamma_{2s-2} = \theta_2 \Theta_2$ $\gamma_{2s-1} = \theta_1 \Theta_2(1 - \theta_2)$ $\gamma_{2s} = -\Theta_2(1 + \theta_1^2 + \theta_2^2)$ $\gamma_{2s+1} = \gamma_{2s-1}$ $\gamma_{2s+2} = \gamma_{2s-2}$ All other autocovariances are zero	(a) $\gamma_{s-2} = \gamma_{s+2}$ (b) $\gamma_{s-1} = \gamma_{s+1}$ (c) $\gamma_{2s-2} = \gamma_{2s+2}$ (d) $\gamma_{2s-1} = \gamma_{2s+1}$

(3a) *Special Case of Model 3*

$w_t = (1 - \theta_1 B - \theta_2 B^2)(1 - \Theta B^s) a_t$

$w_t = a_t - \theta_1 a_{t-1} - \theta_2 a_{t-2} - \Theta a_{t-s} + \theta_1 \Theta a_{t-s-1} + \theta_2 \Theta a_{t-s-2}$

$s \geq 5$

$\gamma_0 = (1 + \theta_1^2 + \theta_2^2)(1 + \Theta^2)$

$\gamma_1 = -\theta_1(1 - \theta_2)(1 + \Theta^2)$

$\gamma_2 = -\theta_2(1 + \Theta^2)$

$\gamma_{s-2} = \theta_2 \Theta$

$\gamma_{s-1} = \theta_1 \Theta(1 - \theta_2)$

$\gamma_s = -\Theta(1 + \theta_1^2 + \theta_2^2)$

$\gamma_{s+1} = \gamma_{s-1}$

$\gamma_{s+2} = \gamma_{s-2}$

All other autocovariances are zero

(a) $\gamma_{s-2} = \gamma_{s+2}$

(b) $\gamma_{s-1} = \gamma_{s+1}$

(3b) *Special Case of Model 3*

$w_t = (1 - \theta B)(1 - \Theta_1 B^s - \Theta_2 B^{2s}) a_t$

$w_t = a_t - \theta a_{t-1} - \Theta_1 a_{t-s} + \theta \Theta_1 a_{t-s-1} - \Theta_2 a_{t-2s} + \theta \Theta_2 a_{t-2s-1}$

$s \geq 3$

$\gamma_0 = (1 + \theta^2)(1 + \Theta_1^2 + \Theta_2^2)$

$\gamma_1 = -\theta(1 + \Theta_1^2 + \Theta_2^2)$

$\gamma_{s-1} = \theta \Theta_1(1 - \Theta_2)$

$\gamma_s = -\Theta_1(1 + \theta^2)(1 - \Theta_2)$

$\gamma_{s+1} = \gamma_{s-1}$

$\gamma_{2s-1} = \theta \Theta_2$

$\gamma_{2s} = -\Theta_2(1 + \theta^2)$

$\gamma_{2s+1} = \gamma_{2s-1}$

All other autocovariances are zero

(a) $\gamma_{s-1} = \gamma_{s+1}$

(b) $\gamma_{2s-1} = \gamma_{2s+1}$

Table 2.3 (Cont'd)

Model	(Autocovariances of w_t)/σ_a^2	Special characteristics
(4) $w_t = (1 - \theta_1 B - \theta_s B^s - \theta_{s+1} B^{s+1})a_t$ $w_t = a_t - \theta_1 a_{t-1} - \theta_s a_{t-s}$ $\quad - \theta_{s+1} a_{t-s-1}$ $s \geq 3$	$\gamma_0 = 1 + \theta_1^2 + \theta_s^2 + \theta_{s+1}^2$ $\gamma_1 = -\theta_1 + \theta_s \theta_{s+1}$ $\gamma_{s-1} = \theta_1 \theta_s$ $\gamma_s = \theta_1 \theta_{s+1} - \theta_s$ $\gamma_{s+1} = -\theta_{s+1}$ All other autocovariances are zero	(a) In general, $\gamma_{s-1} \neq \gamma_{s+1}$ $\gamma_1 \gamma_s \neq \gamma_{s+1}$
(4a) *Special Case of Model 4* $w_t = (1 - \theta_1 B - \theta_s B^s)a_t$ $w_t = a_t - \theta_1 a_{t-1} - \theta_s a_{t-s}$ $s \geq 3$	$\gamma_0 = 1 + \theta_1^2 + \theta_s^2$ $\gamma_1 = -\theta_1$ $\gamma_{s-1} = \theta_1 \theta_s$ $\gamma_s = -\theta_s$ All other autocovariances are zero	(a) Unlike model 4, $\gamma_{s+1} = 0$
(5) $(1 - \Phi B^s)w_t = (1 - \theta_1 B - \theta_s B^s - \theta_{s+1} B^{s+1})a_t$ $w_t - \Phi w_{t-s} = a_t - \theta_1 a_{t-1} - \theta_s a_{t-s} - \theta_{s+1} a_{t-s-1}$ $s \geq 3$	$\gamma_0 = 1 + \theta_1^2 + \dfrac{(\theta_s - \Phi)^2}{1 - \Theta^2} + \dfrac{(\theta_{s+1} + \theta_1 \Phi)^2}{1 - \Phi^2}$ $\gamma_1 = -\theta_1 + \dfrac{(\theta_s - \Phi)(\theta_{s+1} + \theta_1 \Phi)}{1 - \Phi^2}$ $\gamma_{s-1} = (\theta_s - \Phi)\left[\theta_1 + \Phi \dfrac{(\theta_{s+1} + \Phi\theta_1)}{1 - \Phi^2}\right]$	(a) $\gamma_{s-1} \neq \gamma_{s+1}$ (b) $\gamma_j = \Phi \gamma_{j-s}$ $j \geq s + 2$

$$\gamma_s = -(\theta_s - \Phi)\left[1 - \Phi\frac{(\theta_s - \Phi)}{1-\Phi^2}\right]$$
$$+ (\theta_{s+1} + \theta_1\Phi)\left[\theta_1 + \Phi\frac{(\theta_{s+1} + \theta_1\Phi)}{1-\Phi^2}\right]$$

$$\gamma_{s+1} = -(\theta_{s+1} + \theta_1\Phi)\left[1 - \Phi\frac{(\theta_s - \Phi)}{1-\Phi^2}\right]$$

$$\gamma_j = \Phi\gamma_{j-s} \qquad j \geq s+2$$

For $s \geq 4$, $\gamma_2, \ldots, \gamma_{s-2}$ are all zero

(5a) *Special Case of Model 5*

$$(1 - \Phi B^s)w_t = (1 - \theta_1 B - \theta_s B^s)a_t,$$
$$w_t - \Phi w_{t-s} = a_t - \theta_1 a_{t-1} - \theta_s a_{t-s}$$
$$s \geq 3$$

$$\gamma_0 = 1 + \frac{\theta_1^2 + (\theta_s - \Phi)^2}{1-\Phi^2}$$

$$\gamma_1 = -\theta_1\left[1 - \Phi\frac{(\theta_s - \Phi)}{1-\Phi^2}\right]$$

$$\gamma_{s-1} = \frac{\theta_1(\theta_s - \Phi)}{1-\Phi^2}$$

$$\gamma_s = \frac{\Phi\theta_1^2 - (\theta_s - \Phi)(1 - \Phi\theta_s)}{1-\Phi^2}$$

$$\gamma_j = \Phi\gamma_{j-s} \qquad j \geq s+1$$

For $s \geq 4$, $\gamma_2, \ldots, \gamma_{s-2}$ are all zero

(a) Unlike model 5,

$$\gamma_{s+1} = \Phi\gamma_1$$

Source: Box and Jenkins (1976).

2.6.3 Verification

Given a correct model structure and parameters, the model residuals a_t should form a white normal random sequence. Model verification and diagnosis determine whether actual residuals have this expected theoretical behavior.

Adequacy of model structure is checked by the concept of overfitting. Essentially, a suspected (look at the autocorrelation and partial autocorrelation again) higher-order model of the same nature is fitted. Compute the corrected residual variance for both models, where the corrected variance (Delleur, 1978, and Salas et al., 1980) is

$$\hat{\sigma}_{a_c}^2 = \frac{1}{N-n} S(\phi, \theta), \tag{2.136}$$

with N being the number of data points and n the number of parameters. The original model remains adequate unless the corrected residual variance has been considerably reduced by the higher-order model.

Hippel (1975) also suggests fitting a high-order, $r(20 \leq r \leq 30)$, AR model to the data and computing its residual variance σ_r^2. A statistic is calculated

$$N \ln \frac{\hat{\sigma}_a^2}{\hat{\sigma}_r^2} \approx \chi^2(r-k), \tag{2.137}$$

where k is the order of the original model ($k = p + q + sP + sQ$). A statistic larger than the $\chi^2(r-k)$ for a chosen level of significance indicates that a model with more parameters may be better.

Lack of correlation of residuals can be investigated in several ways. Graphical display may help in detection of persistence. The autocorrelation and partial autocorrelation of the residuals may be investigated for whiteness using techniques from Section 2.6.1. A portmanteau lack-of-fit test looks at the autocorrelation of the residuals, $r_k(a)$, of the fitted ARMA(p, q) model and computes,

$$Q' = N \sum_{k=1}^{L} r_k^2(a), \tag{2.138}$$

where L is a very high lag so that the weights at lag L of the moving-average form of the model $Z_t = \psi(B)a_t$ are small. The Q' statistic is approximately distributed as χ^2 with $(L - p - q - P - Q)$ degrees of freedom. Comparing Q' with the theoretical χ^2 value at a given level of significance, α, will indicate correlation $[Q' > \chi^2(L - p - q - P - Q)]$ or lack of correlation.

The normalized cumulative periodogram $C(f_j)$ is an estimate of the normalized cumulative spectrum of a process (see Chapter 4). It is estimated as

$$I(f_i) = \frac{2}{N}\left[\left(\sum_{t=1}^{N} a_t \cos 2\pi f_i t\right)^2 + \left(\sum_{t=1}^{N} a_t \sin 2\pi f_i t\right)^2\right]$$

$$C(f_j) = \frac{\sum_{i=1}^{j} I(f_i)}{N\hat{\sigma}_a^2},$$

(2.139)

where the a_ts are the residual estimates and $f_i = i/N$ is a frequency; $\hat{\sigma}_a^2$ is an estimate of the residual variance.

If the normalized cumulative periodogram of the residuals a_t lies around a straight line from $(0,0)$ to $(0.5,1)$ (a property of white noise), then the residuals are uncorrelated and have no dominant frequencies. A large deviation of $C(f_j)$ from a straight line indicates a periodicity at the frequency f_j. The significance of a deviation can be tested using the Kolmogorov–Smirnov statistic,

$$K = \max |S(f) - C(f_j)|,$$

where $S(f)$ is a straight line. The distribution of $|S(f) - C(f_j)|$ is known from tables that allow the solution of

$$P\left[|S(f) - C(f_j)| \geq K_\alpha\right] = \alpha,$$

where $P[\cdot]$ implies probability. If $K < K_\alpha$, then the hypothesis of white noise for the residuals is accepted at level α.

Normality of residuals can be corroborated by plotting the residual frequency on normal paper and checking residual skewness and kurtosis.

Homoscedasticity requires that the variance of residuals, $\hat{\sigma}_a^2$, be constant. The simplest tests are graphical (Draper and Smith, 1966). Chronological plots of a_t, as well as a_t vs. Z_t, will indicate nonstationarities in mean or variance. Other mathematical tests are discussed by Hipel (1975).

Lack of homoscedasticity and normality affects the validity of all parameter-estimation procedures. Box and Jenkins (1976) recommend data transformation in such cases. In particular they suggest power transformations:

$$Z_t^\lambda = \begin{cases} \frac{1}{\lambda}[Z_t + \text{constant}]^\lambda - 1 & \lambda \neq 0 \\ \ln[Z_t + \text{constant}] & \lambda = 0. \end{cases}$$

(2.140)

The estimation of λ and the unknown constant can be a major inference problem. Several theoretical maximum likelihood results exist (see Hipel,

1975); nevertheless trial-and-error remains the most popular method. In hydrology, the dominant result is $\lambda = 0$ and zero constant, leading to the popular logarithmic transformation. A nonzero constant with $\lambda = 0$ is also common in hydrology and corresponds to variables distributed with a three-parameter log-normal distribution.

EXAMPLE 2.4

Hipel (1975) fitted the $\text{ARIMA}(1,0,0) \times (0,1,1)_{12}$ model identified for the South Saskatchewan River and found that the residuals failed most diagnostic tests. The portmanteau statistic had a value of 80.84 with 48 degrees of freedom. This is significant even at the 1% level. An $\text{ARIMA}(1,0,5) \times (0,1,1)_{12}$ model was then suggested to account for the decreasing correlations at lags two, three, four, and five. Parameters were estimated that indicated that θ_4 was not significantly different from zero. They were re-estimated constraining θ_4 to zero. Resulting residuals were uncorrelated but skewed and of varying variance. A Box–Cox transformation with $\lambda = -0.16649$ and zero constant was performed. Figures 2.21 and 2.22 illustrate the lack of correlation of resulting residuals. Normality and constant test were also favorable. The final model was

$$(1 - 0.95354B)Z_t' = (1 - 0.37119B - 0.13932B^2 - 0.08882B^3$$

$$- 0.10771B^5)(1 - 0.95646B^{12})a_t,$$

where

$$Z_t' = (1 - B^{12})\left[\frac{1}{-0.16649}\left(Z_t^{-0.16649} - 1\right)\right].$$

A simple logarithmic transformation ($\lambda = 0$, constant $= 0$) yielded reasonable results and the following model parameters

$$\phi_1 = 0.95230, \qquad \theta_1 = 0.35861, \qquad \theta_2 = 0.16503,$$
$$\theta_3 = 0.08624, \qquad \theta_5 = 0.09740, \qquad \theta_1^s = 0.95706.$$

2.7 SIMULATION

As discussed at the beginning of this chapter, Monte Carlo simulation has played a major role in the design, analysis, and operation of water-resources systems. The objective is to solve an otherwise unmanageable derived-distributions problem. The systems, for example a river basin with multiple reservoirs and water uses, is deterministically represented and excited by a stochastic input series (rainfall, streamflow, evapotranspiration). The output and ob-

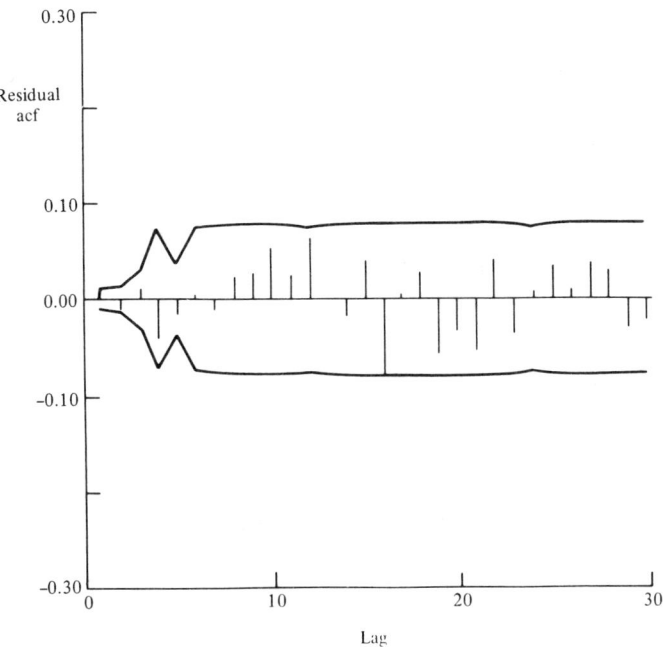

Figure 2.21 Residual acf for the South Saskatchewan River (from Hipel, 1975).

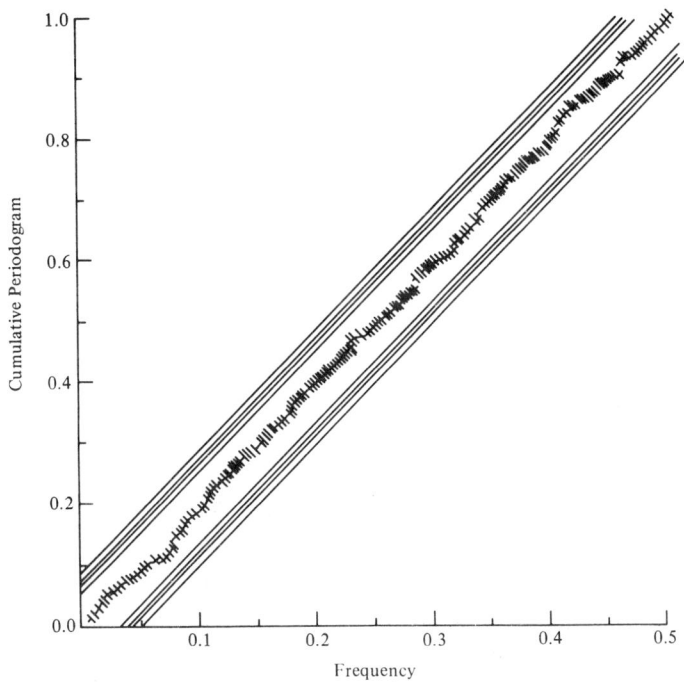

Figure 2.22 Residual cumulative periodogram for the South Saskatchewan River (from Hipel, 1975).

served system behavior is then a stochastic process that can be characterized and studied.

If the identified, estimated, and verified model correctly represents the observed inputs, then it is expected that the sequence of synthetic inputs is well behaved and statistically indistinguishable from the historical series. Nevertheless the user must be aware of the many problems of statistical inference that may seriously affect the adequacy of the model.

Chapter 1 introduced Monte Carlo simulation as the exercise of sampling from a conditional distribution. This is simple when using linear ARIMA models. Given past values of the autoregressive and moving-average terms, simulation requires only sampling a random noise a_t from a normal distribution with zero mean and variance σ_a^2.

Hipel (1975) summarizes the available simulation techniques. His ideas follow.

The simplest way of obtaining an ARMA series of n values is first to express the model in its random-shock form,

$$Z_t = \theta(B)\phi^{-1}(B)a_t$$

$$= \psi(B)a_t.$$

The polynomial $\psi(B)$ is infinite but with decaying weights. It is then possible to approximate it by a truncated polynomial

$$Z_t = \left(1 + \psi_1 B + \psi_2 B^2 + \cdots + \psi_{q'} B^{q'}\right)a_t. \tag{2.141}$$

Generate $(n + q')$ white-noise terms from a normal distribution with mean 0 and variance σ_a^2

$$a_{-q'+1}, a_{-q'+2}, \ldots, a_0, a_1, \ldots, a_n,$$

and use Eq. (2.141) to obtain r values of Z_t, where $r = \max(p, q)$. This set establishes the initial values to generate $n - r$ values based on the exact equation

$$Z_t = \phi_1 Z_{t-1} + \phi_2 Z_{t-2} + \cdots + \phi_p Z_{t-p} + a_t - \theta_1 a_{t-1} - \cdots - \theta_q a_{t-q}.$$

The possible errors of such a procedure are due to the transitory effects of the approximation in Eq. (2.141). Generally these are minimal. The procedure is exact for a pure MA(q) model.

Possible approximation errors can be reduced by generating r ($r > \max(p, q)$) initial Z_t values with the theoretically correct (corresponding to the model at hand) autocovariance function, γ_k; $k = 0, 2, \ldots, r - 1$. This is achieved in the following manner. Form an $(r \times r)$ covariance matrix

$$\Gamma_r = \{\gamma_{i-j}\} = \mathbf{BB}^T.$$

Following procedures to decompose the matrix Γ_r find the matrix $\mathbf{B}$. (Chapter 3 discusses these methods in detail.) A sequence of random variables with the correct correlation can be generated as

$$\begin{bmatrix} Z_1 \\ \vdots \\ Z_r \end{bmatrix} = \mathbf{B}\boldsymbol{\varepsilon}, \tag{2.142}$$

where $\boldsymbol{\varepsilon}$ is an $(r \times 1)$ vector of normal deviates with 0 mean and unit variance. Given the Z_i $(i = 1, \ldots, r)$ the corresponding a_i $(i = 1, \ldots, r)$ are solved for, using the reversed-process expression, Eq. (2.128). Now, the available initial Z_i and a_i values are used in the exact ARMA-model equation to generate as many sequences as desired by sampling a new $a_i \sim N(0, \sigma_a^2)$ at each time step.

The above procedure, suggested by Hipel, can be numerically difficult and unstable if r is very large, because the decomposition of Γ_r commonly exhibits numerical errors. The added expected accuracy may not be worth the extra computations and difficulties.

By differencing, ARIMA models filter unknown random trends and essentially represent an equivalent random process, with the correct probabilistic behavior up to the unknown trend. Therefore the value of simulation of such processes is unclear, since the underlying level of the result may be incorrect. Nevertheless, Hipel (1975) suggests a simulation scheme that requires known starting values to set the initial level. After a long simulation, though, the absolute magnitude of the results may be in question. The value of ARIMA models is mostly in forecasting, as discussed in the next section.

2.8 FORECASTING

Chapter 1 defined forecasting as the computation of the distribution of future events conditional on observed past behavior. Due to the linearity of ARIMA models and the normality of added residuals, the conditional distribution is Gaussian and fully defined by the conditional expected value of the process and the conditional variance.

The general ARIMA$(p, d, q) \times (P, D, Q)$ can be written for time $t + L$ as

$$\phi(B)\phi(B^s)\nabla^d\nabla_s^D X_{t+L} = \theta(B)\theta(B^s)a_{t+L}$$

or

$$\phi'(B)X_{t+L} = \theta'(B)a_{t+L}, \tag{2.143}$$

where $\phi'(B)$ is a polynomial of degree $p + sP + d + sD$, and $\theta'(B)$ has degree $q + sQ$.

If the time series X_t is known up to time t, the conditional expected value of X_{t+L} is given by

$$E[X_{t+L}] = \phi'_1 E[X_{t+L-1}] + \phi'_2 E[X_{t+L-2}]$$
$$+ \cdots + \phi'_{p+sP+d+sD} E[X_{t+L-p-sP-d-sD}] + E[a_{t+L}]$$
$$- \theta'_1 E[a_{t+L-1}] - \cdots - \theta'_{q+sQ} E[a_{t+L-q-sQ}], \qquad (2.144)$$

where the expectation of known values is simply the value:

$$E[X_{t-i}] = x_{t-i} \qquad i \geq 0. \qquad (2.145)$$

Define $\hat{X}(t + L|t)$ as $E[X_{t+L}]$ or the conditional expected value of X at time $t + L$ given observations up to time t. Using the above definitions the expected value of shocks occurring up to time t is the observed value computed from

$$a_{t-j} = x_{t-j} - \hat{X}(t - j|t - j - 1), \qquad j \geq 0. \qquad (2.146)$$

Note that the residual is simply the difference between observed and predicted value. The expected value of shocks that have not occurred, a_{t+L}, for $L \geq 1$ is zero.

In summary, a lead-one prediction (conditional mean) of model (2.143) is

$$\hat{X}(t + 1|t) = \phi'_1 x_t + \phi'_2 x_{t-1}$$
$$+ \cdots + \phi'_{p+sP+d+sD} x_{t+1-p-sP-d-sD}$$
$$- \theta'_1 E[a_t] - \cdots - \theta'_{q+sQ} E[a_{t+1-q-sQ}],$$

where $E[a_t] = x_t - \hat{X}(t|t - 1)$ and so on.

The lead-two prediction would be a function of the lead-one prediction

$$\hat{X}(t + 2|t) = \phi'_1 \hat{X}(t + 1|t) + \phi'_2 x_t + \phi'_3 x_{t-1} +$$
$$\cdots + \phi'_{p+sP+d+sD} x_{t+2-p-sP-d-sD} - \theta'_2 E[a_t] -$$
$$\cdots - \theta'_{q+sQ} E[a_{t+2-q-sQ}].$$

As in Eq. (2.88), any ARIMA model can be written in terms of past random shocks,

$$X_{t+L} = a_{t+L} + \psi_1 a_{t+L-1} + \cdots. \qquad (2.147)$$

The lead L forecast based on the above formulation would be

$$\hat{X}(t+L|t) = \psi_L a_t + \psi_{L+1} a_{t-1} + \cdots . \qquad (2.148)$$

Subtracting Eq. (2.148) from (2.147) yields the lead L prediction error,

$$e(t+L|t) = a_{t+L} + \psi_1 a_{t+L-1} + \cdots + \psi_{L-1} a_{t+1}. \qquad (2.149)$$

Squaring and taking the expected value of the above yields the conditional prediction variance,

$$\sigma_x^2(t+L|t) = \sigma_a^2 \sum_{i=0}^{L-1} \psi_i^2, \qquad (2.150)$$

where $\psi_0 = 1$. Note that the future value, lead L, of X is Gaussian with mean given by Eq. (2.148) and variance by Eq. (2.150). It is then possible to develop confidence limits on the forecast. As the lead L goes to infinity, Eq. (2.150) approaches the finite variance of a stationary ARMA model. On the other hand, an ARIMA model, has nonconvergent ψ_is, which implies that as L goes to infinity the variance of forecast goes to infinity.

Forecasts can be recursively obtained. Using Eq. (2.147), the lead $L-1$ forecast from information up to time $t+1$ is

$$\hat{X}(t+L|t+1) = \psi_{L-1} a_{t+1} + \psi_L a_t + \cdots . \qquad (2.151)$$

From Eq. (2.148), the lead L forecast from time t is

$$\hat{X}(t+L|t) = \psi_L a_t + \psi_{L+1} a_{t-1} + \cdots . \qquad (2.152)$$

Subtracting Eq. (2.152) from (2.151) yields the convenient recursive relationship,

$$\hat{X}(t+L|t+1) = \psi_{L-1} a_{t+1} + \hat{X}(t+L|t), \qquad (2.153)$$

where $a_{t+1} = x(t+1) - \hat{X}(t+1|t)$.

So a new prediction is a weighted sum of new observations and previous predictions.

2.9 A CASE STUDY: STREAMFLOW FORECASTING IN THE ORINOCO RIVER

The Orinoco River is one of the giant fluvial systems in the world. Its basin covers an area of 950,000 km², draining about two thirds of the territory of Venezuela, and about 350,000 km² of Colombia. Its annual mean flow is 28,000 m³/s with maximum flows reaching 100,000 m³/s. In discharge it ranks second only to its neighboring basin, the Amazon River. It is one of the longest rivers in the world (2063 km) and ranks seventh in terms of drainage area.

The Orinoco basin is a major internal waterway for Venezuela as well as a potential provider of enormous quantities of hydropower. It is also the guardian of and access to tremendous mineral and oil resources, which will support large industrial developments on the river banks.

For navigation, flood control, and planning and design, forecasting of Orinoco River streamflows is extremely important. Sancholuz, Carrasquel, and Vargas (1981) report on a stage forecasting model for the Orinoco River based on the time-series–analysis theory seen in this chapter. Their results will illustrate the techniques presented in this chapter.

Figure 2.23 shows 8 years (1943–1950) of daily levels in Palua, a downstream station in the Orinoco River. There is no doubt of the Orinoco's annual cycle and reasonably simple annual hydrograph. Only points corresponding to flow on the 15th of each month are shown. To identify a model capable of a forecasting stage at Palua, it is necessary to follow the identification, estimation, and verification steps outlined in this chapter.

Based on spectral analyses and sampling properties of the spectrum (Chapter 4) Sancholuz et al. (1981) concluded that using instantaneous river levels every five days (i.e., sampling the record every five days) did not result in loss of information relative to using the daily stage records. Given the size and large response times of the Orinoco, their well-documented conclusion is easily accepted. Figure 2.24 shows the means and variances of stage every five days (73 points in total, for 365 days) obtained from 37 years of data, from 1943 through 1979. Figure 2.25 shows the autocorrelation function of the data for the same period; each lag is five days. The periodicity and nonstationarity of the series is obvious, the autocorrelation does not decay in time and has a definite one-year cycle.

For reasons of homogeneity of the series, Sancholuz et al. used the 11 years from January 1, 1969, through December 31, 1979, to identify and fit several possible forecasting models. One alternative was to study the standardized series. This was described in Section 2.5.2, Eq. (2.102). Each of the 803 values (11 × 73) of the series was standardized by subtracting the every five-day mean over the 11 years of data and dividing by the corresponding standard deviation. The autocorrelation of the resulting series is shown in Fig. 2.26. The observed nonstationarity and periodicity of the original series disappeared. The autocorrelation exhibits a dominant "exponential-like" decay typical of AR models, in fact very close to the $\rho^{|\tau|}$ form of an AR(1) model. The decay is

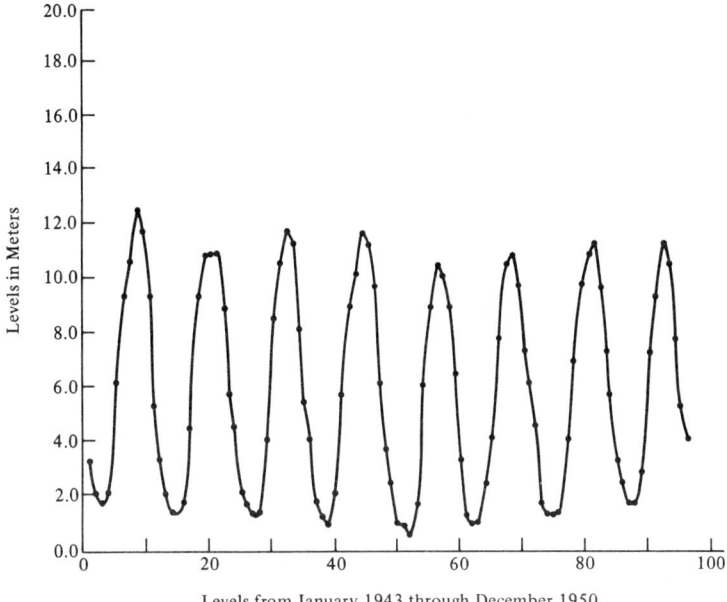

Levels from January 1943 through December 1950

Figure 2.23 Daily levels of the Orinoco River at Palua, points corresponding to the 15th day of each month. (A. Sancholuz, S. Carrasquel, D. Vargas, Laboratorio Nacional de Hidráulica, 1981. Under contract by INC, Caracas, Venezuela.)

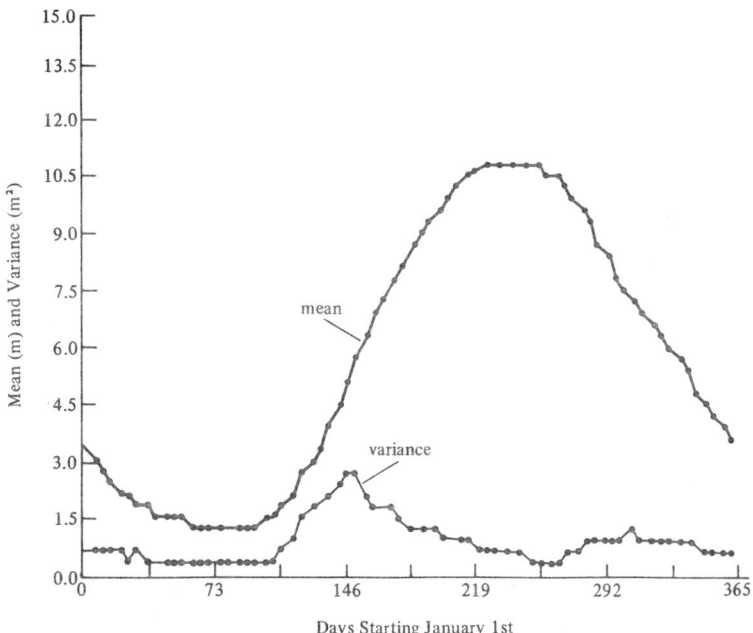

Days Starting January 1st

Figure 2.24 Means and variances of daily levels for the Orinoco River at Palua. The data covers the period 1943–1979, points are plotted every five days. (A. Sancholuz, S. Carrasquel, D. Vargas, Laboratorio Nacional de Hidráulica, 1981. Under contract by INC, Caracas, Venezuela.)

81

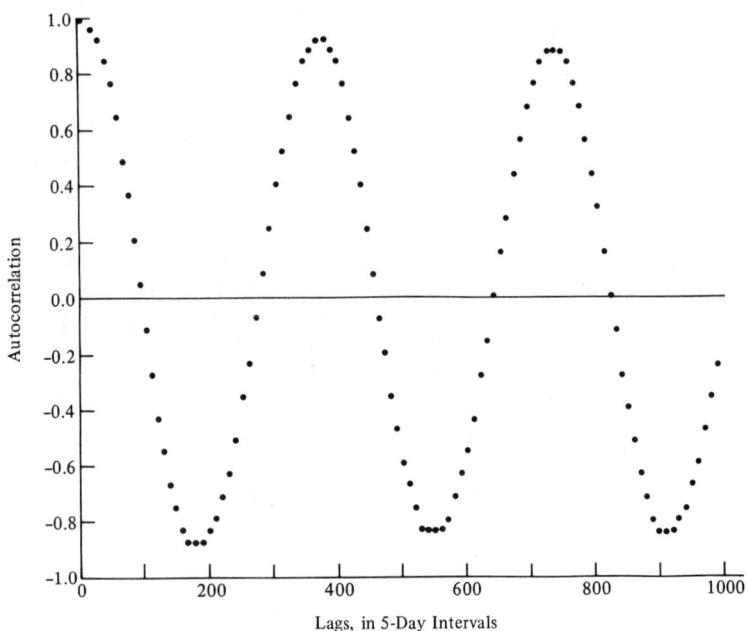

Figure 2.25 Autocorrelation function for the Orinoco River at Palua. (A. Sancholuz, S. Carrasquel, D. Vargas, Laboratorio Nacional de Hidráulica, 1981. Under contract by INC, Caracas, Venezuela.)

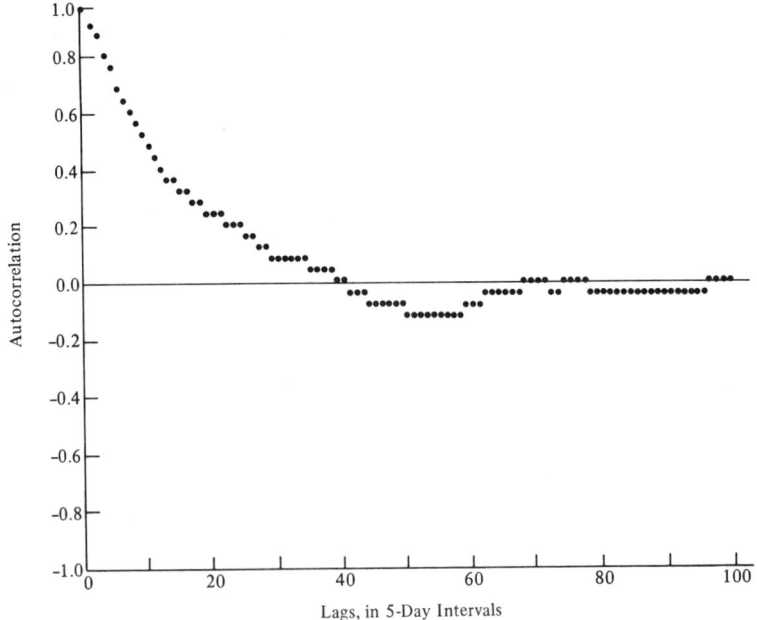

Figure 2.26 Autocorrelation function for the standardized series of the Orinoco River at Palua (A. Sancholuz, S. Carrasquel, D. Vargas, Laboratorio Nacional de Hidráulica, 1981. Under contract by INC, Caracas, Venezuela.)

nevertheless slightly slower than would be expected from a pure AR(1) model. For example at lag 12 the autocorrelation is 0.4 when a pure AR(1) model would imply about $0.37(0.92^{12})$. Based on this slight evidence the model developers also suspected a moving average term of order 1. The partial autocorrelation of the standardized series, shown in Fig. 2.27, supported the AR(1) hypothesis with a significant ϕ_{11} value. A few of the other partial autocorrelations are over or close to the two-standard-deviation lines computed with Eq. (2.115), i.e., S.D. $=1/\sqrt{843} = 0.035$.

The developers first tested an ARMA(1,1) model. Parameter identification was done by minimizing the sum of squared residuals as explained in Section 2.6.2, Eq. (2.135). The parameters of the ARMA(1,1) obtained for the standardized series were $\phi_1 = 0.929$, $\theta_1 = -0.09$, $\sigma_a^2 = 0.1$. The results confirm the dominance of the AR(1) term.

The residuals of the model were studied during a verification step. Both the autocorrelation and the partial autocorrelation of residuals were computed. They both exhibited slight violations of the $2(1/\sqrt{N})$ limits that would imply uncorrelated residuals. The portmanteau lack-of-fit test (Section 2.6.3, Eq. 2.138) indicated slight correlation at a 0.01 level of significance ($Q \approx 74$, $\chi^2_{0.01} \approx 72$). A chi-square test of the normality of residuals also failed at the 0.01 level of significance.

The model developers finally converged on an ARMA(2,73) model estimated as

$$(1-0.59B-0.325B^2)Z_t = (1+0.444B-0.033B^2-0.199B^{73})a_t$$

with $\sigma_a^2 = 0.099$. (*Note:* remember that there are 73 five-day intervals in a year.)

Figures 2.28 and 2.29 show the autocorrelations and partial autocorrelations of the residuals of the above model, respectively. Again, the correlations show some marginally significant values that lie on or slightly above the two-standard-deviation limits. The portmanteau test indicated uncorrelated noise at a 0.048 level of significance. The chi-square test did not reject the normality hypothesis. A histogram of residuals is shown in Fig. 2.30.

At this point it is important to add that the improvements achieved by the ARMA(2,73) model over the ARMA(1,1) or the AR(1) were relatively minor and other analysts may have justifiably stopped with the more parsimonious models.

Sancholuz et al. (1981) also studied the applicability of a seasonal multiplicative model of the type discussed in Section 2.5.2, Eq. (2.100). A first-order seasonal difference of the original time series was taken to yield

$$Z_t = X_t - X_{73}$$
$$= \nabla_{73} X_t.$$

The autocorrelation and partial autocorrelation of the differenced series are

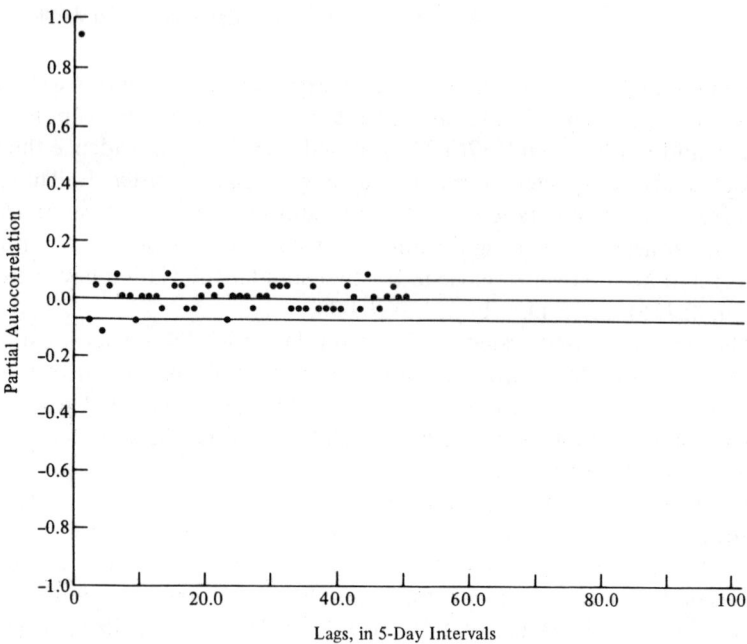

Figure 2.27 Partial autocorrelation function for the standardized series of the Orinoco River at Palua. (A. Sancholuz, S. Carrasquel, D. Vargas, Laboratorio Nacional de Hidráulica, 1981. Under contract by INC, Caracas, Venezuela.)

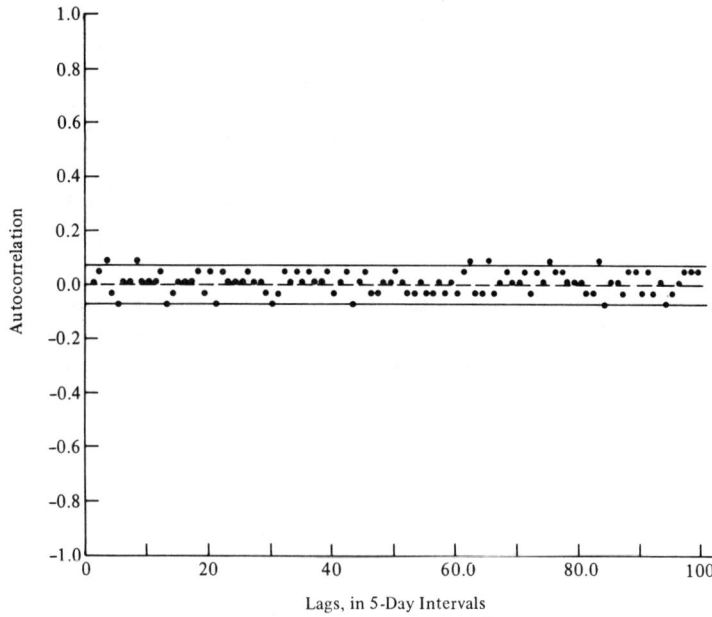

Figure 2.28 Autocorrelation function for the residuals of an ARMA(2, 73) model adjusted to the standardized series of the Orinoco River at Palua. (A. Sancholuz, S. Carrasquel, D. Vargas, Laboratorio Nacional de Hidráulica, 1981. Under contract by INC, Caracas, Venezuela.)

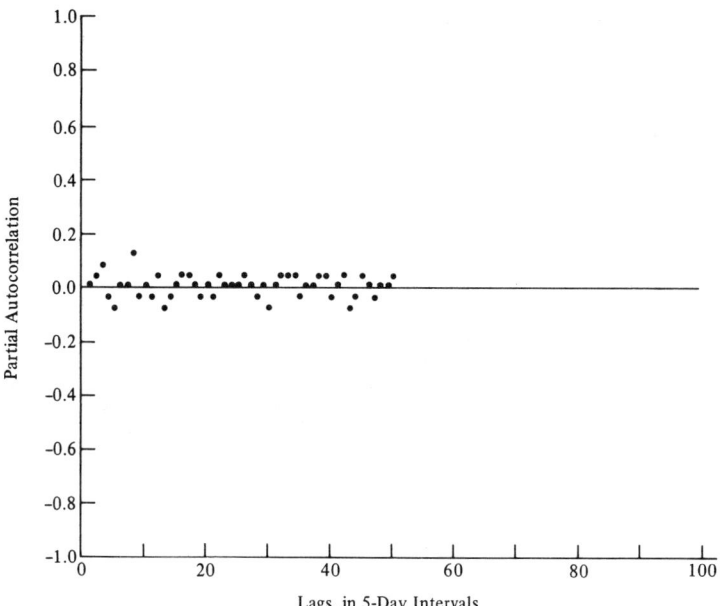

Figure 2.29 Partial autocorrelation function of the residuals of an ARMA(2,73) model adjusted to the standardized series of the Orinoco River at Palua. (A. Sancholuz, S. Carrasquel, D. Vargas, Laboratorio Nacional de Hidráulica, 1981. Under contract by INC, Caracas, Venezuela.)

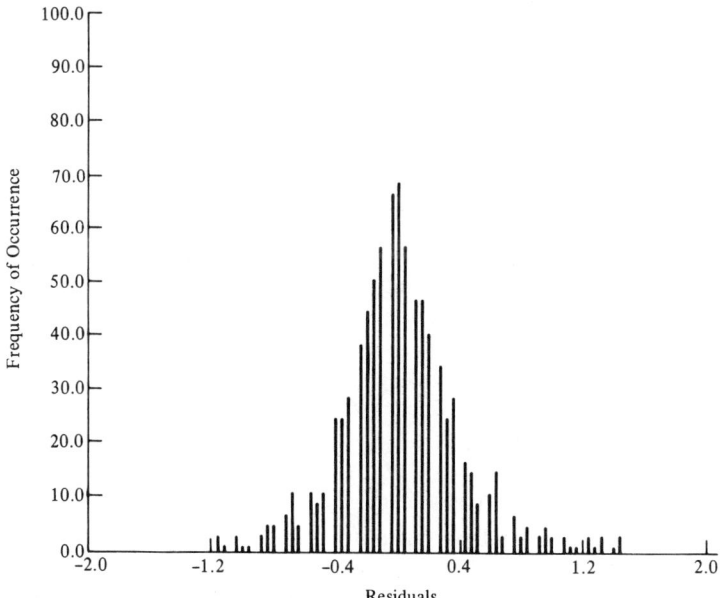

Figure 2.30 Histogram of residuals for an ARMA(2,73) adjusted to the standardized series of the Orinoco River at Palua. (A. Sancholuz, S. Carrasquel, D. Vargas, Laboratorio Nacional de Hidráulica, 1981. Under contract by INC, Caracas, Venezuela.)

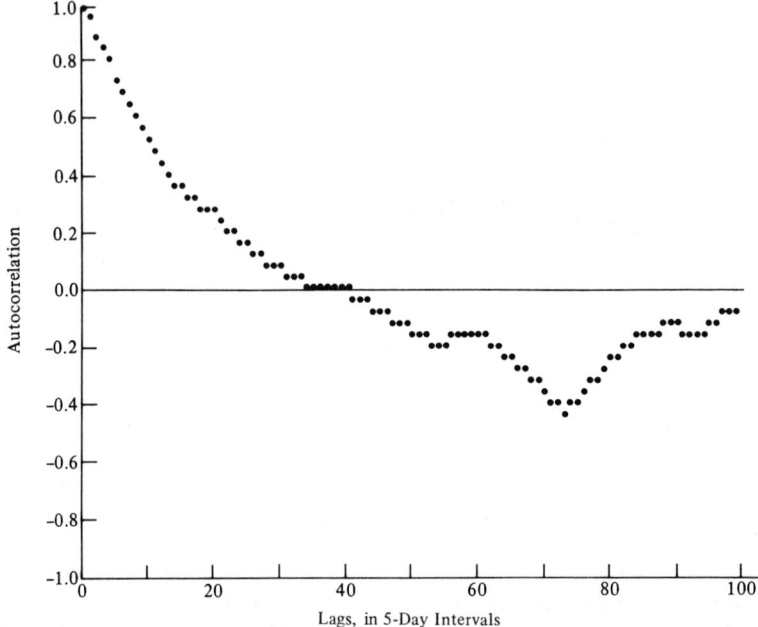

Figure 2.31 Autocorrelation function for the differenced series $Z_t = \nabla_{73} X_t$ of the Orinoco River at Palua. (A. Sancholuz, S. Carrasquel, D. Vargas, Laboratorio Nacional de Hidráulica, 1981. Under contract by INC, Caracas, Venezuela.)

given in Figs. 2.31 and 2.32, respectively. From the autocorrelation, it is clear that the seasonal nonstationarities are eliminated. There is a typical autoregressive decay at low lags, and a significant negative correlation at around lag 73. The partial autocorrelation shows a significant ϕ_{11} value and marginal ϕ_{22} and ϕ_{33} values. The model proponents suggested a seasonal moving-average term to account for the autocorrelation at lag 73. They also converged on an order-5 autoregressive term. Their final model was then an ARIMA$(5,0,0)\times(0,1,1)_{73}$ of the form

$$(1 - 1.136B + 0.238B^2 - 0.129B^3 + 0.069B^4 + 0.11B^5) \nabla_{73} X_t$$
$$= (1 - 0.894B^{73}) a_t,$$

with residual variance $\sigma_a^2 = 0.085$.

The autocorrelation and partial autocorrelation of residuals are given in Figs. 2.33 and 2.34, respectively. The portmanteau test did not reject the nonautocorrelation hypothesis at the 0.05 level of significance. The normality hypothesis on residuals was not rejected using a χ^2 test, also at the 0.05 level of significance.

The two identified models, one based on the standardized series (model 1) and the other on the original series (model 2), were used to make lead-12

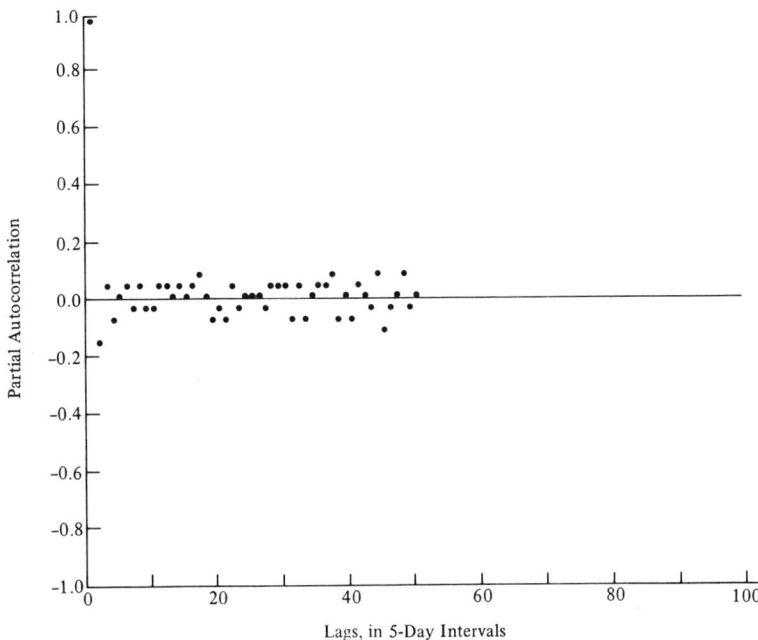

Figure 2.32 Partial autocorrelation function for the differenced series $Z_t = \nabla_{73} X_t$ of the Orinoco River at Palua. (A. Sancholuz, S. Carrasquel, D. Vargas, Laboratorio Nacional de Hidráulica, 1981. Under contract by INC, Caracas, Venezuela.)

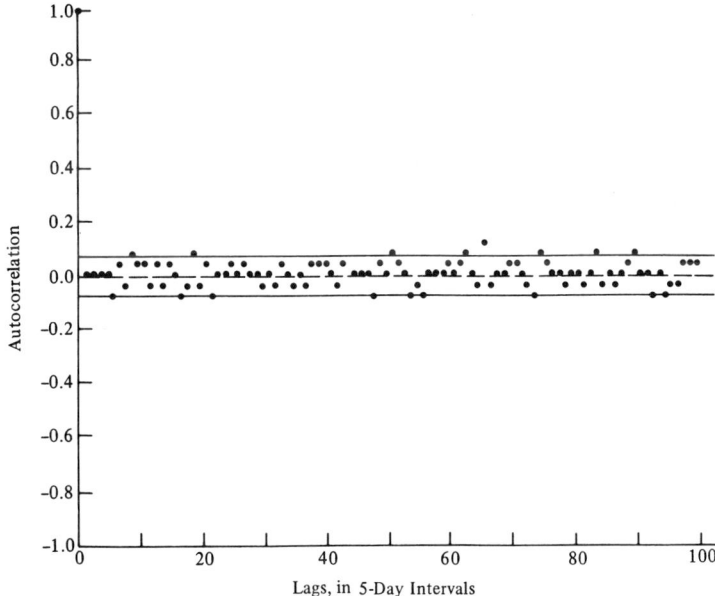

Figure 2.33 Autocorrelation function of the residuals of an ARIMA(5,0,0) $\times (0,1,1)_{73}$ adjusted to the original series of the Orinoco River at Palua. (A. Sancholuz, S. Carrasquel, D. Vargas, Laboratorio Nacional de Hidráulica, 1981. Under contract by INC, Caracas, Venezuela.)

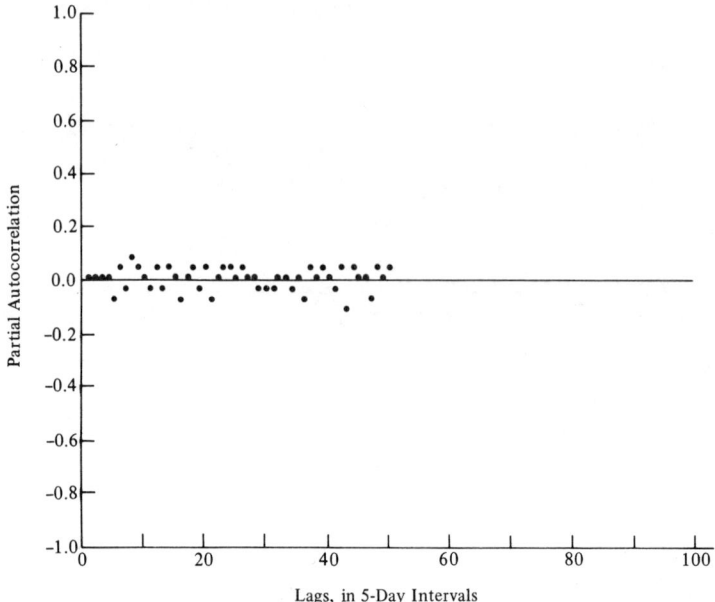

Figure 2.34 Partial autocorrelation function of the residuals of an ARIMA $(5,0,0) \times (0,1,1)_{73}$ adjusted to the original series of the Orinoco River at Palua. (A. Sancholuz, S. Carrasquel, D. Vargas, Laboratorio Nacional de Hidráulica, 1981. Under contract by INC, Caracas, Venezuela.)

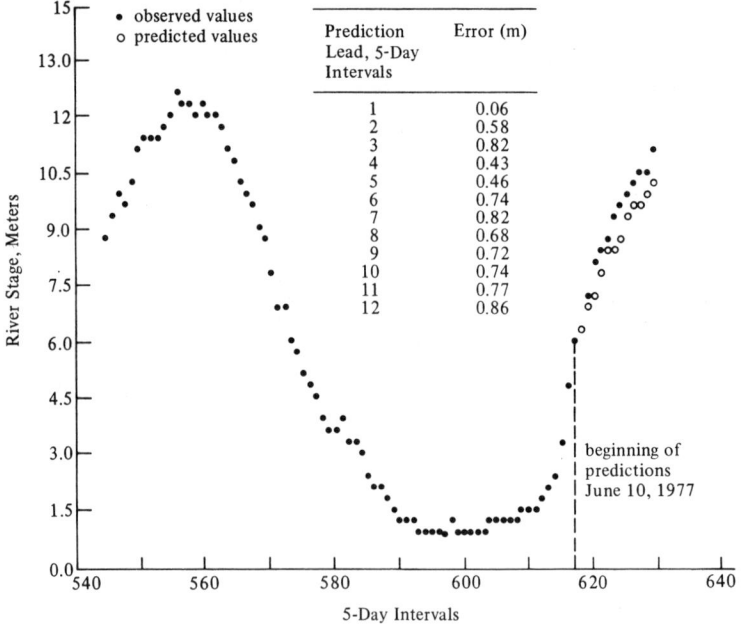

Prediction Lead, 5-Day Intervals	Error (m)
1	0.06
2	0.58
3	0.82
4	0.43
5	0.46
6	0.74
7	0.82
8	0.68
9	0.72
10	0.74
11	0.77
12	0.86

Figure 2.35 Predictions of model 1 for the Orinoco River at Palua starting June 10, 1977. (A. Sancholuz, S. Carrasquel, D. Vargas, Laboratorio Nacional de Hidráulica, 1981. Under contract by INC, Caracas, Venezuela.)

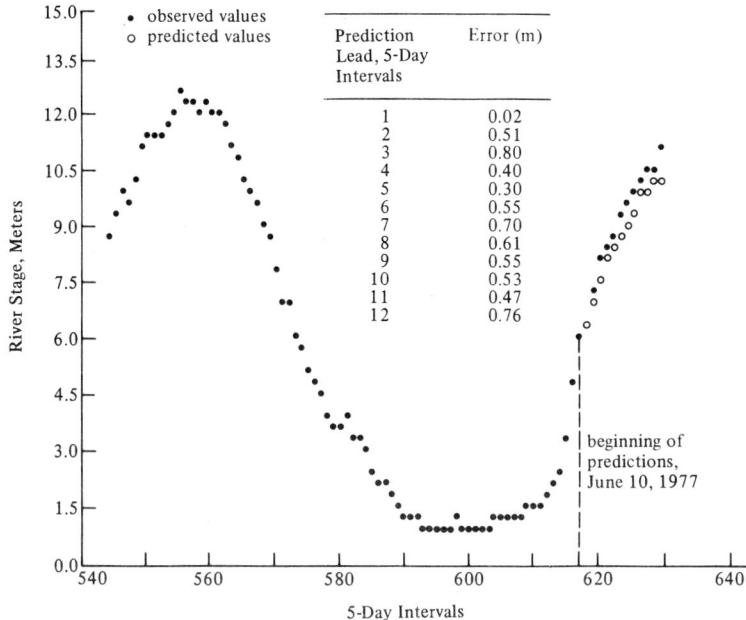

Figure 2.36 Predictions of model 2 for the Orinoco River at Palua starting June 10, 1977. (A. Sancholuz, S. Carrasquel, D. Vargas, Laboratorio Nacional de Hidráulica, 1981. Under contract by INC, Caracas, Venezuela.)

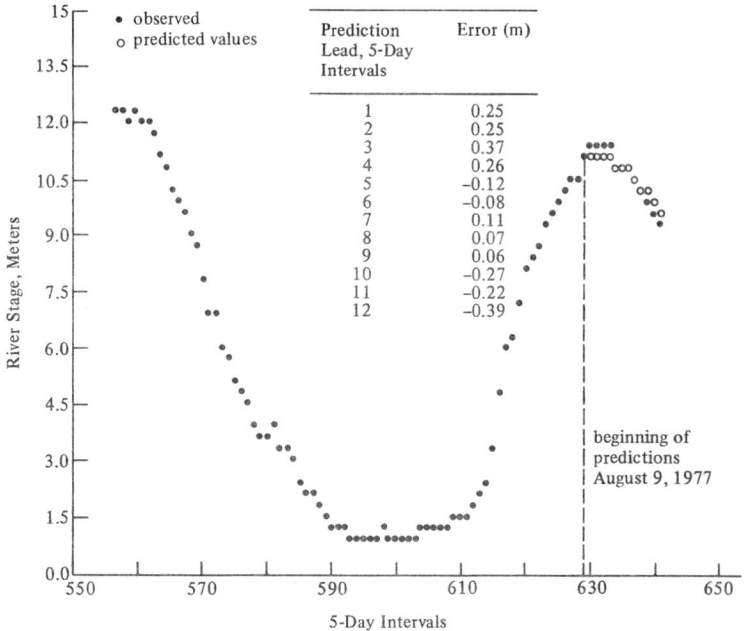

Figure 2.37 Predictions of model 1 for the Orinoco River at Palua starting August 9, 1977. (A. Sancholuz, S. Carrasquel, D. Vargas, Laboratorio Nacional de Hidráulica, 1981. Under contract by INC, Caracas, Venezuela.)

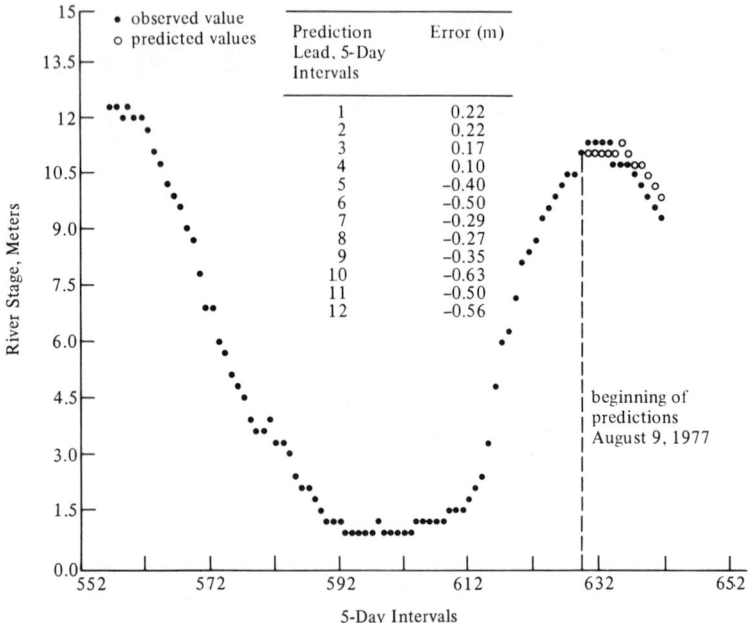

Figure 2.38 Predictions of model 2 for the Orinoco River at Palua starting August 9, 1977. (A. Sancholuz, S. Carrasquel, D. Vargas, Laboratorio Nacional de Hidráulica, 1981. Under contract by INC, Caracas, Venezuela.)

(60 days) forecasts in the Orinoco over several periods. Unfortunately, the examples include periods used during calibration. Behavior over only two periods are shown here, all involving predictions of the rising limb and peak of the hydrograph. Figures 2.35 and 2.36 show 12-lead predictions of models 1 and 2, respectively, for the period starting June 10, 1977. Similarly Figs. 2.37 and 2.38 are predictions for the period starting August 9, 1977. That is, the figures show two consecutive prediction periods of 60 days each. Both models performed relatively well; error magnitudes are also given in the figures. The model developers concluded that of all tests, model 1 had a 50% probability of making an error of less than ± 0.47 m at the one-month (lead-6) forecast and of ± 0.57 m at the two-month (lead-12) forecast. Model 1, though, required standardizing the series by first estimating 73 daily flow means and 73 corresponding variances. Lead-6 predictions with model 2 were in error by less than ± 0.49 m 50% of the time. The equivalent lead-12 error statistic was ± 0.61 m. Nevertheless, no standardization or estimation of means and variances were required by this more parsimonious model.

CHAPTER 3

Multivariate Time-Series Analysis

3.1 INTRODUCTION

Hydrologic studies are commonly multivariate in nature. For example, river-basin planning usually involves development of multiple sites, all of which are naturally related. Reservoir operation cannot be independent of other impoundments in the same river network. Rainfall is usually sampled at discrete, related locations. Therefore it is necessary to consider jointly data from the various rain gages.

Although conceptually multivariate time-series analysis follows the same ideas presented in Chapter 2, in practice the mathematics and the theory lag behind. This lag in development responds to computational and theoretical difficulties. Of the latter, most important is the lack of a unified approach to represent jointly each of the random processes described by a different stochastic-process model. The most serious computational difficulty is parameter estimation. The quantities of data required for adequate parameter estimation in a multivariate model can be unmanageable. The mathematics of sophisticated estimation procedures also become burdensome.

Multivariate stochastic hydrologic modeling mostly has followed the philosophy of fitting limited moments of historical time series. This is in contrast to the philosophy of extensive identification, estimation, and verification of Chapter 2, which emphasized more detailed reproduction of the properties of the original time series. This chapter will first discuss multivariate models of the ARMA type. Next, seasonal multivariate models will be discussed. Among these, a general ARMA-type seasonal model will be developed in detail, following the necessary steps of identification, estimation, and verification. Finally, as an alternative to seasonal simulation, the popular disaggregation model will be presented.

3.2 MULTIVARIATE STATIONARY MODELS

3.2.1 Multivariate AR(1)

By far the most popular multivariate hydrologic streamflow model is the autoregressive lag-one. Its use in hydrology was first suggested by Matalas (1967). Its simplest form is

$$\mathbf{Z}(t) = \mathbf{A}\mathbf{Z}(t-1) + \mathbf{B}\boldsymbol{\varepsilon}(t), \tag{3.1}$$

where

$$\mathbf{Z}^T(t) = \left[Z_1(t) \dots Z_n(t) \right]$$
$$\boldsymbol{\varepsilon}^T(t) = \left[\varepsilon_1(t) \dots \varepsilon_n(t) \right]$$

and $\mathbf{A}$ and $\mathbf{B}$ are $n \times n$ parameter matrices. The vector $\mathbf{Z}(t)$ is composed of n different but interdependent, zero-mean time series. For example, they may be runoff or rainfall at n different locations. They could also be related series of rainfall and runoff. The vector, $\boldsymbol{\varepsilon}(t)$ consists of n uncorrelated shocks (white noise) of zero mean and unit variance. The vector covariance of $\boldsymbol{\varepsilon}(t)$ is the identity matrix,

$$E\left[\boldsymbol{\varepsilon}(t)\boldsymbol{\varepsilon}^T(t) \right] = \mathbf{I}.$$

Vector $\boldsymbol{\varepsilon}(t)$ is uncorrelated with $\mathbf{Z}(\tau)$, for $\tau < t$:

$$E\left[\mathbf{Z}(\tau)\boldsymbol{\varepsilon}^T(t) \right] = 0.$$

Note that following the convention of Chapter 2, $\mathbf{Z}(t)$ is a zero-mean vector. Generally it would be possible to state that

$$\mathbf{X}(t) = \mathbf{Z}(t) + \mathbf{m}, \tag{3.2}$$

where $\mathbf{m}$ is an $n \times 1$ vector of constant means of the original process $\mathbf{X}(t)$.

The parameter matrices $\mathbf{A}$ and $\mathbf{B}$ are generally obtained using the method of moments. Therefore they are estimated in order to preserve explicitly a limited number of moments of the original time series. Analogous to the Thomas–Fiering model described in Chapter 2 the preserved moments are the covariance and lag-one covariance of the process.

Equation (3.1) can be post-multiplied by the transpose of $\mathbf{Z}(t)$ and expectations taken. This amounts to computing the covariance of $\mathbf{Z}(t)$,

$$E\left[\mathbf{Z}(t)\mathbf{Z}^T(t)\right] = \mathbf{A}E\left[\mathbf{Z}(t-1)\mathbf{Z}^T(t)\right]$$
$$+\mathbf{B}E\left[\boldsymbol{\varepsilon}(t)\left[\mathbf{Z}^T(t-1)\mathbf{A}^T + \boldsymbol{\varepsilon}^T(t)\mathbf{B}^T\right]\right]. \qquad (3.3)$$

Defining the covariance

$$\mathbf{M}_0 = E\left[\mathbf{Z}(t)\mathbf{Z}^T(t)\right]$$

and the covariance at lag one

$$\mathbf{M}_1^T = E\left[\mathbf{Z}(t-1)\mathbf{Z}^T(t)\right]$$

and exploiting the white-noise properties of $\boldsymbol{\varepsilon}(t)$, reduces Eq. (3.3) to

$$\mathbf{M}_0 = \mathbf{A}\mathbf{M}_1^T + \mathbf{B}\mathbf{B}^T. \qquad (3.4)$$

Equation (3.4) relates moments $\mathbf{M}_0$ and $\mathbf{M}_1$ to parameters $\mathbf{A}$ and $\mathbf{B}$ through a quadratic matrix equation. In order to specify the system completely another equation is required.

Equation (3.1) can now be post-multiplied by $\mathbf{Z}^T(t-1)$ and expected values taken to define the lag-one covariance,

$$E\left[\mathbf{Z}(t)\mathbf{Z}^T(t-1)\right] = \mathbf{A}E\left[\mathbf{Z}(t-1)\mathbf{Z}^T(t-1)\right] + \mathbf{B}E\left[\boldsymbol{\varepsilon}(t)\mathbf{Z}^T(t-1)\right].$$

Again using the white-noise properties of $\boldsymbol{\varepsilon}(t)$, the above reduces to

$$\mathbf{M}_1 = \mathbf{A}\mathbf{M}_0, \qquad (3.5)$$

where system stationarity was invoked to state

$$E\left[\mathbf{Z}(t)\mathbf{Z}^T(t)\right] = E\left[\mathbf{Z}(t-1)\mathbf{Z}^T(t-1)\right] = \mathbf{M}_0. \qquad (3.6)$$

Equation (3.5) can be solved for $\mathbf{A}$ if $\mathbf{M}_0$ is a valid covariance matrix, i.e., it is positive definite.

$$\mathbf{A} = \mathbf{M}_1\mathbf{M}_0^{-1}. \qquad (3.7)$$

Given Eq. (3.7), Eq. (3.4) reduces to a quadratic expression for $\mathbf{B}$,

$$\mathbf{B}\mathbf{B}^T = \mathbf{M}_0 - \mathbf{M}_1\mathbf{M}_0^{-1}\mathbf{M}_1^T. \qquad (3.8)$$

The form of Eq. (3.8) will recur throughout the formulation of multivariate ARMA models. There are several possible solutions to **B**, given its quadratic form $\mathbf{BB}^T$. Two of these solutions are discussed in the next subsection and their use will be required whenever equations of the form of Eq. (3.8) appear.

Assuming, as the next subsection will prove, that Eq. (3.8) has a solution, then expressions to obtain **A** and **B** will be available. Note that the above model collapses to the Thomas–Fiering model (AR(1)) when only one time series is involved. In that case, letting $n = 1$,

$$\mathbf{M}_0 = \sigma_z^2$$
$$\mathbf{M}_1 = \mathrm{cov}\left[Z_t Z_{t-1}\right]$$

and

$$\rho_1 = \frac{\mathbf{M}_1}{\mathbf{M}_0}.$$

Therefore

$$\mathbf{A} = \frac{\mathbf{M}_1}{\mathbf{M}_0} = \rho_1$$

and

$$\mathbf{BB}^T = \sigma_z^2 - \sigma_z^2 \rho_1^2$$
$$= \sigma_z^2\left(1 - \rho_1^2\right)$$

or

$$\mathbf{B} = \left(1 - \rho_1^2\right)^{1/2}\sigma_z.$$

The above expressions for **A** and **B** correspond to those obtained in Chapter 2.

Note that a typical element of vector $\mathbf{Z}(t)$ is generated from an equation of the form

$$Z_i(t) = \sum_{j=1}^n a_{ij} Z_j(t-1) + \sum_{j=1}^n b_{ij}\varepsilon_j(t) \qquad \forall_i, \tag{3.9}$$

which is similar to a regression equation with correlated residuals. In fact the elements of the matrix $\mathbf{A}_{ij}$, a_{ij}, are generalized least-square regression coefficients.

Operationally, estimates of **A** and **B** are obtained from Eqs. (3.7) and (3.8) by using sample estimates $\hat{\mathbf{M}}_0$ and $\hat{\mathbf{M}}_1$ of the covariance and lag-one covariance matrices, respectively. These estimates have the forms:

$$
\hat{\mathbf{M}}_0 =
\begin{bmatrix}
S_{z_1}^2 & \cdots & r_{z_1 z_n}^0 S_{z_1} S_{z_n} \\
\vdots & \ddots & \vdots \\
r_{z_1 z_n}^0 S_{z_1} S_{z_n} & \cdots & S_{z_n}^2
\end{bmatrix}
\tag{3.10}
$$

$$
\hat{\mathbf{M}}_1 =
\begin{bmatrix}
r_{z_1 z_1}^1 S_{z_1}^2 & \cdots & r_{z_1 z_n}^1 S_{z_1} S_{z_n} \\
\vdots & \ddots & \\
r_{z_n z_1}^1 S_{z_1} S_{z_n} & \cdots & r_{z_n z_n}^1 S_{z_n}^2
\end{bmatrix},
\tag{3.11}
$$

where S_{z_i} is the sample standard deviation of variable Z_i, and $r_{z_i z_j}^k$ is the correlation between variables Z_i and Z_j at lag k.

The elements of the above matrices can be computed using traditional estimators. As discussed in the subsection below, numerical stability and consistency require some care in defining sample statistics. The following estimators are recommended.

$\overline{\mathbf{X}}$ is the matrix of historical (observed) values of the vector $\mathbf{Z}(t)$, each column is an observed year of $\mathbf{Z}(t)$. If there are N observed years, then $\overline{\mathbf{X}}$ is an $n \times N$ matrix. $\overline{\mathbf{Y}}$ is the matrix of historical (observed) values of the vector $\mathbf{Z}(t-1)$, each column being an observed year of $\mathbf{Z}(t-1)$. Given N years of observations $\overline{\mathbf{Y}}$ is $n \times N$. Note that in this case $\overline{\mathbf{Y}}$ is the same as $\overline{\mathbf{X}}$ lagged by one year. Therefore, $N+1$ years of data are actually required. The sample estimates of $\mathbf{M}_0$ and $\mathbf{M}_1$ are then,

$$
\hat{\mathbf{M}}_0 = \frac{1}{N} \overline{\mathbf{X}} \overline{\mathbf{X}}^T
\tag{3.12}
$$

$$
\hat{\mathbf{M}}_1 = \frac{1}{N} \overline{\mathbf{X}} \overline{\mathbf{Y}}^T.
\tag{3.13}
$$

Finally, it is useful to point out that $\mathbf{M}_0$ is a symmetrical matrix, but $\mathbf{M}_1$ is not.

Decomposition of $\mathbf{BB}^T$

Matrices of the form $\mathbf{D} = \mathbf{BB}^T$ are called Gramian matrices, $\mathbf{D}$ being the Gramian of $\mathbf{B}$. It can be shown that $\mathbf{D}$ must be positive or positive-semidefinite, and in fact that any positive or positive-semidefinite matrix is a Gramian of some matrix (Valencia and Schaake, 1972). Given $\mathbf{D}$, there are infinite solutions for $\mathbf{B}$, since the above equation is satisfied by any matrix of the form $\mathbf{B} \cdot \mathbf{C}$, where $\mathbf{C}$ is orthogonal, implying $\mathbf{CC}^T = \mathbf{I}$. For any such $\mathbf{C}$,

$$
\mathbf{D} = \mathbf{BCC}^T \mathbf{B}^T = \mathbf{BB}^T.
\tag{3.14}
$$

There are two popular solutions for **B** in Eq. (3.14). The first uses the properties of principal components and is able to handle the case when **D** is positive semidefinite, implying that **B** has rank less than n, if $n \times n$ are the dimensions of **D**. This is useful in dealing with a particular class of multivariate models—the Disaggregation Model—which will be discussed in Section 3.5.

The procedure uses the properties of eigenvalues and eigenvectors. Call,

$$\mathbf{D} = \mathbf{BB}^T. \tag{3.15}$$

D is always a positive-semidefinite matrix.

Define a matrix

$$\mathbf{P} = \begin{bmatrix} \mathbf{P}_1 & \vdots & \cdots & \vdots & \mathbf{P}_n \end{bmatrix},$$

where $\mathbf{P}_i = i$th eigenvector of matrix **D**.

The matrix **P** has the property of orthogonality.

$$\mathbf{P}^T\mathbf{P} = \mathbf{PP}^T = \mathbf{I}.$$

Define $e_1, \ldots, e_n$, the eigenvalues of matrix **D**, and using the properties of eigenvalues and eigenvectors,

$$\mathbf{DP}_i = \mathbf{P}_i e_i. \tag{3.16}$$

Defining a diagonal matrix of eigenvalues,

$$\mathbf{E} = \begin{bmatrix} e_1 & \cdots & 0 \\ & \ddots & \\ 0 & \cdots & e_n \end{bmatrix},$$

the system of equations given by Eq. (3.16) can be represented as

$$\mathbf{DP} = \mathbf{PE} \tag{3.17}$$

or

$$\mathbf{D} = \mathbf{PEP}^{-1} = \mathbf{BB}^T. \tag{3.18}$$

Therefore,

$$\mathbf{B} = \mathbf{PE}^{1/2}, \tag{3.19}$$

where

$$\mathbf{E}^{1/2} = \begin{bmatrix} e_1^{1/2} & \cdots & 0 \\ \vdots & \ddots & \vdots \\ 0 & \cdots & e_n^{1/2} \end{bmatrix}.$$

The **B** matrix obtained in this manner is of rank $n - \ell$, where ℓ is the number of zero eigenvalues in **D**.

Numerically, the above procedure is limited by the algorithms used in finding eigenvalues and eigenvectors. For large matrices such procedures may result in errors and instabilities.

A second procedure, valid for positive-definite matrices, suggests a lower triangular form for matrix **B**,

$$\mathbf{B} = \begin{bmatrix} b_{11} & 0 & \cdot & \cdots & 0 \\ b_{21} & b_{22} & 0 & \cdots & 0 \\ \vdots & \vdots & & & \vdots \\ b_{n1} & \cdot & \cdot & \cdots & b_{nn} \end{bmatrix}.$$

Multiplying **B** by its transpose,

$$\mathbf{D} = \mathbf{B}\mathbf{B}^T$$

$$= \begin{bmatrix} b_{11}^2 & b_{11}b_{21} & b_{11}b_{31} & \cdots & b_{11}b_{n1} \\ b_{21}b_{11} & b_{21}^2 + b_{22}^2 & b_{21}b_{31} + b_{22}b_{32} & \cdots & b_{21}b_{n1} + b_{22}b_{n2} \\ \vdots & \vdots & \vdots & & \vdots \\ b_{n1}b_{11} & b_{n1}b_{21} + b_{22}b_{n2} & \cdots & \cdots & \sum_{i=1}^{n} b_{nn}^2 \end{bmatrix}.$$

From the above matrix equation, the elements of matrix **B** can be obtained through sequential algebraic equations. For example, defining d_{ij} as an element of **D**,

$$b_{11} = \sqrt{d_{11}}$$
$$b_{21} = d_{12} / \sqrt{d_{11}}$$
$$\vdots$$
$$b_{n1} = d_{1n} / \sqrt{d_{11}}.$$

$$(3.20)$$

In general multiplying the ith row of **B** by the ith column of $\mathbf{B}^T$ yields

$$b_{i1}^2 + b_{i2}^2 + \cdots + b_{ii}^2 = d_{ii}$$

on solving for b_{ii}:

$$b_{ii} = \left(d_{ii} - \left[b_{i1}^2 + \cdots + b_{ii-1}^2 \right] \right)^{1/2}, \qquad (3.21)$$

assuming the quantity in parenthesis is positive, which is the positive-definite condition.

Given the first element of column i, b_{ii}, the remaining parameters result from

$$b_{ji} = \frac{d_{ij} - \left[b_{i1}b_{j1} + b_{i2}b_{j2} + \cdots + b_{ii-1}b_{ji-1}\right]}{b_{ii}}, \qquad (3.22)$$

with $j > i$ and $b_{ii} \neq 0$.

The above procedure was first suggested for use in hydrology by Young and Pisano (1968). It is computationally fast and accurate for small matrices. When dealing with large matrices, numerical accuracy deteriorates due to the many sequential multiplications and additions required. In such cases the first method is as fast as and more accurate than the second, assuming that a reasonable algorithm to obtain eigenvalues and eigenvectors is used. Note that the first method is computationally feasible even if zero or negative eigenvalues result from numerical errors. The second method will fail any time any b_{ii} is zero or if the term in parenthesis in Eq. (3.21) is negative, a possible numerical anomaly in practice.

In multivariate AR(1), $\mathbf{BB}^T$ (Eq. 3.8) is a covariance matrix; in fact it is the conditional covariance of $\mathbf{Z}(t)$ given past observations $\mathbf{Z}(t-1)$. As such, $\mathbf{BB}^T$ must be positive-definite, assuming we are dealing with stationary processes. This will be the case any time matrices of the $\mathbf{BB}^T$ form appear in the definition of multivariate ARMA models.

Values of $\mathbf{BB}^T$ resulting from sample statistics, i.e., $\hat{\mathbf{M}}_0$ and $\hat{\mathbf{M}}_1$, can nevertheless in fact be nonpositive-definite. Such unexpected results are either of statistical origin or related to numerical accuracy of computational algorithms. They will commonly occur if the elements of $\hat{\mathbf{M}}_0$ and $\hat{\mathbf{M}}_1$ are calculated from data sets of different lengths; this yields sample statistics of differing accuracy and variability. A useful reference in this case is the paper of Crosby and Maddock (1970). The use of different statistical procedures in defining the elements of the necessary covariance matrices can also result in inconsistent $\mathbf{BB}^T$. To avoid such difficulties the estimators given in Eqs. (3.12) and (3.13) are recommended.

It is still possible to obtain inconsistent $\mathbf{BB}^T$ matrices. Such failures can be attributed to data transformations, as will be discussed later, or more commonly to numerical errors in the extensive and repetitive calculations required to obtain $\mathbf{BB}^T$ estimates. Particularly if the elements of the vector $\mathbf{Z}$ are highly correlated among themselves at lag zero or lag one. In such cases the only alternative is to adjust elements of $\mathbf{BB}^T$ in order to obtain the desired positive definiteness. A popular and easily used adjustment was proposed by Mejía and Millán (1974).

A new $\mathbf{BB}^T$ matrix is defined as follows:

$$\mathbf{B'B'}^T = \mathbf{BB}^T + \lambda_j, \qquad (3.23)$$

where

$$\lambda_j = \begin{bmatrix} |\lambda| & & 0 \\ & \ddots & \\ 0 & & |\lambda| \end{bmatrix},$$

and λ is the most negative eigenvalue of the original $\mathbf{BB}^T$ matrix. This adjustment is repeated, obtaining λ_j from the latest $\mathbf{B'B}'^T$, until the new $\mathbf{B'B}'^T$ matrix is positive-definite. Subscript j is used to keep track of the number of iterations.

Obtaining $\mathbf{B'}$ from $\mathbf{B'B}'^T$, the equation of the new multivariate AR(1) model will be

$$\mathbf{Z}(t) = \frac{1}{\sqrt{1 + \sum\limits_{j=1}^{m} \lambda_j}} (\mathbf{AZ}(t-1) + \mathbf{B'\varepsilon}(t)), \tag{3.24}$$

where m is the number of iterations needed to achieve positive semidefiniteness.

This new model preserves the mean and variance of the historical data, affecting only the correlation coefficients by a factor equal to

$$\frac{1}{\sqrt{1 + \sum\limits_{j=1}^{m} \lambda_j}}.$$

Thus, if $\sum_{j=1}^{m} \lambda_j$ is small, the estimation error can be neglected.

3.2.2 Multivariate AR(2)

Higher-order multivariate autoregressive models are studied similarly to the AR(1). Of particular interest is the AR(2)

$$\mathbf{Z}(t) = \mathbf{AZ}(t-1) + \mathbf{BZ}(t-2) + \mathbf{C\varepsilon}(t) \tag{3.25}$$

with parameter matrices $\mathbf{A}$, $\mathbf{B}$, and $\mathbf{C}$. Clarke (1973), Salas and Pegram (1977), Puente (1978), and Deeb and Puente (1979) used this model and the method-of-moments approach to identify $\mathbf{A}$, $\mathbf{B}$, and $\mathbf{C}$. This would require defining $\mathbf{M}_0$, $\mathbf{M}_1$, and $\mathbf{M}_2$, where the latter is now the covariance matrix at lag two.

Post-multiplying Eq. (3.25) by $\mathbf{Z}^T(t)$ and taking expected values yields

$$\mathbf{M}_0 = \mathbf{AM}_1^T + \mathbf{BM}_2^T + \mathbf{CC}^T, \tag{3.26}$$

where the unit variance and white-noise properties of $\varepsilon(t)$ have been exploited.

The lag-one covariance $\mathbf{M}_1$ results from post-multiplying Eq. (3.25) by $\mathbf{Z}^T(t-1)$ and performing expectations. It is useful to illustrate this:

$$E\left[\mathbf{Z}(t)\mathbf{Z}^T(t-1)\right] = \mathbf{A}E\left[\mathbf{Z}(t-1)\mathbf{Z}^T(t-1)\right] + \mathbf{B}E\left[\mathbf{Z}(t-2)\mathbf{Z}^T(t-1)\right]$$
$$+ \mathbf{C}E\left[\boldsymbol{\varepsilon}(t)\{\mathbf{Z}^T(t-2)\mathbf{A}^T + \mathbf{Z}^T(t-3)\mathbf{B}^T + \boldsymbol{\varepsilon}^T(t-1)\mathbf{C}^T\}\right].$$

$$(3.27)$$

Using the stationary properties of covariance and again the noncorrelation of $\boldsymbol{\varepsilon}(t)$ with past values of itself and of $\mathbf{Z}(t)$ reduces the above to

$$\mathbf{M}_1 = \mathbf{A}\mathbf{M}_0 + \mathbf{B}\mathbf{M}_1^T. \tag{3.28}$$

On post-multiplying by $\mathbf{Z}^T(t-2)$ a similar exercise yields

$$\mathbf{M}_2 = \mathbf{A}\mathbf{M}_1 + \mathbf{B}\mathbf{M}_0. \tag{3.29}$$

Equations (3.26), (3.28), and (3.29) are three simultaneous matrix equations on three unknowns. Combining Eqs. (3.28) and (3.29) yields

$$\mathbf{B} = \left(\mathbf{M}_2 - \mathbf{M}_1\mathbf{M}_0^{-1}\mathbf{M}_1\right)\left(\mathbf{M}_0 - \mathbf{M}_1^T\mathbf{M}_0^{-1}\mathbf{M}_1\right)^{-1}. \tag{3.30}$$

The resulting value of $\mathbf{B}$ can be used to obtain $\mathbf{A}$ from Eq. (3.28):

$$\mathbf{A} = \left(\mathbf{M}_1 - \mathbf{B}\mathbf{M}_1^T\right)\mathbf{M}_0^{-1}. \tag{3.31}$$

Using $\mathbf{A}$ and $\mathbf{B}$ in Eq. (3.26) results in the quadratic form of $\mathbf{C}$:

$$\mathbf{C}\mathbf{C}^T = \mathbf{M}_0 - \mathbf{A}\mathbf{M}_1^T - \mathbf{B}\mathbf{M}_2^T, \tag{3.32}$$

which can be decomposed to obtain $\mathbf{C}$ using either of the two procedures described above.

3.2.3 Multivariate MA(1)

The form of the multivariate moving-average model of order 1 is

$$\mathbf{Z}(t) = \mathbf{A}\boldsymbol{\varepsilon}(t) - \mathbf{B}\boldsymbol{\varepsilon}(t-1). \tag{3.33}$$

Parameters are obtained in terms of the lag-zero and lag-one covariances. Post-multiplying by $\mathbf{Z}^T(t)$ and $\mathbf{Z}^T(t-1)$ and taking expectations yields the following respective expressions:

$$\mathbf{M}_0 = \mathbf{A}\mathbf{A}^T + \mathbf{B}\mathbf{B}^T \tag{3.34}$$

$$\mathbf{M}_1 = -\mathbf{B}\mathbf{A}^T. \tag{3.35}$$

As in univariate moving-average models, the above two equations are nonlinear on parameters **A** and **B**. A possible solution is an iterative search as proposed by O'Connell (1974).

Equation (3.35) can be solved for **B**, leading to

$$\mathbf{B} = -\mathbf{M}_1(\mathbf{A}^T)^{-1} \tag{3.36}$$

and

$$\mathbf{B}^T = -\mathbf{A}^{-1}\mathbf{M}_1^T. \tag{3.37}$$

Substituting the above in Eq. (3.34) yields

$$\mathbf{A}\mathbf{A}^T + \mathbf{M}_1(\mathbf{A}^T)^{-1}\mathbf{A}^{-1}\mathbf{M}_1^T = \mathbf{M}_0$$

or

$$\mathbf{A}\mathbf{A}^T + \mathbf{M}_1(\mathbf{A}\mathbf{A}^T)^{-1}\mathbf{M}_1^T = \mathbf{M}_0. \tag{3.38}$$

With $\mathbf{U} = \mathbf{A}\mathbf{A}^T$, Eq. (3.38) represents an iterative algorithm

$$\mathbf{U}_j = \mathbf{M}_0 - \mathbf{M}_1\mathbf{U}_{j-1}^{-1}\mathbf{M}_1^T, \tag{3.39}$$

where subscript j stands for the jth iteration. The solution starts with an assumption for $\mathbf{U}_j$, possibly the identity matrix. Once $\mathbf{U}_j$ converges to a solution, it can be decomposed to obtain an **A** value, which in turn is used in Eq. (3.36) to find **B**. Note that **A** could be assumed to be lower triangular, but **B** will not be so.

Successful convergence of the iterative procedure ensures the preservation of $\mathbf{M}_0$ and $\mathbf{M}_1$ values used in the fitting exercise. O'Connell (1974) states that the convergence of such algorithms is satisfactory. In some cases though convergence may not occur and the values of **U** may oscillate. It is then possible to suggest a filtered solution

$$\mathbf{U}_j = \mathbf{M}_0 - \lambda\mathbf{M}_1\mathbf{U}_{j-1}^{-1}\mathbf{M}_1^T, \tag{3.40}$$

where λ is a value between 0 and 1. When using this equation you must keep in mind that introducing λ values different from 1 implies distortions in the preservation of $\mathbf{M}_0$ and $\mathbf{M}_1$ by the model.

An alternative approximate solution involves combining Eqs. (3.34) and (3.35) to yield

$$(\mathbf{A}+\mathbf{B})(\mathbf{A}+\mathbf{B})^T = \mathbf{M}_0 - \mathbf{M}_1 - \mathbf{M}_1^T = \mathbf{F} \tag{3.41}$$

$$(\mathbf{A}-\mathbf{B})(\mathbf{A}-\mathbf{B})^T = \mathbf{M}_0 + \mathbf{M}_1 + \mathbf{M}_1^T = \mathbf{G}. \tag{3.42}$$

The above are in the quadratic $(\mathbf{B}\mathbf{B}^T)$ form and can be decomposed into

$$(\mathbf{A}+\mathbf{B}) = \mathbf{P} \tag{3.43}$$

$$(\mathbf{A}-\mathbf{B}) = \mathbf{Q}, \tag{3.44}$$

which in turn can be solved for $\mathbf{A}$ and $\mathbf{B}$

$$\mathbf{A} = \frac{\mathbf{P} + \mathbf{Q}}{2} \tag{3.45}$$

$$\mathbf{B} = \frac{\mathbf{P} - \mathbf{Q}}{2}. \tag{3.46}$$

Since $\mathbf{M}_0$ is a symmetrical, $n \times n$ matrix, and $\mathbf{M}_1$ is also $n \times n$ but not symmetrical, there are

$$n^2 + \frac{n(n+1)}{2}$$

conditions available to define matrices $\mathbf{A}$ and $\mathbf{B}$. However, the symmetry of $\mathbf{F}$ and $\mathbf{G}$ in Eqs. (3.41) and (3.42) implies that only $n(n+1)$ conditions are being imposed in the approximate solutions of $\mathbf{A}$ and $\mathbf{B}$. The implication is lack of assurance that $\mathbf{M}_1$ will in fact be exactly preserved.

Puente (1978) found that the quadratic form (Eq. 3.41) commonly is nonpositive-definite in practice, and is inconsistent with the theory and solution methods. This is because the parameters of the model are not on its feasible region. Equation (3.42) is rarely problematic, as expected from its additive nature. Modifying Eq. (3.41) using the method of Mejía and Millán implies defining

$$(\mathbf{A}' + \mathbf{B}')(\mathbf{A}' + \mathbf{B}')^T = (\mathbf{A} + \mathbf{B})(\mathbf{A} + \mathbf{B})^T + \lambda \mathbf{I}$$
$$= \mathbf{F} + \lambda \mathbf{I}. \tag{3.47}$$

Sufficient conditions for satisfying Eq. (3.42) are

$$\mathbf{A}'\mathbf{A}'^T + \mathbf{B}'\mathbf{B}'^T = \mathbf{A}\mathbf{A}^T + \mathbf{B}\mathbf{B}^T + [\lambda/2]\mathbf{I} \tag{3.48}$$

$$\mathbf{B}'\mathbf{A}'^T = \mathbf{B}\mathbf{A}^T + [\lambda/4]\mathbf{I}, \tag{3.49}$$

which leaves Eq. (3.42) unchanged:

$$(\mathbf{A}' - \mathbf{B}')(\mathbf{A}' - \mathbf{B}')^T = (\mathbf{A} - \mathbf{B})(\mathbf{A} - \mathbf{B})^T = \mathbf{G}. \tag{3.50}$$

If and only if Eqs. (3.48) and (3.49) are satisfied, the basic multivariate MA(1) is modified to

$$\mathbf{Z}(t) = \frac{1}{\sqrt{1 + \lambda/2}} \mathbf{A}'\boldsymbol{\varepsilon}(t) + \frac{1}{\sqrt{1 + \lambda/2}} \mathbf{B}'\boldsymbol{\varepsilon}(t-1), \tag{3.51}$$

which can be shown to lead to,

$$\mathbf{M}_0' = \frac{1}{1 + \lambda/2} \left[\mathbf{M}_0 + (\lambda/2)\mathbf{I} \right] \tag{3.52}$$

$$\mathbf{M}_1' = \frac{1}{1 + \lambda/2} \left[\mathbf{M}_1 - (\lambda/4)\mathbf{I} \right]. \tag{3.53}$$

If λ is small, acceptable distortions are introduced in $\mathbf{M}_0$ and $\mathbf{M}_1$, the latter suffering less.

As previously stated, the above set of equations are valid only for $\mathbf{A}'$ and $\mathbf{B}'$ obeying Eqs. (3.48) and (3.49). Note that those equations can be formulated as

$$\mathbf{A}'\mathbf{A}'^T + \mathbf{B}'\mathbf{B}'^T = \mathbf{M}_0 + (\lambda/2)\mathbf{I} \tag{3.54}$$

$$\mathbf{B}'\mathbf{A}'^T = -\mathbf{M}_1 + (\lambda/4)\mathbf{I}, \tag{3.55}$$

which in principle can be solved using an iterative scheme such as that discussed in Eqs. (3.36) to (3.39).

Usually, practitioners simply use the approximate scheme of decomposing Eqs. (3.47) and (3.50):

$$(\mathbf{A}' + \mathbf{B}') = \mathbf{P}'$$

$$(\mathbf{A}' - \mathbf{B}') = \mathbf{Q}'$$

and

$$\mathbf{A}' = \frac{\mathbf{P}' + \mathbf{Q}'}{2}$$

$$\mathbf{B}' = \frac{\mathbf{P}' - \mathbf{Q}'}{2}.$$

The above solution most probably will not satisfy Eqs. (3.48), (3.49), and (3.51). However, experience indicates that for small λs the distortions of the resulting models remain acceptable.

3.2.4 Multivariate MA(2)

The solution of

$$\mathbf{Z}(t) = \mathbf{A}\boldsymbol{\varepsilon}(t) - \mathbf{B}\boldsymbol{\varepsilon}(t-1) - \mathbf{C}\boldsymbol{\varepsilon}(t-2) \tag{3.56}$$

proceeds in a manner analogous to that in the previous section. Finding $E[\mathbf{Z}(t)\mathbf{Z}^T(t)]$, $E[\mathbf{Z}(t)\mathbf{Z}^T(t-1)]$, and $E[\mathbf{Z}(t)\mathbf{Z}^T(t-2)]$ results in

$$\mathbf{M}_0 = \mathbf{A}\mathbf{A}^T + \mathbf{B}\mathbf{B}^T + \mathbf{C}\mathbf{C}^T \tag{3.57}$$

$$\mathbf{M}_1 = -\mathbf{B}\mathbf{A}^T + \mathbf{C}\mathbf{B}^T \tag{3.58}$$

$$\mathbf{M}_2 = -\mathbf{C}\mathbf{A}^T \tag{3.59}$$

An approximate alternative to a nonlinear iterative solution of the above is to use the following valid expressions:

$$(\mathbf{A}-\mathbf{B}-\mathbf{C})(\mathbf{A}-\mathbf{B}-\mathbf{C})^T = \mathbf{M}_0 + \mathbf{M}_1 + \mathbf{M}_1^T + \mathbf{M}_2 + \mathbf{M}_2^T \quad (3.60)$$

$$(-\mathbf{A}-\mathbf{B}+\mathbf{C})(-\mathbf{A}-\mathbf{B}+\mathbf{C})^T = \mathbf{M}_0 + \mathbf{M}_2 + \mathbf{M}_2^T - \mathbf{M}_1 - \mathbf{M}_1^T \quad (3.61)$$

Solving for $(\mathbf{A}-\mathbf{B}-\mathbf{C}) = \mathbf{R1}$ and $(-\mathbf{A}-\mathbf{B}+\mathbf{C}) = \mathbf{R2}$ and combining yields

$$\mathbf{B} = -\frac{\mathbf{R1}+\mathbf{R2}}{2}. \quad (3.62)$$

Using the resulting $\mathbf{B}$ in Eq. (3.57) and combining with Eq. (3.59) also implies

$$(\mathbf{A}-\mathbf{C})(\mathbf{A}-\mathbf{C})^T = \mathbf{M}_0 - \mathbf{BB}^T + \mathbf{M}_2 + \mathbf{M}_2^T \quad (3.63)$$

$$(\mathbf{A}+\mathbf{C})(\mathbf{A}+\mathbf{C})^T = \mathbf{M}_0 - \mathbf{BB}^T - \mathbf{M}_2 - \mathbf{M}_2^T, \quad (3.64)$$

which after decomposition $(\mathbf{A}-\mathbf{C} = \mathbf{R3}, \ \mathbf{A}+\mathbf{C} = \mathbf{R4})$ can be solved for,

$$\mathbf{A} = \frac{\mathbf{R3}+\mathbf{R4}}{2} \quad (3.65)$$

$$\mathbf{C} = \frac{\mathbf{R4}-\mathbf{R3}}{2}. \quad (3.66)$$

Puente (1978) encountered frequent inconsistencies (negative definiteness) in Eq. (3.64). Treatment of these numerical problems using Mejía and Millán's formulation again requires the solution of equations corresponding to Eqs. (3.48) and (3.49). The resulting modified model is

$$\mathbf{Z}(t) = \frac{1}{\sqrt{1+\lambda/2}}\mathbf{A}'\boldsymbol{\varepsilon}(t) - \frac{1}{\sqrt{1+\lambda/2}}\mathbf{B}\boldsymbol{\varepsilon}(t-1) - \frac{1}{\sqrt{1+\lambda/2}}\mathbf{C}'\boldsymbol{\varepsilon}(t-2),$$

$$(3.67)$$

where $\mathbf{A}'$ and $\mathbf{C}'$ correspond to

$$(\mathbf{A}'+\mathbf{C}')(\mathbf{A}'+\mathbf{C}')^T = (\mathbf{A}+\mathbf{C})(\mathbf{A}+\mathbf{C})^T + \lambda\mathbf{I}. \quad (3.68)$$

Corresponding distortions on $\mathbf{M}_0$, $\mathbf{M}_1$, and $\mathbf{M}_2$ are

$$\mathbf{M}_0' = \frac{1}{(1+\lambda/2)}\left[\mathbf{M}_0 + (\lambda/2)\mathbf{I}\right]$$

$$\mathbf{M}_1' = \frac{1}{(1+\lambda/2)}\left[-\mathbf{BA}^T + \mathbf{C}'\mathbf{B}^T\right]$$

$$\mathbf{M}_2' = \frac{1}{(1+\lambda/2)}\left[\mathbf{M}_2 - (\lambda/4)\mathbf{I}\right].$$

3.2.5 Multivariate ARMA(1,1)

O'Connell (1974) did much to popularize the use of the ARMA(1,1) model in hydrology. Some of his univariate results will be discussed in some detail in the next chapter. He was also the first to suggest the multivariate ARMA(1,1)

$$\mathbf{Z}(t) = \mathbf{A}\mathbf{Z}(t-1) + \mathbf{B}\boldsymbol{\varepsilon}(t) - \mathbf{C}\boldsymbol{\varepsilon}(t-1). \tag{3.69}$$

Parameter matrices $\mathbf{A}$, $\mathbf{B}$, and $\mathbf{C}$ derive from the theoretical expressions for $\mathbf{M}_0$, $\mathbf{M}_1$, and $\mathbf{M}_2$, i.e., $E[\mathbf{Z}(t)\mathbf{Z}^T(t)]$, $E[\mathbf{Z}(t)\mathbf{Z}^T(t-1)]$, and $E[\mathbf{Z}(t)\mathbf{Z}^T(t-2)]$:

$$\mathbf{M}_0 = \mathbf{A}\mathbf{M}_1^T + \mathbf{B}\mathbf{B}^T - \mathbf{C}\mathbf{B}^T\mathbf{A}^T + \mathbf{C}\mathbf{C}^T \tag{3.70}$$

$$\mathbf{M}_1 = \mathbf{A}\mathbf{M}_0 - \mathbf{C}\mathbf{B}^T \tag{3.71}$$

$$\mathbf{M}_2 = \mathbf{A}\mathbf{M}_1. \tag{3.72}$$

From Eq. (3.72)

$$\mathbf{A} = \mathbf{M}_2\mathbf{M}_1^{-1}, \tag{3.73}$$

which on substitution in Eqs. (3.70) and (3.71) yields

$$\mathbf{B}\mathbf{B}^T + \mathbf{C}\mathbf{C}^T = \mathbf{M}_0 - \mathbf{M}_2\mathbf{M}_1^{-1}\mathbf{M}_1^T + \left[\mathbf{M}_2\mathbf{M}_1^{-1}\mathbf{M}_0 - \mathbf{M}_1\right]\left[\mathbf{M}_2\mathbf{M}_1^{-1}\right]^T = \mathbf{F} \tag{3.74}$$

$$\mathbf{C}\mathbf{B}^T = \mathbf{M}_2\mathbf{M}_1^{-1}\mathbf{M}_0 - \mathbf{M}_1 = \mathbf{G}, \tag{3.75}$$

which in turn can be combined to yield

$$(\mathbf{B}+\mathbf{C})(\mathbf{B}+\mathbf{C})^T = \mathbf{F} + \mathbf{G} + \mathbf{G}^T \tag{3.76}$$

$$(\mathbf{B}-\mathbf{C})(\mathbf{B}-\mathbf{C})^T = \mathbf{F} - \mathbf{G} - \mathbf{G}^T. \tag{3.77}$$

Decomposition of the above into $\mathbf{B}+\mathbf{C}=\mathbf{R}$ and $\mathbf{B}-\mathbf{C}=\mathbf{S}$ permits the approximate solution

$$\mathbf{B} = \frac{\mathbf{R}+\mathbf{S}}{2} \tag{3.78}$$

$$\mathbf{C} = \frac{\mathbf{R}-\mathbf{S}}{2}. \tag{3.79}$$

The approximate nature of the above must be re-emphasized. The original Eqs. (3.70) to (3.72) impose $2n^2 + n(n-1)/2$ conditions on the matrices $\mathbf{A}$, $\mathbf{B}$, and $\mathbf{C}$. The solution of $\mathbf{A}$ uses n^2 of those conditions, leaving $n^2 + n(n-1)/2$ conditions for full definition of $\mathbf{B}$ and $\mathbf{C}$. The approximate procedure uses only $n(n-1)$ of these due to symmetry considerations. As a consequence exact preservation of $\mathbf{M}_0$ and $\mathbf{M}_1$ cannot be ensured; only their diagonal values will be preserved.

An exact iterative nonlinear solution for Eqs. (3.74) and (3.75) is possible. From Eq. (3.75),

$$C = GB^{T^{-1}}$$
$$C^T = (GB^{T^{-1}})^T = B^{-1}G^T. \tag{3.80}$$

Substitution in Eq. (3.74) leads to

$$BB^T + G(BB^T)^{-1}G^T = F.$$

Defining $U_j = BB^T$ as the j estimate of BB^T, the iterative form is

$$U_j = F - GU_{j-1}^{-1}G^T, \tag{3.81}$$

whose solution can begin with $U_j = I$. After a BB^T and corresponding B are found, Eq. (3.80) can then be used for a full definition of C.

A Simple Alternate Form to the ARMA(1,1)

O'Connell (1974) suggests limiting matrix A to diagonal form,

$$A = \begin{bmatrix} a_{11} & \cdots & & 0 \\ & a_{22} & & \\ \vdots & & \ddots & \vdots \\ 0 & \cdots & & a_{nn} \end{bmatrix},$$

where the elements a_{ii} $i = 1, \ldots, n$ obey the relationship common to univariate models

$$a_{ii} = \rho_i(2)/\rho_i(1), \tag{3.82}$$

with $\rho_i(j)$ being the j lag correlation of site i. Provided that matrices B and C are obtained using the iterative procedures of Eqs. (3.80) and (3.81), the model will correctly represent all elements of M_0 and M_1. Only the diagonals of M_2 will be correct, with the nondiagonals taking the value

$$\rho_{ij}(2) = a_{ii}\rho_{ij}(1). \tag{3.83}$$

Effectively, the simplification allows the independent assessment for each station of the autoregressive term in the ARMA(1,1) multivariate model.

3.2.6 An Example of Multivariate Simulation

Puente (1978) studied flows at five stations in a river basin in Colombia. Each station had 23 years of records. Figures 3.1 to 3.5 show the historical data, plotted at monthly intervals. The seasonality was treated by standardizing

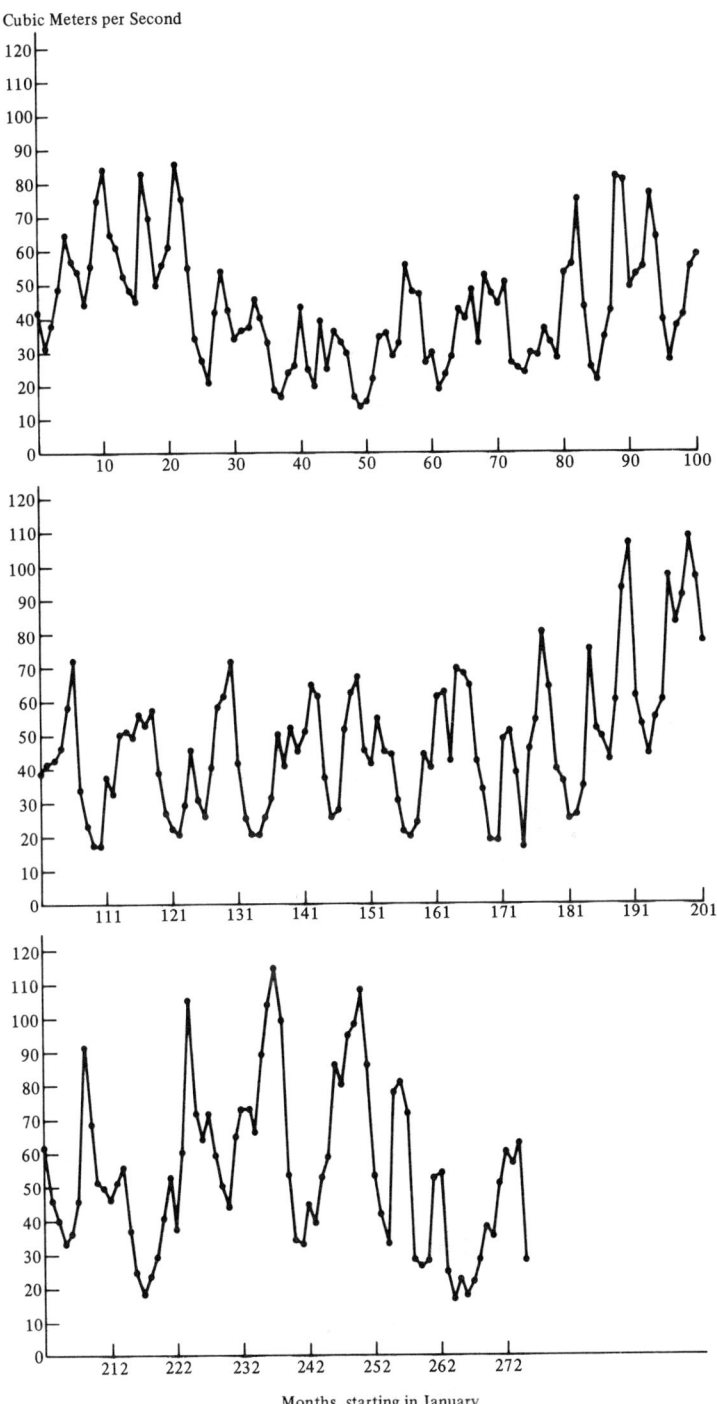

Figure 3.1 Monthly streamflows at station 1, Colombian river-basin example.

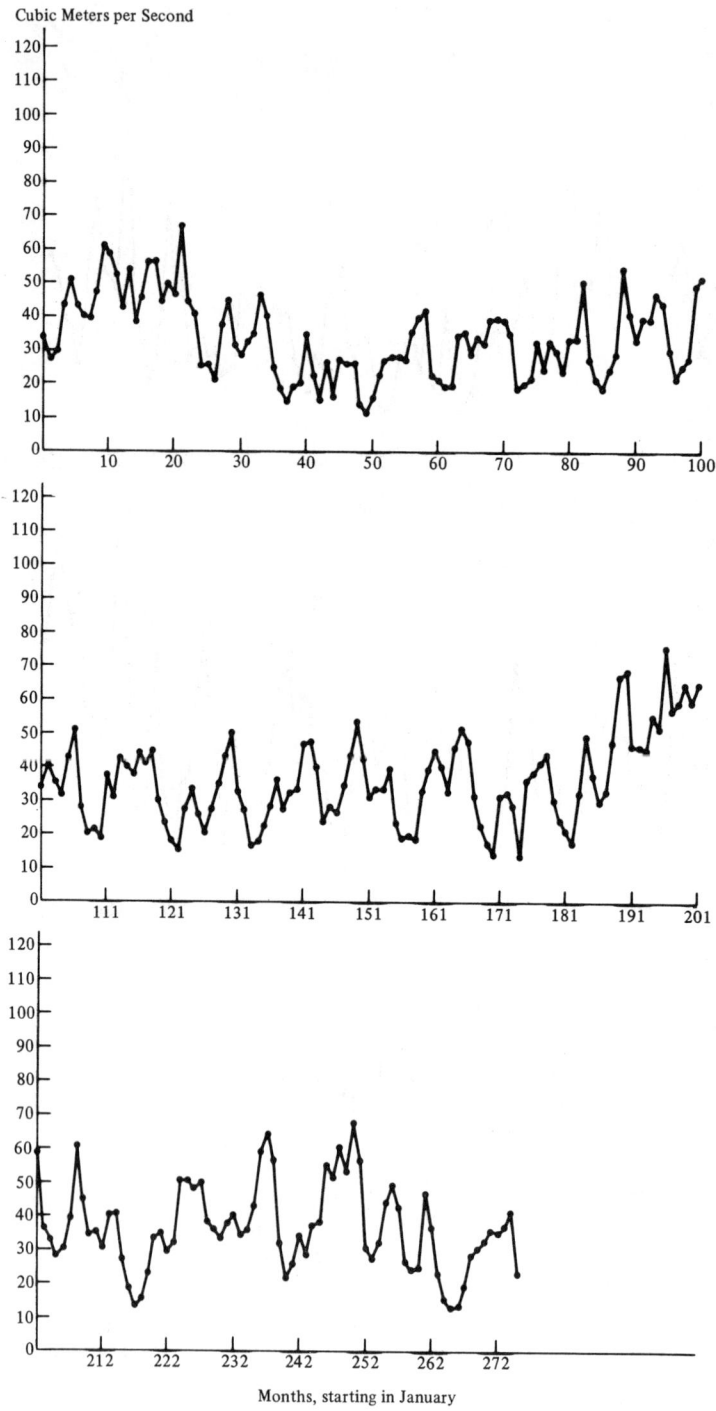

Figure 3.2   Monthly streamflows at station 2, Colombian river-basin example.

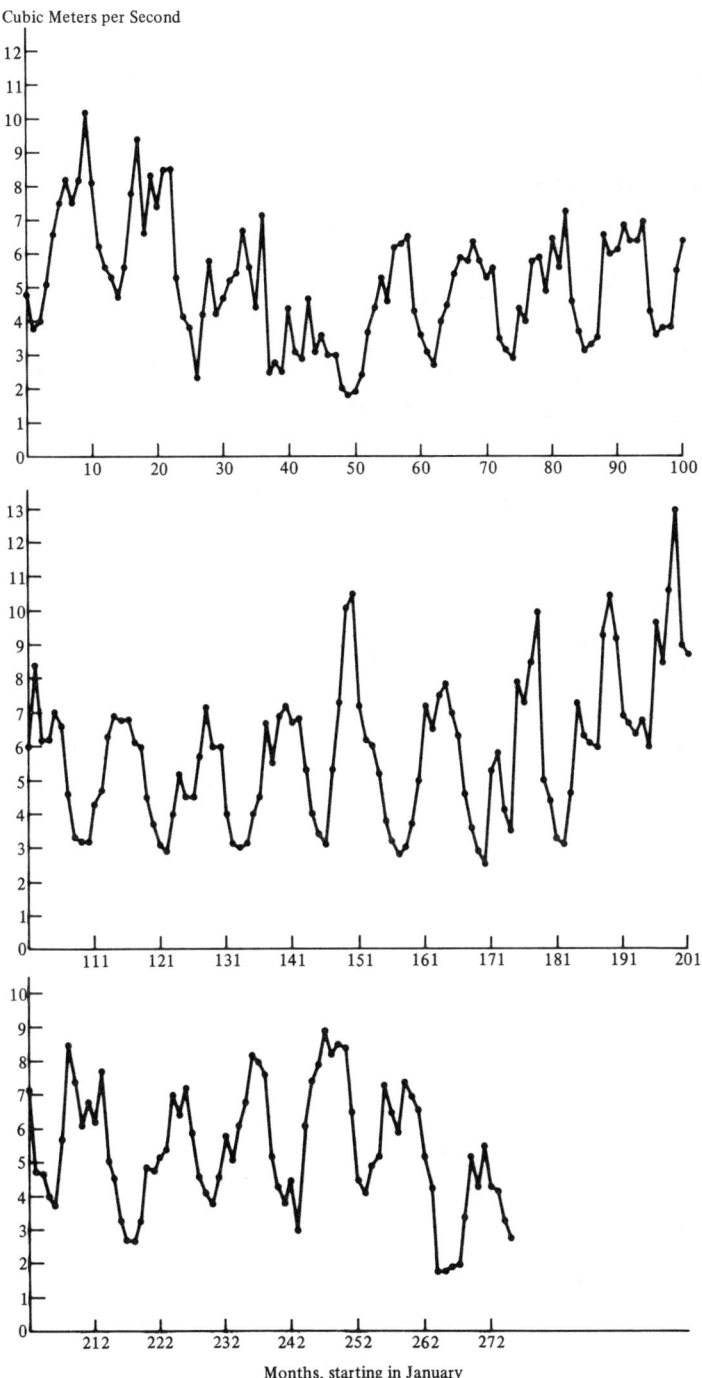

Figure 3.3 Monthly streamflows at station 3, Colombian river-basin example.

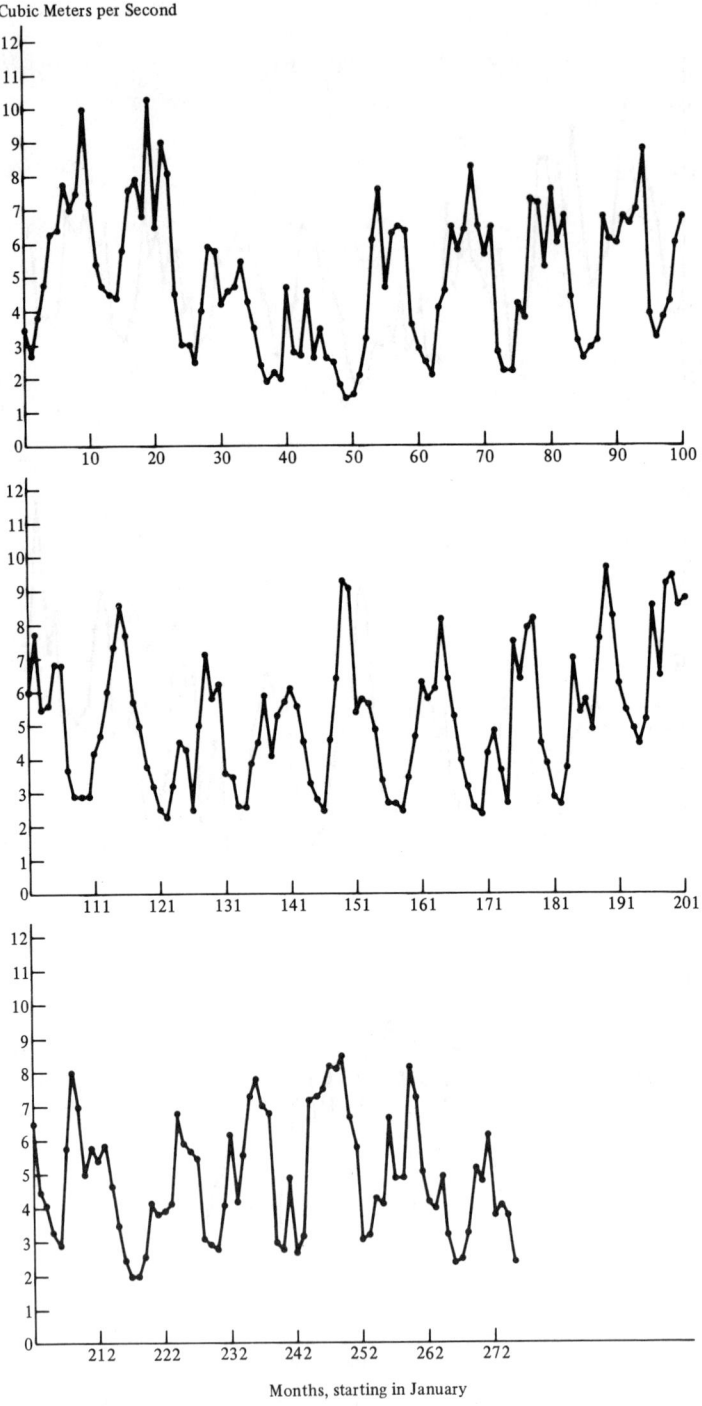

Figure 3.4 Monthly streamflows at station 4, Colombian river-basin example.

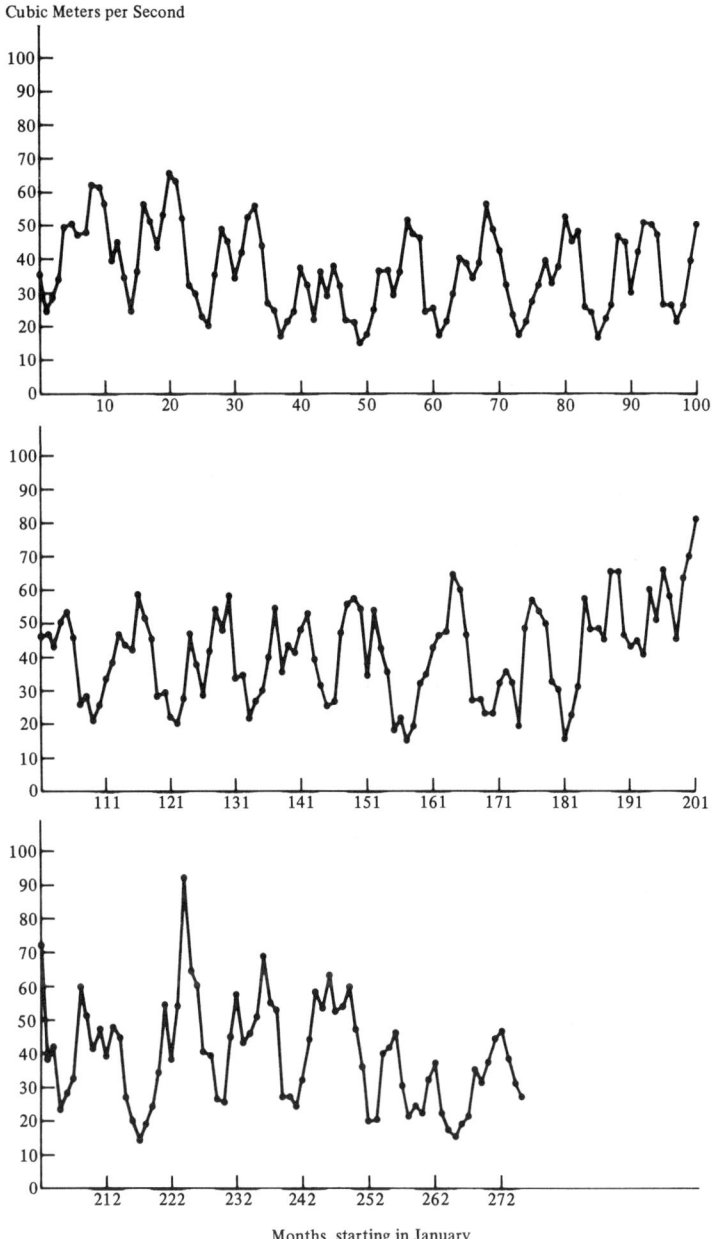

Figure 3.5   Monthly streamflows at station 5, Colombian river-basin example.

each record—subtracting its monthly mean and dividing by the sample monthly standard deviation. It was then assumed that the resulting time series was stationary not only in terms of mean and variance but in terms of constant correlation structure from month to month. That is, $\mathbf{M}_0$, $\mathbf{M}_1$, and $\mathbf{M}_2$ were assumed to be the same for every month. The computed sample covariances were

$$\hat{\mathbf{M}}_0 = \begin{bmatrix} 1.000 & 0.880 & 0.760 & 0.631 & 0.703 \\ 0.880 & 1.000 & 0.860 & 0.749 & 0.780 \\ 0.760 & 0.860 & 1.000 & 0.836 & 0.727 \\ 0.631 & 0.749 & 0.836 & 1.000 & 0.632 \\ 0.703 & 0.780 & 0.727 & 0.632 & 1.000 \end{bmatrix}$$

$$\hat{\mathbf{M}}_1 = \begin{bmatrix} 0.760 & 0.719 & 0.660 & 0.549 & 0.619 \\ 0.706 & 0.761 & 0.722 & 0.625 & 0.664 \\ 0.629 & 0.686 & 0.764 & 0.629 & 0.609 \\ 0.485 & 0.555 & 0.602 & 0.596 & 0.475 \\ 0.547 & 0.621 & 0.582 & 0.487 & 0.708 \end{bmatrix}$$

$$\hat{\mathbf{M}}_2 = \begin{bmatrix} 0.652 & 0.627 & 0.573 & 0.465 & 0.584 \\ 0.638 & 0.697 & 0.655 & 0.556 & 0.632 \\ 0.588 & 0.663 & 0.697 & 0.554 & 0.601 \\ 0.474 & 0.556 & 0.570 & 0.493 & 0.455 \\ 0.491 & 0.565 & 0.549 & 0.456 & 0.625 \end{bmatrix}$$

Two multivariate models were investigated, AR(1) and ARMA(1,1). The AR(1) was identified with parameters

$$\hat{\mathbf{A}} = \begin{bmatrix} 0.590 & 0.008 & 0.136 & -0.010 & 0.105 \\ 0.150 & 0.322 & 0.204 & 0.029 & 0.141 \\ 0.091 & -0.019 & 0.675 & -0.037 & 0.093 \\ 0.016 & 0.076 & 0.253 & 0.297 & 0.032 \\ 0.046 & 0.174 & 0.108 & -0.062 & 0.566 \end{bmatrix}$$

$$\hat{\mathbf{B}} = \begin{bmatrix} 0.625 & 0.000 & 0.000 & 0.000 & 0.000 \\ 0.472 & 0.409 & 0.000 & 0.000 & 0.000 \\ 0.354 & 0.287 & 0.446 & 0.000 & 0.000 \\ 0.342 & 0.315 & 0.363 & 0.507 & 0.000 \\ 0.362 & 0.232 & 0.121 & 0.041 & 0.532 \end{bmatrix}$$

Figures 3.6 through 3.10 show the historical correlations at each station, together with the sample correlation resulting from the generation of 30 years of monthly streamflows with the AR(1) model.

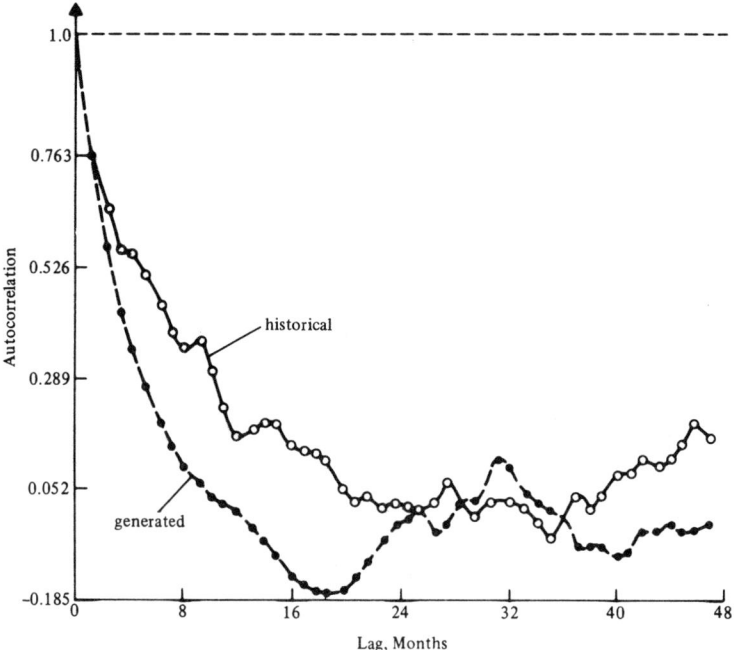

Figure 3.6 Historical correlation function of station 1 versus sample correlation function from series generated with the AR(1) model, Colombian river-basin example.

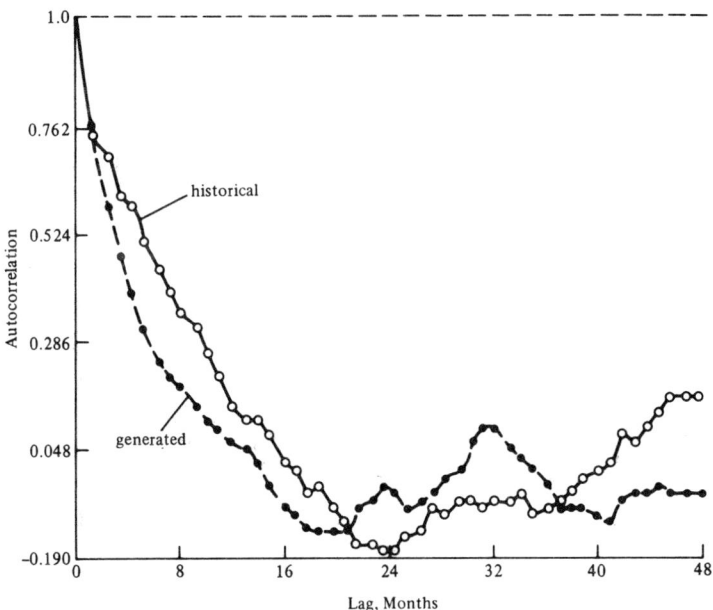

Figure 3.7 Historical correlation function of station 2 versus sample correlation function from series generated with the AR(1) model, Colombian river-basin example.

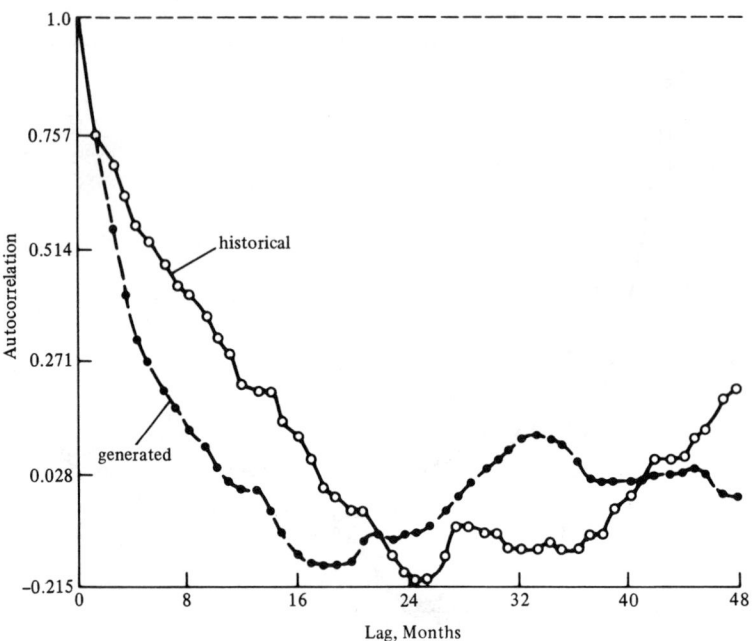

Figure 3.8 Historical correlation function of station 3 versus sample correlation function from series generated with the AR(1) model, Colombian river-basin example.

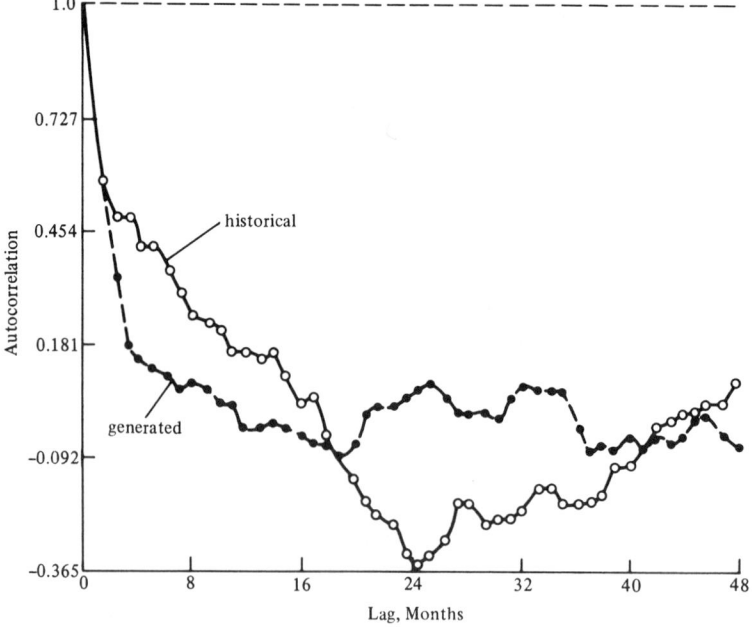

Figure 3.9 Historical correlation function of station 4 versus sample correlation function from series generated with the AR(1) model, Colombian river-basin example.

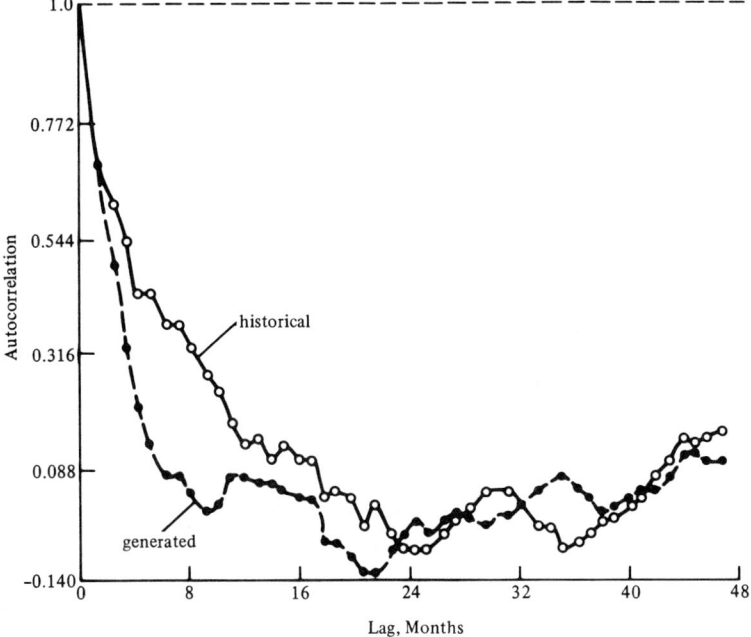

Figure 3.10 Historical correlation function of station 5 versus sample correlation function from series generated with the AR(1) model, Colombian river-basin example.

The matrices $\hat{\mathbf{A}}$, $\hat{\mathbf{B}}$, and $\hat{\mathbf{C}}$ for the fitted ARMA(1,1) model were

$$\hat{\mathbf{A}} = \begin{bmatrix} 0.732 & -0.025 & 0.047 & -0.119 & 0.247 \\ -0.104 & 1.019 & 0.011 & -0.145 & 0.115 \\ -0.299 & 0.671 & 0.800 & -0.383 & 0.048 \\ -0.458 & 1.112 & 0.303 & -0.048 & -0.230 \\ -0.899 & 0.077 & 0.172 & -0.038 & 0.775 \end{bmatrix}$$

$$\hat{\mathbf{B}} = \begin{bmatrix} 0.633 & 0.000 & 0.000 & 0.000 & 0.000 \\ 0.467 & 0.414 & 0.000 & 0.000 & 0.000 \\ 0.354 & 0.809 & 0.460 & 0.000 & 0.000 \\ 0.334 & 0.497 & 0.396 & 0.481 & 0.000 \\ 0.337 & 0.209 & 0.079 & 0.047 & 0.502 \end{bmatrix}$$

$$\hat{\mathbf{C}} = \begin{bmatrix} 0.120 & 0.000 & 0.000 & 0.000 & 0.000 \\ 0.114 & 0.249 & 0.000 & 0.000 & 0.000 \\ 0.037 & 0.146 & 0.075 & 0.000 & 0.000 \\ 0.007 & 0.104 & 0.113 & -0.003 & 0.000 \\ 0.181 & 0.015 & 0.189 & -0.047 & 0.162 \end{bmatrix}$$

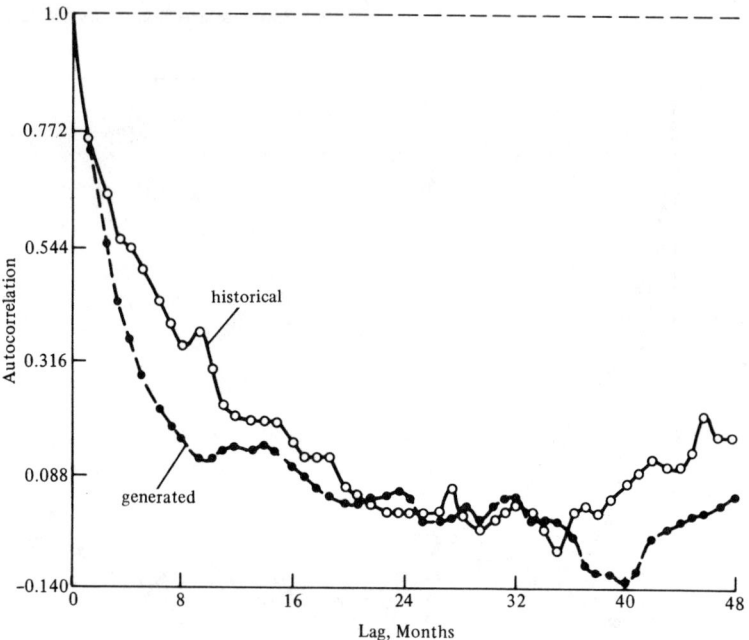

Figure 3.11 Historical correlation function of station 1 versus sample correlation function from series generated with the ARMA(1,1) model, Colombian river-basin example.

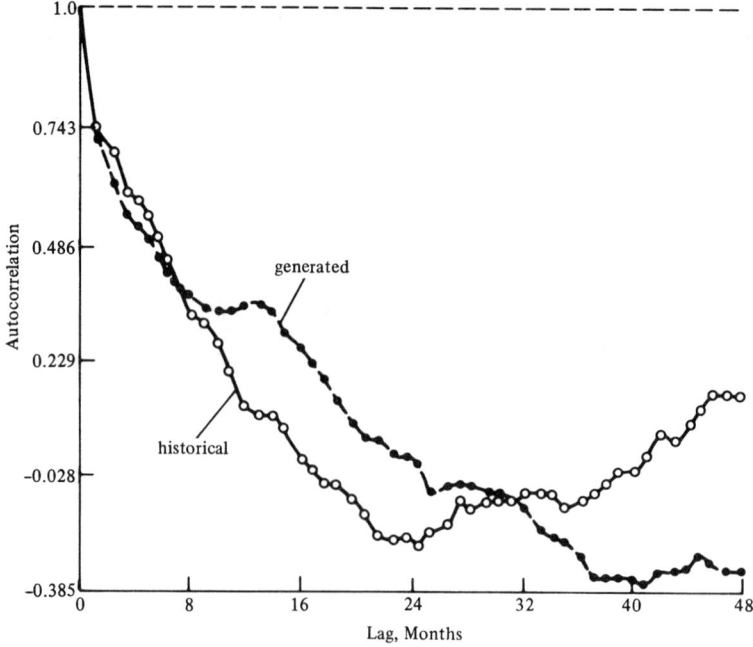

Figure 3.12 Historical correlation function of station 2 versus sample correlation function from series generated with the ARMA(1,1) model, Colombian river-basin example.

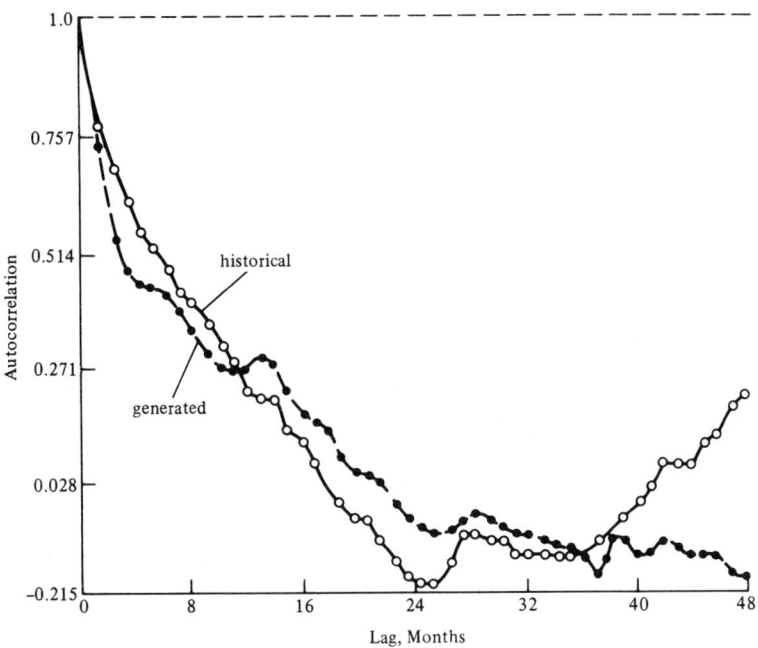

Figure 3.13 Historical correlation function of station 3 versus sample correlation function from series generated with the ARMA(1,1) model, Colombian river-basin example.

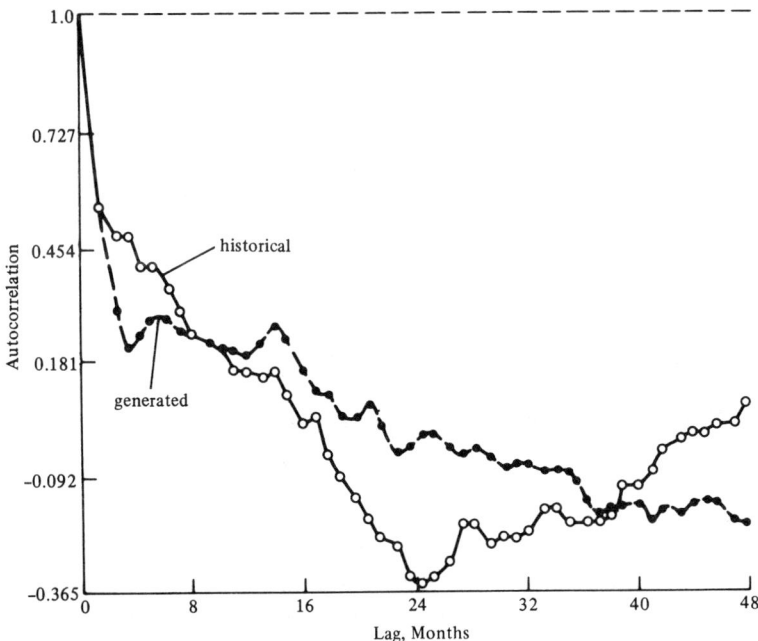

Figure 3.14 Historical correlation function of station 4 versus sample correlation function from series generated with the ARMA(1,1) model, Colombian river-basin example.

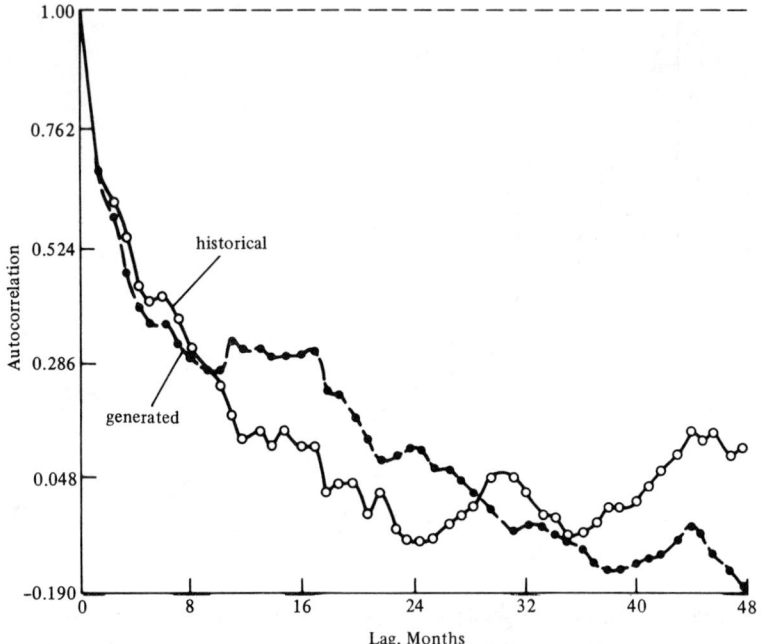

Figure 3.15 Historical correlation function of station 5 versus sample correlation function from series generated with the ARMA(1,1) model, Colombian river-basin example.

Note that the approximate procedure, yielding lower triangular matrices $\hat{\mathbf{B}}$ and $\hat{\mathbf{C}}$, was used for parameter estimation. The historical and generated (30 years) sample correlations are shown in Figs. 3.11 through 3.15 for all five stations. It is apparent that the overall behavior of the ARMA(1,1) is preferable to that of the AR(1) model.

3.3 SEASONAL MULTIVARIATE MODELS

3.3.1 The Seasonal Multivariate Autoregressive Model

The above example dealt with seasonality by standardizing flows but ignored nonstationarity in the correlation structure. Time-varying parameters are required to include seasonal variability in the correlation structure. In theory, it is possible to reformulate all the models in Section 3.2 in this nonstationarity mode. Most widely used is the seasonal AR(1) (Salas and Pegram, 1977; Salas et al, 1980).

The multivariate seasonal autoregressive model of order 1 will preserve all seasonal means of all variables in the state vector, all seasonal variances, all correlations among all elements of the state vector, and lag-one correlations between adjacent seasons and between all variables. As formulated here, the model will be able to handle normal and log-normal variables as well as mixed normal and log-normal elements in the state vector.

The Multivariate Normal Case

Let

$$(\mathbf{X}_{ij} - \mathbf{m}_j) = \mathbf{A}_j(\mathbf{X}_{i,j-1} - \mathbf{m}_{j-1}) + \mathbf{B}_j\boldsymbol{\varepsilon}_{ij}, \tag{3.84}$$

where $\mathbf{X}_{ij}$ is the state vector $(n \times 1)$ of random variables x_{ij}^ℓ, during the year i and season j at location ℓ with mean $\mathbf{m}_j$.

In detail,

$$\mathbf{X}_{ij} = \begin{bmatrix} x_{ij}^1 \\ x_{ij}^2 \\ \vdots \\ x_{ij}^n \end{bmatrix} \quad \text{and} \quad \mathbf{m}_j = \begin{bmatrix} m_{x_j}^1 \\ m_{x_j}^2 \\ \vdots \\ m_{x_j}^n \end{bmatrix}.$$

For example, x_{ij}^1 can be interpreted as the monthly streamflow at station 1 during year i and month j (season) and $m_{x_j}^1$ as the mean value of this variable during the month j.

$\mathbf{A}_j$ and $\mathbf{B}_j$ are parameter matrices $(n \times n)$, one for each season. The $(n \times 1)$ vector of standard normal deviates is $\boldsymbol{\varepsilon}_{ij}$, for year i and season j.

Notation for Eq. (3.84) is simplified by introducing the zero-mean vector, $\mathbf{Z}_{ij} = (\mathbf{X}_{ij} - \mathbf{m}_j)$:

$$\mathbf{Z}_{ij} = \mathbf{A}_j\mathbf{Z}_{i,j-1} + \mathbf{B}_j\boldsymbol{\varepsilon}_j. \tag{3.85}$$

The similarity between Eqs. (3.85) and (3.1) is obvious; the only difference is the seasonal dependence of $\mathbf{A}_j$ and $\mathbf{B}_j$. The lag-zero covariance for season j is $_j\mathbf{M}_0$. The covariance matrix between vectors $\mathbf{Z}_{ij}$ and $\mathbf{Z}_{i,j-1}$ is $_j\mathbf{M}_1$. Following a derivation analogous to that of Section 3.2.1, it should be easy to arrive at the following results:

$$\mathbf{A}_j = {}_j\mathbf{M}_{1\,j-1}\mathbf{M}_0^{-1} \tag{3.86}$$

$$\mathbf{B}_j\mathbf{B}_j^T = {}_j\mathbf{M}_0 - {}_j\mathbf{M}_{1\,j-1}\mathbf{M}_0^{-1}{}_j\mathbf{M}_1^T. \tag{3.87}$$

The decomposition of $\mathbf{B}_j\mathbf{B}_j^T$ is accomplished using the procedures described in Section 3.2.1. Since every covariance function is seasonally dependent, so are the resulting parameters.

It is now convenient notationally to redefine the vectors $\mathbf{Z}_{ij}$ and $\mathbf{Z}_{i,j-1}$ as $\mathbf{Y}$ and $\mathbf{X}$, respectively. That is, vector $\mathbf{Y}$, with n elements y_i, $i = 1, \ldots, n$, will represent the state vector at season j; and vector $\mathbf{X}$, with elements x_i, $i = 1, \ldots, n$, will represent the state vector in the previous season $j - 1$. The model is now

$$\mathbf{Y} = \mathbf{A}_j \mathbf{X} + \mathbf{B}_j \boldsymbol{\varepsilon}_j. \tag{3.88}$$

The covariances of interest are now,

$$_j\mathbf{M}_0 = \mathbf{S}_{yy} = E[\mathbf{YY}^T]$$

$$_j\mathbf{M}_1 = \mathbf{S}_{yx} = E[\mathbf{YX}^T]$$

$$_{j-1}\mathbf{M}_0 = \mathbf{S}_{xx} = E[\mathbf{XX}^T]$$

$$\mathbf{S}_{xy} = \mathbf{S}_{yx}^T$$

Matrices $\mathbf{S}_{xx}$, $\mathbf{S}_{xy}$, $\mathbf{S}_{yx}$, and $\mathbf{S}_{yy}$ can be represented in terms of variances, standard deviations, and correlations as

$$\mathbf{S}_{xx} = \begin{bmatrix} S_{x_1}^2 & r_{x_1x_2}S_{x_1}S_{x_2} & \cdots & r_{x_1x_n}S_{x_1}S_{x_n} \\ r_{x_2x_1}S_{x_2}S_{x_1} & S_{x_2}^2 & & r_{x_2x_n}S_{x_2}S_{x_n} \\ \vdots & & \ddots & \vdots \\ r_{x_nx_1}S_{x_n}S_{x_1} & r_{x_nx_2}S_{x_n}S_{x_2} & \cdots & S_{x_n}^2 \end{bmatrix} \tag{3.89}$$

$$\mathbf{S}_{yy} = \begin{bmatrix} S_{y_1}^2 & r_{y_1y_2}S_{y_1}S_{y_2} & & r_{y_1y_n}S_{y_1}S_{y_n} \\ \vdots & \ddots & & \vdots \\ r_{y_ny_1}S_{y_n}S_{y_1} & r_{y_ny_2}S_{y_n}S_{y_2} & \cdots & S_{y_n}^2 \end{bmatrix} \tag{3.90}$$

$$\mathbf{S}_{yx} = \begin{bmatrix} r_{y_1x_1}S_{x_1}S_{y_1} & r_{y_2x_1}S_{x_1}S_{y_2} & \cdots & r_{y_nx_1}S_{x_1}S_{y_n} \\ \vdots & \ddots & & \vdots \\ r_{y_1x_n}S_{x_n}S_{y_1} & r_{y_2x_n}S_{x_n}S_{y_2} & \cdots & r_{y_nx_n}S_{x_n}S_{y_n} \end{bmatrix}, \tag{3.91}$$

where S_{x_i} is the standard deviation of variable x_i, $r_{x_ix_j}$ is the lag-zero correlation between stations (variables) x_i and x_j, and $r_{y_ix_j}$ is the lag-one correlation between variables y_i and x_j.

The estimation of sample covariances should again follow Eqs. (3.12) and (3.13) to minimize the occurrences of inconsistent (nonpositive definite) $\mathbf{B}_j\mathbf{B}_j^T$ matrices.

Multivariate Log-Normal Case

Consider the case where the elements of vectors X and Y, x_i and y_i, are random variables following a two-parameter log-normal distribution. Define the new variables, x_i' and y_i', as follows:

$$x_i' = \ln(x_i) \tag{3.92}$$

$$y_i' = \ln(y_i). \tag{3.93}$$

Thus, the original variables x_i and y_i are log-normally distributed with means m_{x_i} and m_{y_i}, standard deviations S_{x_i} and S_{y_i}, and the correlation coefficient among them given by $r_{x_i y_i}$.

Therefore, the transformed variables x_i' and y_i' are normally distributed with means $m_{x_i'}$ and $m_{y_i'}$, standard deviations $S_{x_i'}$ and $S_{y_i'}$, and the correlation coefficient among them equal to $r_{x_i' y_i'}$.

To preserve without bias the statistics of the original variables instead of the statistics of the transformed variables it is necessary to compute parameters of the distribution of x_i' and y_i' based on the parameters of the distribution of the original variables x_i and y_i, using the expressions given by Matalas (1967):

$$m_{x_i} = \exp\left\{ S_{x_i'}^2/2 + m_{x_i'} \right\}$$

$$m_{y_i} = \exp\left\{ S_{y_i'}^2/2 + m_{y_i'} \right\}$$

$$S_{x_i}^2 = \exp\left\{ 2\left[S_{x_i'}^2 + m_{x_i'} \right] \right\} - \exp\left[S_{x_i'}^2 + 2m_{x_i'} \right]$$

$$S_{y_i}^2 = \exp\left\{ 2\left[S_{y_i'}^2 + m_{y_i'} \right] \right\} - \exp\left[S_{y_i'}^2 + 2m_{y_i'} \right]$$

$$r_{x_i y_i} = \frac{\exp\left\{ S_{x_i'} S_{y_i'} r_{x_i' y_i'} \right\} - 1}{\sqrt{\left\{ \exp\left(S_{x_i'}^2 \right) - 1 \right\} \left\{ \exp\left(S_{y_i'}^2 \right) - 1 \right\}}}.$$

Solving the above system of simultaneous equations for $m_{x_i'}$, $m_{y_i'}$, $S_{x_i'}$, $S_{y_i'}$, and $r_{x_i' y_i'}$ yields

$$m_{x_i'} = \ln(m_{x_i}) - S_{x_i'}^2/2 \tag{3.94}$$

$$S_{x_i'}^2 = \ln\left(\frac{S_{x_i}^2}{m_{x_i}^2} + 1 \right) \tag{3.95}$$

$$m_{y_i'} = \ln(m_{y_i}) - S_{y_i'}^2/2 \tag{3.96}$$

$$S_{y_i'}^2 = \ln\left(\frac{S_{y_i}^2}{m_{y_i}^2} + 1 \right) \tag{3.97}$$

$$r_{x_i' y_i'} = \frac{\ln\left\{ 1 + r_{x_i y_i} \sqrt{(e^{S_{x_i}} - 1)(e^{S_{y_i}} - 1)} \right\}}{S_{x_i'} S_{y_i'}}. \tag{3.98}$$

Thus, having the sample variance and correlation coefficient from the historical records, it is possible to compute the values of the transformed variables using Eqs. (3.94) through (3.98).

The parameters of the transformed variables can be used to build the necessary autocovariance and cross-covariance matrices using the definitions given in Eqs. (3.89), (3.90), and (3.91). Generation matrices $\mathbf{A}_j$ and $\mathbf{B}_j$ are then available from the previously derived equations.

In order to get the original variables from generated synthetic data, the user must perform the inverse transformation

$$x_i = \exp\left(x_i'\right)$$

$$y_i = \exp\left(y_i'\right).$$

Mixture of Normal and Log-Normal Variables in the Autoregressive Model

Variables such as streamflow are best described by different marginal distributions at each season. For example, the Nile River (Curry and Bras, 1978) exhibits alternating normal and log-normal distributions each month of the year. The multivariate autoregressive model allows the mixing of normal and log-normal variables.

Let x_i and y_i be the original random variables. Assume x_i is log-normally distributed and y_i is normally distributed:

$$x_i' = \ln(x_i)$$

$$x_i' \sim N\left(m_{x_i'}, S_{x_i'}^2\right)$$

$$y_i \sim N\left(m_{y_i}, S_{y_i}^2\right).$$

The procedure to obtain the parameters $m_{x_i'}$ and $S_{x_i'}^2$ of the log-normal variables was described in the last section and made use of Eqs. (3.94) to (3.98). In this case, the problem is to obtain the correlation coefficient between the log-normal and the normal variables while preserving the parameters of the untransformed data. The necessary expression is given by Mejía et al. (1974).

If $r_{x_i y_j}$ is the correlation coefficient among the original variables and $r_{x_i' y_j}$ the correlation coefficient between the transformed log-normal and the normal variables, the following relationship holds,

$$r_{x_i' y_j} = \frac{r_{x_i y_j}\sqrt{\left(e^{S_{x_i'}^2} - 1\right)}}{S_{x_i'}}. \tag{3.99}$$

In order to illustrate how to build the required covariance matrices, consider the following example.

Define the random vectors $\mathbf{X}$ and $\mathbf{Y}$ with three random variables each,

$$\mathbf{X} = \begin{bmatrix} x_1 \\ x_2' \\ x_3' \end{bmatrix} \qquad \mathbf{Y} = \begin{bmatrix} y_1 \\ y_2 \\ y_3' \end{bmatrix}, \qquad (3.100)$$

where the random variables x_1, y_1, y_2 are normally distributed, and x_2', x_3', y_3' are log-normally distributed. Then the corresponding covariance matrices will be

$$\mathbf{S}_{xx} = \begin{bmatrix} \left(S_{x_1}\right)^2 & r_{x_1 x_2'} S_{x_1} S_{x_2'} & r_{x_1 x_3'} S_{x_1} S_{x_3'} \\ r_{x_2' x_1} S_{x_2'} S_{x_1} & \left(S_{x_2'}\right)^2 & r_{x_2' x_3'} S_{x_2'} S_{x_3'} \\ r_{x_3' x_1} S_{x_3'} S_{x_1} & r_{x_3' x_2'} S_{x_3'} S_{x_2'} & \left(S_{x_3'}\right)^2 \end{bmatrix}$$

$$\mathbf{S}_{yy} = \begin{bmatrix} \left(S_{y_1}\right)^2 & r_{y_1 y_2} S_{y_1} S_{y_2} & r_{y_1 y_3'} S_{y_1} S_{y_3'} \\ r_{y_2 y_1} S_{y_2} S_{y_1} & \left(S_{y_2}\right)^2 & r_{y_2 y_3'} S_{y_2} S_{y_3'} \\ r_{y_3' y_1} S_{y_3'} S_{y_1} & r_{y_3' y_2} S_{y_3'} S_{y_2} & \left(S_{y_3'}\right)^2 \end{bmatrix} \qquad (3.101)$$

$$\mathbf{S}_{xy} = \begin{bmatrix} r_{x_1 y_1} S_{x_1} S_{y_1} & r_{x_1 y_2} S_{x_1} S_{y_2} & r_{x_1 y_3'} S_{x_1} S_{y_3'} \\ r_{x_2' y_1} S_{x_2'} S_{y_1} & r_{x_2' y_2} S_{x_2'} S_{y_2} & r_{x_2' y_3'} S_{x_2'} S_{y_3'} \\ r_{x_3' y_1} S_{x_3'} S_{y_1} & r_{x_3' y_2} S_{x_3'} S_{y_2} & r_{x_3' y_3'} S_{x_3'} S_{y_3'} \end{bmatrix}$$

The above matrices are again routinely used in the existing expressions to obtain $\mathbf{A}_j$ and $\mathbf{B}_j$. The user must remember to add means and perform required inverse transformations at the end of the computations. Since the elements of the sample covariance matrices are now individually evaluated and result from different nonlinear transformations, it is not unusual to obtain inconsistent estimates for the $\mathbf{B}_j \mathbf{B}_j^T$ matrix.

3.4 A GENERAL MULTIVARIATE SEASONAL STREAMFLOW MODEL

This section generalizes seasonal multivariate streamflow models. In contrast to the well-defined structure discussed in the past sections, the model presented here can be manipulated to a variety of forms. In the simplest case it will collapse to the multivariate AR(1) model previously seen.

This presentation follows the work of Curry and Bras (1980) and in a sense is a generalization to the multivariate case of the univariate model suggested by Rao and Kashyap (1973, 1974) and Kashyap and Rao (1976).

3.4.1 General Model Form and Identification

The structure of the general model is

$$
\mathbf{Y}(k,i) = \mathbf{\Phi}_i
\begin{bmatrix}
\mathbf{Y}(k,i-1) \\
\mathbf{Y}(k,i-2) \\
\vdots \\
\mathbf{Y}(k,i-n_1(i))
\end{bmatrix}
+ \mathbf{U}(i) + \mathbf{V}(k,i) + \mathbf{\Gamma}_i
\begin{bmatrix}
\mathbf{V}(k,i-1) \\
\mathbf{V}(k,i-2) \\
\vdots \\
\mathbf{V}(k,i-n_2(i))
\end{bmatrix},
$$

$$(3.102)$$

where $\mathbf{Y}(k,i)$ is a vector of streamflows (or other stochastic variables) at year k, season i at n_0 locations or gaging stations

$$
\mathbf{Y}(k,i) =
\begin{bmatrix}
Y_1(k,i) \\
Y_2(k,i) \\
\vdots \\
Y_{n_0}(k,i)
\end{bmatrix}.
$$

The $(n_0 \times 1)$ vector $\mathbf{U}(i)$ represents a deterministic term to account for seasonal variation in the mean of the process. This term is periodic, with a basic cycle of one year in the case of streamflows.

The $(n_0 \times 1)$ vector $\mathbf{V}(k,i)$ represents a random component or noise

$$
\mathbf{V}(k,i) =
\begin{bmatrix}
V_1(k,i) \\
V_2(k,i) \\
\vdots \\
V_{n_0}(k,i)
\end{bmatrix}.
$$

Each of the elements of the above vector can be correlated. Correlation is introduced by operating on uncorrelated white noise to form each $V_j(k,i)$

$$
V_j(k,i) = \sum_{m=1}^{n_0} C_i(j,m) W_m(k,i),
$$

where $C_i(j,m)$ is an element of an $(n_0 \times n_0)$ matrix of coefficients, and $W_m(k,i)$ is a white-noise input at station m, year k, season i.

Matrix $\mathbf{\Gamma}_i$, of dimensions $n_0 \times n_0 n_2(i)$, induces autocorrelation of noise terms affecting the present discharge at all locations.

Matrix $\mathbf{\Phi}_i$, with dimensions $n_0 \times n_0 n_1(i)$, controls the transitions to the month i discharges from past occurrences.

The above model is clearly not parsimonious. Besides the obvious parameter matrices $\mathbf{\Phi}_i$, $\mathbf{\Gamma}_i$, and vector $\mathbf{U}(i)$ (note the seasonal dependence of all of the above), you must also fix the autoregressive lag $n_1(i)$, the noise lag $n_2(i)$, and matrix $\mathbf{C}_i$, for all seasons. However, as will be seen, the estimation of all these parameters is not necessary or is simplified in many cases.

The peculiarity of the model represented in Eq. (3.102) is that each station or location is potentially dependent on every other station at several time lags. The model allows for seasonally varying mean, variances, and correlation structures. Finally, noise terms are not only cross-correlated between stations but may be autocorrelated through the introduction of a moving average of past shocks.

A preliminary identification of structure and parameters is sometimes possible. Such decisions should be guided by knowledge of the process being modeled. The concept of causality may be very valuable in this step. Granger and Newbold (1977) discuss it in detail; Curry and Bras (1980) illustrate it together with the model being studied here.

Essentially, in most geophysical and natural processes, it is valid to state that the future cannot cause the past.

Using the above concepts, it may be possible to determine *a priori* that many of the elements in matrix $\mathbf{\Phi}_i$ are zero. For example, in a river basin, flow in upstream stations cannot be caused by past flows in downstream stations. Nevertheless, if the set of possibly causal variables is not complete, spurious relationships may occur. For example, past downstream stations may show some explanatory power relative to upstream streamflow stations if they are in any way acting as surrogates for past rainfall on upstream locations. Care must then be exercised in interpreting causality.

When using the above model for streamflow, it is also possible to state that $\mathbf{U}(i)$ has a yearly period and to limit $n_1(i)$ and $n_2(i)$ to a reasonable range. Structural simplifications are also possible by eliminating the autocorrelated residuals, i.e., using a model with $\mathbf{\Gamma}_i$ equal to zero. In a forecasting application it will be seen that the need to estimate $\mathbf{C}_i$ does not arise.

Another structural simplification arises by assuming that matrices $\mathbf{\Phi}_i$, $\mathbf{\Gamma}_i$, and parameters $n_1(i)$ and $n_2(i)$ are in fact seasonally invariant, and not a function of i.

The estimation of the nonautocorrelated and constant parameter models are the subject of the next subsections.

3.4.2 Parameter Estimation

Two possible simplifications of the generalized model were suggested previously. One implied elimination of the autocorrelation of residuals, leading to the following form,

$$\mathbf{Y}(k,i) = \mathbf{\Phi}_i \begin{bmatrix} \mathbf{Y}(k,i-1) \\ \mathbf{Y}(k,i-2) \\ \vdots \\ \mathbf{Y}(k,i-n_1(i)) \end{bmatrix} + \mathbf{U}(i) + \mathbf{V}(k,i) \qquad \begin{array}{l} k = 0,1,\ldots, \\ i = 0,1,\ldots,11, \end{array}$$

$$(3.103)$$

where for simplicity of discussion a season has been considered equal to a month.

The other alternative was to assume seasonally invariant parameter matrices,

$$\mathbf{Y}(k,i) = \mathbf{\Phi} \begin{bmatrix} \mathbf{Y}(k,i-1) \\ \mathbf{Y}(k,i-2) \\ \vdots \\ \mathbf{Y}(k,i-n_1) \end{bmatrix} + \mathbf{U}(i) + \mathbf{V}(k,i) + \mathbf{\Gamma} \begin{bmatrix} \mathbf{V}(k,i-1) \\ \mathbf{V}(k,i-2) \\ \vdots \\ \mathbf{V}(k,i-n_2) \end{bmatrix} \tag{3.104}$$

$$k = 0,1,\ldots,$$
$$i = 0,1,2,\ldots,11$$

Parameter Estimation of the Nonautocorrelated Residual Model

Parameter estimation in this case relies heavily on Generalized Least Squares techniques. Excellent and complete references on the subject are Goldberger (1964) and Johnston (1972), and the reader is referred there for review, if necessary. The following discussion assumes knowledge of linear regression and closely follows Curry and Bras (1980).

From Eq. (3.103), the expression for any particular station j and month i is

$$Y_j(k,i) = \sum_{\ell=1}^{n_1(i)} \sum_{m=1}^{n_0} \left[\Phi_i(j,(m-1)n_1(i)+\ell) \cdot Y_m(k,i-\ell) \right]$$
$$+ U_j(i) + V_j(k,i). \tag{3.105}$$

As previously discussed, some elements of $\mathbf{\Phi}_i$ may be assumed *a priori* to be zero. However, to simplify the following discussion, it will be assumed that this information is not available.

If there are N total months of observations, or $N/12$ observations of month i, Eq. (3.105) may be written in the notation of a general linear model [let $n_1(i)=1$ and $j=1$ for ease of notation]:

$$\mathbf{Z}_1 = \mathbf{X}_1\mathbf{\beta}_1 + \mathbf{\epsilon}_1, \tag{3.106}$$

where

$$\mathbf{Z}_1 = \begin{bmatrix} Y_1(1,i) \\ Y_1(2,i) \\ \vdots \\ Y_1(N/12,i) \end{bmatrix}$$

$$\mathbf{X}_1 = \begin{bmatrix} 1 & Y_1(1, i-1) & Y_2(1, i-1) & \cdots & Y_{n_0}(1, i-1) \\ 1 & Y_1(2, i-1) & Y_2(2, i-1) & \cdots & Y_{n_0}(2, i-1) \\ \vdots & \vdots & \vdots & & \vdots \\ 1 & Y_1(N/12, i-1) & Y_2(N/12, i-1) & \cdots & Y_{n_0}(N/12, i-1) \end{bmatrix}$$

$$\beta_1 = \begin{bmatrix} U_1(i) \\ \Phi_i(1,1) \\ \Phi_i(1,2) \\ \vdots \\ \Phi_i(1, n_0) \end{bmatrix}$$

$$\varepsilon_1 = \begin{bmatrix} V_1(1, i) \\ V_1(2, i) \\ \vdots \\ V_1(N/12, i) \end{bmatrix}.$$

It would be possible to use Ordinary Least Squares techniques to estimate the coefficients in Eq. (3.105) (as represented in Eq. 3.106) if the error term between different equations (locations) were not correlated, i.e., $E[V_j(k,i)V_\ell(k,i)] = 0$, $\forall j, \ell$. In fact, by construction, they will undoubtedly be correlated. In such cases, Ordinary Least Squares estimators, although consistent, are not asymptotically efficient. The coefficients of a single equation of the multivariate model (Eq. 3.105) should be estimated simultaneously with all other coefficients. This may be accomplished as follows.

The $N/12$ observations of each equation of the multivariate model may be written together in the notation of a general linear model as

$$\begin{bmatrix} \mathbf{Z}_1 \\ \mathbf{Z}_2 \\ \vdots \\ \mathbf{Z}_{n_0} \end{bmatrix} = \begin{bmatrix} \mathbf{X}_1 & & & \mathbf{0} \\ & \mathbf{X}_2 & & \\ & & \ddots & \\ \mathbf{0} & & & \mathbf{X}_{n_0} \end{bmatrix} \begin{bmatrix} \beta_1 \\ \beta_2 \\ \vdots \\ \beta_{n_0} \end{bmatrix} + \begin{bmatrix} \varepsilon_1 \\ \varepsilon_2 \\ \vdots \\ \varepsilon_{n_0} \end{bmatrix}, \tag{3.107}$$

where $\mathbf{Z}_i$, $\mathbf{X}_i$, and ε_i, $i = 1, 2, \ldots, n_0$, are as defined in Eq. (3.106). Equation (3.107) may be written as

$$\mathbf{Z}' = \mathbf{X}'\beta' + \varepsilon'. \tag{3.108}$$

The error term in a single station equation (Eq. 3.106) is homoscedastic (constant variance) and not cross-correlated, implying

$$E\left[\varepsilon_i \varepsilon_i^T\right] = \sigma_{\varepsilon_i \varepsilon_i} \mathbf{I}$$

$$E\left[\varepsilon_i \varepsilon_j^T\right] = \sigma_{\varepsilon_i \varepsilon_j} \mathbf{I},$$

where $\mathbf{I}$ is the identity matrix. By definition, the variance–covariance matrix of $\boldsymbol{\varepsilon}'$ is then,

$$E[\boldsymbol{\varepsilon}'\boldsymbol{\varepsilon}'^T] = \begin{bmatrix} E[\boldsymbol{\varepsilon}_1\boldsymbol{\varepsilon}_1^T] & E[\boldsymbol{\varepsilon}_1\boldsymbol{\varepsilon}_2^T] & \cdots & E[\boldsymbol{\varepsilon}_1\boldsymbol{\varepsilon}_{n_0}^T] \\ E[\boldsymbol{\varepsilon}_2\boldsymbol{\varepsilon}_1^T] & E[\boldsymbol{\varepsilon}_2\boldsymbol{\varepsilon}_2^T] & \cdots & E[\boldsymbol{\varepsilon}_2\boldsymbol{\varepsilon}_{n_0}^T] \\ \vdots & \vdots & & \vdots \\ E[\boldsymbol{\varepsilon}_{n_0}\boldsymbol{\varepsilon}_1^T] & E[\boldsymbol{\varepsilon}_{n_0}\boldsymbol{\varepsilon}_2^T] & \cdots & E[\boldsymbol{\varepsilon}_{n_0}\boldsymbol{\varepsilon}_{n_0}^T] \end{bmatrix}$$

$$= \begin{bmatrix} \sigma_{\varepsilon_1\varepsilon_1} & \sigma_{\varepsilon_1\varepsilon_2} & \cdots & \sigma_{\varepsilon_1\varepsilon_{n_0}} \\ \sigma_{\varepsilon_2\varepsilon_1} & \sigma_{\varepsilon_2\varepsilon_2} & \cdots & \sigma_{\varepsilon_2\varepsilon_{n_0}} \\ \vdots & \vdots & & \vdots \\ \sigma_{\varepsilon_{n_0}\varepsilon_1} & \sigma_{\varepsilon_{n_0}\varepsilon_2} & \cdots & \sigma_{\varepsilon_{n_0}\varepsilon_{n_0}} \end{bmatrix} \otimes \mathbf{I} \equiv \Sigma_{\varepsilon'\varepsilon'} \otimes \mathbf{I}, \quad (3.109)$$

where $\otimes$ denotes Kronecker multiplication of matrices.*

The homoscedasticity and noncorrelation necessary for Ordinary Least Squares (OLS) are then violated by the complete model. However, using Generalized Least Squares (GLS) estimators, the coefficients of Eq. (3.107) can be consistently and efficiently estimated.

Let

$$E[\boldsymbol{\varepsilon}'\boldsymbol{\varepsilon}'^T] = \Omega,$$

where Ω is the matrix defined in Eq. (3.109). Assuming that Ω is a positive-definite matrix, there exists a nonsingular $\mathbf{P}$ that satisfies

$$\Omega = \mathbf{P}\mathbf{P}^T$$

so that

$$\mathbf{P}^{-1}\Omega(\mathbf{P}^{-1})^T = \mathbf{I}$$

and

$$(\mathbf{P}^{-1})^T\mathbf{P}^{-1} = \Omega^{-1}.$$

---

*

$$\mathbf{A} \otimes \mathbf{B} = \begin{bmatrix} a_{11}\mathbf{B} & a_{12}\mathbf{B} & \cdots & a_{1m}\mathbf{B} \\ a_{21}\mathbf{B} & a_{22}\mathbf{B} & \cdots & a_{2m}\mathbf{B} \\ \vdots & \vdots & & \vdots \\ a_{n1}\mathbf{B} & a_{n2}\mathbf{B} & \cdots & a_{nm}\mathbf{B} \end{bmatrix}$$

where $\mathbf{A}$ is an $(n \times m)$ matrix and $\mathbf{B}$ is of arbitrary dimension.

Pre-multiplying Eq. (3.108) by $\mathbf{P}^{-1}$ gives

$$\mathbf{P}^{-1}\mathbf{Z}' = \mathbf{P}^{-1}\mathbf{X}'\boldsymbol{\beta}' + \mathbf{P}^{-1}\boldsymbol{\varepsilon}'$$

or

$$\mathbf{Z}'' = \mathbf{X}''\boldsymbol{\beta}' + \boldsymbol{\varepsilon}'', \qquad (3.110)$$

where

$$\mathbf{Z}'' = \mathbf{P}^{-1}\mathbf{Z}'$$
$$\mathbf{X}'' = \mathbf{P}^{-1}\mathbf{X}'$$
$$\boldsymbol{\varepsilon}'' = \mathbf{P}^{-1}\boldsymbol{\varepsilon}'.$$

The covariance of $\boldsymbol{\varepsilon}''$ is given by

$$E[\boldsymbol{\varepsilon}''\boldsymbol{\varepsilon}''^{T}] = \mathbf{P}^{-1}\boldsymbol{\Omega}(\mathbf{P}^{-1})^{T}$$
$$= \mathbf{I}.$$

Thus, the transformed model of Eq. (3.110) satisfies the assumptions required for OLS estimation so that

$$\hat{\boldsymbol{\beta}}' = (\mathbf{X}''^{T}\mathbf{X}'')^{-1}\mathbf{X}''^{T}\mathbf{Z}''$$
$$= (\mathbf{X}'^{T}\boldsymbol{\Omega}^{-1}\mathbf{X}')^{-1}\mathbf{X}'^{T}\boldsymbol{\Omega}^{-1}\mathbf{Z}' \qquad (3.111)$$

is a consistent and efficient estimator of the coefficients $\boldsymbol{\beta}'$. A consistent estimate of the variance–covariance matrix of the estimated coefficients is given by Goldberger (1964)

$$\mathbf{S}_{\hat{\beta}'\hat{\beta}'} = E\left[(\hat{\boldsymbol{\beta}}' - \boldsymbol{\beta}')(\hat{\boldsymbol{\beta}}' - \boldsymbol{\beta}')^{T}\right] = [\mathbf{X}'^{T}\boldsymbol{\Omega}^{-1}\mathbf{X}']^{-1}. \qquad (3.112)$$

The difficulty with implementing Eqs. (3.111) and (3.112) is that the matrix $\boldsymbol{\Omega}^{-1}$ is unknown. However, the OLS estimator may be used on each individual station equation (Eq. 3.106) of the multivariate model to obtain estimates of the coefficients (Johnston, 1972). Using these coefficients, the residuals of each individual equation may be computed and subsequently used to estimate $\boldsymbol{\Omega}^{-1}$.
From Eq. (3.109)

$$\hat{\boldsymbol{\Omega}} = \mathbf{S}_{\varepsilon'\varepsilon'} \bigotimes I. \qquad (3.113)$$

where $\mathbf{S}_{\varepsilon'\varepsilon'}$ is an estimate of $\boldsymbol{\Sigma}_{\varepsilon'\varepsilon'}$. A given element of $\mathbf{S}_{\varepsilon'\varepsilon'}$ is estimated from

$$S_{\varepsilon'\varepsilon'}(i, j) = \frac{(\mathbf{Z}_i - \mathbf{X}_i\hat{\boldsymbol{\beta}}_i)^{T}(\mathbf{Z}_j - \mathbf{X}_j\hat{\boldsymbol{\beta}}_j)}{((N/12) - \mathrm{rank}(\mathbf{X}_i))^{1/2}((N/12) - \mathrm{rank}(\mathbf{X}_j))^{1/2}}, \qquad (3.114)$$

where $\hat{\beta}$ is the OLS estimate of β, and $\mathbf{Z}$ and $\mathbf{X}$ are as defined in Eq. (3.106). Thus, using Eqs. (3.113) and (3.114), $\hat{\Omega}$ may be found.

Parameter Estimation of the Autocorrelated Residuals Model

The expression of one element (one station) of the vector $\mathbf{Y}(k, i)$ in Eq. (3.104) can be expressed in terms of a continuous index t, by assuming a 12-season (months) year:

$$Y_j(t) = \sum_{\ell=1}^{n_1} \sum_{m=1}^{n_0} \Phi(j,(m-1)n_0+\ell)Y_m(t-\ell) + U_j(\text{mod}_{12}(t))$$

$$+ V_j(t) + \sum_{\ell=1}^{n_2} \sum_{m=1}^{n_0} \Gamma(j,(m-1)n_0+\ell)V_m(t-\ell), \tag{3.115}$$

where $t = 12k + i$; $k = 0,1,2,\ldots$; $i = 0,1,2,\ldots,11$; and $\text{mod}_{12}(t)$ is the modulus, base 12, of t.

Keep in mind that the terms U_j and V_m remain seasonally dependent. In fact $U_j(i)$ can be expressed as

$$U_j(i) = \alpha(j,1) + \sum_{\ell=1}^{n_3} \left[\alpha(j,2\ell)\cos\omega_\ell(i+1) + \alpha(j,2\ell+1)\sin\omega_\ell(i+1) \right]$$

$$i = 0,1,\ldots,11; \; \omega_\ell = \frac{2\pi\ell}{12}; \; n_3 \leq 6 \tag{3.116}$$

The use of OLS estimators with Eq. (3.115) is inappropriate because the explanatory variables $Y_m(t-\ell)$ are contemporaneously correlated with the disturbance terms.

Panuska (1969) derived a consistent estimation algorithm for the coefficients of a particular case of the stochastic difference equation given in Eq. (3.115). The algorithm was developed for the univariate case (i.e., only one Y sequence) and for $V(\cdot)$ with constant variance. In Kashyap (1971), Kashyap and Rao (1976), and Rao and Kashyap (1973, 1974), theoretical properties of the algorithm are discussed in great detail.

An algorithm, essentially like Panuska's, but capable of estimating the parameters of the multivariate Eq. (3.104) is derived in the following paragraphs. The algorithm is closely related to GLS estimation as discussed in the past section.

The single element Eq. (3.115) of the multivariate model can be written in the form of a linear model (for convenience, let $n_1 = 1$ and $n_2 = 1$, and assume that no elements of Φ or Γ are a priori assumed zero)

$$Z_j(t) = \mathbf{X}_j(t)\beta_j + \varepsilon_j(t), \tag{3.117}$$

where

$$Z_j = Y_j(t)$$

$$\varepsilon_j(t) = V_j(t)$$

$$\mathbf{X}_j(t) = \big[Y_1(t-1)Y_2(t-1),\ldots, Y_{n_0}(t-1), 1,$$

$$\cos \omega_1(\mathrm{mod}_{12}(t)), \sin \omega_1(\mathrm{mod}_{12}(t)),\ldots,$$

$$\cos \omega_{n_3}(\mathrm{mod}_{12}(t)), \sin \omega_{n_3}(\mathrm{mod}_{12}(t)),$$

$$\varepsilon_1(t-1), \varepsilon_2(t-1),\ldots, \varepsilon_{n_0}(t-1)\big]$$

$$\boldsymbol{\beta}_j^T = \big[\Phi(j,1), \Phi(j,2),\ldots, \Phi(j,n_0), \alpha(j,1), \alpha(j,2),\ldots,$$

$$\alpha(j,2n_3+1), \Gamma(j,1), \Gamma(j,2),\ldots, \Gamma(j,n_0)\big].$$

Given N observations of $Y_j(\cdot)$ $\{ Y_j(i),\ i = 0,1,2,\ldots, N-1 \}$, the station equation may be written as

$$\mathbf{Z}_j^N = \mathbf{X}_j^N \boldsymbol{\beta}_j + \boldsymbol{\varepsilon}_j^N, \tag{3.118}$$

where

$$\mathbf{Z}_j^N = \begin{bmatrix} Z_j(0) \\ Z_j(1) \\ \vdots \\ Z_j(N-1) \end{bmatrix} = \begin{bmatrix} Y_j(0) \\ Y_j(1) \\ \vdots \\ Y_j(N-1) \end{bmatrix}$$

$$\mathbf{X}_j^N = \begin{bmatrix} \mathbf{X}_j(0) \\ \mathbf{X}_j(1) \\ \vdots \\ \mathbf{X}_j(N-1) \end{bmatrix}$$

$$\boldsymbol{\varepsilon}_j^N = \begin{bmatrix} \varepsilon_j(0) \\ \varepsilon_j(1) \\ \vdots \\ \varepsilon_j(N-1) \end{bmatrix}$$

The superscript N is used to call attention to the number of observations, or row dimension, of each vector or matrix. This notation will simplify the subsequent derivation.

The variance of $\varepsilon_j(t), V_j(t)$, is a function of season and hence the variance of $\varepsilon_j(\cdot)$ has period 12 when modeling monthly discharges. For now, assume that

$$\text{var}\big[\varepsilon_j(t)\big] = \Psi_j\big(\text{mod}_{12}(t)\big)^2 \tag{3.119}$$

Using Eq. (3.119) and noting that $\varepsilon_j(\cdot)$ is not autocorrelated (i.e., $E[\varepsilon_j(i)\varepsilon_j(\ell)] = 0 \ \forall i \neq \ell$) the noise covariance matrix is

$$E\Big[\varepsilon_j^N\big(\varepsilon_j^N\big)^T\Big] = \begin{bmatrix} \Psi_j(0)^2 & & & & & & 0 \\ & \Psi_j(1)^2 & & & & & \\ & & \ddots & & & & \\ & & & \Psi_j(11)^2 & & & \\ & & & & \Psi_j(0)^2 & & \\ & & & & & \ddots & \\ 0 & & & & & & \Psi_j\big(\text{mod}_{12}(N)\big)^2 \end{bmatrix}$$

$$= \Omega_j^N. \tag{3.120}$$

The use of the Generalized Least Squares estimator of β_j in Eq. (3.118) will yield consistent estimates since all elements of X_j^N are contemporaneously uncorrelated with ε_j^N (each sequence $\{V_j(i), \ i = 0, 1, \ldots, N-1\} \ \forall j$ is a white-noise process) and other necessary assumptions are satisfied. Thus,

$$\hat{\beta}_j^N = \Big(\big(X_j^N\big)^T\big(\Omega_j^N\big)^{-1}X_j^N\Big)^{-1}\big(X_j^N\big)^T\big(\Omega_j^N\big)^{-1}Z_j^N. \tag{3.121}$$

Of course this does not solve the estimation problem since the $\varepsilon(\cdot)$s (and hence X_j^N) and $\Psi_j(\cdot)$s (and hence Ω_j^N) are unknown.

Note, however, that given initial estimates of parameter vectors $\{\hat{\beta}_j^1, \ j = 1, 2, \ldots, n_0\}$ the residuals $\{\hat{\varepsilon}_j(1), \ j = 1, 2, \ldots, n_0\}$ may be estimated as

$$\hat{\varepsilon}_j(1) = Z_j(1) - X_j(1)\hat{\beta}_j^1 \qquad j = 1, 2, \ldots, n_0, \tag{3.122}$$

where the elements $\{\varepsilon_j(0), \ j = 1, 2, \ldots, n_0\}$ in $X_j(1)$ are replaced with their expected value (zero). The estimated residuals $\{\hat{\varepsilon}_j(1), \ j = 1, 2, \ldots, n_0\}$ may then be used to form estimated matrices $\{X_j, \ j = 1, 2, \ldots, n_0\}$. Assuming that estimates exist of the functions $\{\Psi_j(\cdot), \ j = 1, 2, \ldots, n_0\}$ (this point will be discussed later) the matrices $\{\Omega_j, \ j = 1, 2, \ldots, n_0\}$ may be formed and $\hat{\beta}_j$ for all j may be estimated using Eq. (3.121). This process can be repeated until the parameters $\{\hat{\beta}_j, \ j = 1, 2, \ldots, n_0\}$ are found after N iterations. Keep in mind that although not explicitly indicated, matrices $\mathbf{X}_j$, $\mathbf{\Omega}_j^i$, and $\mathbf{S}_j^i$ are estimates in Eqs. (3.121), (3.122), and, following expressions.

As formulated, the above estimation algorithm is computationally inefficient. The addition of a new observation requires the problem to be completely reworked and no use is made of the previous estimate of β_j. This seems to waste effort. A sequential form for the estimate can be determined so that new observations can be incorporated without completely reworking the problem. The details are found in Curry and Bras (1980); the final forms of the equations follows.

The $(i+1)$th estimate of the coefficient vector β is given by

$$\hat{\beta}_j^{i+1} = \hat{\beta}_j^i + \mathbf{K}_j^{i+1}\left(Z_j(i+1) - \mathbf{X}_j(i+1)\hat{\beta}_j^i\right). \qquad (3.123)$$

The vector $\mathbf{K}_j^{i+1}$ takes the form

$$\mathbf{K}_j^{i+1} = \frac{\mathbf{S}_j^{i+1}\left(\mathbf{X}_j(i+1)\right)^T}{\Psi_j\left(\mathrm{mod}_{12}(i+1)\right)^2}. \qquad (3.124)$$

The variance of estimation is updated by

$$\mathbf{S}_j^{i+1} = \mathbf{S}_j^i - \frac{\mathbf{S}_j^i\left(\mathbf{X}_j(i+1)\right)^T\mathbf{X}_j(i+1)\mathbf{S}_j^i}{\left[\Psi_j\left(\mathrm{mod}_{12}(i+1)\right)^2 + \mathbf{X}_j(i+1)\mathbf{S}_j^i\left(\mathbf{X}_j(i+1)\right)^T\right]}. \qquad (3.125)$$

Two methods may be used to assign initial values to $\hat{\beta}_j$ and $\mathbf{S}_j$. In the first (Panuska, 1969)

$$\hat{\beta}_j^1 = \text{element of } B_j \qquad \forall j,$$

where B_j is a set known to contain β_j, and

$$\mathbf{S}_j^1 = \mathbf{I} \qquad \forall j.$$

The second method is

$$\mathbf{S}_j^1 = \left[\left(\mathbf{X}_j^1\right)^T\left(\mathbf{\Omega}_j^1\right)^{-1}\mathbf{X}_j^1\right]^{-1} \qquad \forall j$$

$$\hat{\beta}_j^1 = \mathbf{S}_j^1\left(\mathbf{X}_j^1\right)^T\left(\mathbf{\Omega}^1\right)^{-1}\mathbf{Z}_j^1 \qquad \forall j.$$

The matrices $\mathbf{X}^1$ are constructed using the available observations $Y_j(\cdot)$, $j = 1, \ldots, n_0$ and the indicated deterministic functions. The residuals $\{\varepsilon_j(\cdot)$, $j = 1, 2, \ldots, n_0\}$ are replaced by randomly generated Gaussian deviates with zero mean and variance equal to $\Psi_j(\ell)^2$ (where $\ell = \mathrm{mod}_{12}(t)$), if available and otherwise the sample variance of the observations, $Y_j(\cdot)$, is used. If the functions $\{\Psi_j(\ell), \forall j, \ell\}$ are not available, the Ω_j^1 is assumed to be $\mathbf{I}$ for all j.

The above estimation procedure deals with parameters of a single-station equation, independently of the others. Not accounting for the cross-correlated error between equations leads to inefficiency in the estimation. As in the algorithm defined on page 126, asymptotically efficient estimates of β may be obtained by extension to the simultaneous estimation of $\{\beta_j, \ j=1,2,\ldots,n_0\}$ rather than the estimation of each β_j separately. However, the computational burden (memory and execution time) would be greatly increased and the algorithm will not be recursive.

3.4.3 Tests of Model Adequacy

Chapter 2 emphasized that model building consists of repeatedly exercising identification, estimation, and verification steps. Verification of the generalized multivariate model follows the pattern of their univariate cousins.

Testing for parameter significance in the model with nonautocorrelated noise follows techniques common to OLS or linear regression. The availability of many books on that subject makes a detailed discussion here superfluous. (The reader is referred to Johnston [1972] and Draper and Smith [1966].) In estimating parameters for the nonautocorrelated-residuals model, it is recommended that a stepwise regression algorithm be used. This algorithm uses a F-test in selecting explanatory variables that yield significant coefficients according to the user's criteria.

The partial F-test and stepwise regression is not applicable to the model with autocorrelated noise, particularly since the explanatory variables include past residuals. The parameter error covariance matrix, $\mathbf{S}_j^N$ (Eq. 3.125) may be used to define reasonable ranges for the parameters. As a rule of thumb, parameters within the range $[-S_j^{1/2}(i,i), S_j^{1/2}(i,i)]$ may be considered insignificant. Another aid in verification is to study the evolution of parameter estimates. Parameters that exhibit wildly fluctuating behavior as new data points are processed imply that the model structure or the corresponding explanatory variable is not useful.

The properties of residuals are the best clue to model adequacy. By construction, the residual terms $\mathbf{V}(k,i)$, both of models with correlated and models with uncorrelated residuals, should have zero mean and zero autocorrelations and cross-correlations at lags larger than one. Ideally the distribution of estimated residuals (obtained from the difference of observed and predicted values) should also be Gaussian. A few possible tests for residuals follow.

Durbin–Watson Test

Durbin and Watson (1950, 1951) derived a widely used test for autocorrelation of residuals. Although the test was derived for nonstochastic regression variables, it is applicable to a model with stochastic regression variables so long as they are serially independent and do not contain lagged values of the regressand or lagged noise terms (i.e., no autoregressive or moving-average terms). This being the case, the application of the test is only theoretically appropriate for the residuals of the model with nonautocorrelated residuals.*

The Durbin–Watson test is applied to the OLS residuals of each equation j of the monthly multivariate model with nonautocorrelated residuals by calculating the d statistic as

$$d_{ji} = \sum_{k=1}^{N/12} \left(\hat{V}_j(k,i) - \hat{V}_j(k-1,i) \right)^2 \bigg/ \sum_{k=0}^{N/12} \hat{V}_j(k,i)^2. \qquad (3.126)$$

Upper (d_U) and lower (d_L) limits, dependent on the number of explanatory variables, observations $(N/12 + 1)$, and chosen significance level, are used to test the hypothesis of zero autocorrelation against the alternative hypothesis of positive first-order autocorrelation.

If $d_{ji} < d_L$, reject the hypothesis of nonautocorrelated residuals in favor of the hypothesis of positive autocorrelation.

If $d_{ji} > d_U$, do not reject the null hypothesis.

If $d_L < d_{ji} < d_U$, the test is inconclusive.

Values for d_L and d_U, tabulated against their arguments, may be found in Durbin and Watson (1951) and Johnston (1972).

To test the alternative hypothesis of negative first-order autocorrelation, compute $(4 - d_{ji})$ and compare this value to d_L and d_U as if testing for positive autocorrelation.

The above test is not appropriate for the residuals obtained from the GLS estimation of each multivariate monthly model with nonautocorrelated residuals. However, if these GLS residuals are grouped according to j and the Durbin–Watson statistic is calculated as indicated in Eq. (3.126), so that each monthly multivariate equation yields n_0 statistics, the test may be assumed to be approximately correct.

Durbin (1970) suggested a large-sample $(N > 30)$ procedure to test for autocorrelated residuals obtained from a model whose explanatory variables contain some autoregressive terms but no lagged noise (i.e., no moving-average terms).

Thus, the application of the test to the residuals of the model with autocorrelated noise may be assumed approximately correct when $n_1 \geq 1$ and

---

*Assuming $n_1(i)$ is less than 12 for all i, the regression variables of nonautocorrelated noise models are not lagged values of the regressand, since they are different months. If $n_1(i)$ is greater than or equal to 12, the test may not be applicable theoretically. Serial independence of the regression variables is assumed approximately to hold, since they are separated by a lag of 12 months.

$n_2 = 0$, and when estimated by either least squares or weighted least squares. (Also, the test may be applicable to some residuals of the nonautocorrelated noise model if $n_1(i) \geq 12$ for any i.)

The test is applied to each equation j of the multivariate model (Eq. 3.104) by computing the statistics h_j, where

$$h_j = \frac{\displaystyle\sum_{k=0}^{N/12} \sum_{i=0}^{11} \hat{V}_j(k, i+1)\hat{V}_j(k, i)}{\displaystyle\sum_{k=0}^{N/12} \sum_{i=0}^{11} \hat{V}_j(k, i+1)^2} \sqrt{\frac{N}{1 - N \operatorname{var}\left(\hat{\beta}_j(1)\right)}} \qquad (3.127)$$

and $\operatorname{var}(\hat{\beta}_j(1))$ equals the variance for the estimated coefficient of the first autoregressive term $(Y_j(k, i-1))$ (see Eq. 3.125).

Under the hypothesis of nonautocorrelated noise, the statistic h_j is distributed as a standard normal deviate. Thus, if $|h_j| > 1.96$, the hypothesis of zero autocorrelation is rejected at the 5% level of significance and if $|h_j| > 2.58$, the hypothesis is rejected at the 1% level of significance. Note that the test breaks down if $N \operatorname{var}(\hat{\beta}_j(1))$ is greater than one.

The tests of correlation of residuals discussed in Chapter 2 are also valid guidelines in the multivariate case.

Tests of Normality

The partial F-test used in the previously mentioned stepwise-regression procedure depends on the assumption of normally distributed residuals. As discussed in Chapter 2, normality can be tested for by studying skewness and kurtosis properties as well as χ^2 tests or visual inspection of residuals.

A coordinate system may be defined such that the cumulative distribution function (CDF) of any normal density function plots linearly. If the estimated CDF of a set of residuals plots approximately linearly in this coordinate system, it is a good indication that the residuals are normally distributed. Paper with this special coordinate system predefined may be obtained and is generally referred to as normal-probability paper.

The estimated CDF of a set of residuals $\{\hat{V}_j(k, i), \ k = 0, 1, 2, \ldots, N/12\}$ can be found. First, the residuals are ordered with respect to magnitude so that $\hat{V}_j^m(k, i)$ denotes the mth smallest value. Values of the CDF are then estimated by

$$\operatorname{prob}\left(V_j(\cdot) \leq \hat{V}_j^m(k, i)\right) \approx \frac{m}{(N/12)+1}.$$

Performance Evaluation

Many times, the ultimate use of the stochastic model is in the multi-lead forecast of monthly discharge. Thus, it is prudent to test the forecasting ability of a few candidate models before making a selection.

Let $\hat{Y}_j(k, i|k, i - \ell)$ denote a forecast of discharge at station j, year k, and month i, with a lead of ℓ months. The bias (B) and mean square error (MSE) of the forecasts may be tabulated over ℓ, i, and j, using the following relationships

$$B(\ell, i, j) = \frac{1}{(N/12)} \sum_{k=0}^{N/12} \left[Y_j(k, i) - \hat{Y}_j(k, i|k, i - \ell) \right] \qquad \forall j, i, \ell;$$

$$(3.128)$$

$$\text{MSE}(\ell, i, j) = \frac{1}{(N/12)} \sum_{k=0}^{N/12} \left[Y_j(k, i) - \hat{Y}_j(k, i|k, i - \ell) \right]^2 \qquad \forall j, i, \ell.$$

$$(3.129)$$

Using the historical standard deviation of the sequence $\{Y_j(k, i), \; k = 0, 1, \ldots, N/12\}$, denoted by $S^2(j, i)$, a scaled statistic $R^2(\ell, i, j)$ may be computed from $\text{MSE}(\ell, i, j)$ as

$$R^2(\ell, i, j) = \frac{S^2(j, i) - \text{MSE}(\ell, i, j)}{S^2(j, i)}, \qquad (3.130)$$

where

$$S^2(j, i) = \frac{1}{(N/12)} \sum_{k=0}^{N/12} \left(Y_j(k, i) - \overline{Y}_j(i) \right)^2$$

and

$$\overline{Y}_j(i) = \frac{1}{(N/12) + 1} \sum_{k=0}^{N/12} Y_j(k, i).$$

The statistics $B(\ell, i, j)$ and $R^2(\ell, i, j)$, $\forall i, j$, and $\ell = 1, 2, 3, \ldots$, are useful in evaluating the forecasting performance of a given model. Clearly, an unbiased forecast is desirable so that the closer the various values of B are to zero, the better the performance. Similarly, it is desirable to minimize the mean square error of forecasts and, thus, values of $R^2(\ell, i, j)$ close to 1 are desired.

3.4.4 An Example: Forecasting Flows in the Nile River

Curry and Bras (1980) evaluated several alternative forms of the models discussed in the previous section. The goal was to predict discharges in the Nile River, Egypt, and ultimately to use monthly forecasted flows to control releases from the High Aswan Dam (Bras et al., 1983). The most successful of the models corresponds to Eq. (3.103), a multivariate model with nonautocorrelated residuals.

The model was calibrated using monthly data from 1912 to 1965 inclusive, in all eight stations schematized in Fig. 3.16. Models for all eight stations were evaluated simultaneously using the algorithm discussed on pages 126–130. For the sake of expediency only results of the equation for Wadi Halfa, at the entrance to Lake Nasser, are given.

Table 3.1 gives the selected explanatory variables and estimated coefficients for all 12 months in Wadi Halfa. For comparison, both Ordinary Least Squares (OLS) and Generalized Least Squares (GLS) results are given. Of interest is the different nature of each monthly equation; each has very different explanatory variables and coefficients.

Table 3.2 gives the results of a Durbin–Watson test on residuals. The hypothesis of uncorrelated residuals could not be rejected in the great majority of the equations. In some the test was inconclusive or not applicable. In four cases, the hypothesis was rejected.

The calibrated model was used for forecasting at various leads. Forecasting is accomplished using the expected value of Eq. (3.103). For example, a forecast one month ahead (lead one) is

$$\hat{\mathbf{Y}}(k, i) = \hat{\boldsymbol{\Phi}}_i \begin{bmatrix} \mathbf{Y}(k, i-1) \\ \mathbf{Y}(k, i-2) \\ \vdots \\ \mathbf{Y}(k, i - n_1(i)) \end{bmatrix} + \hat{\mathbf{U}}(i), \qquad (3.131)$$

where the symbol "^" designates an estimate. At lead one all explanatory variables are actual observations. If multi-lead forecasts are required, Eq. (3.131) is used recursively. For example, a lead-two forecast for month i (from the end of month $i-2$) utilizes the lead-one forecasting equation for month i, replacing lag-one explanatory variables (which have not been observed at the time of the lead-two forecast) with the lead-one predictions generated from the end of month $i-2$. Using both the lead-one and lead-two predictions, as well as all observations prior to and including month $i-2$, lead-three forecasts can be generated, and so forth. Further details of the forecasting procedures for all models are found in Curry and Bras (1980).

The resulting R^2 statistics of predicting all months at various lags were computed using Eq. (3.130) and are given in Table 3.3. Considerable variation exists in the accuracy of forecasts for any given lead. For example, the dry month of April at Wadi Halfa can be forecast in March (lead 1) with a variance reduction of 96%, while the 12-month lead forecast has a variance reduction of 58%. On the other hand, the flood month of August at Wadi Halfa is forecast with a 75% variance reduction at lead 1, but only a 1% variance reduction with a six-month lead.

Tables 3.4 through 3.6 (Bras et al., 1983) give examples of actual predictions and comparisons to observed discharges. Forecasts are given for three years from the end of January, April, and August, for various leads. For example, from the end of January 1922, the model predicts reasonably well

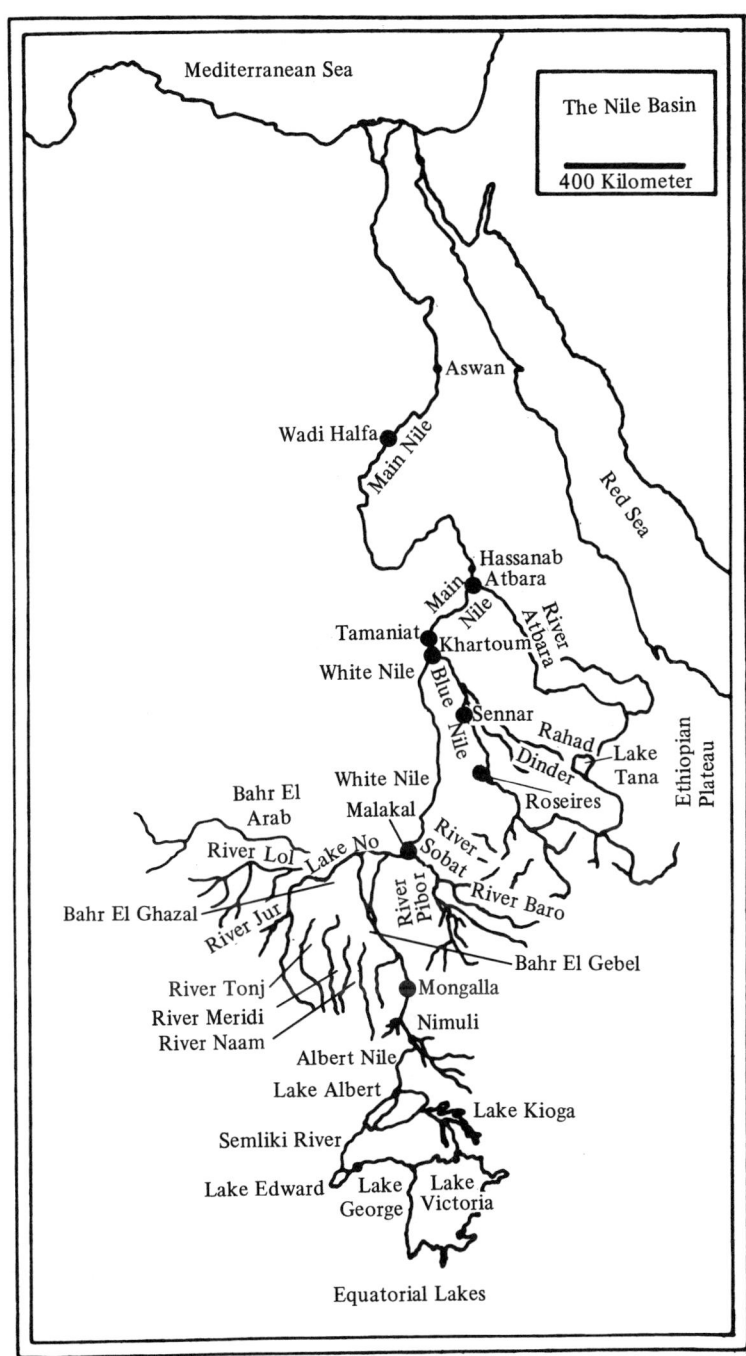

Figure 3.16 Map of Nile River Basin.

Table 3.1

Results for Ordinary Least Squares (OLS) and Generalized Least Squares (GLS) estimation of multivariate model 3.103; monthly equations for the Nile River at Wadi Halfa

Model of	Variable	Lag	OLS estimated coefficient	OLS standard error	Partial F statistic	Degrees of freedom	Significance level	GLS estimated coefficient	GLS standard error
January	Wadi Halfa	December 1	0.457	0.064	51.549	47	100.0	0.468	0.055
	Tamaniat	December 1	0.273	0.071	14.949	47	100.0	0.0268	0.062
	Malakal	December 1	0.393	0.039	103.792	47	100.0	0.390	0.034
	Sennar	November 2	-0.180	0.025	49.853	47	100.0	-0.189	0.023
	Roseires	June 7	0.083	0.024	11.894	47	99.9	0.078	0.021
	Tamaniat	March 10	-0.141	0.025	31.896	47	100.0	-0.145	0.022
	Khartoum	March 10	0.257	0.107	5.753	47	98.0	0.269	0.095
	Constant		-372.251	108.342	11.805	47	99.9	-357.115	96.522
February	Wadi Halfa	January 1	0.419	0.043	92.821	52	100.0	0.405	0.036
	Malakal	January 1	0.420	0.034	152.367	52	100.0	0.435	0.030
	Constant		-9.647	103.870	0.009	52	97.4	4.554	89.308
March	Atbara	February 1	-3.611	1.548	5.441	46	97.6	-3.562	1.185
	Tamaniat	February 1	0.922	0.092	101.589	46	100.0	0.960	0.071
	Malakal	February 1	0.507	0.113	20.045	46	100.0	0.456	0.092
	Tamaniat	January 2	-0.437	0.058	57.020	46	100.0	-0.417	0.046
	Mongalla	November 4	-0.503	0.077	43.232	46	100.0	-0.389	0.058
	Roseires	November 4	0.225	0.057	15.583	46	100.0	0.164	0.044
	Atbara	October 5	-0.218	0.097	5.010	46	97.0	-0.132	0.074
	Mongalla	October 5	0.316	0.077	16.910	46	100.0	0.237	0.061
	Constant		62.061	133.407	21.953	46	100.0	588.348	111.500
April	Tamaniat	March 1	0.860	0.047	334.866	46	100.0	0.868	0.041
	Malakal	March 1	0.160	0.046	12.289	46	99.9	0.147	0.040
	Khartoum	March 1	-1.714	0.202	72.104	46	100.0	-1.518	0.176

Month	Location	Date							
	Roseires	March 1	2.113	0.198	113.347	46	100.0	1.896	0.174
	Atbara	February 2	−7.616	1.323	33.149	46	100.00	−6.370	1.121
	Roseires	May 10	−0.422	0.124	11.628	46	99.9	0.389	0.171
	Khartoum	May 10	0.441	0.203	4.714	46	96.5	−0.349	0.104
	Khartoum	April 12	0.480	0.154	9.772	46	99.7	0.389	0.130
	Constant		−400.029	111.060	12.974	46	99.9	−390.171	96.824
May	Tamaniat	April 1	0.593	0.035	287.089	45	100.0	0.618	0.029
	Malakal	April 1	0.764	0.118	41.885	45	100.0	0.806	0.094
	Khartoum	April 1	0.385	0.172	5.014	45	97.0	0.447	0.138
	Khartoum	March 2	−1.729	0.347	24.813	45	100.0	−1.280	0.272
	Sennar	March 2	0.919	0.272	11.399	45	99.8	0.613	0.213
	Tamaniat	February 3	−0.308	0.063	24.092	45	100.0	−0.272	0.049
	Roseires	February 3	0.685	0.208	10.881	45	99.8	0.453	0.162
	Malakal	January 4	0.347	0.052	44.175	45	100.0	0.279	0.041
	Mongalla	October 7	−0.123	0.037	11.274	45	99.8	−0.107	0.029
	Constant		−302.098	97.250	9.650	45	99.7	−354.327	80.777
June	Tamaniat	May 1	0.703	0.046	236.361	50	100.0	0.768	0.038
	Roseires	May 1	0.655	0.093	49.095	50	100.0	0.547	0.080
	Atbara	February 4	11.982	2.963	16.352	50	100.0	9.218	2.201
	Atbara	January 5	−4.743	1.847	6.596	50	98.7	−4.583	1.386
	Constant		164.115	96.321	2.903	50	90.5	104.608	84.533
July	Atbara	June 1	7.344	1.137	41.715	44	100.0	5.537	0.841
	Sennar	June 1	−1.230	0.378	10.571	44	99.8	−0.909	0.280
	Roseires	June 1	2.103	0.412	26.009	44	100.0	1.854	0.309
	Atbara	May 2	−9.597	4.618	4.319	44	95.6	−6.662	3.415
	Mongalla	May 2	0.374	0.103	13.046	44	99.9	0.370	0.077
	Khartoum	May 2	1.561	0.469	11.097	44	99.8	1.243	0.347
	Atbara	April 3	−23.426	7.361	10.127	44	99.7	−16.482	5.444
	Roseires	December 7	1.952	0.355	30.303	44	100.0	1.542	0.262
	Khartoum	November 8	−0.501	0.126	15.953	44	100.0	−0.364	0.093
	Atbara	July 12	−0.497	0.124	16.005	44	100.0	−0.401	0.092
	Constant		715.067	446.619	2.563	44	88.3	995.165	341.805

Table 3.1 *(cont'd)*

Model of	Variable	Lag	OLS estimated coefficient	OLS standard error	Partial F statistic	Degrees of freedom	Significance level	GLS estimated coefficient	GLS standard error
August	Atbara	July 1	1.410	0.383	13.534	50	99.9	1.282	0.249
	Sennar	July 1	1.762	0.211	69.519	50	100.0	1.411	0.151
	Sennar	June 2	−1.305	0.545	5.739	50	98.0	−0.302	0.346
	Sennar	February 6	−4.799	1.829	6.888	50	98.9	−2.738	1.213
	Constant		10102.008	1175.671	73.832	50	100.0	10150.064	902.921
September	Atbara	August 1	0.949	0.211	20.286	50	100.0	0.771	0.111
	Sennar	August 1	0.591	0.145	16.538	50	100.0	0.571	0.085
	Khartoum	February 7	12.715	3.380	14.152	50	100.0	4.596	1.834
	Khartoum	November 10	−1.004	0.448	5.019	50	97.0	−0.310	0.247
	Constant		4584.462	1737.609	6.961	50	98.9	7657.633	1172.785
October	Sennar	September 1	1.073	0.086	156.919	51	100.0	1.040	0.071
	Atbara	April 6	−34.304	10.616	10.442	51	99.8	−18.518	5.760
	Sennar	April 6	−2.864	1.193	5.758	51	98.0	−1.536	0.730
	Constant		1891.081	1210.145	2.442	51	87.6	1760.970	968.057
November	Sennar	October 1	0.677	0.033	422.734	50	100.0	0.622	0.025
	Wadi Halfa	December11	0.612	0.142	18.451	50	100.0	0.609	0.083
	Atbara	December11	−5.709	2.186	6.820	50	98.8	−4.255	1.277
	Roseires	November 12	−0.460	0.183	6.291	50	98.5	−0.485	0.106
	Constant		1326.222	506.258	6.863	50	98.8	1687.365	310.970
December	Tamaniat	November 1	0.480	0.032	229.332	48	100.0	0.502	0.028
	Malakal	October 2	0.203	0.067	9.223	48	99.6	0.160	0.059
	Tamaniat	September 3	0.121	0.035	12.099	48	99.9	0.134	0.029
	Khartoum	September 3	−0.086	0.034	6.362	48	98.5	−0.100	0.029
	Atbara	May 7	2.526	1.042	5.878	48	98.1	2.240	0.877
	Atbara	March 9	−18.341	7.871	5.429	48	97.6	−17.660	6.623
	Constant		426.739	234.214	3.320	48	92.5	427.699	208.979

Curry and Bras, 1980 (with the permission of Kevin Curry, Sperry Flight Systems, Phoenix, Ariz.).

Table 3.2

Results of Durbin–Watson test on residuals obtained from Generalized Least Squares (GLS) estimated multivariate model 3.103

	January	February	March	April	May	June	July	August	September	October	November	December
Wadi Halfa $H: \rho = 0$	2.293 I	1.738 NR	2.006 NR*	1.876 NR*	2.065 NR*	2.188 NR	1.872 NR*	2.016 NR	2.108 NR	2.272 NR	2.615 I	2.310 I
Atbara $H: \rho = 0$	NA	1.125 R	1.928 NR	NA	NA	NA	—	1.962 NR	2.360 I	1.570 I	2.230 NR	NA
Tamaniat $H: \rho = 0$	2.029 NR*	1.732 NR	NA	2.297 I	1.742 I	2.307 I	1.568 I	2.299 NR	1.989 NR	2.375 I	2.222 NR*	NA
Malakal $H: \rho = 0$	2.106 NR	1.670 I	2.535 I	1.719 I	2.277 NR	2.036 NR*	1.786 NR	2.197 NR*	2.386 I	1.956 NR	1.421 I	1.757 I
Mongalla $H: \rho = 0$	2.479 I	1.577 I	1.736 NR	1.583 I	1.965 NR	1.745 NR	2.543 I	2.145 NR	1.878 NR	NA	1.631 I	1.931 NR
Khartoum $H: \rho = 0$	1.366 R*	1.791 NR*	1.551 I	1.783 NR	1.877 NR	1.916 NR	1.818 NR	2.264 NR	2.314 I	2.405 I	2.078 NR	NA
Sennar $H: \rho = 0$	1.282 R	1.296 R	1.665 I	1.673 I	1.998 NR	1.864 NR	1.688 NR	2.180 NR	2.026 NR	2.362 I	2.080 NR	1.923 NR
Roseires $H: \rho = 0$	2.160 NR	NA	2.033 NR	1.757 NR	1.957 —	1.824 NR	1.767 NR	2.231 NR	2.015 NR	2.412 I	2.100 I	NA

	5% Level	
K	d_L	d_U
2	1.49	1.64
3	1.45	1.68
4	1.41	1.72
5	1.38	1.77
> 5	< 1.38	> 1.77

K = number of explanatory variables; N = number of observations (55); NR = hypothesis not rejected ($H: \rho = 0$); I = inconclusive results; R = hypothesis rejected ($H: \rho = 0$); NA = test not applicable (due to lagged autoregressive term); and * = conclusion is tentative since upper and lower limits are not precisely known.

Curry and Bras, 1980 (with the permission of Kevin Curry, Sperry Flight Systems, Phoenix, Ariz.).

Table 3.3
R^2 statistics for multi-lead forecasts at Wadi Halfa

	Month											
Lead	1	2	3	4	5	6	7	8	9	10	11	12
1	0.9864	0.9569	0.9281	0.9589	0.9822	0.9255	0.8358	0.7502	0.7180	0.7724	0.8918	0.9282
2	0.9507	0.9089	0.7298	0.8299	0.9238	0.6928	0.4002	0.1037	0.4571	0.4262	0.4285	0.8707
3	0.8992	0.8891	0.6603	0.8166	0.9013	0.5959	0.2291	0.0279	0.0533	0.3944	0.2111	0.6297
4	0.8174	0.8637	0.6679	0.8289	0.8649	0.6018	0.2045	0.0132	0.0451	0.1818	0.1881	0.4598
5	0.6503	0.7822	0.6191	0.8162	0.8284	0.5423	0.2132	0.0140	0.0466	0.1345	0.0826	0.3796
6	0.5565	0.6277	0.6113	0.7985	0.7989	0.5154	0.2202	0.0146	0.0466	0.1343	0.0426	0.2725
7	0.4489	0.5273	0.4940	0.7609	0.7962	0.4942	0.1986	0.0214	0.0466	0.0541	0.0465	0.2191
8	0.3617	0.4194	0.4218	0.6992	0.7261	0.4791	0.1872	0.0107	0.0326	0.0475	0.0353	0.2049
9	0.3355	0.3379	0.3595	0.7034	0.6512	0.4440	0.1678	0.1827	0.0080	0.0471	0.0298	0.2139
10	0.3118	0.3137	0.2344	0.6788	0.6207	0.4066	0.1822	0.1759	0.0916	0.0349	0.0301	0.1997
11	0.3035	0.2628	0.1861	0.6281	0.5835	0.3774	0.1734	0.1677	0.0897	0.1027	0.0130	0.1913
12	0.2779	0.2684	0.1481	0.5849	0.5394	0.3631	0.1792	0.1651	0.0875	0.0989	0.0915	0.1700

Curry and Bras, 1980 (with the permission of Kevin Curry, Sperry Flight Systems, Phoenix, Ariz.).

Table 3.4
Forecasts and actual observations from the end of January for the years 1917, 1922, and 1965 in 10^9 m^3

| | Forecasts, observations | | | | | | Forecast | Sample statistics | |
| | 1917 | | 1922 | | 1965 | | Standard | | Standard |
Lead	Forecast	Observation	Forecast	Observation	Forecast	Observation	deviation	Mean	deviation
1	4.03	4.13	2.04	1.96	4.97	5.08	0.14	2.45	0.69
2	3.52	3.79	1.76	1.38	4.55	4.39	0.35	2.28	0.67
3	2.56	2.41	1.59	1.05	3.31	3.58	0.30	2.04	0.69
4	2.49	1.76	1.17	0.88	4.07	4.17	0.28	1.92	0.76
5	2.58	2.08	1.54	1.00	4.17	4.52	0.53	2.07	0.78
6	4.14	5.83	4.58	3.82	6.61	5.82	1.35	5.17	1.53

Bras et al., 1983.

Table 3.5
Forecasts and actual observations from the end of April for the years 1922, 1951, and 1965 in 10^9 m^3

| | Forecasts, observations | | | | | | Forecast | Sample statistics | |
| | 1922 | | 1951 | | 1965 | | Standard | | Standard |
Lead	Forecast	Observation	Forecast	Observation	Forecast	Observation	deviation	Mean	deviation
1	0.72	0.88	2.17	1.99	4.18	4.17	0.10	1.92	0.76
2	1.16	1.00	1.80	1.30	4.54	4.52	0.43	2.07	0.78
3	4.64	3.82	4.59	3.40	6.54	5.82	1.34	5.17	1.53

Bras et al., 1983.

Table 3.6

Forecasts and actual observations from the end of August for the years 1913, 1941, and 1964 in 10^9 m^3

| | Forecasts, observations | | | | | | Forecast | Sample statistics | |
| | 1913 | | 1941 | | 1964 | | | | |
Lead	Forecast	Observation	Forecast	Observation	Forecast	Observation	Standard deviation	Mean	Standard deviation
1	14.92	13.40	17.57	15.10	24.52	25.70	2.01	21.99	3.78
2	10.72	7.86	13.34	11.10	16.10	16.40	2.39	14.61	3.15
3	6.11	4.14	6.26	7.43	7.47	9.45	1.57	7.17	1.77
4	3.83	3.02	3.92	3.99	6.00	5.85	0.64	4.54	0.87
5	2.08	2.10	2.10	2.99	4.05	5.75	0.44	3.51	0.74
6	1.57	1.42	1.59	2.08	3.89	5.08	0.42	2.45	0.69
7	2.02	1.29	2.04	2.54	3.90	4.39	0.48	2.28	0.67
8	1.42	1.10	1.92	2.52	3.32	3.58	0.38	2.04	0.69
9	1.33	1.18	1.56	1.81	3.50	4.17	0.45	1.92	0.76
10	1.75	1.14	1.78	1.92	3.53	4.52	0.60	2.07	0.78

Bras et al., 1983.

(up through July) the below-average flows. Similarly, the 1965 high flows are well predicted from the end of January. Note that as the lead increases, the forecast deteriorates and approaches the historical mean. On June 1965 the observed flow was 4.52×10^9 m^3. At lead two, from the end of April, this month was forecast as 4.54×10^9 m^3. At lead 5, from the end of January, the forecast was 4.17×10^9 m^3, still very good. From the end of August 1964 (lead 10) the forecast is 3.53×10^9 m^3. This quantity has a much higher error; nevertheless, the model still recognizes flows well above the 2.07×10^9 m^3 historical mean.

3.5 THE DISAGGREGATION MODEL

Section 2.5.2 discussed seasonal univariate models. It should be evident by now that model building to preserve multi-lag correlations among the various seasons can be fairly complicated and tedious. Failure to represent these correlations adequately can lead to serious errors in analysis. For example, reservoirs may be underdesigned if seasonal correlations at lags higher than one are significant and they are not taken into account. Blind use of the seasonal lag-one (Thomas–Fiering) model does indeed commonly result in this type of error. Valencia and Schaake (1972, 1973) suggested preserving all interseasonal correlations by appropriately disaggregating annual (aggregated) variables. Throughout this section, then, assume that a set of aggregated random variables is available (i.e., annual streamflow at n locations) and it is desired to obtain a corresponding series of seasonal variables (i.e., monthly streamflows at all n locations). This disaggregation must be performed so that all correlations among seasons (months) and between variables (sites) will be preserved. Logically, seasonal values must add up to the original aggregated variables.

Assume you have a vector of annual values of discharge at different locations:

$$\mathbf{X} = \begin{bmatrix} x_1 \\ \vdots \\ x_n \end{bmatrix}, \tag{3.132}$$

where n is the number of sites. Define a vector of seasonal (monthly, quarterly, etc.) values of discharge at the n sites

$$\mathbf{Y} = \begin{bmatrix} y_{11} \\ y_{21} \\ \vdots \\ y_{m1} \\ \vdots \\ y_{1n} \\ \vdots \\ y_{mn} \end{bmatrix}, \tag{3.133}$$

where y_{ij} is the discharge at location j during season i. There are m seasons.

Note that

$$x_j = \sum_{i=1}^{m} y_{ij} \tag{3.134}$$

or

$$X = CY,$$

where

$$C = \begin{bmatrix} 1 \cdots 1 & 0 \cdots 0 & \cdots & 0 \cdots 0 \\ 0 \cdots 0 & 1 \cdots 1 & \cdots & 0 \cdots 0 \\ \vdots & & & \\ 0 \cdots & & \cdots & 1 \cdots 1 \\ & & & \overset{\longleftrightarrow}{m} \end{bmatrix} \tag{3.135}$$

The disaggregation model takes the form

$$Y = AX + BW, \tag{3.136}$$

where Y and X are zero mean vectors (mean subtracted), A is an $nm \times n$ matrix, B is an $nm \times nm$ matrix, W is an $nm \times 1$ vector of standard normal deviates, and $E[WW^T] = I$, $E[BWY^T] = 0$.

Parameters A and B will be selected to preserve the correct cumulative relation (Eq. 3.134), the spatial correlation between all stations and months, and the correlation between the aggregated (X) and disaggregated (Y) vectors.

The above equation is clearly analogous to the expression for the multivariate AR(1) model (Eq. 3.1). The parameter matrices A and B are then obtained according to equations equivalent to Eqs. (3.7) and (3.8)

$$A = S_{yx} S_{xx}^{-1} \tag{3.137}$$

$$BB^T = S_{yy} - S_{yx} S_{xx}^{-1} S_{xy}, \tag{3.138}$$

where $S_{xx} = E[XX^T]$, $S_{yy} = E[YY^T]$, and $S_{yx} = E[YX^T]$. The resulting BB^T matrix is always positive semidefinite, with rank $nm - n$. The rank is not full because there must be n dependent columns in the matrix. This results from the fact the sum of the seasons must add up to the annual values, leaving only $m - 1$ degrees of freedom per station. The decomposition of BB^T is then usually done using the principal-components method discussed on pages 95–96.

Sample estimates of S_{yx}, S_{xx}, and S_{yy} are obtained with equations equivalent to Eqs. (3.12) and (3.13).

As long as **A** and **B** are obtained as previously described, the disaggregation model will have the following properties:

1. Preserve mean of vector **Y**
2. Preserve variance of elements of **Y**
3. Preserve the cross-correlation matrix between **Y** and **X**, $\mathbf{S}_{yx}$
4. Preserve the correlation among the elements of **Y**, $\mathbf{S}_{yy}$
5. Preserve the cumulative relation, $\mathbf{X} = \mathbf{CY}$, therefore, the disaggregated values add up to the annual components.

The last property is easily shown. Equation (3.136) will satisfy any linear relation

$$\mathbf{X} = \mathbf{CY}. \tag{3.139}$$

Using Eq. (3.139)

$$\mathbf{S}_{yx} = \mathbf{S}_{yy}\mathbf{C}^T$$
$$\mathbf{S}_{xx} = E[\mathbf{XX}^T] = \mathbf{CS}_{yy}\mathbf{C}^T \tag{3.140}$$
$$\mathbf{S}_{xy} = \mathbf{CS}_{yy}.$$

Substituting into the expressions for **A** and $\mathbf{BB}^T$ (Eqs. 3.137 and 3.138),

$$\mathbf{A} = \mathbf{S}_{yy}\mathbf{C}^T\left(\mathbf{CS}_{yy}\mathbf{C}^T\right)^{-1} \tag{3.141}$$

$$\mathbf{BB}^T = \mathbf{S}_{yy} - \mathbf{S}_{yy}\mathbf{C}^T\left(\mathbf{CS}_{yy}\mathbf{C}^T\right)^{-1}\mathbf{CS}_{yy}. \tag{3.142}$$

Pre-multiplying **A** by **C** leads to

$$\mathbf{CA} = \mathbf{CS}_{yy}\mathbf{C}^T\left(\mathbf{CS}_{yy}\mathbf{C}^T\right)^{-1} = \mathbf{I}. \tag{3.143}$$

Pre-multiplying $\mathbf{BB}^T$ by **C** yields

$$\mathbf{CBB}^T = \mathbf{CS}_{yy} - \mathbf{CS}_{yy}\mathbf{C}^T\left(\mathbf{CS}_{yy}\mathbf{C}^T\right)^{-1}\mathbf{CS}_{yy} = \mathbf{0}. \tag{3.144}$$

Eq. (3.144) implies

$$\mathbf{CB} = \mathbf{0}. \tag{3.145}$$

It is then easy, using Eqs. (3.143) and (3.145), to premultiply Eq. (3.136) by **C**, leading to

$$\mathbf{CY} = \mathbf{CAX} + \mathbf{CBW}$$
$$= \mathbf{X},$$

which is the desired linear relation.

The conditions

$$CA = I$$
$$CB = 0$$

are good checks for the correct estimation of **A** and **B**.

3.5.1   Comments and an Example on Disaggregation

The disaggregation model is possibly one of the most widely accepted tools in stochastic hydrology. Conceptually it is simple, but nevertheless it accomplishes a considerable amount of otherwise very difficult tasks. In practice, the model can lead to various numerical problems. Essentially, the decomposition of the matrix BB^T can rapidly deteriorate numerically, particularly if its dimensions ($mn \times mn$) are large. Similarly, the necessary inversion of matrix S_{xx} can lead to numerical errors. Computer storage can also be a problem. Note that in a not uncommon problem of 6 stations and 12 seasons, the **Y** vector consists of 72 elements and the largest matrix will be of dimensions 72×72, a considerable storage demand. For accuracy, double precision is always recommended. In order to save space, it is possible to disaggregate in various stages, as shown in Fig. 3.17. Note, for example, that quarterly values are disaggregated into months independently. In such a case the correlation of the last month of every quarter and the first month of the next quarter will not be explicitly preserved. This behavior is shown in Fig. 3.18. Shown are lag-one correlations in four streamflow stations in the Nile River, Egypt. Compared to the historical values are those obtained by synthetic generation with a seasonal multivariate AR(1) model (Section 3.3.1) and by disaggregation of annual values generated by a multivariate broken-line model, to be described in Chapter 5. The nature of the approximation schematized in Fig. 3.17 is obvious. The disaggregated values fail to preserve the correlation between adjacent months in different seasons. Nevertheless, since correlations between and within quarters will be correctly handled, the errors incurred should be minimal. Alternatively, the simple step-disaggregation proposed by Santos and Salas (1983) may be useful here.

The reader should be able to recognize that the general disaggregation scheme also ignores seasonal correlations between different years, in particular the correlation between the last and first seasons of adjacent years. If necessary, this can be completely solved by adding additional elements to vector **Y** (state augmentation) or, more simply, by modifying Eq. (3.136) to include an additional linear dependency between vector **Y** and seasons corresponding to the previous year (previous **X** value). Mejía and Rouselle (1976) discuss such an approach in detail, although their particular approach is flawed (Valencia et al., 1983).

The disaggregation model operates on aggregated values, regardless of their origin. Typical applications would be the generation of seasonal stream-

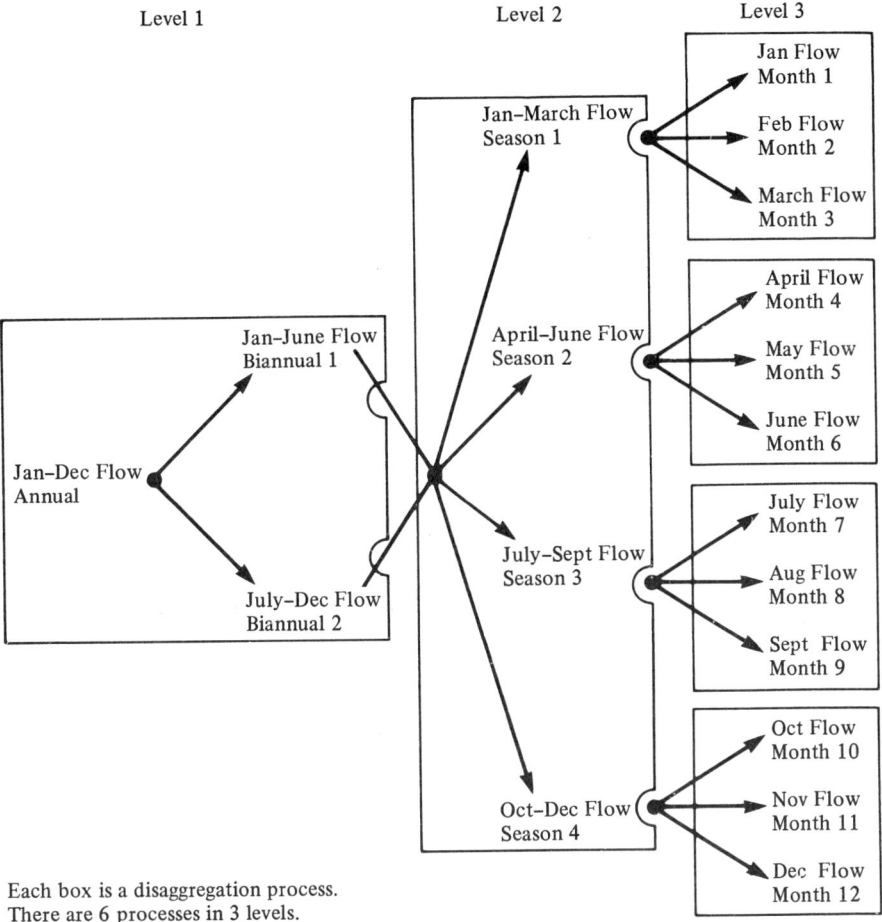

Figure 3.17   Disaggregation scheme in various stages (from Curry and Bras, 1978 [with the permission of Kevin Curry, Sperry Flight Systems, Phoenix, Ariz.]).

flows from an available annual record; or the disaggregation of annual streamflows generated using another time-series model. (Chapter 2 discussed univariate models capable of annual streamflow generation.)

The multivariate seasonal AR(1) (mixed distributions) proved to be a very good model of the Nile River. The correlation structure at lags higher than the explicitly preserved lag-one was reasonably well reproduced, at least down to the quarterly level. Even then, the performance of the AR(1) in preserving statistics at aggregated levels, such as the annual level, is not the best. For example, Table 3.7 gives historical lag-one correlations of annual flows in four stations in the Nile. Given for comparison are the annual lag-one correlations resulting from the broken-line model, which are unaltered by disaggregation,

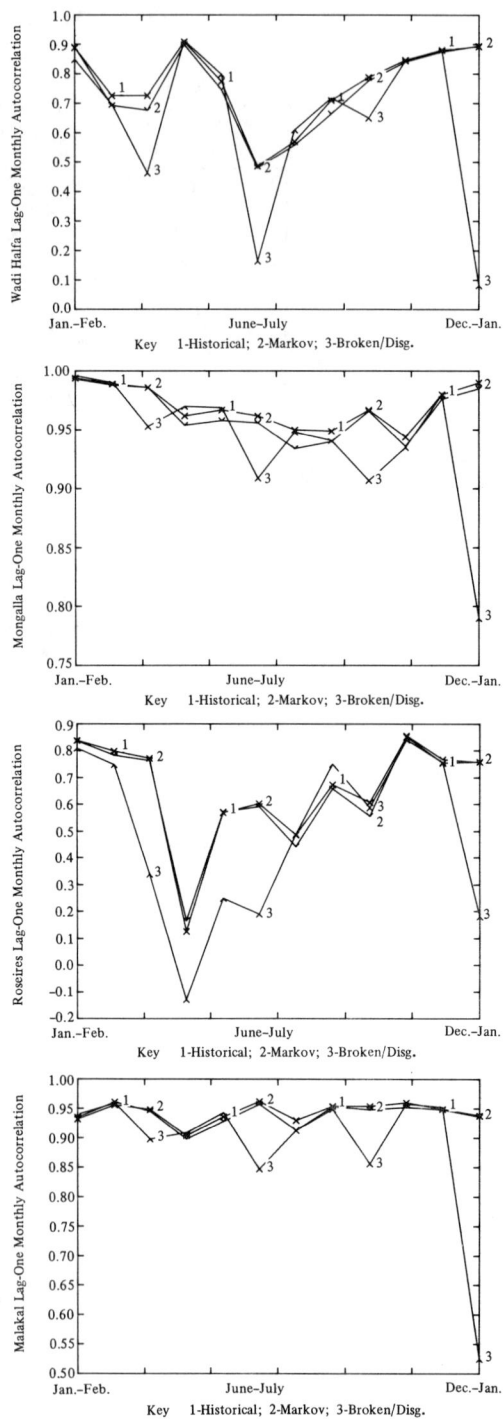

Figure 3.18 Lag-one monthly autocorrelations for the Nile River at Wadi Halfa, Mongalla, Roseires, and Malakal: (1) historical values, (2) values from a seasonal multivariate AR(1) model, (3) values from the disaggregation of annual flows generated by a multivariate broken-line model (from Curry and Bras, 1978 [with the permission of Kevin Curry, Sperry Flight Systems, Phoenix, Ariz.]).

TABLE 3.7
Comparison of annual Nile River lag-one correlations as obtained by the multivariate seasonal AR(1) model and the broken-line model

	Lag-one autocorrelations of annual streamflow		
	Historical	*AR(1)*	*Broken line*
Wadi Halfa	0.184	0.089	0.207
Mongalla	0.886	0.825	0.881
Roseires	0.162	0.067	0.146
Malakal	0.801	0.752	0.786

Curry and Bras, 1978 (with the permission of Kevin Curry, Sperry Flight Systems, Phoenix, Ariz.).

Table 3.8
Storage required* to meet monthly target releases where annual target is 70% of mean annual flow[†]

40-year reliability (%)	Thomas–Fiering monthly flow model	AR(1) annual flow generator	ARMA(1,1) annual flow generator
80	0.31	0.37	0.38
95	0.34	0.50	0.61
98	0.36	0.62	0.83

*Storage reported as a fraction of mean annual flow.
[†]Values for Ishikari River, Japan, based on study reported in Hoshi et al. (1978).

Louks, D. P. et al., 1981, *Water Resource Systems Planning and Analysis*, Englewood Cliffs, N.J.: Prentice-Hall).

and those resulting from the aggregated monthly streamflows generated by the seasonal AR(1). Even in this favorable example the AR(1) is clearly inferior in this respect.

The possible value of the disaggregation model is well illustrated by Hoshi et al. (1978), as reviewed by Loucks et al. (1981). Table 3.8 summarizes a storage-yield analysis performed with three different models: a monthly lag-one autoregressive model (Section 2.5.2, Eq. 2.103), an AR(1), and an ARMA(1,1). The last two nonseasonal models were then coupled with a disaggregation model to obtain monthly streamflows. For high-release targets, such as the 70% of mean annual flow shown, the disaggregation-based procedures always yielded larger reservoirs. The required reservoir size increased significantly with reliability for both the AR(1) and ARMA(1,1) models. This behavior is what motivated Valencia and Schaake to formulate the disaggregation model. Monthly correlations at large lags can be significant enough to affect reservoir design and operation.

Finally, you should realize that the disaggregation model is not parsimonious. It requires a lot of parameters and data. Improvements requiring less parameters have been presented by Santos and Salas (1983), Stedinger (1983), and Lane (1983).

CHAPTER 4

Frequency-Domain Analysis of Hydrologic Processes

4.1 INTRODUCTION

In the past three chapters we have discussed random processes in the time domain. All properties, such as correlation, were studied in terms of the original argument of the process, or space of definition, generally taken as time in hydrology. Most people prefer to think in terms of the daily concept of time, but the fact is that transformation of the process domain to frequency can yield considerable insight. Historically the analysis of signals, including random signals, was considerably advanced in the frequency domain. The works of Wiener (1949) and his student Lee (1960), dealing with transformed processes defined in the frequency domain, form the basis of the material in this chapter. Emphasis on time-domain analysis is relatively recent, probably because of the availability and popularity of numerical, computer-based solutions to differential equations, eliminating the need for the analytic integrals in frequency-domain work. This newly discovered freedom has also led to advances and solutions that were stifled by the old approach.

This chapter will familiarize the reader with the traditional concepts of frequency-domain analysis. We intend not only to provide the right historical perspective, but to illustrate how new insights into random-process behavior can be obtained. We will also show how some common hydrologic problems are in fact easier after the frequency transformation. The concepts presented in this chapter are necessary for a better understanding of Chapters 5 and 6.

4.2 DETERMINISTIC PERIODIC SIGNALS

Analysis of random series in the frequency domain is simpler with a good understanding of decomposition of deterministic signals in harmonic constituents. The simplest deterministic signal and the one most hydrologists seem to be familiar with is the periodic one, defined as

$$X(t) = X(t + T), \tag{4.1}$$

where T is the period. The frequency f (in cycles per unit time) is defined as

$$f = \frac{1}{T} \tag{4.2}$$

and the angular frequency as

$$\omega_0 = 2\pi f \tag{4.3}$$

in radians per unit time.

Any periodic continuous function can be represented by an infinite sum of trigonometric terms, usually called a Fourier series representation:

$$
\begin{aligned}
X(t) &= \frac{1}{2}a_0 + a_1\cos \omega_0 t + a_2\cos 2\omega_0 t + \cdots + b_1\sin \omega_0 t + b_2\sin 2\omega_0 t + \cdots \\
&= \frac{1}{2}a_0 + \sum_{n=1}^{\infty}\left(a_n\cos n\omega_0 t + b_n\sin n\omega_0 t \right).
\end{aligned}
\tag{4.4}
$$

The above series expansion of $X(t)$ exists as long as the following integral is finite,

$$\int_{-T/2}^{T/2} |X(t)|\,dt < \infty, \tag{4.5}$$

which implies that the integration of $X(t)$ is bounded within one period; this type of signal is sometimes referred to as energy bounded.

Note that Eq. (4.4) is a series of sinusoidal components with increasing frequencies, integral multiples of the fundamental frequency ω_0. Figure 4.1 illustrates the increasing accuracy of approximating a periodic signal with sinusoids of increasing frequency (up to five harmonics).

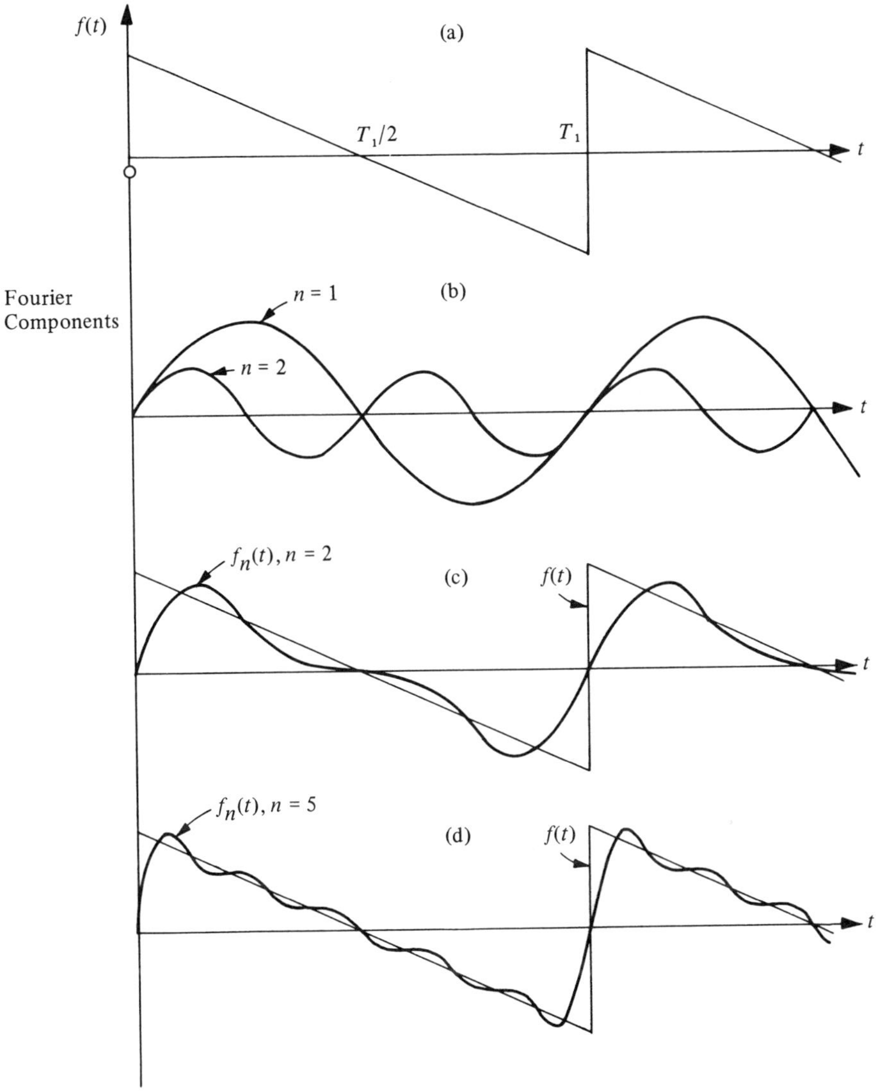

$f(t)$ (a)

$T_1/2$ T_1 t

Fourier Components

$n = 1$ (b)

$n = 2$ t

$f_n(t), n = 2$ (c) $f(t)$

t

$f_n(t), n = 5$

(d) $f(t)$

t

Figure 4.1 Fourier series approximation of periodic function (from Eagleson, 1969). Temporal and Spatial Definition in Hydrologic Modelling. Cambridge, Mass.: Massachusetts Institute of Technology (unpublished class notes).

Using algebra and trigonometric identities, a term of Eq. (4.4) can be expressed as

$$a_n\cos n\omega_0 t + b_n\sin n\omega_0 t = \sqrt{a_n^2 + b_n^2}\left[\frac{a_n}{\sqrt{a_n^2 + b_n^2}}\cos n\omega_0 t + \frac{b_n}{\sqrt{a_n^2 + b_n^2}}\sin n\omega_0 t\right]$$

$$= C_n\cos(n\omega_0 t + \theta_n), \tag{4.6}$$

where

$$C_n = \sqrt{a_n^2 + b_n^2}$$

$$\theta_n = \tan^{-1}\left(-\frac{b_n}{a_n}\right).$$

If $C_0 = \frac{1}{2}a_0$, Eq. (4.4) becomes

$$X(t) = C_0 + \sum_{n=1}^{\infty} C_n \cos(n\omega_0 t + \theta_n). \tag{4.7}$$

The coefficients a_n and b_n are obtained using the orthogonality property of the sine and cosine functions. Multiplying Eq. (4.4) by $\cos m\omega_0 t$, where m is an integer, and integrating over the period yields,

$$\int_{-T/2}^{T/2} X(t)\cos m\omega_0 t\, dt = \frac{1}{2}a_0 \int_{-T/2}^{T/2} \cos m\omega_0 t\, dt$$

$$+ \int_{-T/2}^{T/2} \left[\sum_{n=1}^{\infty} a_n \cos n\omega_0 t\right] \cos m\omega_0 t\, dt$$

$$+ \int_{-T/2}^{T/2} \left[\sum_{n=1}^{\infty} b_n \sin n\omega_0 t\right] \cos m\omega_0 t\, dt. \tag{4.8}$$

Interchanging summations and integrations,

$$\int_{-T/2}^{T/2} X(t)\cos m\omega_0 t\, dt = \frac{1}{2}a_0 \int_{-T/2}^{T/2} \cos m\omega_0 t\, dt$$

$$+ \sum_{n=1}^{\infty} a_n \int_{-T/2}^{T/2} \cos n\omega_0 t \cos m\omega_0 t\, dt$$

$$+ \sum_{n=1}^{\infty} b_n \int_{-T/2}^{T/2} \sin n\omega_0 t \cos m\omega_0 t\, dt. \tag{4.9}$$

The orthogonality and periodic properties state

$$\int_{-T/2}^{T/2} \cos m\omega_0 t\, dt = 0$$

for $m \neq 0$, and

$$\int_{-T/2}^{T/2} \sin n\omega_0 t \cos m\omega_0 t\, dt = 0$$

for all m and n. Using the above properties, Eq. (4.9) reduces to

$$\int_{-T/2}^{T/2} X(t)\cos m\omega_0 t\, dt = \sum_{n=1}^{\infty} a_n \int_{-T/2}^{T/2} \cos n\omega_0 t \cos m\omega_0 t\, dt. \quad (4.10)$$

The integral on the right of Eq. (4.10) is zero except when $n = m \neq 0$, in which case,

$$\int_{-T/2}^{T/2} \cos^2 n\omega_0 t\, dt = \frac{T}{2}. \quad (4.11)$$

With the above results, it is clear that

$$\int_{-T/2}^{T/2} X(t)\cos n\omega_0 t\, dt = \frac{T}{2} a_n$$

or

$$a_n = \frac{2}{T} \int_{-T/2}^{T/2} X(t)\cos n\omega_0 t\, dt. \quad (4.12)$$

If Eq. (4.4) is multiplied by $\sin m\omega_0 t$ and integrated over the period, a similar argument will yield

$$b_n = \frac{2}{T} \int_{-T/2}^{T/2} X(t)\sin n\omega_0 t\, dt. \quad (4.13)$$

When $n = 0$, Eq. (4.12) gives

$$a_0 = \frac{2}{T} \int_{-T/2}^{T/2} X(t)\, dt. \quad (4.14)$$

Therefore $a_0/2$ is the average over the period of the function $X(t)$.

Fourier series can also be expressed as a summation of complex numbers. Using the following identities,

$$\cos n\omega_0 t = \frac{1}{2}\left(e^{jn\omega_0 t} + e^{-jn\omega_0 t}\right)$$

$$\sin n\omega_0 t = \frac{1}{2j}\left(e^{jn\omega_0 t} - e^{-jn\omega_0 t}\right)$$

and substituting into Eq. (4.4) yields

$$X(t) = \frac{1}{2}a_0 + \sum_{n=1}^{\infty} \left[a_n \frac{1}{2}(e^{jn\omega_0 t} + e^{-jn\omega_0 t}) + b_n \frac{1}{2j}(e^{jn\omega_0 t} - e^{-jn\omega_0 t}) \right].$$

(4.15)

Using the identity $1/j = -j$,

$$X(t) = \frac{1}{2}a_0 + \sum_{n=1}^{\infty} \left[\frac{1}{2}(a_n - jb_n)e^{jn\omega_0 t} + \frac{1}{2}(a_n + jb_n)e^{-jn\omega_0 t} \right]. \quad (4.16)$$

If

$$X(0) = \frac{1}{2}a_0$$

$$X(n) = \frac{1}{2}(a_n - jb_n)$$

$$X(-n) = \frac{1}{2}(a_n + jb_n)$$

Eq. (4.16) becomes

$$X(t) = X(0) + \sum_{n-1}^{\infty} X(n)e^{jn\omega_0 t} + \sum_{n=-1}^{-\infty} X(n)e^{jn\omega_0 t}$$

$$= \sum_{n=-\infty}^{\infty} X(n)e^{jn\omega_0 t}.$$

(4.17)

The function $X(n) = \frac{1}{2}(a_n - jb_n)$ is called the complex spectrum of $X(t)$ and is also given by

$$X(n) = \frac{1}{2}\sqrt{a_n^2 + b_n^2} \exp\left[j \tan^{-1}\left(-\frac{b_n}{a_n} \right) \right]$$

$$= |X(n)| \exp[j\theta(n)]$$

(4.18)

$|X(n)| = \frac{1}{2}\sqrt{a_n^2 + b_n^2}$ is the amplitude spectrum of $X(t)$. By definition, the amplitude spectrum is real and positive, as well as an even function. The amplitude spectrum takes a real value only for integer values of n; for this reason it is also called a line spectrum.

The function

$$\theta(n) = \tan^{-1}\left(-\frac{b_n}{a_n} \right)$$

(4.19)

is the phase spectrum of $X(t)$.

If integral expressions of a_n and b_n (Eqs. 4.12 and 4.13) were substituted in the expression for the complex spectrum $X(n)$, the result would indicate

$$X(n) = \frac{1}{T} \int_{-T/2}^{T/2} X(t) e^{-jn\omega_0 t} \, dt. \tag{4.20}$$

Equations (4.17) and (4.20) are called a Fourier-transform pair.

Figure 4.2 illustrates the line and phase spectra of a periodic train of rectangular pulses with period $T_1 = 4b$. Using Eq. (4.20) the complex spectrum is given by

$$X(n) = \frac{1}{T_1} \int_0^b P e^{-jn\omega_0 t} \, dt = \left| \frac{Pb}{T_1} \frac{\sin\left(n\pi\dfrac{b}{T_1}\right)}{\dfrac{n\pi b}{T_1}} \right| e^{-jn\pi b/T_1}.$$

The normalized amplitude spectrum is defined as

$$^\circ|X(n)| = \frac{|X(n)|}{|X(0)|}.$$

In the case of Fig. 4.2 using L'Hospital's rule we get

$$|X(0)| = \frac{Pb}{T_1} \lim_{n \to 0} \left| \frac{\sin\left(n\pi\dfrac{b}{T_1}\right)}{n\pi b/T_1} \right| = \frac{Pb}{T_1},$$

whereupon

$$^\circ|X(n)| = \left| \frac{\sin\left(n\pi\dfrac{b}{T_1}\right)}{n\pi b/T_1} \right|.$$

The phase spectrum is in this case

$$\theta(n) = -n\pi\frac{b}{T_1}.$$

Of considerable importance are the concepts of autocorrelation and cross-correlation of periodic signals. Analogously to time averaging of stationary

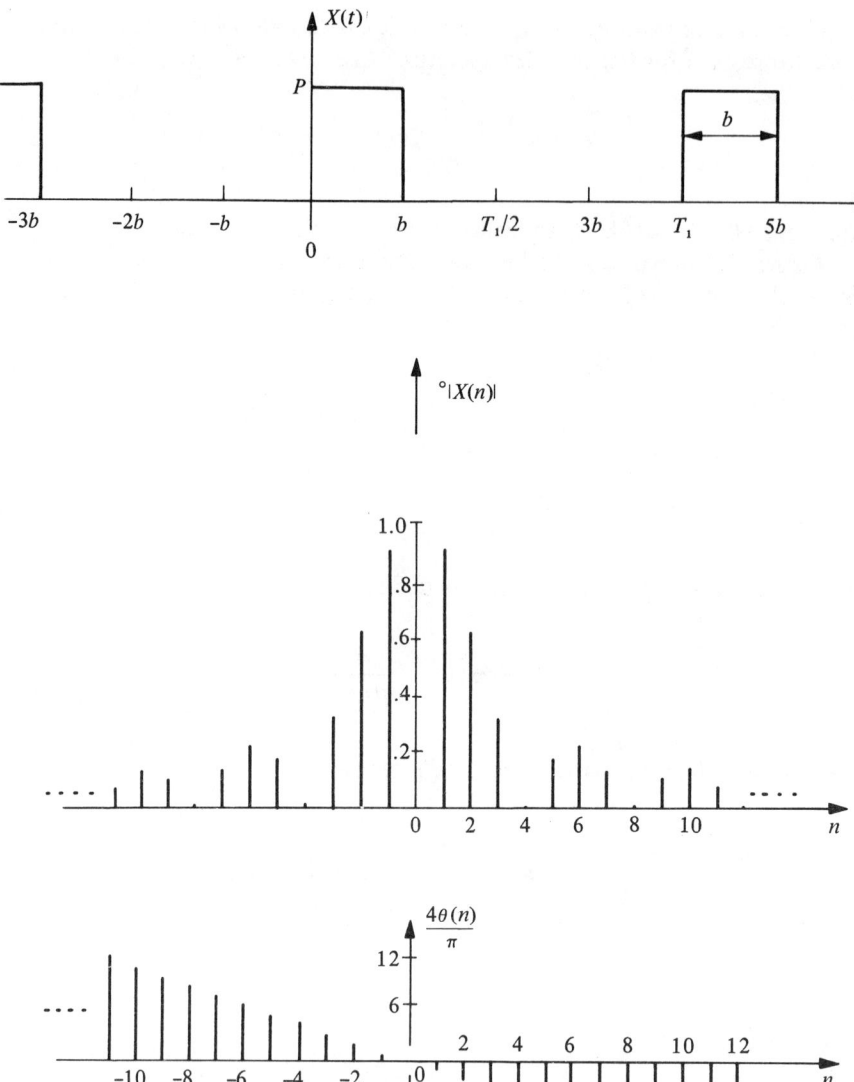

Figure 4.2 Amplitude and phase spectra of periodic train of rectangular pulses ($T_1 = 4b$) (from Eagleson, 1969). Temporal and Spatial Definition in Hydrologic Modelling. Cambridge, Mass.: Massachusetts Institute of Technology (unpublished class notes).

random series, the autocorrelation is a function of the lag τ,

$$\psi_{11}(\tau) = \frac{1}{T} \int_{-T/2}^{T/2} X_1(t) X_1(t+\tau) \, dt. \tag{4.21}$$

The cross-correlation between two periodic signals is

$$\psi_{12}(\tau) = \frac{1}{T} \int_{-T/2}^{T/2} X_1(t) X_2(t+\tau) \, dt. \tag{4.22}$$

Expansion of $X_1(t)$ and $X_2(t)$ in Fourier series and substitution in Eq. (4.22) yields

$$\begin{aligned}
\psi_{12}(\tau) &= \frac{1}{T} \int_{-T/2}^{T/2} X_1(t) \, dt \left(\sum_{n=-\infty}^{\infty} X_2(n) e^{jn\omega_0(t+\tau)} \right) \\
&= \sum_{n=-\infty}^{\infty} X_2(n) e^{jn\omega_0\tau} \frac{1}{T} \int_{-T/2}^{T/2} X_1(t) e^{jn\omega_0 t} \, dt \qquad (4.23) \\
&= \sum_{n=-\infty}^{\infty} X_2(n) \overline{X}_1(n) e^{jn\omega_0\tau},
\end{aligned}$$

where $\overline{X}_1(n)$ is the complex conjugate of $X_1(n)$.

Equation (4.23) indicates that $\psi_{12}(\tau)$ is the Fourier transform pair of $X_2(n)\overline{X}_1(n)$. Similarly, $X_1(n)\overline{X}_2(n)$ is related to $\psi_{21}(\tau)$ by a Fourier transformation such as Eq. (4.20).

If $X_1(t) = X_2(t)$, the above result implies that

$$\psi_{11}(\tau) = \sum_{n=-\infty}^{\infty} |X_1(n)|^2 e^{jn\omega_0\tau} \tag{4.24}$$

and

$$\Phi_{11}(n) = |X_1(n)|^2 = \frac{1}{T} \int_{-T/2}^{T/2} \psi_{11}(\tau) e^{-jn\omega_0\tau} \, d\tau, \tag{4.25}$$

where $\Phi_{11}(n)$ is the power spectrum of the function $X_1(t)$ and is mathematically equivalent to the Fourier transform of the autocorrelation function.

When $\tau = 0$, Eq. (4.24) yields

$$\psi_{11}(0) = \sum_{n=-\infty}^{\infty} |X_1(n)|^2, \tag{4.26}$$

which demonstrates that the mean square value of the function over period T, $\psi_{11}(0)$, is equivalent to the sum of the power spectra over all frequencies; each frequency contributes to the mean square value of the function. This result is called the Parseval Theorem.

There are several important properties of the autocorrelation and power spectrum. Using Eq. (4.21) and a change of variables, it is easy to prove that the autocorrelation is an even function,

$$\psi_{11}(\tau) = \psi_{11}(-\tau). \tag{4.27}$$

This implies that its Fourier expansion will not have any sine terms, as all coefficients b_n will be zero; this is easily rationalized from Eq. (4.13). Therefore

$$\psi_{11}(\tau) = \psi_{11}(-\tau) = \sum_{n=-\infty}^{\infty} \Phi_{11}(n)\cos n\omega_0\tau$$

$$= \Phi_{11}(0) + 2\sum_{n=1}^{\infty} \Phi_{11}(n)\cos n\omega_0\tau \tag{4.28}$$

and

$$\Phi_{11}(n) = \frac{1}{T}\int_{-T/2}^{T/2}\psi_{11}(\tau)\cos n\omega_0\,d\tau. \tag{4.29}$$

Note that the lack of sine terms implies no phase angles in the expansion. Conversely, $\Phi_{11}(n)$ contains no phase information on the original signal. This is easy to accept recognizing (from Eq. 4.26) that the power spectrum is the square of the line spectra, which has no phase information.

The properties of the cross-correlation function are different. Simple use of the definitions will yield

$$\psi_{12}(-\tau) = \psi_{21}(\tau)$$
$$\Phi_{12}(n) = \overline{\Phi}_{21}(n), \tag{4.30}$$

where $\Phi_{12}(n)$ is the cross-power spectrum and $\overline{\Phi}_{21}(n)$ is the complex conjugate of $\Phi_{21}(n)$. It can be shown (Lee, 1960) that the cross-power spectrum contains phase information in terms of the differences of the phase angles of the original functions. Also, the coefficients of the Fourier expansion of the cross-power spectrum are products of the coefficients of the expansions of the corresponding functions. If a frequency is not present in either $X_1(t)$ or $X_2(t)$, it will be lacking in $\Phi_{12}(n)$.

4.3 EXAMPLES AND HYDROLOGIC APPLICATIONS OF FREQUENCY-DOMAIN ANALYSIS

Possibly one of the most popular hydrologic concepts is the linear representation of the rainfall–runoff process. A linear system is represented by a convolution integral (summation for discrete processes)

$$Q(t) = \int_0^t i(\tau)h(t-\tau)\,d\tau, \tag{4.31}$$

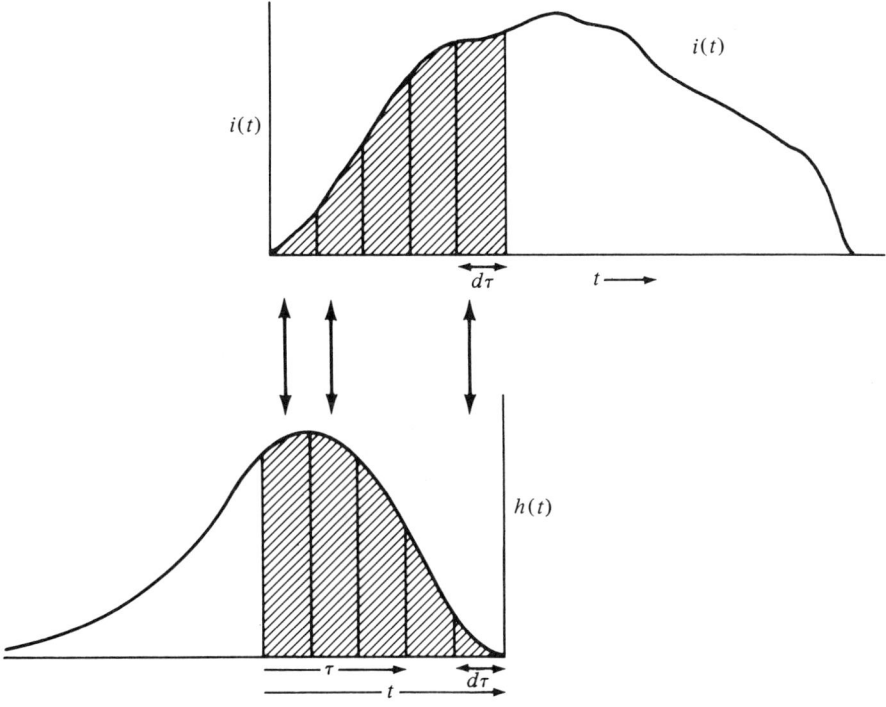

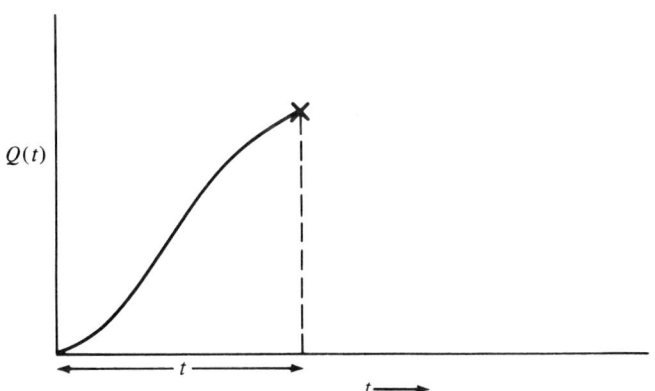

Figure 4.3 The convolution operation between the rainfall input $i(t)$ and the instantaneous unit hydrograph $h(t)$ to yield the streamflow output $Q(t)$.

where $Q(t)$ is discharge at time t; $i(t)$ is a rainfall excess input function; and $h(t)$ is an instantaneous unit hydrograph (IUH), or hydrologic response function. The integration limits are due to the causality of natural systems. The output at time t cannot depend on future inputs. Similarly, negative time has little meaning in nature.

The convolution operation is illustrated in Fig. 4.3. Each point of $Q(t)$ is the result of the integration (summation) of the product of the input (up to time t) and the folded IUH translated by t.

Equation (4.31) represents a relation between transient deterministic events, and so the theory of Fourier series is not strictly applicable. A rainfall event and its resultant runoff hydrograph would not, at first sight, appear to lend themselves to Fourier-series analysis—a technique usually associated with repeating events. However, as O'Donnell (1960) postulated, it is possible to conceive of a rainfall event that repeats itself. The IUH and the runoff output are also interpreted as periodic series. The duration of the surface runoff, T, is the sum of the duration of rainfall excess, T_i, and the duration of the IUH, T_u. The processes of rainfall, IUH, and runoff are thus periodic series with period no less than $T = T_u + T_i$ as shown in Fig. 4.4. Fourier series can then be applied to such a train of periodic signals.

The periodic assumption implies that

$$Q(t+T) = \int_0^t i(\tau+T)h(t-\tau)\,d\tau.$$

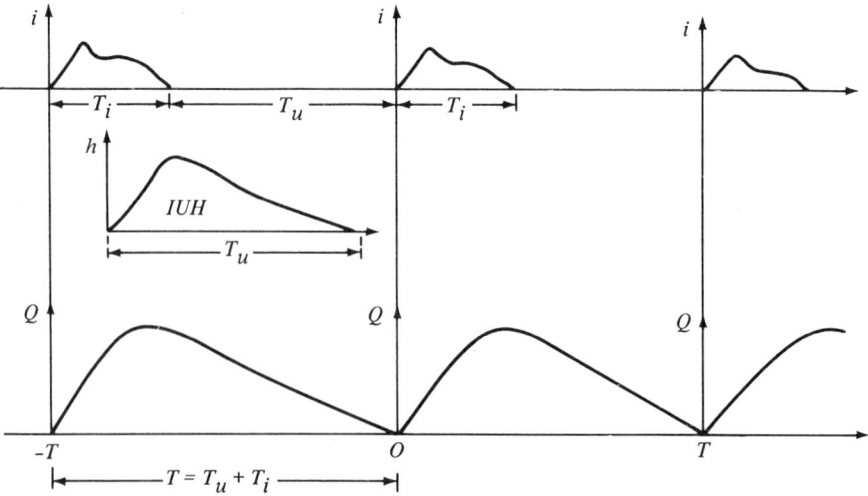

Figure 4.4 Representation of a periodic rainfall event causing a periodic runoff signal (after O'Donnell, 1960).

Letting $\tau + T = s$ and changing integration limits yields

$$Q(t+T) = \int_T^{t+T} i(s)h(t-s+T)\,ds.$$

Defining $t' = t + T$, becomes

$$Q(t') = \int_{t'-t}^{t'} i(s)h(t'-s)\,ds,$$

where t is the effective memory of the process. Since the IUH is zero beyond that memory, it is possible to let $t = T$ and write

$$Q(t) = \int_{t-T}^{t} i(\tau)h(t-\tau)\,d\tau. \tag{4.32}$$

Each of the components of Eq. (4.32) can be expanded in Fourier series,

$$Q(t) = \sum_{r=0}^{\infty} A_r \cos r \frac{2\pi t}{T} + \sum_{r=1}^{\infty} B_r \sin r \frac{2\pi t}{T} \tag{4.33}$$

$$i(\tau) = \sum_{n=0}^{\infty} a_n \cos n \frac{2\pi\tau}{T} + \sum_{n=1}^{\infty} b_n \sin n \frac{2\pi\tau}{T} \tag{4.34}$$

$$h(t-\tau) = \sum_{m=0}^{\infty} \alpha_m \cos m \frac{2\pi(t-\tau)}{T} + \sum_{m=1}^{\infty} \beta_m \sin m \frac{2\pi(t-\tau)}{T}. \tag{4.35}$$

Any fundamental base time greater than T could be used in the above series, merely resulting in different values for the harmonic coefficients.

Substitution of the above Fourier expansions on the convolution integral, multiplying out, and exchanging integral and summation signs, leads to $Q(t)$ as a sum of an infinite number of integrals all of which belong to one of the following four types,

$$a_n \alpha_m \int_{t-T}^{t} \cos n \frac{2\pi\tau}{T} \cos m \frac{2\pi(t-\tau)}{T}\,d\tau$$

$$a_n \beta_m \int_{t-T}^{t} \cos n \frac{2\pi\tau}{T} \sin m \frac{2\pi(t-\tau)}{T}\,d\tau$$

$$b_n \alpha_m \int_{t-T}^{t} \sin n \frac{2\pi\tau}{T} \cos m \frac{2\pi(t-\tau)}{T}\,d\tau$$

$$b_n \beta_m \int_{t-T}^{t} \sin n \frac{2\pi\tau}{T} \sin m \frac{2\pi(t-\tau)}{T}\,d\tau.$$

The previously encountered orthogonal properties of the sine and cosine imply that the above equal zero except when $m = n$ where they become

$$\frac{T}{2} a_n \alpha_n \cos n \frac{2\pi t}{T}$$

$$\frac{T}{2} a_n \beta_n \sin n \frac{2\pi t}{T}$$

$$\frac{T}{2} b_n \alpha_n \sin n \frac{2\pi t}{T}$$

$$-\frac{T}{2} b_n \beta_n \cos n \frac{2\pi t}{T} \,.$$

Equating coefficients of the above expressions to those of the series expansion of the discharge yields

$$\frac{A_0}{2} = \frac{T}{2} a_0 \alpha_0$$

$$A_n = \frac{T}{2}(a_n \alpha_n - b_n \beta_n); \quad n \geq 1$$

$$B_n = \frac{T}{2}(a_n \beta_n + b_n \alpha_n).$$

The previous equations can be useful in identifying response functions of basins (IUH) from observed rainfall and discharge events. The harmonic coefficients of the IUH are obtained from the above equations,

$$\alpha_0 = \frac{1}{T} \frac{A_0}{a_0}$$

$$\alpha_n = \frac{2}{T} \cdot \frac{a_n A_n + b_n B_n}{a_n^2 + b_n^2} \qquad \text{(for } n \geq 1) \tag{4.36}$$

$$\beta_n = \frac{2}{T} \cdot \frac{a_n B_n - b_n A_n}{a_n^2 + b_n^2}$$

O'Donnell's procedure was applied to two rainfall–runoff events from the North Branch Potomac River near Cumberland, Maryland. The data are shown in Table 4.1 and the corresponding IUHs are shown in Fig. 4.5. Two common occurrences are observed: (1) there are oscillations in the IUH tails indicative of high-frequency errors in data, and (2) two different storms yield different IUHs for the same basin, a result that questions the validity of the time-invariant linearity assumption (Eq. 4.31).

It is interesting to note that harmonic analysis is one of the best available methods to estimate the IUH in the presence of errors in the basic data

Table 4.1
**Data from the North Branch Potomac River near Cumberland,
Maryland (2266 km^2).**

Time	Storm 1				Storm 2			
	Rainfall		*Discharge*		*Rainfall*		*Discharge*	
($\Delta t = 4$ hr)								
hrs	*cm*	*in.*	*cm/hr*	*in./hr*	*cm*	*in.*	*cm/hr*	*in./hr*
4	0.31	0.12	0.008	0.003	0.51	0.20	0.0038	0.0015
8	2.24	0.88	0.028	0.011	0.10	0.04	0.0127	0.0050
12	2.03	0.80	0.091	0.036	0.81	0.32	0.0203	0.0080
16	2.54	1.00	0.229	0.090	0.20	0.08	0.0241	0.0095
20	0.61	0.24	0.356	0.140	1.83	0.72	0.0279	0.0110
24			0.356	0.140	1.12	0.44	0.0686	0.0270
28			0.279	0.110	0.31	0.12	0.1270	0.0500
32			0.208	0.082	2.13	0.84	0.1829	0.0720
36			0.145	0.057	0.31	0.12	0.2159	0.0850
40			0.094	0.037	1.32	0.52	0.2261	0.0890
44			0.064	0.025			0.2146	0.0845
48			0.036	0.014			0.2108	0.0830
52			0.020	0.008			0.2121	0.0835
56			0.010	0.004			0.1753	0.0690
60			0.005	0.002			0.1321	0.0520
64			0.003	0.001			0.0914	0.0360
68							0.0660	0.0260
72							0.0483	0.0190
76							0.0356	0.0140
80							0.0254	0.0100
84							0.0178	0.0070
88							0.0102	0.0040
92							0.0025	0.0010
96							0.0064	0.0025
100							0.0013	0.0005

Source: Singh (1976).

(Dooge, 1977). As shown by Laurenson and O'Donnell (1969), the direct solution of the discrete version of Eq. (4.31) is highly sensitive to errors in input and output data. Harmonic analysis works well even in the presence of contaminated data because of its inherent filtering mechanism, which has the effect of subtracting high frequencies containing only a small part of the signal and most of the error.

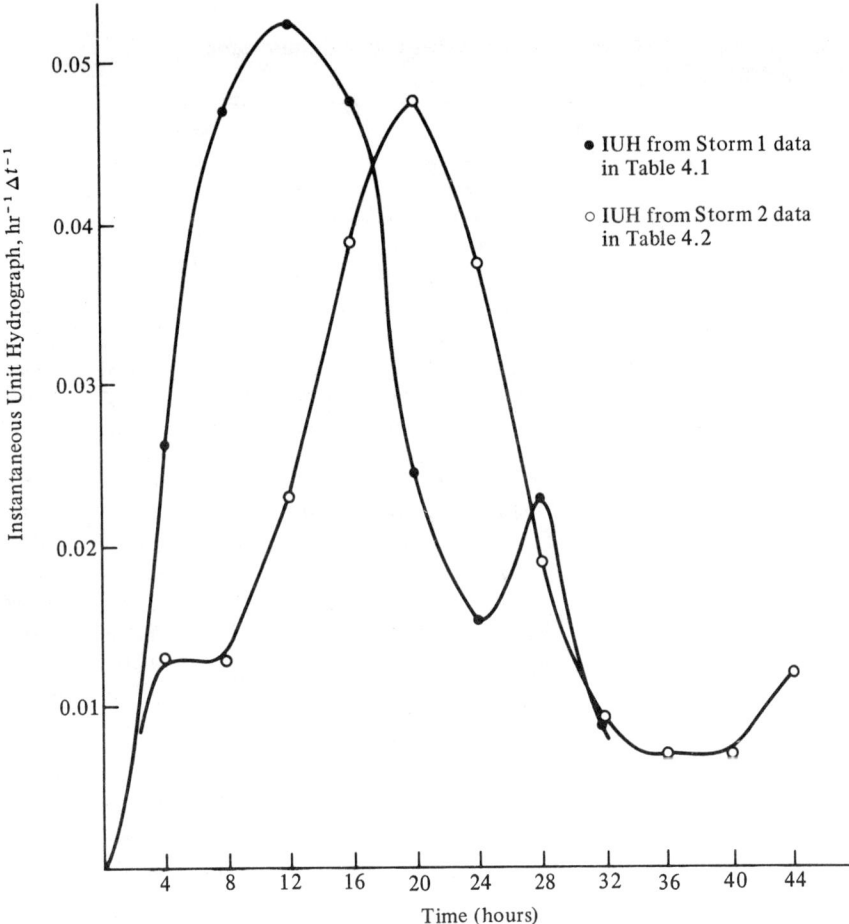

Figure 4.5 Instantaneous unit hydrographs obtained by O'Donnell's procedure for the North Branch Potomac River near Cumberland, Maryland.

4.4 FOURIER SERIES OF DISCRETE DATA AND THE SAMPLING THEOREM

Assume that a continuous and periodic signal is sampled in intervals Δt leading to

$$N = \frac{T}{\Delta t}$$

samples per period. The observed values being

$$X(t = r\Delta t).$$

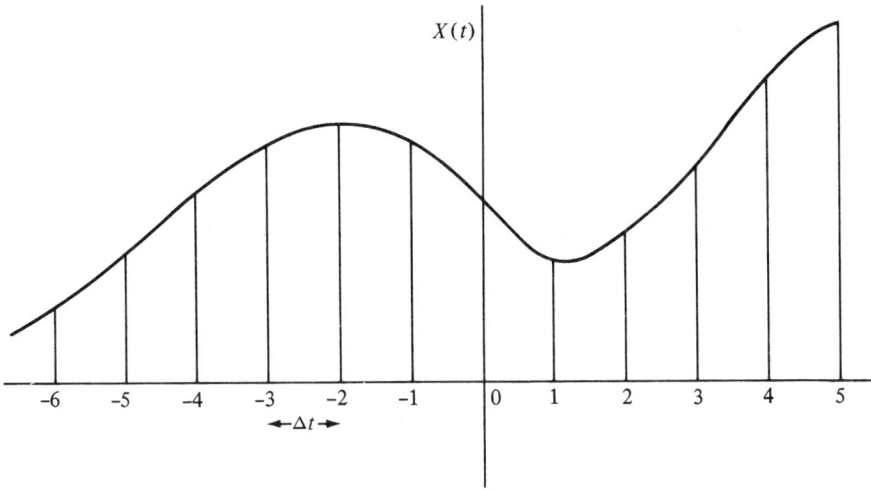

Figure 4.6   Discrete sampling of a continuous function.

If N is even and equal to $2n$, r may take integer values (Fig. 4.6)

$$-n,\ldots,0,1,\ldots,n-1.$$

The function $X(t)$ can be approximated by

$$X(t) = A_0 + 2 \sum_{m=1}^{n-1} \left(A_m \cos 2\pi m f_1 t + B_m \sin 2\pi m f_1 t \right)$$

$$+ A_n \cos 2\pi n f_1 t \tag{4.37}$$

where

$$f_1 = \frac{1}{T} = \frac{1}{N\Delta t}.$$

The coefficients A_m and B_m can be obtained so that they exactly fit the observed points $X(r\Delta t)$. Defining $t = r\Delta t$, the coefficients are:

$$A_m = \frac{1}{N} \sum_{r=-n}^{n-1} X(r\Delta t) \cos \frac{2\pi m r}{N} \tag{4.38}$$

$$B_m = \frac{1}{N} \sum_{r=-n}^{n-1} X(r\Delta t) \sin \frac{2\pi m r}{N}, \tag{4.39}$$

where A_0 is the average of the observed points. If N is odd, there are $2n-1$ terms, and the last value A_n will not exist. Note from Eqs. (4.38) and (4.39) that the highest frequency detected by this procedure is

$$\frac{n}{N\Delta t} = \frac{n}{2n\Delta t} = \frac{1}{2\Delta t}. \tag{4.40}$$

The sampling theorem states that, the maximum detectable frequency is one cycle every two time periods. A reduction of Δt increases the observable frequencies. The above concept is useful in the design of experiments, and will be used later in this book in the sampling of hydrologic signals.

4.5 TRANSIENT DETERMINISTIC SIGNALS

A transient signal can be interpreted as an event with an infinite period. As $T \to \infty$ the fundamental frequency reduces to a differential frequency

$$\omega_0 = \frac{2\pi}{T} \underset{T \to \infty}{\to} d\omega. \tag{4.41}$$

Therefore the discrete line spectra, $X(n\omega_0)$ becomes a continuous function on ω, $X(nd\omega) \approx X(\omega) d\omega$ for infinitesimal $d\omega$. Then, examining the Fourier-series–transform pair as T approaches infinity,

$$X(t) = \lim_{T \to \infty} \sum_{n=-\infty}^{\infty} X(n\omega_0) e^{jn\omega_0 t} = \int_{-\infty}^{\infty} X(\omega) e^{j\omega t} d\omega \tag{1.12}$$

and

$$X(n\omega_0) \underset{T \to \infty}{=} X(\omega) d\omega \underset{T \to \infty}{=} \frac{1}{T} \int_{-T/2}^{T/2} X(t) e^{-j\omega t} dt.$$

Using Eq. (4.41)

$$X(\omega) d\omega = \frac{d\omega}{2\pi} \int_{-\infty}^{\infty} X(t) e^{-j\omega t} dt$$

or

$$X(\omega) = \frac{1}{2\pi} \int_{-\infty}^{\infty} X(t) e^{-j\omega t} dt. \tag{4.43}$$

Equations (4.42) and (4.43) are Fourier-transform pairs for transient functions. $X(\omega)$ is now a continuous function of angular frequency, which is generally complex,

$$X(\omega) = P(\omega) + jQ(\omega).$$

The amplitude-density spectrum is then defined as

$$|X(\omega)| = \sqrt{P^2(\omega) + Q^2(\omega)} \tag{4.44}$$

and the phase-density spectrum is given by

$$\theta(\omega) = \tan^{-1}\left(-\frac{Q(\omega)}{P(\omega)}\right). \tag{4.45}$$

The correlation and cross-correlation functions are also defined for transient phenomena,

$$\begin{aligned}
\psi_{11}(\tau) &= \int_{-\infty}^{\infty} X_1(t) X_1(t+\tau)\, dt \\
&= \int_{-\infty}^{\infty} 2\pi |X_1(\omega)|^2 e^{j\omega\tau}\, d\omega,
\end{aligned} \tag{4.46}$$

where the energy-density spectrum is

$$\Phi_{11}(\omega) = 2\pi |X_1(\omega)|^2. \tag{4.47}$$

The correlation function can then be written as,

$$\psi_{11}(\tau) = \int_{-\infty}^{\infty} \Phi_{11}(\omega)\cos\omega\tau\, d\omega. \tag{4.48}$$

The cross-correlation,

$$\begin{aligned}
\psi_{12}(\tau) &= \int_{-\infty}^{\infty} X_1(t) X_2(t+\tau)\, dt \\
&= \int_{-\infty}^{\infty} 2\pi X_1(\omega)\overline{X}_2(\omega) e^{j\omega\tau}\, d\omega,
\end{aligned}$$

where $\Phi_{12}(\omega) = 2\pi X_1(\omega)\overline{X}_2(\omega)$ is the cross–energy-density spectrum. Parseval's theorem for transient signals becomes

$$\psi_{11}(0) = \int_{-\infty}^{\infty} \Phi_{11}(\omega)\, d\omega,$$

which implies that the total power or energy of the function $X(t)$ is distributed among the contributing frequencies according to a density $\Phi_{11}(\omega)$.

The theory of transient signals is convenient when dealing with linear systems,

$$Q(t) = \int_{-\infty}^{\infty} I(\tau)h(t-\tau)\, d\tau. \tag{4.49}$$

Taking transforms on both sides of the equation,

$$\frac{1}{2\pi}\int_{-\infty}^{\infty} Q(t)e^{-j\omega t}\,dt = \frac{1}{2\pi}\int_{-\infty}^{\infty} e^{-j\omega t}\,dt\int_{-\infty}^{\infty} I(\tau)h(t-\tau)\,d\tau.$$

Redefining $s = t - \tau$, $dt = ds$,

$$Q(\omega) = \frac{1}{2\pi}\int_{-\infty}^{\infty} e^{-j\omega(s+\tau)}\,ds\int_{-\infty}^{\infty} I(\tau)h(s)\,d\tau$$

$$= \frac{1}{2\pi}\int_{-\infty}^{\infty} h(s)e^{-j\omega s}\,ds\int_{-\infty}^{\infty} I(\tau)e^{-j\omega \tau}\,d\tau,$$

or

$$Q(\omega) = 2\pi H(\omega)\cdot I(\omega). \tag{4.50}$$

The convolution integral is converted to an algebraic product on transformed functions. Equation (4.50) has been used in hydrology several times, particularly in the identification of system-response functions (Evans et al. 1972) and the design of data collection networks (Eagleson, 1967a).

The linear-systems formulation results in other useful relationships between the output and input correlation structure. Using Eq. (4.49) in Eq. (4.46) to obtain the autocorrelation of the output of a linear system results in

$$\psi_{QQ}(\tau) = \int_{-\infty}^{\infty} Q(t)Q(t+\tau)\,dt$$

$$= \int_{-\infty}^{\infty} dt\int_{-\infty}^{\infty} h(v)I(t-v)\,dv\int_{-\infty}^{\infty} h(\sigma)I(t+\tau-\sigma)\,d\sigma.$$

Changing the order of integration

$$\psi_{QQ}(\tau) = \int_{-\infty}^{\infty} h(v)\int_{-\infty}^{\infty} h(\sigma)\int_{-\infty}^{\infty} I(t+\tau-\sigma)I(t-v)\,dt\,d\sigma\,dv$$

$$= \int_{-\infty}^{\infty} h(v)\int_{-\infty}^{\infty} h(\sigma)\psi_{II}(\tau+v-\sigma)\,d\sigma\,dv, \tag{4.51}$$

where $\psi_{II}(\tau+v-\sigma)$ is the autocorrelation of the input.

Equation (4.51) can be converted to a more familiar form. Letting $t = \sigma - v$,

$$\psi_{QQ}(\tau) = \int_{-\infty}^{\infty} h(v)\int_{-\infty}^{\infty} h(t+v)\psi_{II}(\tau-t)\,dt\,dv$$

$$= \int_{-\infty}^{\infty} \psi_{II}(\tau-t)\int_{-\infty}^{\infty} h(v)h(v+t)\,dv\,dt \tag{4.52}$$

$$= \int_{-\infty}^{\infty} \psi_{II}(\tau-t)\psi_{hh}(t)\,dt,$$

where $\psi_{hh}(t)$ is the autocorrelation of the response function.

Equation (4.52) is clearly a convolution between autocorrelations of inputs and systems response. By analogy to previous manipulations with convolutions, it should then be obvious that

$$\Phi_{QQ}(\omega) = 2\pi\Phi_{hh}(\omega)\Phi_{II}(\omega) \tag{4.53}$$

and using Eq. (4.47)

$$\Phi_{QQ}(\omega) = 4\pi^2|H(\omega)|^2\Phi_{II}(\omega), \tag{4.54}$$

where

$$\Phi_{hh}(\omega) = \frac{1}{2\pi}\int_{-\infty}^{\infty}\psi_{hh}(\tau)\cos\omega\tau\,d\tau$$

and

$$H(\omega) = \frac{1}{2\pi}\int_{-\infty}^{\infty}h(t)e^{-j\omega t}\,dt = |H(\omega)|e^{j\theta(\omega)},$$

where $|H(\omega)|$ is sometimes called the gain of the linear system and $\theta(\omega)$ is the phase-density spectrum defined in Eq. (4.45).

Simple relationships also exist between the cross-correlation of inputs and outputs. Defining the cross-correlation as

$$\psi_{IQ}(\tau) = \int_{-\infty}^{\infty}I(t)Q(t+\tau)\,dt$$

and substituting the convolution equation for $Q(t)$,

$$\begin{aligned}
\psi_{IQ}(\tau) &= \int_{-\infty}^{\infty}I(t)\int_{-\infty}^{\infty}h(\sigma)I(t+\tau-\sigma)\,d\sigma\,dt \\
&= \int_{-\infty}^{\infty}h(\sigma)\int_{-\infty}^{\infty}I(t)I(t+\tau-\sigma)\,dt\,d\sigma \tag{4.55} \\
&= \int_{-\infty}^{\infty}h(\sigma)\psi_{II}(\tau-\sigma)\,d\sigma.
\end{aligned}$$

Again, Eq. (4.55) is the familiar convolution, which logically leads to

$$\Phi_{IQ}(\omega) = 2\pi H(\omega)\Phi_{II}(\omega). \tag{4.56}$$

4.6 DISCRETE SAMPLING OF CONTINUOUS TRANSIENT SIGNALS

Section 4.4 indicated how the highest observable frequency in a discrete sampling at Δt intervals is

$$f_N = \frac{1}{2\Delta t} \qquad \text{or} \qquad \omega_N = \frac{\pi}{\Delta t}. \tag{4.57}$$

The subscript N alludes to the name Nyquist, which is given to the highest frequency observed. If a continuous process is sampled at discrete points, the spectral estimate of the signal is truncated, leading to some bias (called aliasing). A sampling scheme should avoid cutting important frequencies and minimize aliasing. The idea is to choose Δt so that variation in the continuous series is negligible at frequencies higher than $\pi/\Delta t$.

Following Chatfield (1975), the effect of aliasing is easily derived.

The autocorrelation of the continuous transient process is defined as $\psi(\tau)$. The autocorrelation of its discrete samples is $\psi(k)$, where k is an integer value. For integer values $\tau = k$ the two autocorrelations must be the same. Using Eq. (4.48) integrated to the maximum observable frequency (assuming $\Delta t = 1$), you obtain

$$\psi(k) = \int_0^\pi 2\Phi_d(\omega)\cos\omega k\, d\omega. \tag{4.58}$$

Using the continuous spectrum

$$\psi(\tau) = \int_0^\infty 2\Phi_c(\omega)\cos\omega\tau\, d\omega, \tag{4.59}$$

where $\Phi_d(\omega)$ stands for the power-density spectrum resulting from discrete sampling and $\Phi_c(\omega)$ its continuous counterpart.

Equating the above two equations, for $\tau = k = 0, +1, \ldots$

$$\int_0^\pi 2\Phi_d(\omega)\cos\omega k\, d\omega = \int_0^\infty 2\Phi_c(\omega)\cos\omega k\, d\omega. \tag{4.60}$$

The righthand integral is equivalent to

$$\int_0^\infty \Phi_c(\omega)\cos\omega k\, d\omega = \sum_{s=0}^\infty \int_{2\pi s}^{2\pi(s+1)} \Phi_c(\omega)\cos\omega k\, d\omega. \tag{4.61}$$

Further expanding Eq. (4.61),

$$
\begin{aligned}
\int_0^\infty \Phi_c(\omega)\cos\omega k\, d\omega &= \sum_{s=0}^\infty \int_0^{2\pi} \Phi_c(\omega+2\pi s)\cos\omega k\, d\omega \\
&= \sum_{s=0}^\infty \int_0^\pi \{\Phi_c(\omega+2\pi s)+\Phi_c[2\pi(s+1)-\omega]\}\cos\omega k\, d\omega \\
&= \int_0^\pi \left\{ \sum_{s=0}^\infty \Phi_c(\omega+2\pi s)+\sum_{s=1}^\infty \Phi_c(2\pi s-\omega)\right\}\cos\omega k\, d\omega.
\end{aligned}
\tag{4.62}
$$

Generalizing for general values of Δt and letting $\omega_N = \pi/\Delta t$, it is clear that

$$\Phi_d(\omega) = \sum_{s=0}^\infty \Phi_c(\omega+2\omega_N s)+\sum_{s=1}^\infty \Phi_c(2\omega_N s-\omega). \tag{4.63}$$

Equation (4.63) shows that if the continuous series contains no variations at frequencies above the Nyquist frequency, so that

$$\Phi_c(\omega) = 0 \qquad \text{for} \qquad \omega > \pi/\Delta t,$$

then

$$\Phi_d(\omega) = \Phi_c(\omega).$$

In this case, no information is lost by sampling. In cases where $\Phi_c(\omega) \neq 0$ for $\omega > \pi/\Delta t$, the effect of sampling is that the power-density function is overestimated because the power at frequencies larger than the Nyquist frequency is "folded back" and added to frequencies smaller than the Nyquist frequency in $\Phi_d(\omega)$. If we denote the Nyquist frequency $\pi/\Delta t$ by ω_N, then the frequencies ω, $2\omega_N - \omega$, $2\omega_N + \omega$, $4\omega_N - \omega, \ldots$, are called aliases of one another. Power at all these frequencies in the continuous series will appear as power at frequency ω in the sampled series (Chatfield, 1975). The aliasing effect is illustrated in Fig. 4.7.

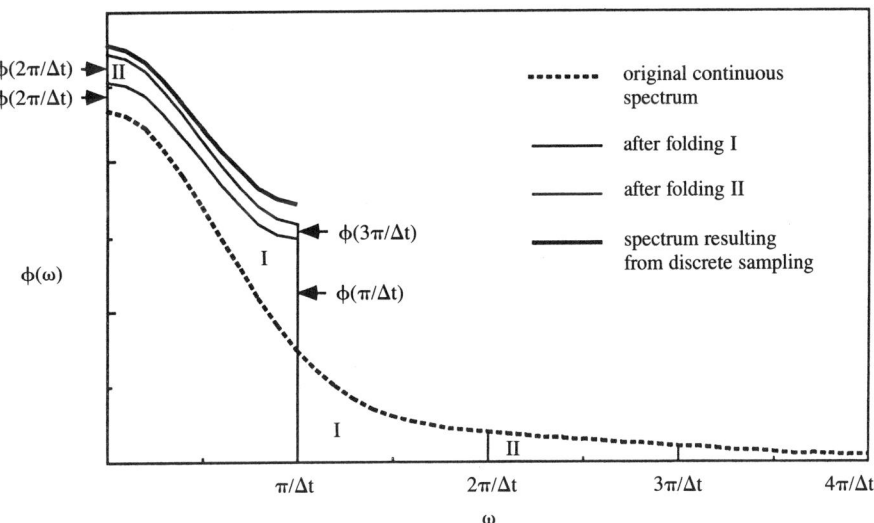

Figure 4.7   Illustration of aliasing effects on power-spectrum estimation.

4.7 AN EXAMPLE OF NETWORK DESIGN

Eagleson (1967a) approached the problem of determining the optimal density of rain gages, required for accurate discharge forecasts, using deterministic linear-system theory in the frequency domain.

A catchment was hypothesized to consist of an infinite array of lumped, stationary, independent linear subsystems leading to an equation of the form

$$Q(a,t) = \int_0^a \int_0^t P(x,y,\tau) \cdot h(x,y,t-\tau)\, d\tau\, da \qquad (4.64)$$

(Fig. 4.8a), where $Q(a,t)$ is the discharge at the catchment outlet at time t resulting from rainfall, $P(x,y,t)$, up to time t and distributed over the catchment area, a. The function $h(x,y,t)$ represents the linear response of the elementary area da at time t.

Eagleson decided to focus his interest on the hydrograph peak so he simplified Eq. (4.64) (Eagleson, 1967b, for details) into a time-independent expression for hydrograph peak in a one-dimensional basin representation (Fig. 4.8b).

The spatial distribution of the rainfall input is considered only in the direction of the main stem of the basin. The linear peak-discharge expression is

$$Q = 2\Theta_c \int_0^{L_s} y^{5/3}(x)\, dx, \qquad (4.65)$$

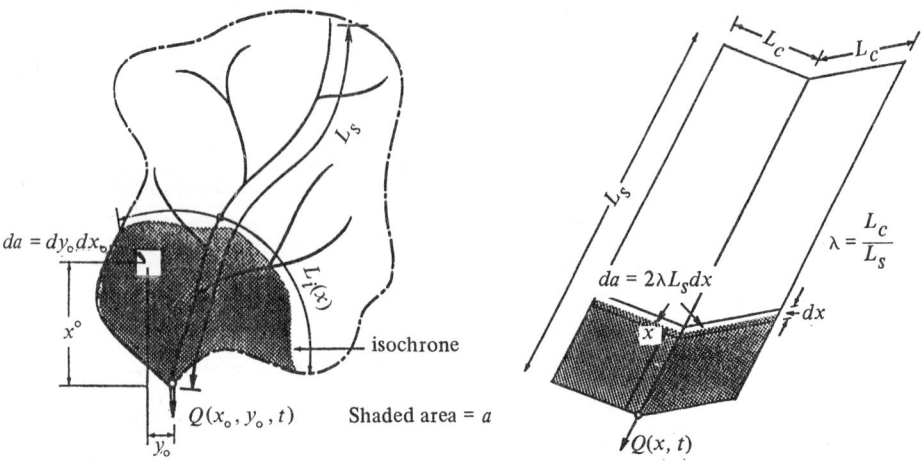

(a) Natural Catchment   (b) One-Dimensional Simplification

Figure 4.8 A one-dimensional simplification of natural catchments (from Eagleson, 1967a).

where Q = peak discharge at basin outlet, $\Theta_c = (G/n_c)S_c^{1/2}$ (G is a constant depending on units, $G = 1$ for the metric system and $G = 1.49$ for the English system), S_c = average slope of overland segments, n_c = average Manning's n of overland segments, L_s = length of mainstream, $P(x, t)$ = rainfall as a function of time and location along the one-dimensional basin x, $t_0 = t_c$ if $T_R \geq t_c$ or $t_0 = T_R/2$ if $T_R < t_c$, T_R = storm duration, and t_c = time of concentration of overland flow segments. Also,

$$y(x) = \int_0^{t_0} P(x, t)\, dt. \tag{4.66}$$

The maximum discharge Q occurs at $t = T_R/2 + t_s$ for $T_R \leq t_c$ and at $t = t_c + t_s$ for $T_R \geq t_c$, where t_s is the time of concentration of the stream element.

Equation (4.65) can be thought of as a convolution integral of the form

$$Q(x) = \int_0^x h(x - s)P(s)\, ds, \tag{4.67}$$

where $h(x) = 2\Theta_c$, a constant for $0 \leq x \leq L_s$ and $P(x) = y^{5/3}(x)$.

In the frequency domain, Eq. (4.67) can be represented as

$$Q(\kappa) = 2\pi H(\kappa)P(\kappa) \tag{4.68}$$

or

$$\Phi_Q(\kappa) = 4\pi^2 |H(\kappa)|^2 \Phi_P(\kappa), \tag{4.69}$$

where $\Phi_x(\kappa)$ is the energy-density spectrum of x, or Fourier transform of the correlation function of x, $\psi_x(s)$,

$$\psi_x(s) = \int_{-\infty}^{\infty} x(y)x(y + s)\, dy \tag{4.70}$$

and κ is the wave number $2\pi/L$, where L is the wavelength of the constituent harmonics.

In order to study the frequency characteristics of the discharge, Eagleson had to hypothesize some one-dimensional storm structures. Two radially symmetric (around storm center) storm structures were used and are illustrated in Fig. 4.9. The selected structures are empirical expressions for the spatial distribution of total storm depths of typical convective storms (in Arizona) and cyclonic storms in the U.S.A.

For convective storms the rainfall average over the area is taken from the storm-centered function of Woolhiser and Schwalen (1959)

$$P_A/P_T(0) = 1 - [0.14/P_T(0)] A_s^{0.6},$$

where $P_T(0)$ is the total depth in inches at the storm center, P_A is the average depth over the circular area A_s (in square miles), and radial symmetry is assumed.

For cyclonic storms the average rainfall in the area is taken from the storm-centered function of Boyer (1957)

$$P_T(r)/P_T(0) = e^{-ar}.$$

Eagleson (1967a) gives the analytical expressions for the correlation functions corresponding to the storm structures given above.

Assuming that rainfall is time synchronized everywhere in space so that the spatial distribution remains constant through time, and assuming that rainfall is one-dimensional and centered at the catchment outlet, then Fig. 4.9 illustrates the spatial distribution of rainfall over Eagleson's one-dimensional basin. The normalized correlation functions, $\psi(x)/\psi(0)$, implied by the storm structures are shown in Fig. 4.10. Note that the spatial coordinate x (distance from basin outlet) is normalized by the "correlation radius," r_0, defined as that distance at which the normalized autocorrelation function takes a value of 0.5.

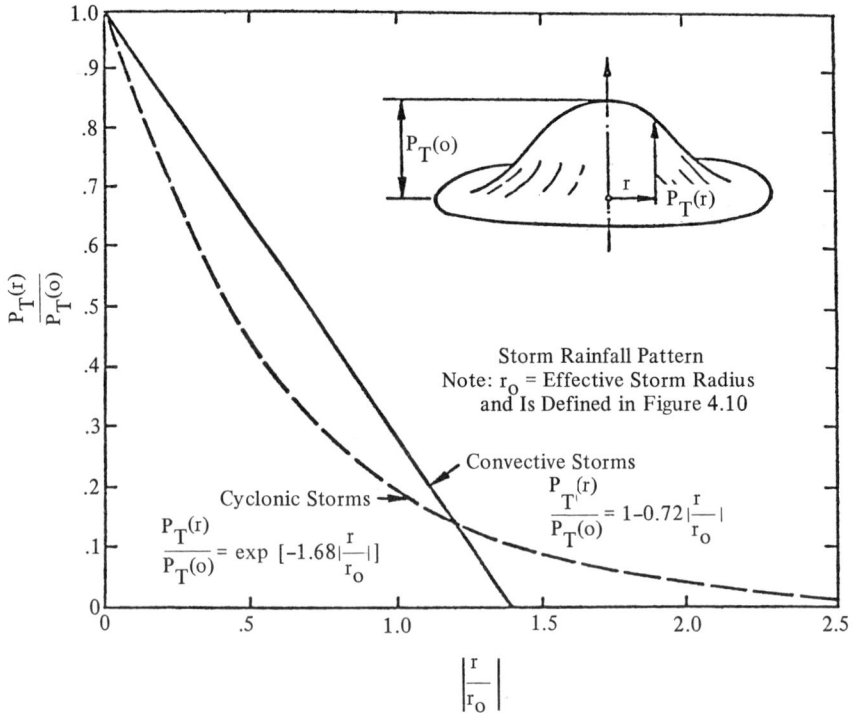

Figure 4.9 Typical storm structures (from Eagleson, 1967a).

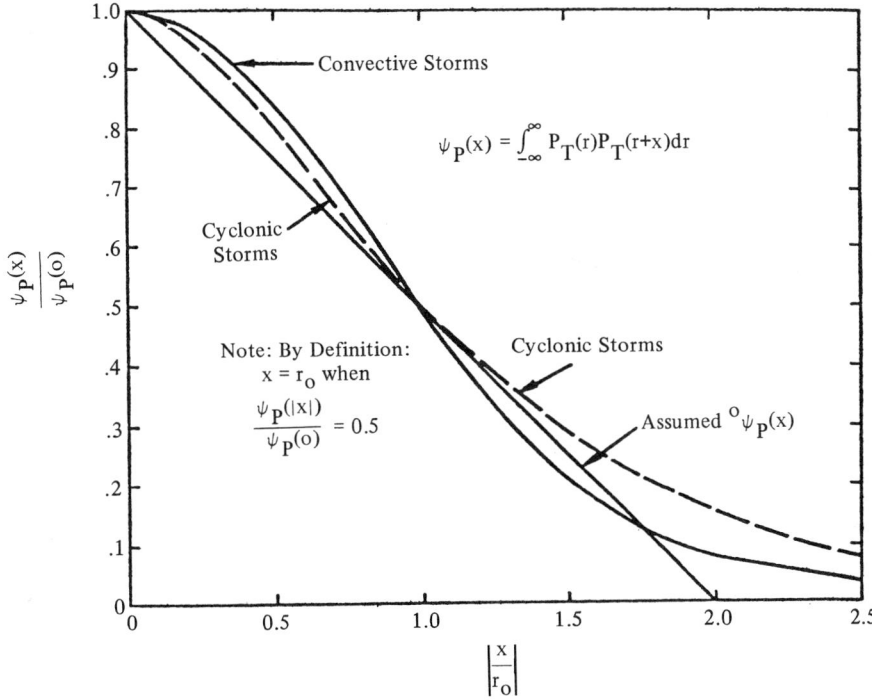

Figure 4.10   Correlation function of typical storms (from Eagleson, 1967a).

The distribution of average rainfall for convective and cyclonic storms can be written as a function of r in the form

$$P_T(r)/P_T(0) = 1 - 0.72(r/r_0) \qquad \text{for convective storms}$$
$$P_T(r)/P_T(0) = e^{-1.68r/r_0} \qquad \text{for cyclonic storms}$$

Figure 4.11 gives the energy-density spectra, $\Phi_p(\alpha)$, of the convective and cyclonic storms (normalized by their value at the origin). These energy spectra are the Fourier transforms of the correlations given in Fig. 4.10. The energy spectrum of rainfall is given in terms of the parameter $\alpha = r_0\kappa$, where r_0 and κ are the correlation radius and wave numbers as previously defined.

The Fourier transform of the response function is given by

$$H(\kappa) = \int_0^{L_s} 2\Theta_c e^{-j\kappa x} \, dx$$

$$= 2\Theta_c L_s \left| \frac{\sin \kappa L_s - j(1 - \cos \kappa L_s)}{\kappa L_s} \right|$$

$$(4.71)$$

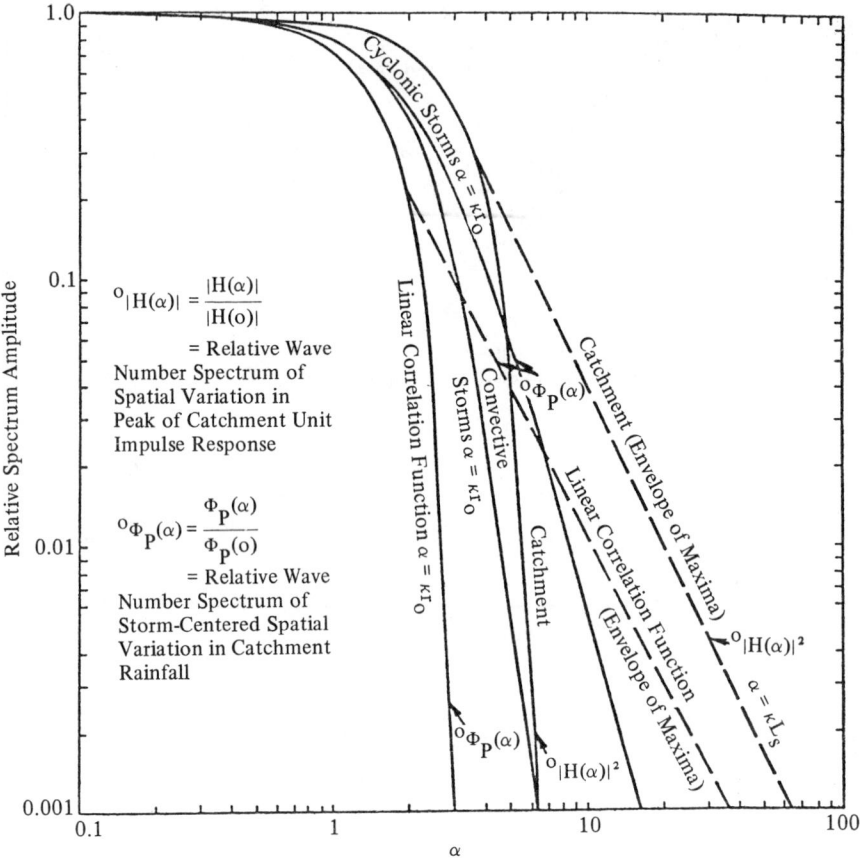

Figure 4.11 Energy density spectra of typical storms and impulse response function of peak discharge (from Eagleson, 1967a).

leading to a normalized square modulus

$$^\circ|H(\kappa)|^2 = \frac{|H(\kappa)|^2}{|H(0)|^2} = \left[4\sin^2(\alpha/2)\right]/\alpha. \qquad (4.72)$$

Here, the parameter $\alpha = \kappa L_s$. Note that it is different from the argument of the rainfall spectrum. Equation (4.72) is also plotted in Fig. 4.11.

From Eq. (4.69), it is clear that the energy spectrum of discharge is given by

$$^\circ\Phi_Q(\alpha, \beta) = \frac{\Phi_Q(\alpha, \beta)}{\Phi_Q(0, \beta)} \qquad (4.73)$$

$$= 4\pi^2 \,^\circ|H(\kappa)|^2 \,^\circ\Phi_p(\alpha/\beta),$$

where $\alpha = \kappa L_s$ and $\beta = L_s/r_0$.

The parameter β is essentially a measure of the size of the basin relative to storm dimensions. As β increases to infinity, the observed effect corresponds to a localized line pulse of rainfall.

The discharge-energy spectrum given in Eq. (4.73) is related to the Fourier transform of the discharge function by Eq. (4.47)

$$\Phi_Q(\alpha, \beta) = 2\pi |Q(\alpha, \beta)|^2. \tag{4.74}$$

Sampling at finite densities (discrete number of points in space) implies truncating the discharge spectra at some nondimensional wave number,

$$\alpha_g = \kappa_g L_s = 2\pi(L_s / L_g), \tag{4.75}$$

where L_g is the smallest constituent wavelength to be measured in the spatial rainfall distribution. The finite sampling implies an error in discharge prediction given by

$$E(\alpha_g) = 1 - \frac{\hat{Q}(L_s)}{Q(L_s)} \tag{4.76}$$

$$\hat{Q}(L_s) = \int_{-\kappa_g}^{\kappa_g} Q(\kappa, \beta) e^{j\kappa L_s} d\kappa \tag{4.77}$$

$$Q(L_s) = \int_{-\infty}^{\infty} Q(\kappa, \beta) e^{j\kappa L_s} d\kappa. \tag{4.78}$$

Using the sampling theorem, Eagleson related the truncation wave number and the network density. Since L_g is the smallest constituent wavelength to be measured and a minimum of two samples per wave are needed to define a given harmonic, the maximum allowable gage spacing x must be

$$x = L_g / 2 = \pi(L_s / \alpha_g). \tag{4.79}$$

Using Fig. 4.8, it is easy to see that the one-dimensional catchment representation implies

$$2L_c = \frac{A}{L_s}, \tag{4.80}$$

where A is the basin area. The ratio of overland catchment length to stream length is defined $\lambda = L_c / L_s$. The number of gages per basin is then

$$G_0 = \frac{A}{x^2} = 2\alpha_g^2 \frac{\lambda}{\pi^2}, \tag{4.81}$$

assuming that all gages are located within the storm area or that the storm essentially covers the whole basin, clearly $G_0 \geq 1$. Given the definition of the correlation radius complete coverage will be generally expected when $\beta < 2$ ($\beta = L_s/r_0$). For $\beta > 2$, partial coverage will occur and the number of rain gages should be increased in proportion to the ratio of catchment to storm area,

$$G = (A/A_0)G_0 \approx (\beta/2)2\alpha_g^2\lambda/\pi^2$$

or

$$2G/\lambda\beta = 2\alpha_g^2/\pi^2 \qquad \text{for } \beta > 2. \qquad (4.82)$$

Equations (4.76), (4.81), and (4.82) are used to produce the graph shown in Fig. 4.12, relating the number of rain gages, G, to the acceptable error, E, and the relative catchment to storm dimensions, β.

Eagleson's method generally requires relatively few rain gages for reasonable forecasting errors. Table 4.2 shows the number of rain gages required for several forecasting errors and values of β. The results correspond to a 540-mi^2 Australian basin with a shape parameter, $\lambda \approx 1$.

Eagleson and Goodspeed (1973) using the same method in other Australian experimental basins found that for areas ranging from 1.19 to 66 mi^2 only one or two gages yielded discharge errors within 5.3%.

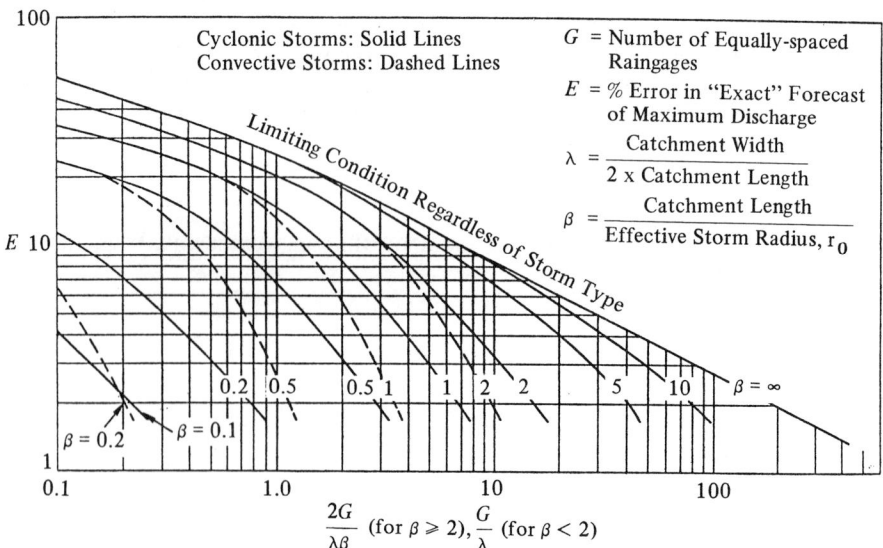

Figure 4.12 Number of equispaced rain gages for flood forecasting (from Eagleson, 1967a).

Table 4.2
Required number of flood-forecasting rain gages for a 540 mi^2 Australian basin (cyclonic storms)

	β			
Error in Q	*1*	*2*	*4*	*10*
5%	4	7	24	46
10%	2	4	10	14
15%	1	2	5	7

Source: Eagleson, P. S. (1967a).

4.8 FREQUENCY-DOMAIN ANALYSIS OF STATIONARY RANDOM SIGNALS

Let $X_1(t)$ represent a stationary random process in continuous time. It is unreasonable to define the Fourier transform of such a realization since any other realization would yield different results. From previous chapters, though, the ergodic autocovariance of a zero-mean process is given by

$$\psi_{11}(\tau) = \lim_{T \to \infty} \frac{1}{T} \int_{-T/2}^{T/2} X_1(t) X_1(t+\tau) \, dt. \tag{4.83}$$

The Wiener–Khintchine theorem states that for a purely random signal,

$$\psi_{11}(\tau) = \int_{-\infty}^{\infty} \Phi_{11}(\omega) e^{j\omega\tau} \, d\omega$$

$$= \int_{-\infty}^{\infty} \Phi_{11}(\omega) \cos \omega\tau \, d\omega,$$

where $\Phi_{11}(\omega)$ is the power-spectrum function.

For $\tau = 0$, Parseval's theorem implies

$$\psi_{11}(0) = \int_{-\infty}^{\infty} \Phi_{11}(\omega) \, d\omega, \tag{4.84}$$

which indicates that $\Phi_{11}(\omega)$ gives the frequency distribution of the mean square value of the function (for a zero-mean process this is exactly the variance, σ^2).

The autocovariance and the power spectrum are a Fourier-transform pair, so

$$\Phi_{11}(\omega) = \frac{1}{2\pi} \int_{-\infty}^{\infty} \psi_{11}(\tau) e^{-j\omega\tau} \, d\tau$$

$$= \frac{1}{2\pi} \int_{-\infty}^{\infty} \psi_{11}(\tau) \cos \omega\tau \, d\tau. \tag{4.85}$$

Normalizing the above by σ^2, yields

$$°\Phi_{11}(\omega) = \frac{\Phi_{11}(\omega)}{\sigma^2} = \frac{1}{2\pi} \int_{-\infty}^{\infty} \rho_{11}(\tau) e^{-j\omega\tau} d\tau, \tag{4.86}$$

where $\rho_{11}(\tau)$ is the autocorrelation of the random process; $°\Phi_{11}(\omega)$ is the spectral-density function of the process $X_1(t)$ and is only a normalized power spectrum.

Similar definitions exist for handling the cross-covariance of two random processes,

$$\psi_{12}(\tau) = \lim_{T \to \infty} \frac{1}{T} \int_{-T/2}^{T/2} X_1(t) X_2(t+\tau) dt$$

$$= \int_{-\infty}^{\infty} \Phi_{12}(\omega) e^{j\omega\tau} d\omega \tag{4.87}$$

and

$$\Phi_{12}(\omega) = \frac{1}{2\pi} \int_{-\infty}^{\infty} \psi_{12}(\tau) e^{-j\omega\tau} d\tau, \tag{4.88}$$

where $\Phi_{12}(\omega)$ is the cross-spectrum between $X_1(t)$ and $X_2(t)$.

By analogy with periodic and aperiodic functions, a stationary random process can be thought of as an infinite summation of sinusoids whose coefficients are considered to be numbers drawn at random from a multivariate normal distribution with zero mean and with variances and covariances that are a function of ω (Pierson, 1960).

The relationships previously derived for linear deterministic systems are still valid for random cases. Given

$$Q(t) = \int_{-\infty}^{\infty} h(\tau) I(t-\tau) d\tau, \tag{4.89}$$

where $I(t)$ and $Q(t)$ are stationary random processes,

$$\Phi_{QQ}(\omega) = 4\pi^2 |H(\omega)|^2 \Phi_{II}(\omega), \tag{4.90}$$

where

$$|H(\omega)|^2 = \bar{H}(\omega) \cdot H(\omega)$$

$$H(\omega) = \frac{1}{2\pi} \int_{-\infty}^{\infty} h(t) e^{-j\omega t} dt = |H(\omega)| e^{j\theta(\omega)},$$

and $|H(\omega)|$ and $\theta(\omega)$ are the gain and the phase of the system.

Similarly, the cross-power spectrum between input and output is given by

$$\Phi_{IQ}(\omega) = 2\pi H(\omega) \Phi_{II}(\omega). \tag{4.91}$$

Completely analogous results exist for the case of discrete-argument random processes. Using the sampling-theorem concepts, it is clear (assuming $\Delta t = 1$) that

$$\psi_{11}(k) = \int_{-\pi}^{\pi} \Phi_{11}(\omega) e^{j\omega k} \, d\omega \tag{4.92}$$

$$\Phi_{11}(\omega) = \frac{1}{2\pi} \sum_{k=-\infty}^{\infty} \psi_{11}(k) e^{-j\omega k}. \tag{4.93}$$

Equations (4.92) and (4.93) are useful in studying the behavior of some previously seen random processes in discrete time. For example, an uncorrelated discrete process (discrete white noise) has covariance

$$\psi_{11}(k) = \begin{cases} \sigma^2 & k = 0 \\ 0 & \text{otherwise} \end{cases}$$

leading to

$$\Phi_{11}(\omega) = \frac{\sigma^2}{2\pi}, \tag{4.94}$$

which implies that the power spectrum is constant over the interval $(0, 2\pi)$.

A first-order moving average MA(1) has autocovariance

$$\psi_{11}(k) = \begin{cases} \sigma^2 & k = 0 \\ -\theta_1 \cdot \sigma^2 / (1 + \theta_1^2) & |k| = 1 \\ 0 & \text{otherwise} \end{cases}$$

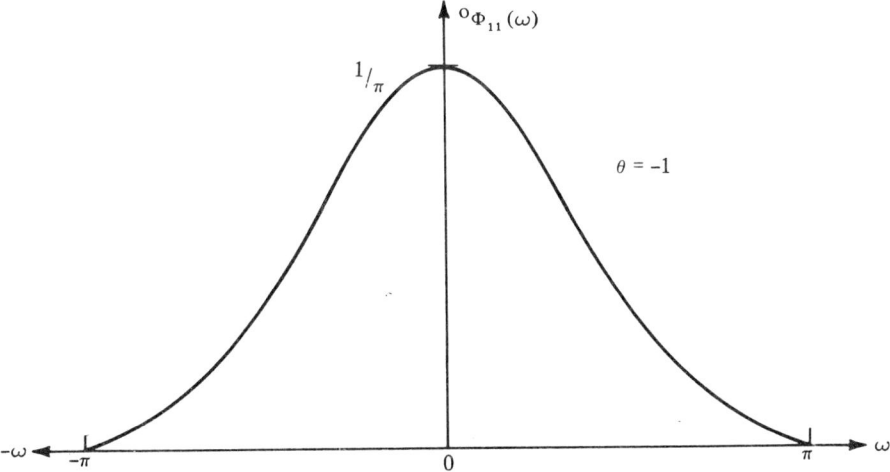

Figure 4.13 Spectral density for a low-frequency MA(1) process corresponding to Eq. (4.95).

leading to the power spectrum

$$\Phi_{11}(\omega) = \frac{\sigma^2}{2\pi}\left[1 - 2\frac{\theta_1}{(1+\theta_1^2)}\cos\omega\right]; \quad -\pi < \omega < \pi, \qquad (4.95)$$

where

$$\sigma^2 = (1+\theta_1^2)\sigma_a^2.$$

Figures 4.13 and 4.14 illustrate low- and high-frequency MA(1), corresponding to negative and positive values of θ, respectively.

The AR(1) model had autocovariance

$$\psi_{11}(k) = \sigma^2\rho^{|k|} \qquad k = (0, \pm 1, \pm 2, \ldots),$$

which using Eq. (4.93) leads to

$$\Phi_{11}(\omega) = \frac{\sigma^2}{2\pi}\left(1 + \sum_{k=1}^{\infty}\rho^k e^{-j\omega k} + \sum_{k=1}^{\infty}\rho^k e^{j\omega k}\right)$$

$$= \frac{\sigma^2}{2\pi}\left(1 + \frac{\rho e^{-j\omega}}{1-\rho e^{-j\omega}} + \frac{\rho e^{j\omega}}{1-\rho e^{j\omega}}\right).$$

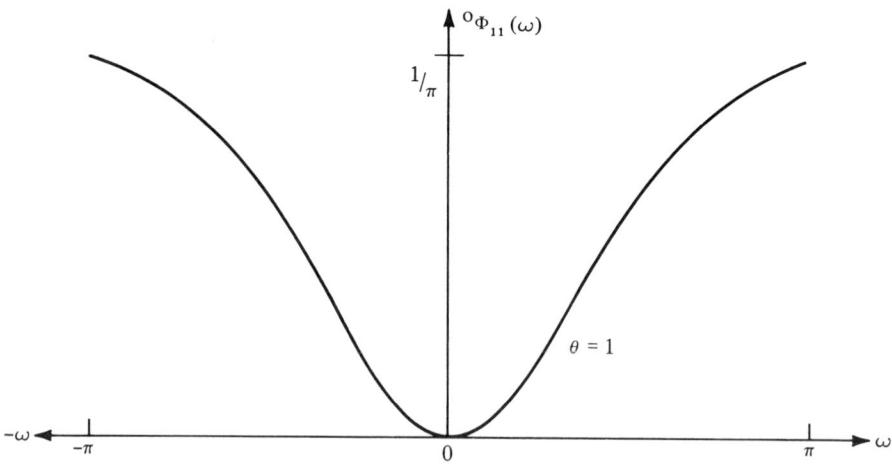

Figure 4.14 Spectral density for a high-frequency MA(1) process corresponding to Eq. (4.95).

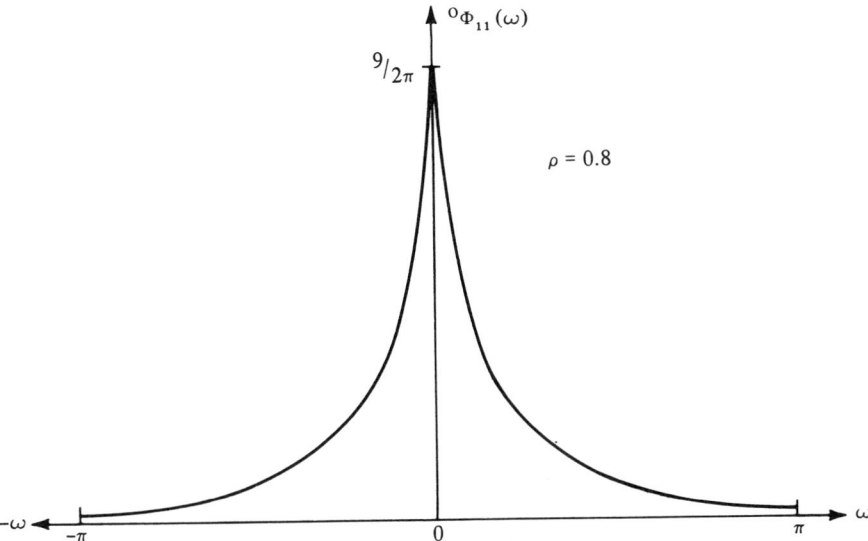

Figure 4.15 Spectral density for a low-frequency AR(1) process corresponding to Eq. (4.96).

This is equivalent to

$$\Phi_{11}(\omega) = \frac{\sigma^2}{2\pi}(1-\rho^2)/(1-2\rho\cos\omega + \rho^2); \quad -\pi < \omega < \pi, \quad (4.96)$$

where ρ is the lag-one autocorrelation and σ^2 is the process variance.

Equation 4.96 is illustrated in Figs. 4.15 and 4.16 for the low-frequency and high-frequency AR(1) models.

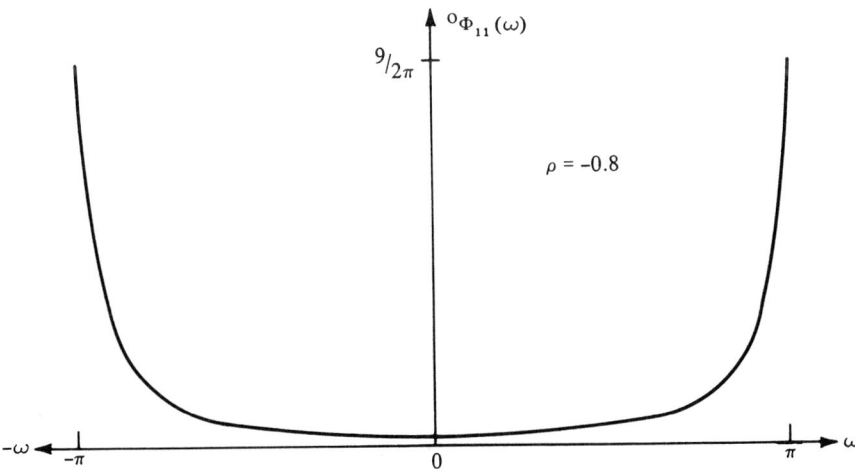

Figure 4.16 Spectral density for a high-frequency AR(1) process corresponding to Eq. (4.96).

4.9 ONE-SIDED SPECTRAL DENSITIES AND COHERENCE

Physical reality makes it difficult to deal with negative angular frequencies. To avoid this problem the one-sided spectral density is defined,

$$G_{11}(\omega) = 2\Phi_{11}(\omega) \qquad \text{for } \omega \geq 0 \qquad (4.97)$$

$$G_{12}(\omega) = 2\Phi_{12}(\omega) \qquad \text{for } \omega \geq 0 \qquad (4.98)$$

Note that since the autocovariance is an even function

$$G_{11}(\omega) = 2\Phi_{11}(\omega) = 2\left(\frac{1}{2\pi}\right) \int_{-\infty}^{\infty} \psi_{11}(\tau) e^{-j\omega\tau} d\tau$$

$$= 4\left(\frac{1}{2\pi}\right) \int_{0}^{\infty} \psi_{11}(\tau) \cos\omega\tau d\tau. \qquad (4.99)$$

The linear-systems relationships discussed in previous sections are valid using one-sided spectral distributions.

Since the cross-covariance $\psi_{12}(\tau)$ between two processes $X_1(t)$ and $X_2(t)$ is not an even function, the cross-spectrum defined in Eq. (4.88) is a complex function

$$G_{12}(\omega) = C_{12}(\omega) + jQ_{12}(\omega), \qquad (4.100)$$

where the real part $C_{12}(\omega)$ is called the cospectrum, a measure of the in-phase covariance, and $Q_{12}(\omega)$ is the quadrature spectrum, a measure of the out-of-phase covariance. The cospectrum $C_{12}(\omega)$ measures the contribution to the total cross-covariance of in-phase elements of $X_1(t)$ and $X_2(t)$ at frequency ω. The quadrature spectrum $Q_{12}(\omega)$ measures the contribution to total cross-covariance of elements of $X_1(t)$ and $X_2(t)$ at frequency ω, which are out of phase by $\pi/2$ (Granger and Hatanaka, 1964).

Using Eq. (4.100) it can be shown that the cross-covariance results from

$$\psi_{12}(\tau) = \int_{0}^{\infty} \left[C_{12}(\omega)\cos\omega\tau - Q_{12}(\omega)\sin\omega\tau \right] d\omega. \qquad (4.101)$$

Since $\psi_{12}(-\tau) = \psi_{21}(\tau)$, the cospectrum may be expressed as

$$C_{12}(\omega) = \frac{1}{2\pi} \int_{0}^{\infty} \left[\psi_{12}(\tau) + \psi_{21}(\tau) \right] \cos\omega\tau d\tau = C_{21}(-\omega) \qquad (4.102)$$

and

$$Q_{12}(\omega) = \frac{1}{2\pi} \int_{0}^{\infty} \left[\psi_{12}(\tau) - \psi_{21}(\tau) \right] \sin\omega\tau d\tau = -Q_{21}(\omega). \qquad (4.103)$$

Similarly,

$$C_{12}(\omega) = \frac{1}{2}[G_{12}(\omega) + G_{21}(\omega)] \tag{4.104}$$

$$Q_{12}(\omega) = \frac{-j}{2}[G_{12}(\omega) - G_{21}(\omega)]. \tag{4.105}$$

In complex polar form

$$G_{12}(\omega) = |G_{12}(\omega)|e^{j\theta_{12}(\omega)} \qquad 0 \le \omega < \infty, \tag{4.106}$$

where the phase function is

$$\theta_{12}(\omega) = \tan^{-1}\left[\frac{-Q_{12}(\omega)}{C_{12}(\omega)}\right]. \tag{4.107}$$

The real quantity, defined as coherence $\gamma_{12}^2(\omega)$, is a direct measure of the square of the correlation of the amplitudes of frequency ω in the processes $X_1(t)$ and $X_2(t)$:

$$\begin{aligned}\gamma_{12}^2(\omega) &= \frac{|G_{12}(\omega)|^2}{G_{11}(\omega)G_{22}(\omega)} \qquad 0 \le \gamma_{12}(\omega) \le 1 \\ &= \frac{|\Phi_{12}(\omega)|^2}{\Phi_{11}(\omega)\Phi_{22}(\omega)}.\end{aligned} \tag{4.108}$$

It should be noted that coherence measures the correlation of elements of the processes at particular frequencies. Important features of the coherence may in fact be masked in the time-domain cross-correlation function. Considering $X_1(t)$ to be the stationary input to a linear system and $X_2(t)$ the output of the system, it is easy to show, through Eqs. (4.90) and (4.91), that

$$\gamma_{12}^2(\omega) = \frac{|H(\omega)|^2 G_{11}^2(\omega)}{G_{11}(\omega)|H(\omega)|^2 G_{11}(\omega)} = 1. \tag{4.109}$$

For two first-order autoregressive processes Rodríguez-Iturbe and Siddiqui (1969) have shown that the coherence function is a constant independent of frequency:

$$\gamma_{12}^2(\omega) = \frac{[1 - \rho_{11}(1)\rho_{22}(1)]^2 \rho_{12}^2(0)}{[1 - \rho_{11}^2(1)][1 - \rho_{22}^2(1)]},$$

where $\rho_{11}(1)$ and $\rho_{22}(1)$ are the autocorrelation of lag one of the two processes and $\rho_{12}(0)$ is the zero-lag cross-correlation.

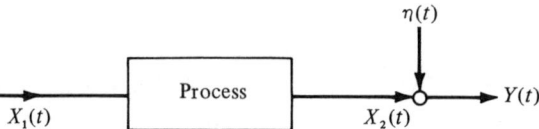

Figure 4.17 Linear system with noise in the measurement of the output (after Goodman et al., 1961).

The coherence then measures the linear relations between frequencies in input and output processes. It can be interpreted as the fractional portion of mean-square value of output $X_2(t)$ contributed by input $X_1(t)$ at frequency ω. Values other than one may be due to basic nonlinearities, extraneous noise in the system, or unknown inputs in the system.

Goodman et al. (1961) examined a single-input linear system with the assumption that there was noise in the measurement of the output (Fig. 4.17). Assume that $\eta(t)$ and $X_1(t)$ are statistically uncorrelated and that all three processes, $X_1(t)$, $Y(t)$, and $\eta(t)$, are stationary Gaussian noises. The effect of the disturbance $\eta(t)$ then appears in the coherence $\gamma^2_{x_1 y}(\omega)$, which is now in the form

$$\gamma^2_{x_1 y}(\omega) = \frac{1}{1 + \dfrac{G_{\eta\eta}(\omega)}{G_{x_2 x_2}(\omega)}}. \tag{4.110}$$

It is seen from Eq. (4.110) that the coherence decreases as the size of the disturbance increases.

Enochson (1964) considered a general case of noise in both input and output measuring devices. Assuming that a measured input $X(t)$ and a measured output $Y(t)$ are composed of true signals $u(t)$ and $v(t)$ and uncorrelated noise components $\eta(t)$ and $m(t)$, respectively, as shown in Fig. 4.18, then the "desired" coherence function is

$$\gamma^2_{xy}(\omega) = \frac{|G_{uv}(\omega)|^2}{G_{uu}(\omega)G_{vv}(\omega)}, \tag{4.111}$$

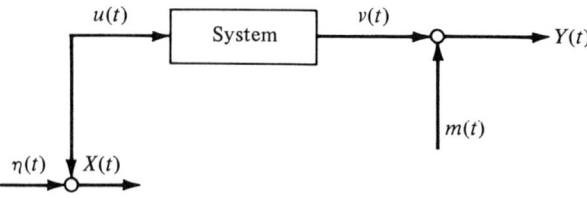

Figure 4.18 Linear system with noise in the measurement of the input and output (after Enochson, 1964).

but the measured coherence function will be

$$\gamma_{xy}^2(\omega) = \frac{|G_{xy}(\omega)|^2}{G_{xx}(\omega)G_{yy}(\omega)}$$

$$= \frac{|G_{uv}(\omega)|^2}{[G_{uu}(\omega)+G_{\eta\eta}(\omega)][G_{vv}(\omega)+G_{mm}(\omega)]}. \qquad (4.112)$$

Thus, theoretically, the measured coherence function will always be less than the desired coherence function.

An analogy can be drawn between coherence and correlation. The following discussion follows Rodríguez-Iturbe (1967).

In statistical analysis of real variables, the correlation coefficient between two variables X and Y with mean values of zero is defined as

$$\rho_{XY} = \frac{E[XY]}{(E[X^2]E[Y^2])^{1/2}}$$

$$= \frac{\text{cov}[X,Y]}{\sigma_x\sigma_y} \qquad (4.113)$$

$$= \frac{\psi_{XY}}{\sigma_X\sigma_Y},$$

where σ_X^2 and σ_Y^2 represent the variances of X and Y, respectively.

Similarly, if complex numbers X and Y are being considered, the square of the correlation coefficient becomes:

$$\rho_{XY} = \frac{E[X\overline{Y}]\overline{E}[X\overline{Y}]}{E[X\overline{X}]E[Y\overline{Y}]}, \qquad (4.114)$$

where the overbar denotes the complex conjugate of the term in question.

Rewriting Eq. (4.114) yields

$$\rho_{XY}^2 = \frac{E[X\overline{Y}]E[\overline{X}Y]}{E[X\overline{X}]E[Y\overline{Y}]} = \frac{|E[X\overline{Y}]|^2}{\sigma_X^2\sigma_Y^2} = \frac{|\psi_{XY}|^2}{\sigma_X^2\sigma_Y^2}. \qquad (4.115)$$

From Eq. (4.108) we can see that the coherence function may be thought of as the square of a correlation coefficient if we replace ψ_{XY} with $\Phi_{12}(\omega)$, σ_X^2 with $\Phi_{11}(\omega)$ and σ_Y^2 with $\Phi_{22}(\omega)$.

Following Cramer's representation of a stationary process with zero mean we have

$$X_1(t) = \int_{-\infty}^{\infty} e^{-i\omega t} dZ_1(\omega), \qquad (4.116)$$

where $dZ_1(\omega)$ approximates $Z_1(d\omega)$ and $Z_1(\omega)$ is an orthogonal set function with

$$E\left|dZ_1(\omega)\right|^2 = d\Phi_{11}(\omega) = \Phi_{11}(\omega)\,d\omega, \qquad (4.117)$$

and $\Phi_{11}(\omega)$ is the power spectrum of the process $X_1(t)$. Similarly,

$$E\left|dZ_2(\omega)\right|^2 = d\Phi_{22}(\omega) = \Phi_{22}(\omega)\,d\omega \qquad (4.118)$$

and

$$E\left[dZ_1(\omega)\cdot d\overline{Z}_2(\omega)\right] = d\Phi_{12}(\omega) = \Phi_{12}(\omega)\,d\omega. \qquad (4.119)$$

Comparing Eqs. (4.117), (4.118), and (4.119) with the expressions

$$\sigma_X^2 = E[\,X\overline{X}\,] = E|X|^2$$
$$\sigma_Y^2 = E[\,Y\overline{Y}\,] = E|Y|^2$$
$$\psi_{XY} = E[\,X\overline{Y}\,]$$

it may be seen that the coherence function can be interpreted as a correlation coefficient squared between the spectral variables $Z_1(\omega)$ and $Z_2(\omega)$ calculated at each frequency ω.

It should now be clear that it is often advantageous to study correlation problems in the frequency domain rather than in the time domain. Working in the frequency domain, any stationary series can be considered as a sum of components or frequency bands, each component being statistically independent of the others. One of the important things that the theory of stationary processes tells us is that not only is the component with center ω_j independent of all the other components of the process, but it is also independent of all components of another process except for the component centered on ω_j. In this manner, when the coherence between two time series is calculated, we look for correlations among them in a very small range of frequencies. On the other hand, with the cross-covariance function we are looking for correlations between the two processes, considering each one as a whole.

4.10 PARTIAL COHERENCE FUNCTIONS

Consider two real-value stationary processes $X(t)$ and $Y(t)$ and assume that the mean values are zero in order to simplify the notation. Define the residual Δy,

$$\Delta Y = Y(t) - \hat{Y}(t), \qquad (4.120)$$

where

$$\hat{Y}(t) = \frac{\psi_{yx}}{\psi_{xx}} X(t). \tag{4.121}$$

Consider now three real-value stationary random processes $X_1(t)$, $X_2(t)$, and $Y(t)$, where the mean values are assumed to be zero. One can define the partial correlation coefficient $\rho_{1y \cdot 2}$ by

$$\rho_{1y \cdot 2}^2 = \frac{\psi_{\Delta x_1 \Delta y}^2}{\psi_{\Delta x_1 \Delta x_1} \psi_{\Delta y \Delta y}} = \frac{\psi_{1y \cdot 2}^2}{\psi_{11 \cdot 2} \psi_{yy \cdot 2}}, \tag{4.122}$$

where

$$\psi_{\Delta x_1 \Delta x_1} = \psi_{11 \cdot 2} = \psi_{11} \left(1 - \rho_{21}^2 \right) \tag{4.123}$$

$$\psi_{\Delta y \Delta y} = \psi_{yy \cdot 2} = \psi_{yy} \left(1 - \rho_{2y}^2 \right) \tag{4.124}$$

$$\psi_{\Delta x_1 \Delta y} = \psi_{1y \cdot 2} = \psi_{1y} \left(1 - \frac{\psi_{12} \psi_{2y}}{\psi_{22} \psi_{1y}} \right). \tag{4.125}$$

Similarly to the partial correlation coefficient in the time domain, it is possible to define in the frequency domain a partial coherence function between $X_1(t)$ and $Y(t)$ with $X_2(t)$ removed at every t, by least-squares prediction, from $X_1(t)$ and $Y(t)$:

$$\gamma_{1y \cdot 2}^2 (\omega) = \frac{\left| \Phi_{1y \cdot 2} (\omega) \right|^2}{\Phi_{11 \cdot 2} (\omega) \Phi_{yy \cdot 2} (\omega)}$$

$$= \frac{\left| G_{1y \cdot 2} (\omega) \right|^2}{G_{11 \cdot 2} (\omega) G_{yy \cdot 2} (\omega)}. \tag{4.126}$$

The terms in Eq. (4.126) are called residual or partial spectra and are defined by

$$\Phi_{1y \cdot 2} (\omega) = \Phi_{1y} (\omega) \left[1 - \frac{\Phi_{12} (\omega) \Phi_{2y} (\omega)}{\Phi_{22} (\omega) \Phi_{1y} (\omega)} \right] \tag{4.127}$$

$$\Phi_{11 \cdot 2} (\omega) = \Phi_{11} (\omega) \left(1 - \gamma_{12}^2 (\omega) \right) \tag{4.128}$$

$$\Phi_{yy \cdot 2} (\omega) = \Phi_{yy} (\omega) \left(1 - \gamma_{2y}^2 (\omega) \right). \tag{4.129}$$

The proof that the partial coherence is nothing but an analogue of the partial correlation coefficient between the spectral variables (calculated at each frequency ω) can be carried out following the same procedure used for the total coherence.

The case of multiple processes is only a generalization of the three-variable case explained before. The partial coherence function between $X_1(t)$ and $Y(t)$ with $X_2(t)$, $X_3(t)$,..., $X_n(t)$ removed at every t, by least-squares prediction, from $X_1(t)$ and $Y(t)$, is defined by

$$\gamma_{1y\cdot 23\ldots n}^2(\omega) = \frac{\left|\Phi_{1y\cdot 23\ldots n}(\omega)\right|^2}{\Phi_{11\cdot 23\ldots n}(\omega)\cdot\Phi_{yy\cdot 23\ldots n}(\omega)}. \qquad (4.130)$$

The definition and calculation of the partial spectra of Eq. (4.127) has been done in matrix form by Goodman (1965), which is suitable for the use of high-speed digital computers.

Similarly to the development made for the partial coherence function it is possible to define the partial phase function between $X_1(t)$ and $Y(t)$ with $X_2(t)$, $X_3(t)$,..., $X_n(t)$ removed at every t, by least-squares prediction, from $X_1(t)$ and $Y(t)$,

$$\theta_{1y\cdot 23\ldots n}(\omega) = \tan^{-1}\left(\frac{\text{imaginary part of } \Phi_{1y\cdot 23\ldots n}(\omega)}{\text{real part of } \Phi_{1y\cdot 23\ldots n}(\omega)}\right). \qquad (4.131)$$

When more than two variables are being considered, the partial coherence function, rather than the ordinary coherence, gives a quantitative indication of the degree of linear dependence between the variables. An example of erroneous high coherence is shown in Fig. 4.19.

Assume that a value of the coherence function near unity is computed between the variables $X_1(t)$ and $Y(t)$. You would be inclined to believe there is linear system relating these two variables. Suppose there is a third variable $X_2(t)$ that is highly coherent with $X_1(t)$ and also passes through a linear system to make up $Y(t)$. In this situation the high coherence computed

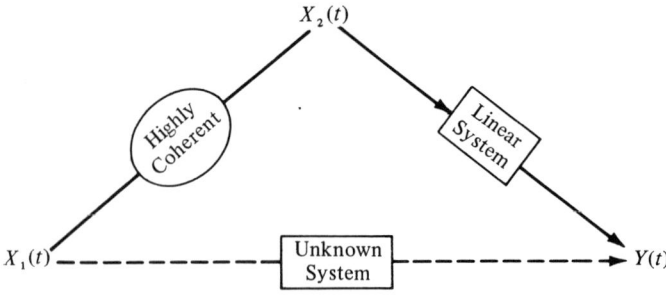

Figure 4.19   Example of erroneous high coherence (after Bendat and Piersol, 1966).

between $X_1(t)$ and $Y(t)$ might be due only to the fact that $X_2(t)$ is highly coherent with $X_1(t)$. If this is in fact the situation, the partial coherence between $X_1(t)$ and $Y(t)$ will be very low.

On the other hand, the opposite situation can exist. If two uncorrelated inputs $X_1(t)$ and $X_2(t)$ pass through existing linear systems to make up the output $Y(t)$, the coherence functions $\gamma_{1y}^2(\omega)$ and $\gamma_{2y}^2(\omega)$ will appear to be less than unity, since a contribution due to the other input will exist; this will appear as noise. If the partial coherences are computed, the effects of the other input will be subtracted and the true coherence will be obtained.

4.11 MULTIPLE-INPUT LINEAR SYSTEMS

Constant-parameter linear systems responding to multiple inputs from stationary random processes will now be considered. It will be assumed that N inputs are occurring with a single output being measured (Fig. 4.20). The output may be considered as the sum of the N partial, not measured, outputs $Y_i(t)$, $i = 1, 2, \ldots, N$. That is,

$$Y(t) = \sum_{i=1}^{N} Y_i(t), \tag{4.132}$$

where $Y_i(t)$ is defined as that part of the output produced by the ith input when all the other inputs are zero. Such a model is a reasonable conceptualization of a river basin with spatially distributed properties.

The cross-spectral relations between the inputs and the output can be expressed concisely with matrix notation. The following formulation of results appears in Enochson (1964) and in Bendat and Piersol (1966).

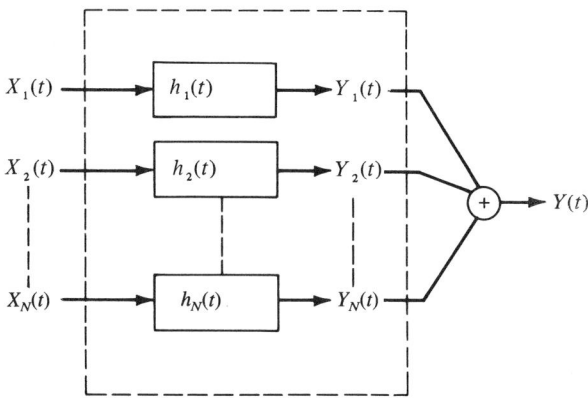

Figure 4.20 Multiple-input linear system.

First define an N-dimensional input vector

$$\mathbf{X}(t) = \left[X_1(t), X_2(t), \ldots, X_N(t) \right].$$ (4.133)

Let $\mathbf{H}(\omega)$ be an N-dimensional frequency-response–function vector

$$\mathbf{H}(\omega) = \left[H_1(\omega), H_2(\omega), \ldots, H_N(\omega) \right].$$ (4.134)

Next, define an N-dimensional cross-spectrum vector of the output $Y(t)$ with the inputs $X_i(t)$,

$$\mathbf{\Phi}_{xy}(\omega) = \left[\Phi_{1y}(\omega), \Phi_{2y}(\omega), \ldots, \Phi_{Ny}(\omega) \right],$$ (4.135)

where

$$\Phi_{iy}(\omega) = \Phi_{x_i y}(\omega) \qquad i = 1, 2, \ldots, N.$$ (4.136)

Finally, define the $N \times N$ cross-spectral matrix of the inputs $X_i(t)$:

$$\mathbf{\Phi}_{xx}(\omega) = \begin{bmatrix} \Phi_{11}(\omega) & \Phi_{12}(\omega) & \cdots & \Phi_{1N}(\omega) \\ \Phi_{21}(\omega) & \Phi_{22}(\omega) & \cdots & \Phi_{2N}(\omega) \\ \vdots & & & \vdots \\ \Phi_{N1}(\omega) & \Phi_{N2}(\omega) & \cdots & \Phi_{NN}(\omega) \end{bmatrix},$$ (4.137)

where

$$\Phi_{ij}(\omega) = \Phi_{x_i x_j}(\omega) \qquad i, j = 1, 2, \ldots, N.$$ (4.138)

The fundamental equation for multiple-input, constant-parameter linear systems can be written as

$$\mathbf{\Phi}_{yy}(\omega) = \mathbf{H}(\omega) \mathbf{\Phi}_{xx}(\omega) \overline{\mathbf{H}}^T(\omega),$$ (4.139)

where $\overline{\mathbf{H}}^T(\omega)$ denotes the complex-conjugate transpose vector of $\mathbf{H}(\omega)$.

The basic equation, which gives the transfer functions $H_j(\omega)$ in the case of multiple correlated inputs, is

$$\mathbf{\Phi}_{xy}(\omega) = \mathbf{\Phi}_{xx}(\omega) \mathbf{H}^T(\omega).$$ (4.140)

Equation (4.140) may be rewritten as the system of equations

$$\Phi_{iy}(\omega) = \sum_{j=1}^{N} H_j(\omega) \Phi_{ij}(\omega).$$ (4.141)

Solving Eq. (4.140) for the transposed row vector $[\mathbf{H}^T(\omega)]$ results in

$$\mathbf{H}^T(\omega) = \Phi_{xx}^{-1}(\omega)\Phi_{xy}^T(\omega). \tag{4.142}$$

Equation (4.142) gives each $H_i(\omega)$ as a function of the input–output cross-spectrum and holds whether or not the inputs are correlated.

The solution of Eq. (4.142) has been presented by Goodman (1965) in the form

$$\left.\begin{aligned}
H_1(\omega) &= \frac{\Phi_{1y \cdot 234\ldots N}(\omega)}{\Phi_{11 \cdot 234\ldots N}(\omega)} \\
&\vdots \\
H_N(\omega) &= \frac{\Phi_{Ny \cdot 1234\ldots N-1}(\omega)}{\Phi_{NN \cdot 1234\ldots N-1}(\omega)}
\end{aligned}\right\}. \tag{4.143}$$

4.12 ESTIMATION OF THE SPECTRAL DENSITY

It would seem logical that given an estimate of the correlation function of a discretely sampled process, the estimated spectral density should be its Fourier transform

$$\hat{\Phi}(\omega) = \frac{1}{2\pi}\left(\hat{\psi}(0) + 2\sum_{k=1}^{N}\hat{\psi}(k)\cos\omega k\right), \tag{4.144}$$

where the symbol "$\hat{\ }$" represents an estimate. The above estimator converges to the true spectral density as N approaches infinity, but the variance of the estimate does not decrease. The estimator is not consistent. Consistency is usually achieved by weighting the estimates of the autocorrelation function

$$\hat{\Phi}(\omega_j) = \frac{1}{2\pi}\left(\lambda_0\hat{\psi}(0) + 2\sum_{k=1}^{M}\lambda_k\hat{\psi}(k)\cos\omega_j k\right), \tag{4.145}$$

where $\omega_j = j\pi/M$; $|j| = 0,\ldots, M$; and $\{\lambda_k\}$ are a set of weights called the lag window (Chatfield, 1975). There are several available expressions for the smoothing weights. The Tukey window is given by

$$\lambda_k = \frac{1}{2}\left(1 + \cos\frac{k\pi}{M}\right) \qquad k = 0,1,\ldots, M. \tag{4.146}$$

The Parzen window takes the form

$$\lambda_k = \begin{cases} 1 - 6\left(\dfrac{k}{M}\right)^2 + 6\left(\dfrac{k}{M}\right)^3 & 0 \le k \le M/2 \\ 2(1 - k/M)^3 & M/2 \le k \le M. \end{cases} \tag{4.147}$$

Table 4.3
Characteristics of the annual data for the Wolf and Fox Rivers used in Figs. 4.21 and 4.22.

	Basin area (mi^2)	Mean annual discharge cfs	Period of record	Coefficient of variation	First autocorrelation coefficient
Wolf	2240	1708	1896–1957	0.278	0.401
Fox	1430	1104	1898–1957	0.225	0.404

Source: Yevjevich (1963).

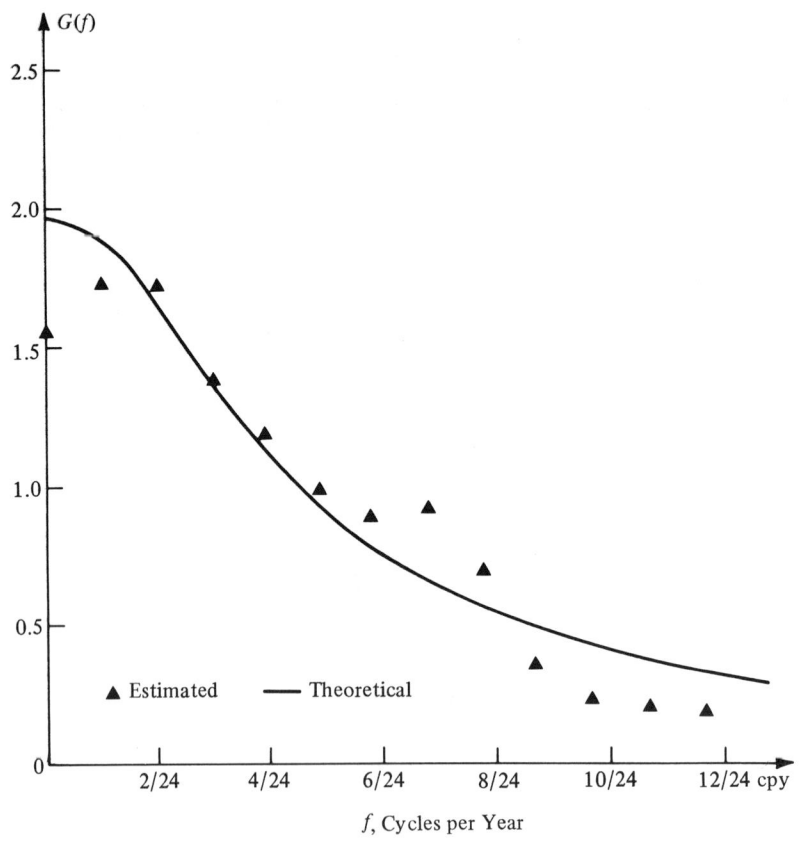

Figure 4.21 Power spectrum of the Wolf River's annual standardized flows (from Rodríguez-Iturbe and Siddiqui, 1969).

The selection of M affects the variance and resolution of the estimator. Small values of M lead to small variance but biased estimates of the spectral density. Large values of M lead to erratic behavior of the estimator, large variance, but increased resolution of frequencies. Values are usually taken in the range $1/20 < M/N < 1/3$, where N is the number of data points.

To estimate the cross-spectrum from discrete estimates of the cross-covariance, we use the discrete analogs of Eqs. (4.102) and (4.103),

$$\hat{C}_{12}(\omega_j) = \frac{\lambda_0}{4\pi}(\hat{\psi}_{12}(0) + \hat{\psi}_{21}(0))$$
$$+ \frac{1}{2\pi} \sum_{k=1}^{M} \lambda_k(\hat{\psi}_{12}(k) + \hat{\psi}_{21}(k))\cos\omega_j k \qquad (4.148)$$

$$\hat{Q}_{12}(\omega_j) = \frac{1}{2\pi} \sum_{k=1}^{M} \lambda_k(\hat{\psi}_{12}(k) - \hat{\psi}_{21}(k))\sin\omega_j k, \qquad (4.149)$$

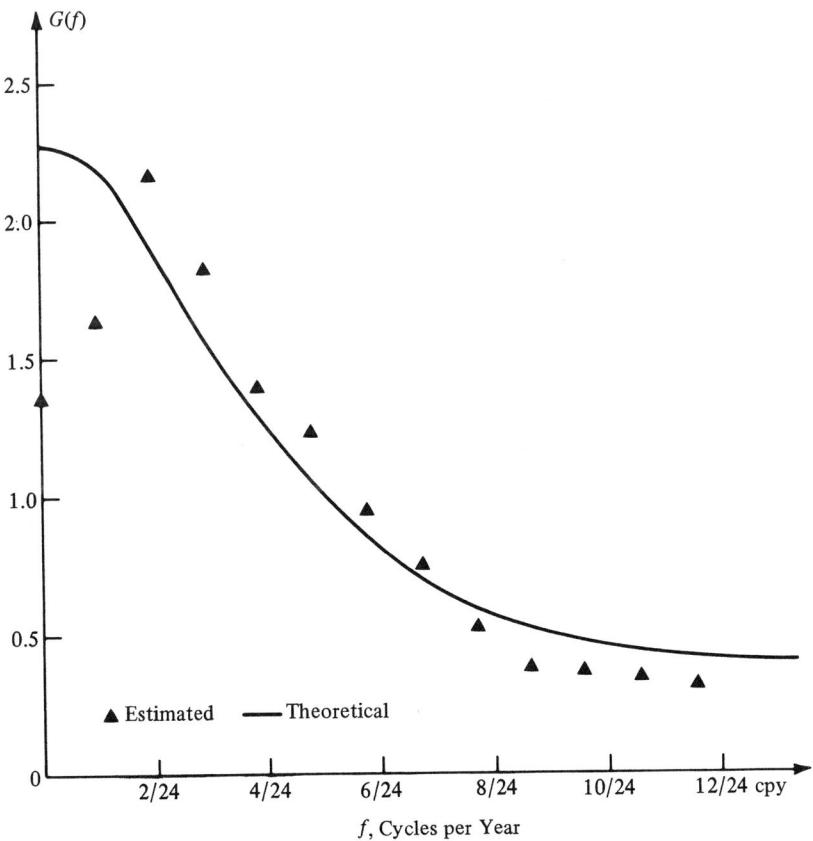

Figure 4.22 Power spectrum of the Fox River's annual standardized flows (from Rodriguez-Iturbe and Siddiqui, 1969).

where $\omega_j = j\pi/M$; $j = 0,\ldots, M$ and the weights, λ_k, are given by Eq. (4.146) or (4.147).

The statistical properties of the spectral estimators are well defined but outside the scope of this work. The reader is referred to Granger and Hatanaka (1964), Bendat and Piersol (1966), and Jenkins and Watts (1968).

The estimation of the power spectra can be illustrated with an example from Rodríguez-Iturbe and Siddiqui (1969). The power spectra of annual standardized flows of the Wolf River at New London, Wisconsin, and the Fox River at Berlin, Wisconsin, were estimated using the Tukey window and a value of M equal to 12. The main characteristics of annual data used in the computation are given in Table 4.3 (from Yevjevich, 1963).

Figures 4.21 and 4.22 show the estimated one-sided spectra (triangles) and in comparison with the theoretical AR(1) spectra, (Eq. 4.96). The frequencies are expressed in cycles per year, since two example points are needed to detect a period and annual data is being used, the highest frequency corresponds to 0.5 cycle/year. The autocovariance was estimated for up to 12 lags in both cases. The reasonable good fit of the AR(1) model to the computed spectra can be used as evidence in a model-identification exercise.

4.13 MODEL INFERENCE USING FREQUENCY-DOMAIN ANALYSIS: A CASE STUDY (after Rodríguez-Iturbe and Nordin, 1968, 1974)

Rodriguez-Iturbe and Nordin (1968, 1974) used the techniques of frequency-domain analysis to study and model the relationship between monthly streamflow and monthly sediment transport in the Rio Grande in the state of New Mexico, United States.

Table 4.4
Summary of the Rio Grande drainage basin characteristics

Station title	Drainage area km^2	Average water discharge m^3/s	Average sediment discharge $metric\ tons/mo$	Median diameter of bed material mm	Approximate channel slope
Otowi	37,000	33.6	214,000	0.45	0.0010
Bernalillo	44,800	28.5	343,000	0.30	0.0008
Bernardo	49,800	24.3	239,000	0.22	0.0006
San Marcial	71,700	23.2	320,000	0.14	0.0004
Major tributaries					
Galisteo	1,650	0.27			
Jemez	2,690	1.30			
Rio Puerco	16,000	1.71			
Rio Salado	3,570	0.39			

Source: Rodríguez-Iturbe and Nordin (1968).

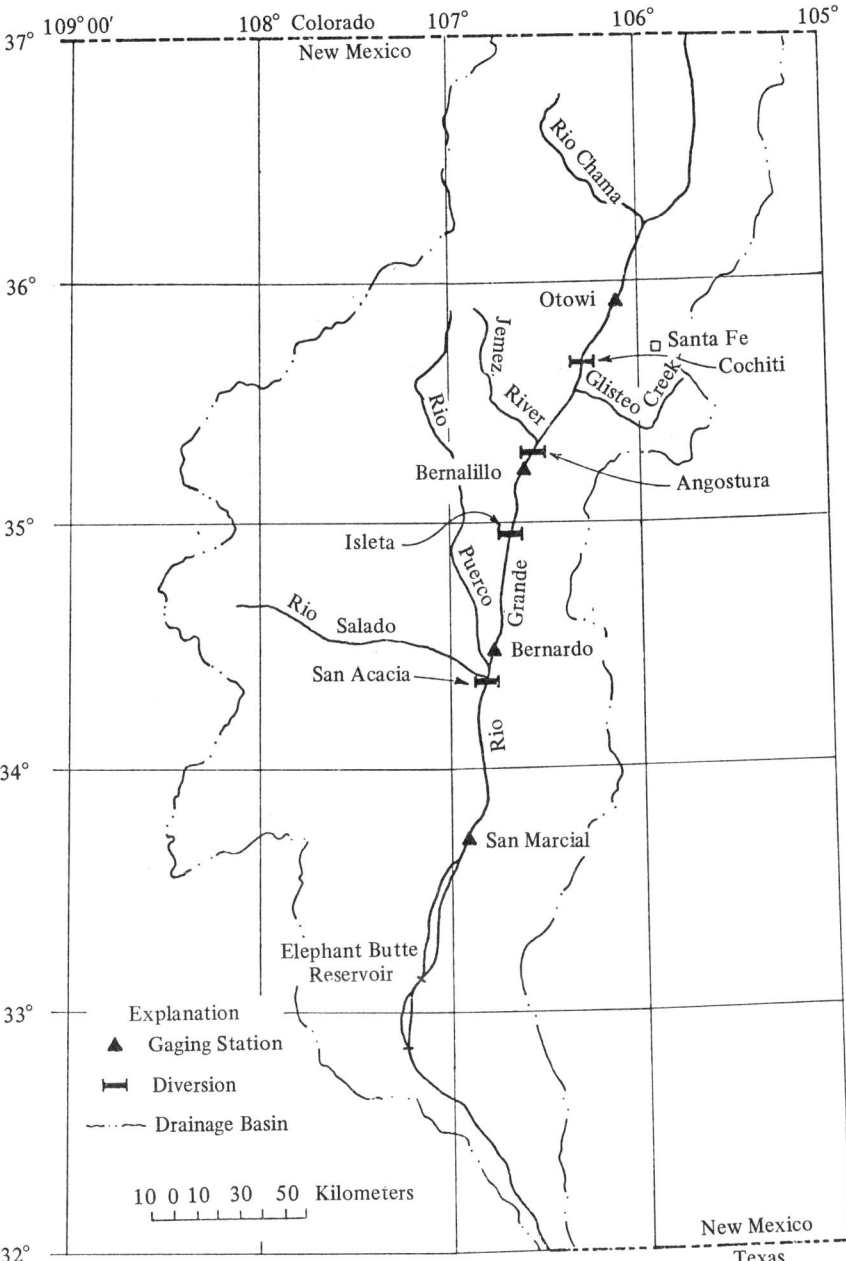

Figure 4.23 Rio Grande drainage basin in New Mexico, U.S.A. (from Rodríguez-Iturbe and Nordin, 1968).

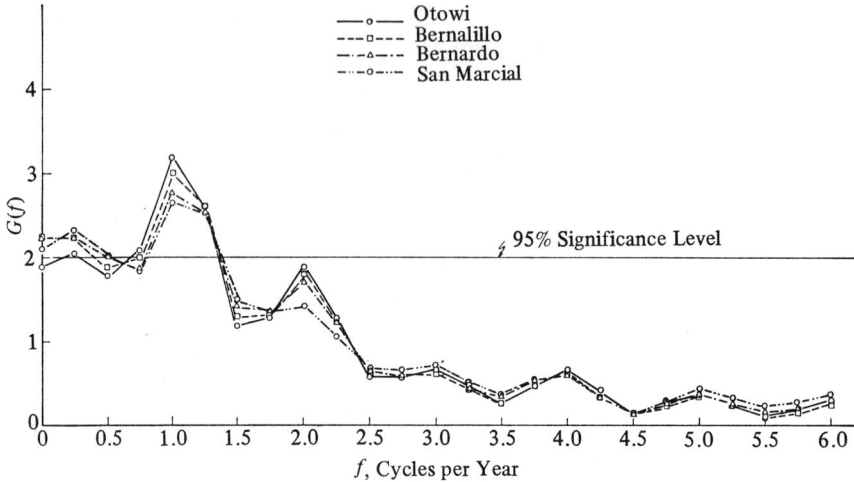

Figure 4.24 Spectra of the Rio Grande series of monthly water discharges (from Rodríguez-Iturbe and Nordin, 1968).

Initially, monthly records of streamflow and suspended sediment at four locations were studied. The locations (Otowi, Bernalillo, San Acacia, and San Marcial) are shown in Fig. 4.23 and a summary of the basic characteristics is presented in Table 4.4. The power spectra of the water and sediment series are shown in Figs. 4.24 and 4.25. From the power spectra, the cross-power spectrum at each station between streamflow and sediment, and coherence

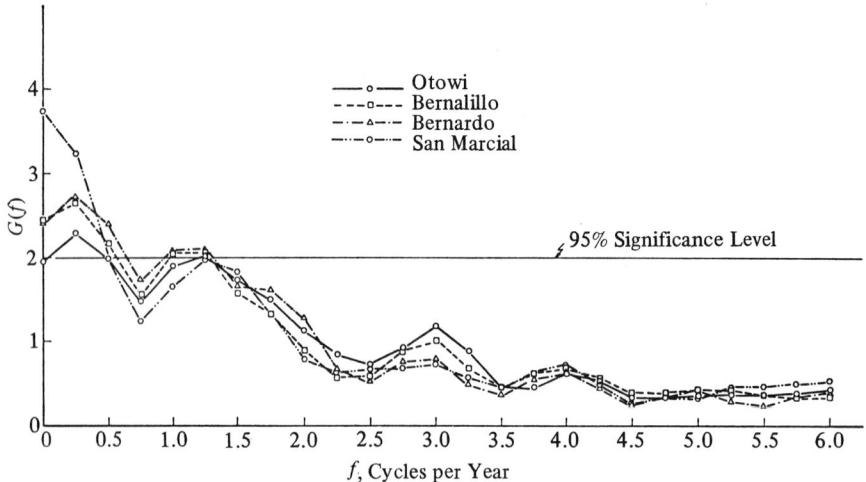

Figure 4.25 Spectra of the Rio Grande series of monthly sediment discharges (from Rodríguez-Iturbe and Nordin, 1968).

analysis between each series, the following main points resulted:

1. The annual and semi-annual cycles explained up to 25% of the variance of the streamflow record, leaving a large portion of variance unexplained by the dominating cyclic components.

2. The power of annual cycles decreases downstream for the streamflow record, a reflection of irrigation withdrawals and thunderstorm activity.

3. The power spectra of the sediment series showed a much less significant influence of the annual cycles, at most explaining 6% of total variance anywhere. No trends were detected.

4. The coherence between sediment and streamflow was significant for all except the most downstream station. Coherence was high, particularly at the low frequencies, at least up to 1 cycle per year (1 cpy) where it was clearly significant, except at San Marcial.

5. The cospectra reinforced that there was high correlation at zero-phase lag between monthly streamflow and sediment series for low frequencies (less than 2 cpy). The cospectrum peaks at 1 cpy, emphasizing the importance of this frequency in correlation analysis.

6. Phase-function diagrams indicated that at Otowi water and sediment were in phase but at Bernalillo the annual cycle of water leads that of sediment by about 18 days (0.6 month). This is computed by using the value of the phase at 1 cpy, 0.31 rad. This is multiplied by 12 months and divided by 2π rad to obtain a 0.6-month phase lag.

Given the dominance of the annual and semi-annual cycles, Rodríguez-Iturbe and Nordin studied the water and sediment series of the Otowi and Bernalillo stations after subtracting the monthly mean values and standardizing the series by dividing by the monthly standard deviations. These will be referred to as standardized series or anomalies series.

The power spectra of the monthly water and sediments standardized series are given in Figs. 4.26 and 4.27, respectively. The main features are the continuing dominance of the low frequencies (below 0.25 cpm) particularly in the water spectra. Shown in Fig. 4.26 is the spectrum of a lag-one autoregressive model fitted to the monthly water series. As previously seen, that spectrum is given by Eq. (4.96). The fit is very reasonable, particularly at the important low frequencies.

Figure 4.28 shows the coherence function at Bernalillo and Otowi, which is always less than one and which drops quickly for frequencies larger than 0.300 cpm. The system contains predictable fluctuations in the longer periods, especially if they are greater than 4 months (0.250 cpm). For frequencies ranging from 0.300 to 0.500 cpm, corresponding to periods between 3.5 and 2 months, the coherence is less than 0.56, which corresponds to the 95% level

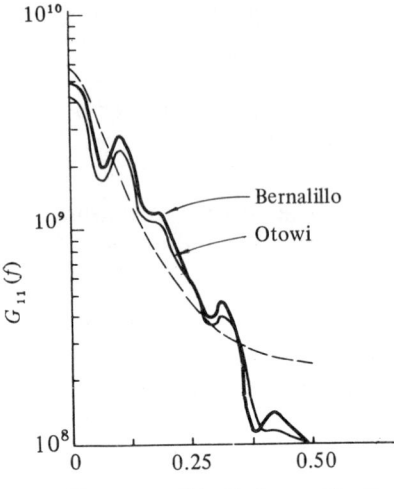

Figure 4.26 Spectra of the Rio Grande standardized series of monthly water discharges. The dashed line is the theoretical spectra for a first-order autoregressive process with parameters estimated from the Bernalillo data (from Rodríguez-Iturbe and Nordin, 1974).

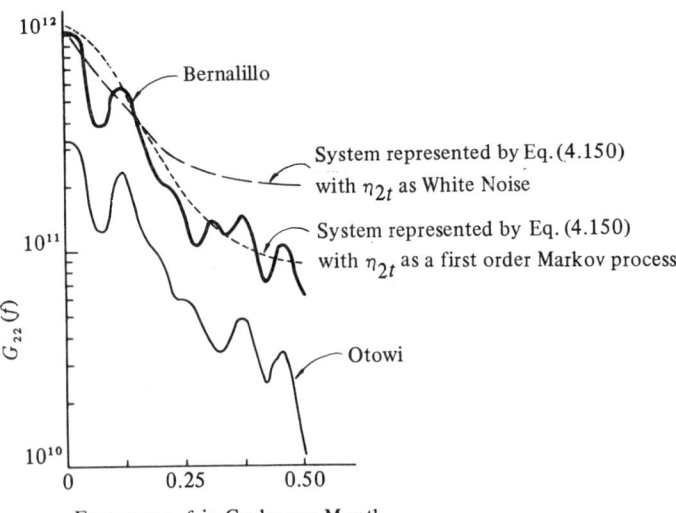

Figure 4.27 Spectra of the Rio Grande standardized series of monthly sediment discharges. The dashed lines are theoretical spectra for a second-order linear system with sediment as output and water as input (from Rodríguez-Iturbe and Nordin, 1974).

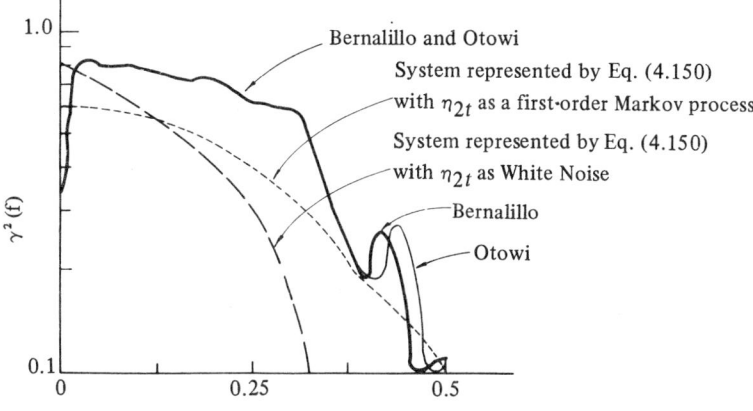

1.0

Bernalillo and Otowi

System represented by Eq. (4.150)
with η_{2t} as a first-order Markov process

System represented by Eq. (4.150)
with η_{2t} as White Noise

Bernalillo

Otowi

$\gamma^2(f)$

0.1

0   0.25   0.5

Frequency, f, in Cycles per Month

Figure 4.28 Coherence functions between the standardized series of water and sediment discharges. The dashed lines are theoretical coherence functions for a second-order linear system with parameters estimated from the Bernalillo data (from Rodríguez-Iturbe and Nordin, 1974).

below which $\gamma^2(\omega)$ is not significantly different from zero, as given by Granger and Hatanaka (1964).

The phase lag between the different frequency components is shown in Fig. 4.29. The standardized water series leads the series of sediment in all the frequency ranges, but the magnitude of the lead decreases with increasing frequency of the fluctuations. For example, the value of the phase for 1/48 cpm is near 175 degrees, which corresponds to $(48 \times 175)/360 = 23$ months. On the other hand, at a frequency of 1/4 cpm, the phase is 156 degrees, which corresponds to 1.7 months. This means that a peak in an oscillation of water anomalies with about a 4-month period is followed about 1.7 months later by a peak in the corresponding oscillation of sediment anomalies. It should be noted that if the coherence is small at frequency f, both $C(f)$ and $Q(f)$ will be small, so the estimate of $Q(f)/C(f)$ is likely to have very large variance. Thus, the points in the phase diagram corresponding to frequencies with low coherence will contain less information than those with high coherence. For this reason, there is not much confidence in the values of $\theta(f)$ for frequencies larger than $f = 0.300$ cpm. Nevertheless, a drop of the phase spectrum is clearly observed.

A great deal might be learned about the relation between the series X_1 and X_2 from the gain function $|H(f)|$, which is similar to a regression coefficient of series X_2 on series X_1 at each frequency f. Figure 4.30 shows the gain function between water and sediment anomalies at Otowi and Bernalillo. The gain estimates for frequencies larger than 0.300 cpm are unreliable because of the

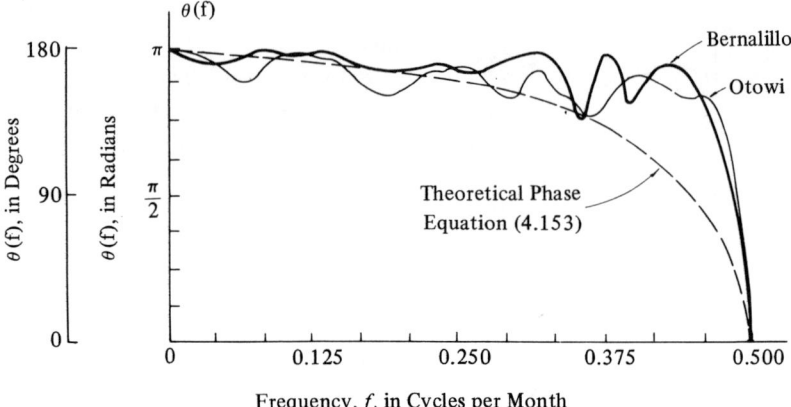

Figure 4.29 Phase diagrams from cross-spectral analysis for the standardized series of water and sediment discharges. The dashed line is the theoretical phase for a second-order linear system with parameters estimated from the Bernalillo data (from Rodríguez-Iturbe and Nordin, 1974).

low coherence existing at those frequencies (Bendat and Piersol, 1966). Nevertheless, a sharp decrease in the gain is observed in the range 0.300 to 0.500 cpm.

For physical reasons, feedback in the water and sediment system is not considered here because the deviation from the mean of the water series at a given month is assumed not to be affected by present or past values of

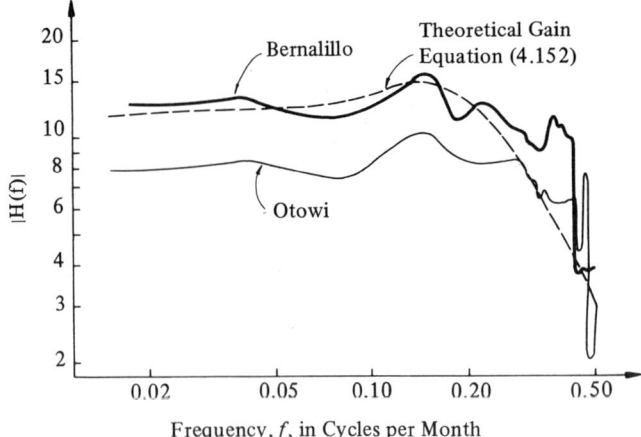

Figure 4.30 Gain functions between the standardized series of water and sediment discharges. The dashed line is theoretical gain function for a second-order linear system with parameters estimated from the Bernalillo data (from Rodríguez-Iturbe and Nordin, 1974).

sediment anomalies. On the other hand, sediment anomalies should be a function of water anomalies and the internal state of the system.

To model the behavior presented in Fig. 4.26 through 4.30, Rodríguez-Iturbe and Nordin (1974) hypothesized the following relationships:

$$x_{2t} = x_t + \eta_{2t}$$
$$x_t = ax_{t-1} + bx_{t-2} + cx_{1t}$$
$$x_{1t} = dx_{1t-1} + \eta_{1t},$$

(4.150)

where x_{2t} is the standardized sediment series and x_{1t} the standardized water series. We will at first assume η_{1t} and η_{2t} to be uncorrelated processes. The state of the system is x_t, and a, b, c, and d are constants.

The theoretical frequency-response function of the second-order linear system described by Eq. (4.150) can be found using Z-transforms (Jenkins and Watts, 1969) and is given by Rodríguez-Iturbe and Nordin (1974) as

$$H(f) = \frac{c}{1 - a\exp(-i2\pi f) - b\exp(-i4\pi f)}.$$

(4.151)

Since $H(f) = |H(f)|\exp[-i\theta(f)]$, one can obtain the theoretical gain and phase of the system:

$$|H(f)| = \frac{c}{\left[1 + a^2 + b^2 - 2b\cos 4\pi f + (2ab - 2a)\cos 2\pi f\right]^{1/2}}$$

(4.152)

$$\theta(f) = \arctan\frac{\sin 2\pi f \cdot (a + 2b\cos 2\pi f)}{1 + b + \cos 2\pi f \cdot (-a - 2b \cdot \cos 2\pi f)}.$$

(4.153)

To obtain the theoretical coherence, use is made of Eq. (4.110)—for the coherence of a linear system when there is noise in the measurement of the output—and Eq. (4.90)—the definition of the power spectra of the output of a linear system.

The theoretical functions obtained with $a = 0.40$, $b = -0.20$, $c = 7.0$, and $d = 0.668$ for Bernalillo are shown with long dashes in Figs. 4.26 through 4.30. In general the fit is good, with the exception of the coherence and the sediment (output) spectrum. Because $G_{x_2x_2}(\omega)$ and $\gamma^2_{x_2x_2}(\omega)$ are directly dependent on $G_{\eta\eta}(\omega)$, the previous result suggests that it is not correct to assume that $\eta_2(t)$ behaves as uncorrelated noise, and furthermore, that a better fit would be obtained if $\eta_2(t)$ is assumed to be a process whose spectrum decreases with frequency. Thus, a first-order Markov model was assumed by Rodríguez-Iturbe and Nordin (1971) for $\eta_2(t)$:

$$\eta_{2t} = 0.4\eta_{2t-1} + \lambda_t,$$

(4.154)

which would not much affect the values of $G_{x_2x_2}(\omega)$ and $\gamma^2_{x_1x_2}(\omega)$ at the low frequencies. The new results for $G_{x_2x_2}(\omega)$ and $\gamma^2_{x_1x_2}(\omega)$ for the station at Bernalillo are shown with short dashes in Figs. 4.27 and 4.28. The fit is greatly improved especially for the high-frequency range.

CHAPTER 5

Long-Term Persistence in Hydrologic Modeling

5.1 INTRODUCTION

One criticism of the ARIMA models, described in Chapter 2, is that it is quite common to simulate streamflow and precipitation series lacking droughts and floods of the magnitude present in historic sequences. The observation that precipitation can be very large or small and that a period of low or high flows can be very long has been designated as the "Noah and Joseph effects" by Mandelbrot and Wallis (1968). These authors note that if an independent or Markovian process is chosen to fit the other aspects of precipitation, it may greatly underestimate the duration of the longest drought. Such processes may be modified by considering more durable aftereffects (for example, through multiple-lag models). However, when the sample duration is further increased unexpected long droughts will again be observed. This shows that the adjusted model attributed a special significance to the sample length available; as the sample length increases the model must be changed.

A possible argument is that Markovian and multiple-lag models fail to exhibit the Noah and Joseph effect when the fit of the independent-component probability distribution is poor. Indeed, this distribution has a large influence in the extreme flow peaks generated by the process, but problems arise even if this component is adequately fitted. Even more important is the fact that although the flood peaks are highly dependent on the distribution of the independent component, other statistics such as the largest-run sum and the range of cumulative departures from the mean are almost independent of the distribution of the independent component, as pointed out be Feller (1951), Thomas and Fiering (1962), and Rodríguez-Iturbe et al. (1971). Therefore, the failure of the Markovian model to reproduce extreme values is not attributable only to the random component. Although there has not been much study about the Noah and Joseph phenomenon as such, the behavior of a related statistic, namely the average range of cumulative departures from the mean, has been an area of active research in hydrology.

5.2 THE HURST PHENOMENON

While studying the design capacity of reservoirs, H. E. Hurst (1951) observed an unexpected behavior of natural time series, which has become known as the Hurst phenomenon.

In a discrete-time series, the range of cumulative departures from the mean value is defined as

$$R(n) = \operatorname{Sup} S(i) - \operatorname{Inf} S(j)$$
$$i \in (0, 1, \ldots, n) \qquad j \in (0, 1, \ldots, n), \tag{5.1}$$

where Sup and Inf are operators defining the largest positive and negative values, respectively, and where

$$S(k) = \sum_{i=1}^{k} \{ X_i - E(X_i) \} \qquad k = 1, 2, \ldots, n. \tag{5.2}$$

The process under consideration (e.g., annual streamflows) is X_i, and n is the sample size. The adjusted range is defined as

$$R^*(n) = \operatorname{Sup} S^*(i) - \operatorname{Inf} S^*(j)$$
$$i \in (0, 1, \ldots, n) \qquad j \in (0, 1, \ldots, n), \tag{5.3}$$

where

$$S^*(k) = \sum_{i=1}^{k} X_i - \frac{k}{n} \sum_{i=1}^{n} X_i, \qquad k = 1, 2, \ldots, n. \tag{5.4}$$

Figure 5.1 shows the accumulated flows and the adjusted range defined by Eq. (5.3). As shown by Borgman and Amorocho (1970), it is possible to give an intuitive meaning to the adjusted range. Think of X_i as the water flow into a reservoir in the ith year and $\sum_{i=1}^{n} X_i$ as the total water inflow in n years. If $(1/n)\sum_{i=1}^{n} X_i$ units are released each year, then $(k/n)\sum_{i=1}^{n} X_i$ will be the amount released in the first k years and $S^*(k)$ is the surplus or deficit relative to the amount released during the kth year and to the starting storage of the reservoir. The difference between the largest surplus and the greatest deficit gives the capacity that a reservoir must have to maintain a constant release equal to the mean of the river without overflows or deficits during the n year period. This concept, as a practical idea for the design of a reservoir, is not of much use, since it is necessary to know the flows that will occur in the next n years. However, the statistical behavior of either $R(n)$ or $R^*(n)$ provides insight into the range of volumes that should be maintained. Hurst and colleagues (1951, 1956, 1965) investigated the behavior of the statistic $R^*(n)/s(n)$, called the rescaled range, where $s(n)$ is the standard deviation of the sample of n values. His studies included various geophysical phenomena in nature, such as tree rings, varves, precipitation series, and streamflow series. He noted that values of n versus $R^*(n)/s(n)$ plotted as straight lines on log–log

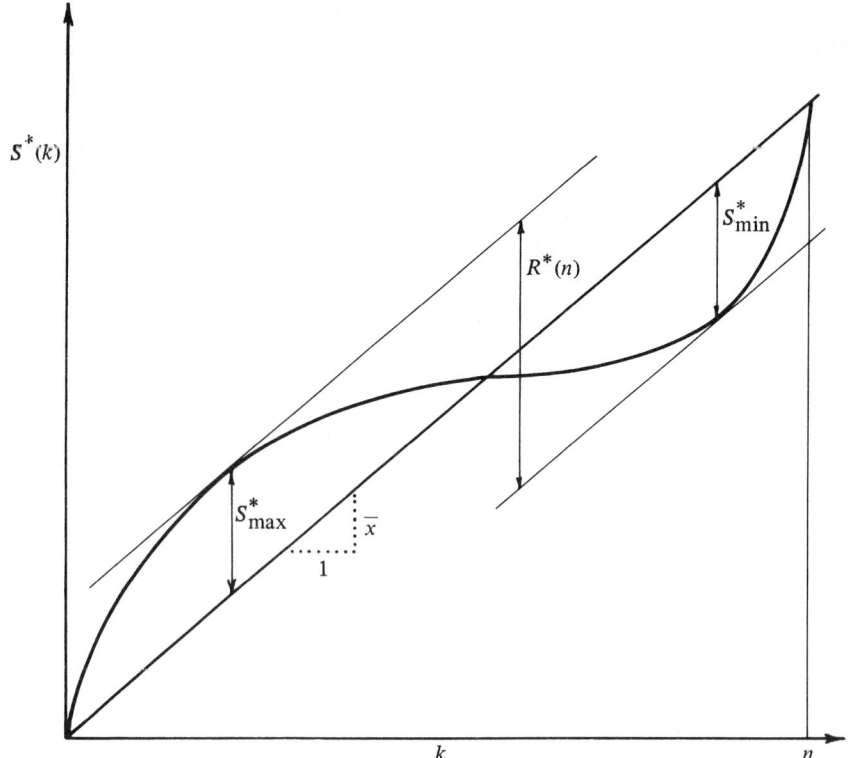

Figure 5.1 Definition of the adjusted range.

graph paper. This indicates that the rescaled range follows an equation of the type

$$\frac{R^*(n)}{s(n)} = Kn^H. \tag{5.5}$$

Hurst found that the average value for H was 0.73 and the standard deviation was 0.08. He also found that for large n a standard normal independent series behaves as

$$E\big[R^*(n)\big] = 1.25n^{0.5}. \tag{5.6}$$

The tendency of geophysical time series to produce values of H greater than 0.5 has become known as Hurst's phenomenon. This behavior of natural time series also interested statisticians at that time and Feller (1951) found the asymptotic distribution of the range (for large n) for both $R(n)$ and $R^*(n)$. He showed that the statistic $R(n)$ has more sampling stability than $R^*(n)$. His result for the asymptotic mean adjusted range $R^*(n)$ agrees with Hurst's result, as given in Eq. (5.6). Feller suggested that the departure of the H exponent from 0.50 could be explained by the use of a Markovian process. However, Barnard (1956) and Moran (1959) showed that even the Markovian models follow the asymptotic behavior with H equal to 0.5. With respect to the Hurst phenomenon, Moran says*:

> This well-established but puzzling phenomenon does not seem easy to explain. It has been suggested that serial correlation or dependence in the sequence $\{X_i\}$ could cause $[E[R(T)]]$ to vary as $[T^H]$ with $[H \neq \frac{1}{2}]$. This however, cannot be true unless the serial dependence is of a very peculiar kind, for with all plausible models of serial dependence the series of values is always approximated by a Bachelier–Wiener process when the time-scale is sufficiently large.

Anis and Lloyd (1953) gave the exact value for the mean of $R(n)$ for standard normal independent variates:

$$E[R(n)] = \sqrt{\frac{2}{\pi}} \sum_{i=1}^{n} i^{-1/2}. \tag{5.7}$$

Anis (1955, 1956) found the moments of the distribution of

$$\operatorname*{Sup}_{1 \leq i \leq n} [S(i)]$$

for standard normal independent variables. However, the moments of the distribution of the range does not follow from this result since $\operatorname{Sup} S(i)$ and $\operatorname{Inf} S(j)$ are dependent.

---

*Bracketed material has been modified.

Yevjevich (1965) found, by the Monte Carlo method, the distribution of the range and the adjusted range of cumulative departure from the mean for values of n up to 50 for a first order Markov process. It is inferred from his plots that both statistics have a large sampling variability. He also suggests (Yevjevich, 1967) that the mean range for linearly dependent normal variables can be expressed as

$$E[R(n)] = \sqrt{\frac{2}{\pi}} \sum_{i=1}^{n} i^{-1}\{\operatorname{var}[S(i)]\}^{1/2} \qquad (5.8)$$

where $\operatorname{var}[S(i)]$ is a function of the parameters and type of the linear dependence model used. However, Salas (1974) showed that Eq. (5.8) is only an approximation. Mandelbrot and Van Ness (1968) showed that the theoretical asymptotic proportionality to $n^{0.5}$ holds for any finite memory process, but a process may exhibit the n^H law (for $H \neq 0.5$) initially as a transient behavior. A question that immediately arises is whether or not hydrologic series are showing the transient behavior as a consequence of the shortness of their records, and if they gradually will shift to the $n^{0.5}$ law in the long run. Hurst's experiments with very long geophysical series (tree rings, varves) tend to invalidate transient behavior as an explanation to the Hurst phenomenon, although Salas, et al. (1979) state that Hurst's series may have been within the transient region of ARMA(1,1) models.

In summary, there are presently three main lines of thought explaining the Hurst phenomenon:

1. The Hurst phenomenon is a transitory behavior. The argument is that our series are simply not long enough to test the steady-state behavior of R, which according to the argument is the square-root law. This period of transition can be reproduced by Markov–autoregressive models. On the basis of a very long time series, Mandelbrot and Wallis (1968) effectively argue against this explanation.

2. The Hurst phenomenon is due to nonstationarities in the underlying mean of the process. This argument claims that a low-frequency, slowly time-varying mean explains the Hurst behavior (Klemeš, 1974; Potter, 1976; Boes and Salas, 1978).

3. The Hurst phenomenon is due to stationary processes with very large memory. That is, stationary processes that have correlation functions that decay very slowly in time, much slower than Markov–Gaussian–autoregressive processes. In the limit, this argument claims infinite memory for natural processes.

In regard to the second explanation above, Bhattacharya et al. (1983) have proved some interesting results. They show that the Hurst effect is present in the case of weakly dependent random variables with a small monotonic trend of the form $f(n) = c(m + n)^\beta$, where m is an arbitrary non-negative parameter

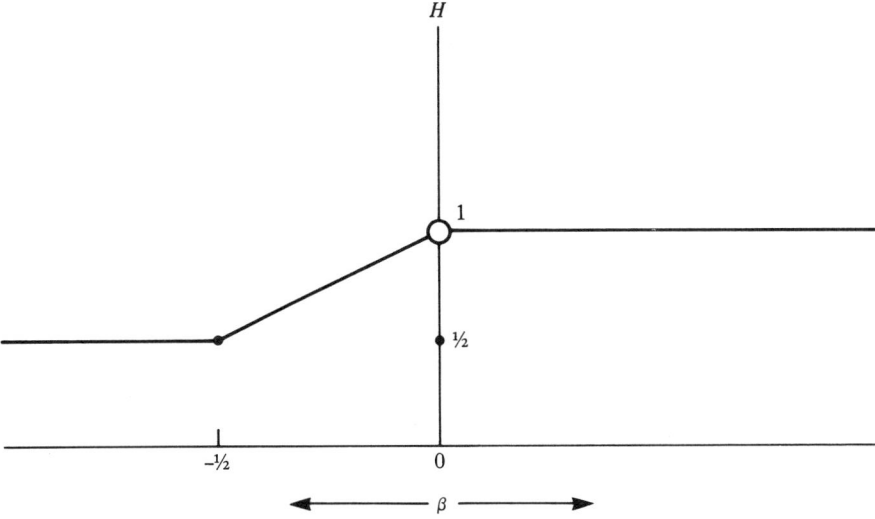

Figure 5.2 The relation between the Hurst exponent H and the trend exponent β (from Bhattacharya et al., 1983. Reproduced from the *Journal of Applied Probability*, by permission of the Applied Probability Trust).

and c is not zero. Thus, let $X_1, X_2, \ldots,$ be a sequence of random variables of the form

$$X_n = Y_n + f(n) \qquad n = 1, 2, \ldots, \tag{5.9}$$

where Y_n is a stationary sequence of weakly dependent random variables having a mean of zero and a variance of one. By weakly dependent we mean that Y_n belongs to the Brownian domain of attraction due to the characteristics of its correlation structure. Bhattacharya et al. show that for $-1/2 < \beta < 0$, the Hurst coefficient is given by $1 + \beta$. For $\beta < -1/2$ the Hurst coefficient is 0.5 and for $\beta > 0$ it is 1.

It is possible for the trend to be so small that it would go unnoticed in any reasonable statistical inspection of the data and yet the Hurst effect would appear asymptotically. Figure 5.2 shows the relation between the Hurst exponent H and the trend exponent β.

Statistically, it cannot be proved that any of the above explanations for the Hurst phenomenon is the correct one. Operationally, though, hydrologists are able to handle only the third argument.

5.3 FRACTIONAL GAUSSIAN NOISE

Mandelbrot (1965) became interested in Hurst's law and jointly with Van Ness (1968) proposed a model for the preservation of the Hurst phenomenon. This model is known as fractional Gaussian noise. Before we describe its use and properties, it is convenient to recall some considerations related to Brownian motion and fractional Brownian motion.

A particle immersed in a liquid or gas exhibits ceaseless irregular motions that are discernible under the microscope. The motion of such a particle is called Brownian, after the English botanist Robert Brown who noticed the phenomenon in 1827. In 1905 Einstein explained the phenomenon of Brownian motion; he showed that the displacement $B(t)$, from the origin and at time t, of a particle in Brownian motion has a Gaussian probability-density function with mean zero and variance $2Dt$, where D is the diffusion coefficient of the surrounding medium.

A model for Brownian motion is provided by a particle undergoing a random walk. Let $X_1, X_2, \ldots, X_n$ be independent and identically distributed random variables with mean $E[X] = 0$ and finite variance $E[X^2]$. The sum $B(n) = X_1 + X_2 + \cdots + X_n$ represents the displacement from its starting position of a particle performing a random walk on a straight line. After n steps, the total displacement $B(n)$ has a mean and variance given by

$$E[B(n)] = 0 \qquad E[B^2(n)] = nE[X^2].$$

Thus if X has a variance of one, we can say that the main property of the process of Brownian motion is that for every $\varepsilon > 0$ the sequence of increments $B(t+\varepsilon) - B(t)$, with t a multiple of ε, is a sequence of independent Gaussian random variables with zero mean and variance equal to ε. This sequence is called discrete-time Gaussian white noise. For the continuous analogue of a random walk we may write

$$B(t) = \int_{-\infty}^{t} X(v) \, dv$$

and the white noise is the derivative of Brownian motion.

Brownian motion is a self-similar process and satisfies $\sqrt{r}$ laws. This means that if we define a new time scale through a ratio r such that the new time u and the old time t are related by $t = ru$, then the functions

$$B(t) - B(0) = B(ru) - B(0)$$

and

$$[B(u) - B(0)] r^{0.5}$$

are generated by the same probabilistic process. Self-similarity implies that the variance of $[B(t+u) - B(t)]/\sqrt{r}$ is equal to 1 and thus the standard deviation of $B(t+u) - B(t) = \sqrt{r}$ for every t and r. Self-similarity also implies that the expected value of the adjusted range is the familiar

$$E\left[R^*(r)\right] = Kr^{0.5} \tag{5.10}$$

where K is a constant.

We now realize that because of the behavior described by Eq. (5.10) the Brownian motion process is unable to reproduce the Hurst phenomenon. As shown by Mandelbrot and Van Ness (1968) this is also the case for all processes whose covariance $C(u)$ is such that $\sum_{u=0}^{\infty} C(u) < \infty$. These processes are said to be in the Brownian domain of attraction.

When $\sum_{u=0}^{\infty} C(u)$ diverges, the $\sqrt{r}$ law expressed in Eq. (5.10) is no longer true. As we already mentioned, it is possible for finite memory processes, $\sum_{u=0}^{\infty} C(u) < \infty$, to have both range and standard deviation proportional to r^H over a finite span of values of r. But the usual $\sqrt{r}$ behavior still applies beyond this transient span. Furthermore, long transients can only be achieved through serious complications or structural deformations in the model. For example, Fiering (1967) experimented with autoregressive processes trying to simulate the Hurst phenomenon. He needed 20 lags in order to ensure that Hurst's law holds over a span of 60 units of time.

There is only one known stationary process with the required infinite memory properties. This process is called fractional Gaussian noise and has a correlation function of the form

$$C(\tau, H) = \tfrac{1}{2}\left[|\tau+1|^{2H} - 2|\tau|^{2H} + |\tau-1|^{2H}\right], \tag{5.11}$$

(Mandelbrot, 1965; Mandelbrot and Van Ness, 1968; Mandelbrot and Wallis, 1968, 1969a,b,c,d,e) where τ is lag and H is the Hurst coefficient. The interested reader can easily verify that the correlation function given above decays a lot more slowly than that of lag-one autoregressive models.

Fractional Gaussian noise (fGn) is intended to simulate $1/f$ noises, so called because its spectrum takes the form f^{1-2H} with f the frequency and $1/2 < H < 1$. This type of noise is present in many communication problems.

Mandelbrot (1977) cites, among others, many natural processes with $1/f$ characteristics: river levels, membrane currents in the nervous systems, coastlines, and variations in sunspots. These noises are not as strongly correlated as Brownian ones, neither are they entirely uncorrelated as white noise. One of the most puzzling properties of these noises is their low-frequency behavior, where most of the power of the spectrum is concentrated. As pointed out by Brophy (1970), the spectrum is extended over the range of all frequencies and the variance is infinite, indicating nonstationarity in the process. Figure 5.3 shows typical patterns of white, $1/f$, and Brownian noise, where the differences are immediately apparent. In order to provide a mathematical model for $1/f$ noises, Mandelbrot and Van Ness (1968) resort to fractional Brownian

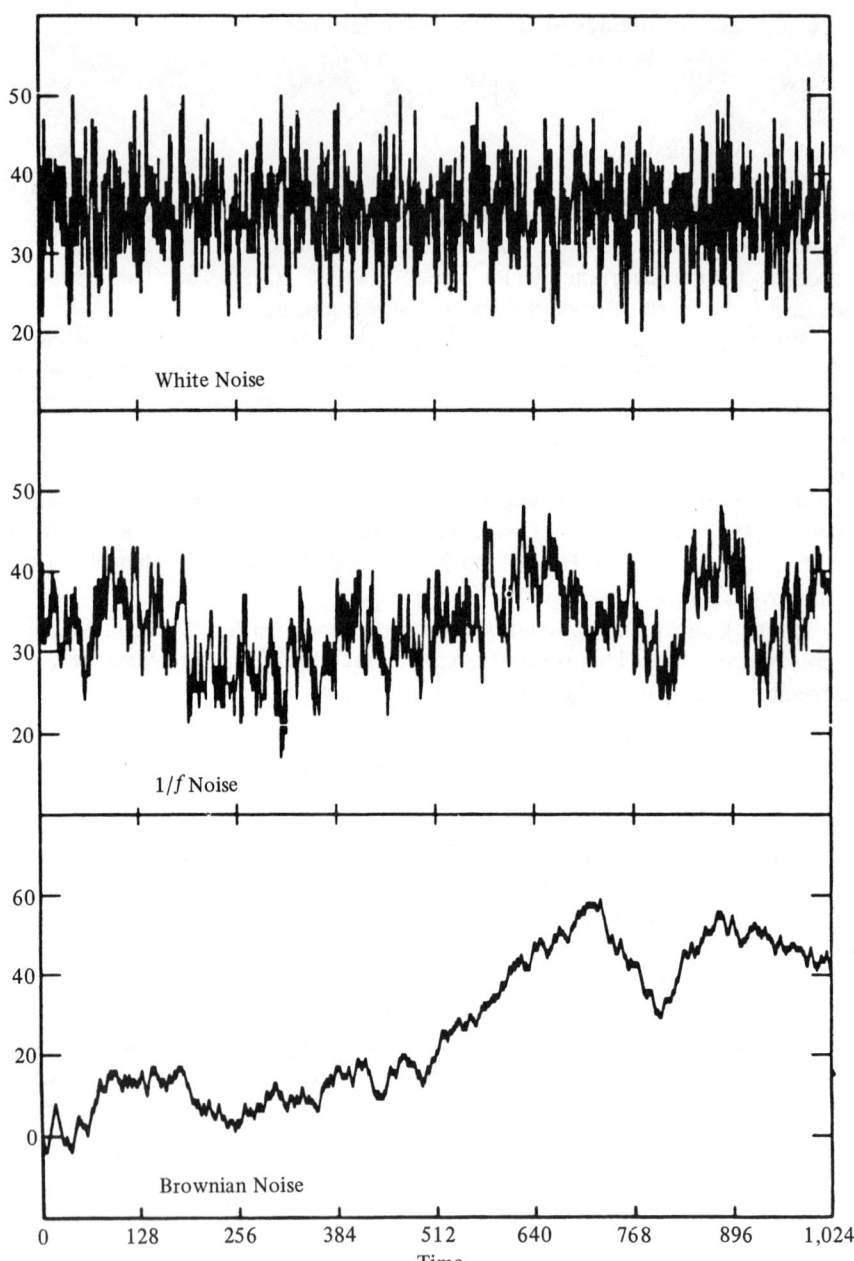

Figure 5.3 Typical patterns of white, $1/f$, and Brownian noise (reproduced with permission of Richard F. Voss/IBM Research).

motion, a type of process previously considered implicitly by Kolmogorov and other investigators.

Fractional Brownian motion (fBm) is defined by Mandelbrot and Van Ness as a moving average of past white noise:

$$B_H(t) - B_H(0) = \frac{1}{\Gamma(H + 0.5)} \left\{ \int_{-\infty}^{0} \left[(t - u)^{H - 0.5} - (-u)^{H - 0.5} \right] dB(u) \right.$$

$$\left. + \int_{0}^{t} (t - u)^{H - 0.5} \, dB(u) \right\},$$

$$0 < H < 1, \quad (5.12)$$

where $dB(u)$ is white noise or an infinitesimal increment of ordinary Brownian motion. $B_H(0)$ is the starting value of the process and may be any arbitrary real number.

Fractional Brownian motion (fBm) has the self-similarity property, making

$$B_H(t) - B_H(0)$$

and

$$T^{-H} [B_H(tT) - B_H(0)]$$

random processes with the same probabilistic behavior.

The variance of fractional Brownian motion is analogously given by

$$\text{var}[B_H(t + T) - B_H(t)] = T^{2H} V_H \qquad 0 < H < 1, \qquad (5.13)$$

where V_H is the variance of unit increments of fBm. For $H = 0.5$, the above converges to the Brownian motion result. The behavior of the expected adjusted range is as desired:

$$E[R^*(T)] = bT^H \qquad 0 < H < 1. \qquad (5.14)$$

Continuous representation of the increments of fractional Brownian motion requires its derivative, $B_H'(t)$, which, by definition, is fractional Gaussian noise. In fact, though, fBm has no derivative. Mandelbrot and Van Ness (1968) then defined a smoothed process,

$$B_H(t, \delta) = \frac{1}{\delta} \int_{t}^{t + \delta} B_H(s) \, ds \qquad \delta > 0, \qquad (5.15)$$

from which a derivative could be defined,

$$B_H'(t, \delta) = \frac{1}{\delta} [B_H(t + \delta) - B_H(t)]. \qquad (5.16)$$

$B_H'(t, \delta)$ is stationary, continuous fGn that is not differentiable.

The autocovariance function of $B_H'(t, \delta)$ is given as

$$C_H(\tau, \delta) = \frac{1}{2} V_H \delta^{2H-2} \left[\left(\frac{|\tau|}{\delta} + 1 \right)^{2H} - 2\left(\frac{\tau}{\delta} \right)^{2H} + \left(\frac{|\tau|}{\delta} - 1 \right)^{2H} \right] \quad (5.17)$$

with finite variance

$$C_H(0, \delta) = V_H \delta^{2H-2}. \quad (5.18)$$

For $0.5 < H < 1$ (cases of hydrologic interest), the covariance function is positive and finite for all τ, but not convergent,

$$\int_0^\infty C_H(\tau, \delta) \, d\tau = \infty. \quad (5.19)$$

Using the basic definition of fBm (Eq. 5.12) and defining

$$[B_H(t + \delta) - B_H(0)] - [B_H(t) - B_H(0)] = [B_H(t + \delta) - B_H(t)] \quad (5.20)$$

it is easy to show

$$B_H'(t, \delta) = \frac{1}{\delta \Gamma(H + 0.5)} \left\{ \int_{-\infty}^t \left[(t + \delta - u)^{H-0.5} - (-u)^{H-0.5} \right] dB(u) \right.$$
$$\left. + \int_t^{t+\delta} (t + \delta - u)^{H-0.5} \, dB(u) \right\}. \quad (5.21)$$

Changing variables, letting $s = u - \delta$, yields

$$B_H'(t, \delta) = \int_{-\infty}^t K_H(t - s, \delta) \, dB(s, \delta), \quad (5.22)$$

where $K_H(t - s, \delta)$ is a weighting function defined by

$$K_H(t, \delta) = \begin{cases} [\delta \Gamma(H + 0.5)]^{-1} t^{H-0.5} & t \le \delta \\ [\delta \Gamma(H + 0.5)]^{-1} \left[t^{H-0.5} - (t - \delta)^{H-0.5} \right] & t > \delta \end{cases}. \quad (5.23)$$

For $0.5 < H < 1$, $K_H(t, \delta)$ is always positive and can always be integrated over t.

Discrete-time fGn is obtained by using integer values of t and $\delta = 1$ in $B_H(t, \delta)$, leading to

$$\Delta B_H(t) = [B_H(t) - B_H(t - 1)]. \quad (5.24)$$

The variance of the discrete process is

$$\text{var}[B_H(t + n) - B_H(t)] = \text{var}[B_H(n) - B_H(0)]$$
$$= n^{2H} V_H \quad (5.25)$$

where t and n are integers and

$$\text{var}\left[B_H(t+1)-B_H(t)\right] = V_H = \sigma^2 \atop = \text{var}[X_t].$$

(5.26)

It can also be shown that the variance of the sample mean of X_t, $\overline{X}$, behaves like

$$\text{var}[\overline{X}] = \sigma^2 n^{2H-2},$$

(5.27)

which for $H = 0.5$ leads to the familiar result of independent random processes and the asymptotic behavior of correlated processes with convergent covariance functions.

The autocovariance function for discrete fGn was derived by Mandelbrot and Wallis (1969c) as:

$$C(\tau, H) = \tfrac{1}{2} V_H\left[|\tau+1|^{2H} - 2|\tau|^{2H} + |\tau-1|^{2H}\right].$$

(5.28)

The expected behavior of the adjusted range is controlled by the coefficient H, as desired,

$$E\left[R^*(n)\right] = bn^H.$$

5.4   SIMULATING FRACTIONAL GAUSSIAN NOISE WITH AN ARIMA(1,0,1)

Analogously to continuous white noise, fractional Gaussian noise cannot be operationally obtained; it can only be approximated. The defining equations do not have analytical solutions. There are few models, three of which we will see, that have been proposed for reproducing fGn behavior. Although chronologically the last one, this section will present the results of O'Connell (1974) in using an ARIMA(1,0,1) for mimicking fGn and preserving the Hurst phenomenon.

The idea is to produce a process with correlation structure similar to Eq. (5.28), for a reasonably long period. The correlation desired must decay slowly, leading to significant higher-lag correlations. This implies that significant low-frequency behavior should be maintained without distorting high-frequency characteristics, represented by the lag-one correlation, ρ_1.

From Chapter 2, the correlation structure of the ARIMA(1,0,1) was given by

$$\rho_1 = \frac{(\phi-\theta)(1-\phi\theta)}{1+\theta^2-2\phi\theta}$$

(5.29)

$$\rho_k = \phi\rho_{k-1} \qquad k \geq 2$$

and the model is represented by

$$(1 - \phi B) X_t = (1 - \theta B) a_t. \tag{5.30}$$

Several interesting conclusions can be drawn from the expressions in Eq. (5.29). The lag-one correlation is controlled by both ϕ and θ. For positive correlations, which are of central interest in hydrology, $\phi \geq \theta$. If $\phi = \theta$, the result is white noise. Furthermore, the decay of the correlation function for lags greater than or equal to 2 is controlled by parameter ϕ. A large ϕ (near the maximum of 1) will imply slow decay. Nevertheless, a large ϕ does not preclude a small ρ_1 since θ could also be large, leading to small differences $(\phi - \theta)$ and $(1 - \phi\theta)$.

The formulation of the ARIMA(1,0,1) model does not permit explicit consideration of the Hurst coefficient as a parameter. For this reason, O'Connell (1974) adopted a Monte Carlo simulation approach to determine pairs of θ and ϕ preserving reasonable values of H and ρ_1. It must be realized that as all Brownian domain processes, the ARIMA(1,0,1) must converge to the 0.5 law. Nevertheless, O'Connell observed that values of $H > 0.5$ could be maintained for very large periods, certainly for periods well above the common hydrologic design horizons of 50 to 100 years.

O'Connell also realized that the available estimators of H are highly variable and biased. To address this issue he performed extensive simulations of the ARIMA(1,0,1) (with different parameter sets) and obtained expected values for H and ρ_1 in realizations of length n. The expected values were obtained by averaging over 10,000 realizations for each n. These results are illustrated in Tables 5.1 through 5.6.

For the hydrologic simulation with the ARIMA(1,0,1), O'Connell then suggests the following procedure.

$$(X_t - \mu_x) = \phi(X_{t-1} - \mu_x) + \sigma_x \sigma_\varepsilon (\varepsilon_t - \theta \varepsilon_{t-1}) \tag{5.31}$$

is the model, where μ_x and σ_x are the mean and standard deviation of the process. The random variate ε_t has variance

$$\sigma_\varepsilon^2 = (1 - \phi^2)/(1 + \theta^2 - 2\phi\theta), \tag{5.32}$$

which ensures that X_t will have variance σ_x^2.

If a skewed variable is desired, then the ε_t should have a skewness related to the skewness of X by

$$\gamma_x = \left(\left[\frac{(1 - \theta^3 + 3\phi\theta^2 - 3\phi^2\theta)}{(1 - \phi^3)} \right] \middle/ \left[\frac{(1 + \theta^2 - 2\phi\theta)^{3/2}}{(1 - \phi^3)} \right] \right) \gamma_\varepsilon. \tag{5.33}$$

Table 5.1
Values of $E[H]_n$ and $E[\hat{\rho}_1]_n$ for selected values of ϕ and θ

		$\phi = 0.96$					
		$E[H]_n$			$E[\hat{\rho}_1]_n$		
θ	ρ_1 ╲ n	25	50	100	25	50	100
0.92	0.058	0.654	0.661	0.657	−0.005	0.005	0.025
0.88	0.146	0.683	0.679	0.698	−0.005	0.038	0.080
0.84	0.250	0.681	0.704	0.724	0.060	0.083	0.132
0.80	0.357	0.693	0.737	0.767	0.069	0.160	0.227
0.76	0.457	0.718	0.752	0.781	0.099	0.196	0.264
0.72	0.545	0.735	0.776	0.811	0.144	0.259	0.366
0.68	0.620	0.756	0.796	0.827	0.219	0.310	0.444
0.64	0.683	0.763	0.818	0.843	0.243	0.382	0.504
0.60	0.734	0.791	0.823	0.855	0.300	0.424	0.567
0.56	0.776	0.803	0.825	0.861	0.353	0.475	0.621
0.52	0.810	0.801	0.852	0.866	0.377	0.565	0.647

Source: O'Connell (1974).

Table 5.2
Values of $E[H]_n$ and $E[\hat{\rho}_1]_n$ for selected values of ϕ and θ

		$\phi = 0.92$					
		$E[H]_n$			$E[\hat{\rho}_1]_n$		
θ	ρ_1 ╲ n	25	50	100	25	50	100
0.88	0.049	0.657	0.664	0.654	0.001	0.012	0.028
0.84	0.114	0.687	0.680	0.686	0.005	0.046	0.079
0.80	0.189	0.686	0.705	0.709	0.072	0.093	0.123
0.76	0.269	0.699	0.735	0.745	0.082	0.160	0.208
0.72	0.349	0.725	0.751	0.756	0.116	0.199	0.240
0.68	0.426	0.740	0.782	0.783	0.169	0.269	0.332
0.64	0.496	0.772	0.783	0.800	0.218	0.309	0.390
0.60	0.560	0.773	0.796	0.803	0.273	0.364	0.437
0.56	0.616	0.774	0.820	0.825	0.285	0.432	0.516
0.52	0.665	0.794	0.825	0.828	0.335	0.467	0.532

Source: O'Connell (1974).

Table 5.3
Values of $E[H]_n$ and $E[\hat{\rho}_1]_n$ for selected values of ϕ and θ

		$\phi = 0.88$					
		$E[H]_n$			$E[\hat{\rho}_1]_n$		
θ	ρ_1 $\diagdown$ n	25	50	100	25	50	100
0.84	0.046	0.659	0.665	0.651	0.005	0.014	0.029
0.80	0.102	0.689	0.678	0.676	0.012	0.050	0.078
0.76	0.176	0.688	0.703	0.696	0.080	0.097	0.117
0.72	0.233	0.703	0.728	0.728	0.091	0.158	0.196
0.68	0.302	0.729	0.745	0.737	0.126	0.198	0.225
0.64	0.370	0.742	0.772	0.764	0.176	0.262	0.309
0.60	0.435	0.771	0.773	0.778	0.223	0.302	0.363
0.56	0.495	0.771	0.788	0.782	0.278	0.356	0.407
0.52	0.550	0.774	0.808	0.802	0.291	0.416	0.481

Source: O'Connell (1974).

Table 5.4
Values of $E[H]_n$ and $E[\hat{\rho}_1]_n$ for selected values of ϕ and θ

		$\phi = 0.84$					
		$E[H]_n$			$E[\hat{\rho}_1]_n$		
θ	ρ_1 $\diagdown$ n	25	50	100	25	50	100
0.80	0.044	0.660	0.663	0.647	0.007	0.016	0.029
0.76	0.096	0.690	0.676	0.668	0.016	0.052	0.077
0.72	0.154	0.688	0.699	0.687	0.086	0.099	0.113
0.68	0.214	0.705	0.720	0.714	0.098	0.156	0.189
0.64	0.277	0.725	0.746	0.733	0.142	0.203	0.237
0.60	0.325	0.753	0.749	0.747	0.180	0.245	0.290
0.56	0.409	0.753	0.766	0.753	0.240	0.298	0.332
0.52	0.454	0.761	0.784	0.773	0.254	0.355	0.406

Source: O'Connell (1974).

Table 5.5
Values of $E[H]_n$ and $E[\hat{\rho}_1]_n$ for selected values of ϕ and θ

		$\phi = 0.80$					
		$E[H]_n$			$E[\hat{\rho}_1]_n$		
θ	ρ_1 $\backslash$ n	25	50	100	25	50	100
0.75	0.055	0.666	0.657	0.649	− 0.006	0.022	0.041
0.70	0.119	0.685	0.694	0.678	0.056	0.079	0.098
0.65	0.188	0.708	0.698	0.708	0.114	0.103	0.166
0.60	0.260	0.736	0.727	0.719	0.139	0.192	0.227
0.55	0.331	0.737	0.750	0.741	0.175	0.247	0.289
0.50	0.400	0.756	0.764	0.746	0.263	0.315	0.343

Source: O'Connell (1974).

The estimates of μ_x, σ_x, γ_x, ρ_x, and H from the historical data are random variables, some of them clearly biased, particularly for the short records available. With valid concern, O'Connell suggests two simulation philosophies.

The first is the traditional approach of estimating $\hat{\mu}_x$, $\hat{\sigma}_x$, $\hat{\rho}_x$, and $\hat{\gamma}_x$, and $\hat{H}$ from the historical data and interpreting those statistics as population parameters to be preserved by the model as the number of generating values goes to infinity. The difficulty lies in the fact that no infinite simulation will be done, and, if it is done, the model H will converge to 0.5. The suggestion is then to use Tables 5.1 through 5.6 and identify a set of parameters ϕ and θ such that $E_n[\hat{H}] = \hat{H}$,

Table 5.6
Values of $E[H]_n$ and $E[\hat{\rho}_1]_n$ for selected values of ϕ and θ

		$\phi = 0.75$					
		$E[H]_n$			$E[\hat{\rho}_1]_n$		
θ	ρ_1 $\backslash$ n	25	50	100	25	50	100
0.70	0.054	0.666	0.655	0.645	− 0.004	0.023	0.041
0.65	0.115	0.684	0.689	0.671	0.059	0.079	0.096
0.60	0.179	0.707	0.691	0.698	0.117	0.102	0.161
0.55	0.246	0.733	0.719	0.707	0.141	0.189	0.219
0.50	0.313	0.734	0.740	0.728	0.177	0.240	0.277
0.45	0.377	0.751	0.754	0.732	0.264	0.307	0.329

Source: O'Connell (1974).

and $\rho_1 \approx \hat{\rho}_x$, where ρ_1 is given by Eq. (5.29) and n is the number of years of simulation.

The second approach is to explicitly account for parameter uncertainty and bias in historic and synthetic sequences of length n. Essentially, historical $\hat{\mu}_x$, $\hat{\sigma}_x^2$, and $\hat{\rho}_x$, and $\hat{H}$ are obtained and Tables 5.1 to 5.6 are used to identify ϕ and θ such that

$$E_n[\hat{H}] \approx \hat{H}$$

and

$$E_n[\hat{\rho}_1] \approx \hat{\rho}_x.$$

Consistent with the above treatment of the lag-one correlation, the second approach also uses an unbiased estimate of σ_x^2 as simulation parameter. O'Connell (1974) states that the sample variance of an n-year simulation of an ARIMA(1,0,1) model satisfies

$$E_n[s^2] = \sigma_x^2 \left[1 - \frac{2\rho_1}{n(n-1)} \left(\frac{n(1-\phi)-(1-\phi^n)}{(1-\phi)^2} \right) \right], \qquad (5.34)$$

where ρ_1 is the lag-one correlation of the ARIMA(1,0,1) in terms of ϕ and θ. Essentially then,

$$E_n[s^2] = \sigma_x^2 f(n, \rho_1, \phi), \qquad (5.35)$$

which equating $E_n[s^2]$ to $\hat{\sigma}_x^2$ leads to an unbiased estimate of σ_x^2:

$$\hat{\sigma}_x^2 = \hat{\sigma}_x^2 / f(n, \rho_1, \phi). \qquad (5.36)$$

Once $\hat{\mu}_x$, $\hat{\sigma}_x^2$, ϕ, and θ are identified, the simulation can proceed. It must be realized that a given realization from the simulation may not exactly exhibit the fitted statistics; they remain random variables. The possibility also exists that estimates of ρ_1 and H may be incompatible with or lie outside the simulation results of Tables 5.1 through 5.6, which then makes the ARIMA(1,0,1) model inadequate.

The main advantage of the use of the ARIMA(1,0,1) to preserve H in the manner presented here is its simplicity. The main disadvantage is the implicit relation of H and the parameters ϕ and θ. The other two models discussed in the following sections explicitly handle the Hurst coefficient.

5.5 FAST FRACTIONAL GAUSSIAN NOISE GENERATOR

Mandelbrot and Wallis (1969a) proposed two approximations of discrete fractional Gaussian noise (called types I and II), which consisted of weighted moving averages of independent Gaussian variables. Because of several drawbacks of the approximations, Mandelbrot (1971) suggested the fast fractional Gaussian noise generator (ffGn). The concept is simple; ffGn consists of a weighted sum of first-order autoregressive models. Mandelbrot suggests that a ffGn process can be divided into high-frequency and low-frequency components,

$$X(t, H) = X_h(t, H) + X_L(t, H). \tag{5.37}$$

It is the low-frequency component, $X_L(t, H)$ that simulates the Hurst phenomenon. $X_L(t, H)$ is designed to mimic the correlation behavior of a discrete fractional Gaussian noise at high lags. This low-frequency element takes the form

$$X_L(t, H) = \sum_{n=1}^{N(T)} W_n X(t, r_n), \tag{5.38}$$

where $X(t, r_n)$ is a lag-one autoregressive model with zero mean, unit variance, and covariance function

$$r_n = \exp(-B^{-n}). \tag{5.39}$$

B is a parameter to be discussed later in greater detail. Each process $X(t, r_n)$ takes the form

$$X(t, r_n) = r_n X(t-1, r_n) + (1 - r_n^2)^{1/2} \varepsilon_t. \tag{5.40}$$

W_n is a weight given by

$$W_n^2 = \frac{H(2H-1)(B^{1-H} - B^{-1+H})}{\Gamma(3-2H)} B^{-2(1-H)n}, \tag{5.41}$$

where H is the desired Hurst coefficient and $\Gamma(\cdot)$ is the gamma function.

The number of added Markov models $N(T)$ is given by

$$N(T) = \| \log(QT)/\log B \|, \tag{5.42}$$

where Q is another specified parameter and T is the number of time periods of simulation desired. The notation $\|y\|$ implies the smallest integer above y. As $B \to 1$ and $Q \to \infty$, the quality of the ffGn approximation to the low-frequency behavior improves [$N(T)$ increases].

As mentioned previously, the model given above simulates the behavior of the higher lags of the discrete ffGn correlation function. This implies that the high-frequency term $X_h(t, H)$ must be used to account for unexplained variance and low-lag correlation behavior, in particular the lag-one correlation.

Mandelbrot suggests representing $X_h(t, H)$ by a lag-one autoregressive model with variance given by

$$\sigma_h^2 = \frac{1 - B^{-(1-H)}H(2H-1)}{\Gamma(3-2H)} \qquad (5.43)$$

and lag-one correlation,

$$\rho_h(1) = 2^{2H-1} - 1 + \sum_{n=1}^{N(T)} W_n(1-r_n) - \frac{B^{-(1-H)}H(2H-1)}{\Gamma(3-2H)}. \qquad (5.44)$$

The above equations were obtained by computing leftover variance not accounted for by the low-frequency term as well as the necessary contribution to the correlation function in order to reproduce the lag-one behavior of the ffGn correlation. The added high frequency Markov term is given a weight of one.

Briefly, the derivation of ffGn follows the ideas described below.

The correlation function of discrete fGn,

$$C(\tau, H) = \tfrac{1}{2}\big[|\tau + 1|^{2H} - 2|\tau|^{2H} + |\tau - 1|^{2H}\big], \qquad (5.45)$$

is first approximated by

$$C(\tau, H) \approx C_1(\tau, H) = H(2H-1)\tau^{2H-2}. \qquad (5.46)$$

Note that in fact the former equation is the second finite difference of $\tau^{2H}/2$, while the latter is the second derivative of the same function. The approximation is very good for high lags. The reader is reminded that the low-frequency term is intended to mimic the process behavior at these high lags.

The correlation $C_1(\tau, H)$ can be expressed as a weighted sum of an infinite number of Markov-models' correlation functions (expressed as exponentials):

$$C_1(\tau, H) = H(2H-1)\tau^{2H-2} = \int_0^\infty e^{-\tau u}W(u)\,du. \qquad (5.47)$$

Equation (5.47) can be interpreted as a Laplace transform of the function $W(u)$. It is known that τ^{-Z} is the Laplace transform of $u^{Z-1}/\Gamma(Z)$. Therefore

$$H(2H-1)\tau^{2H-2} = \frac{\displaystyle\int_0^\infty e^{-\tau u}u^{1-2H}\,du}{\Gamma(-2H+2)/H(2H-1)}, \qquad (5.48)$$

which leads to a square-weights form

$$W(u) = \frac{u^{1-2H}}{2|\Gamma(-2H)|}. \tag{5.49}$$

Equation (5.47) is further approximated by:

a. eliminating the very-high-frequency terms of $C_1(\tau, H)$, which correspond to high values of u. A Markovian process is rapidly varying if its lag-one correlation e^{-u} lies below some threshold (th 1) less than 1. From the viewpoint of simulation, the sum of such Markov components can be approximated by the single process $X_h(t, H)$ of Eq. (5.37).

b. eliminating low values of u, which correspond to Markovian models with very little variance and which are essentially a constant process.

c. discretization of the continuous integral.

In order to discretize, the following change of variables is performed

$$u = B^{-v} \tag{5.50}$$

where $B > 1$, leading to

$$C_1(\tau, H) = \frac{\log B}{|2\Gamma(-2H)|} \int_{-\infty}^{\infty} \exp(-\tau B^{-v}) B^{-2(1-H)v} \, dv. \tag{5.51}$$

The variable v is then discretized in intervals $n - 0.5$ to $n + 0.5$, where n is an integer value. Equation (5.51) is now approximated by evaluating $\exp(-\tau B^v)$ at the middle of the discretization interval, $\exp(-\tau B^n)$, and averaging $B^{2(1-H)v}$ over $n - 0.5$, $n + 0.5$. This leads to

$$
\begin{aligned}
C(\tau, H) &\approx \frac{\log B}{2|\Gamma(-2H)|} \sum_{n=0}^{N(T)} \exp(-\tau B^{-n}) \int_{n-0.5}^{n+0.5} B^{-2(1-H)y} \, dy \\
&= \sum_{n=0}^{N(T)} \exp(-\tau B^n) \cdot \frac{B^{1-H} - B^{1+H}}{4(H-1)|\Gamma(-2H)|} B^{-2(1-H)n}.
\end{aligned}
\tag{5.52}
$$

The lower limit $n = 0$ in Eq. (5.52) implies accounting for lag-one correlations as low as

$$r_{\min} = \exp(-B^{0.5}),$$

which for typical values of $B = 2, 3, 4$ yields $r_{\min}$ of 0.25, 0.18, and 0.14, respectively. At the other end, values of n much above zero (low values of u) correspond to very low frequencies and can be truncated beyond the value of n

given by Eq. (5.42). Q is a quality parameter usually taking values around 4, 5, and 6. Mandelbrot (1971) gives Q as a function of H and two threshold values: th2 and th3; th2 represents the lowest change in variance that will be accounted for during the T years of simulation. Very highly correlated processes add very little variance to the simulation. Threshold th3 weeds out processes with small weights and those with negligible mean sample variance over the T periods. It represents the ratio of sample mean variance of negligible terms to sample mean variance of all terms.

Equation (5.52) indicates that the ffGn approximation to the low-frequency components is a weighted discrete sum of Markovian models with correlation function

$$r_n^\tau = \exp(-\tau B^n).\tag{5.53}$$

The square weights are

$$W_N^2 = \frac{B^{1-H} - B^{-1+H}}{4(H-1)|\Gamma(-2H)|} B^{-2(1-H)n},\tag{5.54}$$

which, using

$$|\Gamma(-2H)| = \Gamma(3-2H)[(2H)(2H-2)(2H-1)]^{-1},$$

leads to Eq. (5.41).

The approximations to Eq. (5.47) lead to an error term whose variance can be approximated by the high-frequency component $X_h(t, H)$ with parameters given in Eqs. (5.43) and (5.44) (Mandelbrot, 1971).

To our knowledge fast fractional Gaussian noise has never been used in practice. An operational model would require the ability to explicitly fit any combination of lag-one correlation and Hurst coefficient. This would require weighting of the added high-frequency component. The mechanics of this procedure have never been published. The extension of the ffGn generator to the multivariate case also remains to be investigated. The next section deals with a model based on ideas similar to those of the fast fractional Gaussian noise and which is fully operational.

5.6 THE BROKEN-LINE PROCESS (adapted from Curry and Bras, 1978)

The broken-line process was introduced by Ditlevsen (1971) to check by simulation some of his results in first-passage theory. Mejia and colleagues (1972, 1974) extended Ditlevsen's work and pioneered its use in hydrology.

The simple broken-line process (Fig. 5.4) results from the linear interpolation of equally spaced independent random variables. The series is made

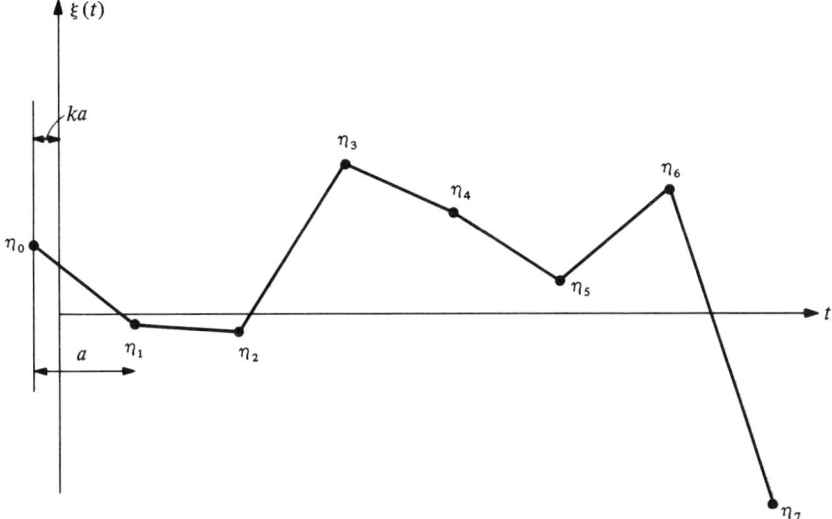

Figure 5.4 Schematic representation of simple broken line (from Mejía et al., 1972).

stationary by imposing a random initial displacement. Formally, a simple broken-line process is given by

$$\xi(t - ka) = \sum_{n=0}^{\infty} \left[\eta_n + \frac{(\eta_{n+1} - \eta_n)(t - na)}{a} \cdot I_{[na,(n+1)a]}(t) \right], \quad (5.55)$$

where $\eta_0, \eta_1, \ldots, \eta_n, \ldots$, are independent identically distributed random variables with zero mean, variance σ_η^2, and skewness $\gamma(\eta)$; k is a random variable uniformly distributed over the interval $(0, 1)$; a is the time distance among the different values of η; and $I(t)$ is the indicator function

$$I_{[na,(n+1)a]}(t) = \begin{cases} 1 & \text{for } na \leq t \leq (n+1)a \\ 0 & \text{otherwise.} \end{cases} \quad (5.56)$$

The stationarity of the process, shown by Ditlevsen (1971), implies

$$E[\xi(t')] = E[\xi(0)]. \quad (5.57)$$

From Eq. (5.55)

$$E[\xi(0)] = E[\eta_0 + (\eta_1 - \eta_0)k] = E[\eta_0(1 - k) + \eta_1 k]$$
$$= \int_0^1 [(1 - x)E[\eta_0] + xE[\eta_1]] f_k(x) \, dx = 0,$$

where $f_k(x)$ is the uniform distribution of k. Thus,

$$E[\xi(t')] = \overline{\xi(t')} = 0. \tag{5.58}$$

The variance of the process is given by

$$
\begin{aligned}
\sigma_\xi^2 = \operatorname{var}[\xi(t')] &= E[\xi(t')^2] - \overline{\xi(t')}^2 = E[\xi(0)^2] \\
&= E[(\eta_0 + (\eta_1 - \eta_0)k)^2] = \int_0^1 [(1-x)^2 E[\eta_0^2] + x^2 E[\eta_1^2]] f_k(x)\, dx \\
&= \int_0^1 [1 - 2x + 2x^2] \sigma_\eta^2\, dx = \tfrac{2}{3}\sigma_\eta^2.
\end{aligned} \tag{5.59}
$$

The skewness of the process is given by

$$
\begin{aligned}
E[[\xi(t') - \bar{\xi}]^3] = E[\xi(0)^3] &= E[[\eta_0 + (\eta_1 - \eta_0)k]^3] \\
&= \int_0^1 (1 - 3x + 3x^2) E[\eta^3] f_k(x)\, dx = \frac{\sigma_\eta^3 \gamma(\eta)}{2},
\end{aligned} \tag{5.60}
$$

where $\gamma(\eta)$ is the skewness coefficient of the ηs.

The derivation of the covariance and correlation function for the simple broken-line process follows. Given the stationarity and zero mean property of the process

$$\operatorname{cov}[\xi(t'), \xi(t' + \tau)] = E[\xi(t')\xi(t' + \tau)] = E[\xi(0)\xi(\tau)], \tag{5.61}$$

by Eq. (5.55) $\xi(0)$ is found to be

$$\xi(0) = \eta_0 + (\eta_1 - \eta_0)k. \tag{5.62}$$

The expression for $\xi(\tau)$ depends on τ, a, and k. For $\tau \le a$

$$\xi(\tau) = \begin{cases} \eta_0 + (\eta_1 - \eta_0)(k + \tau/a) & k \le 1 - \tau/a \\ \eta_1 + (\eta_2 - \eta_1)(k - 1 + \tau/a) & 1 - \tau/a \le k \le 1 \end{cases} \tag{5.63}$$

and for $a < \tau \le 2a$

$$\xi(\tau) = \begin{cases} \eta_1 + (\eta_2 - \eta_1)(k - 1 + \tau/a) & k \le 2 - \tau/a \\ \eta_2 + (\eta_3 - \eta_2)(k - 2 + \tau/a) & 2 - \tau a \le k \le 1. \end{cases} \tag{5.64}$$

When $\tau > 2a$, the form of $\xi(\tau)$ is

$$\xi(\tau) = \eta_i + (\eta_{i+1} - \eta_i)(k + f(a, \tau)) \qquad i = 3, 4, \text{ etc.} \tag{5.65}$$

where $f(a, \tau)$ denotes a function of τ and a.

Therefore, for $\tau \le a$

$$E[\xi(0)\xi(\tau)] = \int_0^{1-\tau/a} E[[\eta_0 + (\eta_1 - \eta_0)x][\eta_0 + (\eta_1 - \eta_0)(x + \tau/a)]]$$

$$\cdot f_k(x)\,dx + \int_{1-\tau/a}^1 E[[\eta_0 + (\eta_1 - \eta_0)x]$$

$$\cdot[\eta_1 + (\eta_2 - \eta_1)(x + \tau/a - 1)]]\cdot f_k(x)\,dx. \qquad (5.66)$$

Since

$$E[\eta_i \eta_j] = \begin{cases} 0 & i \ne j \\ \sigma_\eta^2 & i = j, \end{cases} \qquad (5.67)$$

Eq. (5.66) reduces to

$$E[\xi(0)\xi(\tau)] = \sigma_\eta^2 \left[\frac{2}{3} - \frac{1}{2}(\tau/a)^2\left(2 - \frac{\tau}{a}\right)\right]. \qquad (5.68)$$

For $a < \tau \le 2a$, we get

$$E[\xi(0)\xi(\tau)] = \int_0^{2-\tau/a} E[[\eta_0 + (\eta_1 - \eta_0)x][\eta_1 + (\eta_2 - \eta_1)(x - 1 + \tau/a)]]$$

$$\cdot f_k(x)\,dx + \int_{2-\tau/a}^1 E[[\eta_0 + (\eta_1 - \eta_0)x]$$

$$\cdot[\eta_2 + (\eta_3 - \eta_2)(x - 2 + \tau/a)]]\cdot f_k(x)\,dx. \qquad (5.69)$$

Since all products of η involve different subscripts, the second term of Eq. (5.69) is zero; thus we get

$$E[\xi(0)\xi(\tau)] = \sigma_\eta^2 \left[\frac{(2 - \tau/a)^3}{6}\right]. \qquad (5.70)$$

For $\tau > 2a$, all products of η would involve different subscripts, leading to

$$E[\xi(0)\xi(\tau)] = 0. \qquad (5.71)$$

Summarizing,

$$
\text{cov}\big[\xi(t'),\xi(t'+\tau)\big] =
\begin{cases}
\sigma_\eta^2\!\left[\dfrac{2}{3}-\dfrac{1}{2}\!\left(\dfrac{\tau}{a}\right)^{\!2}\!\left[2-\left(\dfrac{\tau}{a}\right)\right]\right] & 0<\tau<a \\[3mm]
\sigma_\eta^2\!\left[\left(\dfrac{2-\tau/a}{6}\right)^{\!3}\right] & 0<\tau\le a \\[3mm]
0 & \tau>2a
\end{cases}
\tag{5.72}
$$

Using Eqs. (5.72) and (5.59) the correlation function is easily derived.

$$
\rho(\tau,a) = \frac{\text{cov}\big[\xi(t'),\xi(t'+\tau)\big]}{\text{var}\big[\xi(t')\big]^{1/2}\text{var}\big[\xi(t'+\tau)\big]^{1/2}} = \frac{\text{cov}\big[\xi(t'),\xi(t'+\tau)\big]}{\tfrac{2}{3}\sigma_\eta^2}
$$

$$
=
\begin{cases}
1-\dfrac{3}{4}\!\left(\dfrac{\tau}{a}\right)^{\!2}\!\left(2-\dfrac{\tau}{a}\right) & 0\le\tau\le a \\[3mm]
\dfrac{1}{4}\!\left(2-\dfrac{\tau}{a}\right)^{\!3} & a<\tau<2a \\[3mm]
0 & \tau\ge 2a
\end{cases}
\tag{5.73}
$$

The lag-zero cross-correlation between two simple broken-line processes with the same parameter a and initial displacement k is easily obtained. Given the processes

$$
\xi_a^i(t-ka) = \sum_{m=0}^{\infty} \eta_m^i + \frac{\eta_{m+1}^i - \eta_m^i}{a}(t-ma)I_{[ma,(m+1)a]}(t)
$$

$$
\xi_a^j(t-ka) = \sum_{m=0}^{\infty} \eta_m^j + \frac{\eta_{m+1}^j - \eta_m^j}{a}(t-ma)I_{[ma,(m+1)a]}(t).
$$

If the ηs are such that

$$
\left.
\begin{aligned}
E\big[\eta_m^i\big] &= E\big[\eta_m^j\big] = 0 \\
\text{var}\big[\eta_m^i\big] &= \text{var}\big[\eta_m^j\big] = \sigma_\eta^2 \\
\text{skewness coefficient }\big[\eta_m^i\big] &= \gamma(\eta^i) \\
\text{skewness coefficient }\big[\eta_m^j\big] &= \gamma(\eta^j)
\end{aligned}
\right\}
\quad m=0,1,\dots
$$

$$
E\big[\eta_m^i \eta_p^j\big] =
\begin{cases}
0 & i=j, & m\ne p \\
\sigma_\eta^2 & i=j, & m=p \\
0 & i\ne j, & m\ne p \\
\lambda_{ij} & i\ne j, & m=p
\end{cases}
\quad
\begin{array}{l} m=0,1,\dots \\ p=0,1,\dots \end{array}
\tag{5.74}
$$

then

$$E\left[\xi_a^i(t')\xi_a^j(t')\right] = E\left[\xi_a^i(0)\xi_a^j(0)\right]$$
$$= \int_0^1 E\left[\left\{\eta_0^i + (\eta_1^i - \eta_0^i)x\right\}\left\{\eta_0^j + (\eta_1^j - \eta_0^j)x\right\}\right] f_k(x)\, dx$$
$$= \int_0^1 \lambda_{ij}(1 - 2x + 2x^2)\, dx = \tfrac{2}{3}\lambda_{ij} \qquad i \neq j \qquad (5.75)$$

and thus the lag-zero cross-correlation is

$$\frac{\mathrm{cov}\left[\xi_a^i(t')\xi_a^j(t')\right]}{\mathrm{var}\left[\xi_a^i(t')\right]^{1/2}\mathrm{var}\left[\xi_a^j(t')\right]^{1/2}} = \frac{E\left[\xi_a^i(0)\xi_a^j(0)\right]}{\tfrac{2}{3}\sigma_\eta^2}$$
$$= \lambda_{ij}/\sigma_\eta^2 \qquad i \neq j$$

(Mejía, 1971). That is, the lag-zero cross-correlation between two broken lines with the same parameter a is equal to the cross-correlation between the ηs from which each was constructed, divided by the variance of the ηs.

If $\beta_N(t)$ is a broken-line process formed from the addition of N weighted simple broken lines,

$$\beta_N(t) = \sum_{i=1}^{N} W_i\xi_i(t), \qquad (5.76)$$

where the ξ_is are simple broken-line processes with parameters a_i, its correlation function is given by

$$\rho_{\beta_N}(\tau) = \frac{\displaystyle\sum_{i=1}^{N} \sigma_{\xi_i}^2 W_i^2 \rho(\tau, a_i)}{\displaystyle\sum_{i=1}^{N} W_i^2 \sigma_{\xi_i}^2}, \qquad (5.77)$$

where $\rho(\tau, a_i)$ is the correlation function of $\xi_i(t)$. By varying the number, N, of summed simple broken lines, and weighting individual processes, it is possible to generate a process that has as many degrees of freedom as desired. Hence, a broken-line process can be constructed to reproduce properties of a given correlation function. This is the idea underlying the use of broken-line processes to simulate hydrologic records that exhibit the Hurst phenomenon. A broken-line process can be constructed that approximately reproduces the correlation function required to preserve the Hurst phenomenon; this was first proposed by Mejía et al. (1972). Furthermore, by using the relationship expressed in Eq. (5.75), broken-line processes can be constructed to have a given cross-correlation and thus can be used for multivariate simulation of hydrologic events (Mejía et al., 1974). The details of the univariate and multivariate broken-line models that explicitly preserve any given Hurst coefficient are shown in the Appendix. This is the only operational multivariate model that we know of that can explicitly preserve any value of H in several correlated sites simultaneously.

5.7 AN EXAMPLE OF BROKEN-LINE BEHAVIOR

To demonstrate the utility of the broken-line process for simulation purposes, let us consider the case in which the weights of Eq. (5.76) are taken equal to one and the following arbitrary restrictions are applied to the parameters of the process:

$$a_i = a_1 q^{i-1} \qquad q > 1 \qquad (5.78)$$

and

$$\sigma_i = \sigma_1 (a_1/a_i)^{1/2} = \sigma_1 (1/q^{i-1})^{1/2}, \qquad (5.79)$$

where q is a new parameter. The "memory" of the process is defined by the time lag for which the correlation function becomes zero; for the compound broken line, this is given by

$$\text{memory} = 2a_N = 2a_1 q^{N-1}.$$

This equation implies that the memory can be fixed by varying q and N—the number of simple broken lines being added.

The above restrictions are not the only ones possible; others may be used if required by the fitting of the parameters, as shown in the Appendix on the preservation of H. If variance of the compound broken line is fixed at 1, the restrictions of Eqs. (5.78) and (5.79) imply that the standard deviation of the first added broken line must be

$$\sigma_1 = \left[3q^{N-1}(q-1)/2(q^N-1)\right]^{1/2}. \qquad (5.80)$$

Table 5.7
Characteristics of the five processes generated

Process	a_1	q	N	Memory
1	5	1.5	20	22,168
2	5	5	20	1.9×10^{14}
3	2	1.5	10	154
4	1	1.1	10	4.72
5	1	5	5	1250

Source: Mejía et al. (1972).

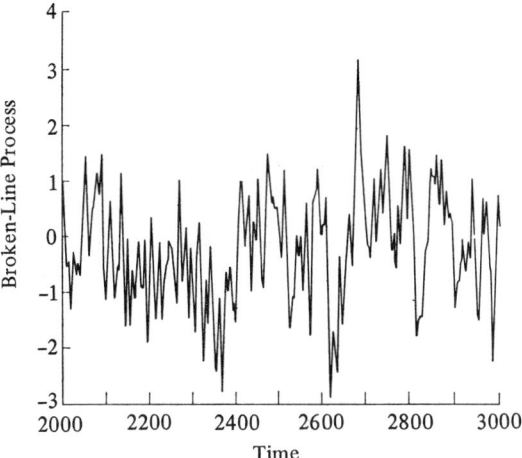

Figure 5.5 Sample realization of process 1, where $a_1 = 5$, $q = 1.5$, and $N = 20$ (from Mejía et al., 1972).

Therefore this construction of parameters depends only on values of a_1, q, and N. Mejía et al. (1972) simulated five sample processes with the characteristics given in Table 5.7. Figures 5.5 to 5.7 show examples of three of these.

Figures 5.8 to 5.10 show the correlation functions, theoretical and estimated, for the simulations for three of the processes. Note that the span of interdependence between the elements of the process can be controlled from

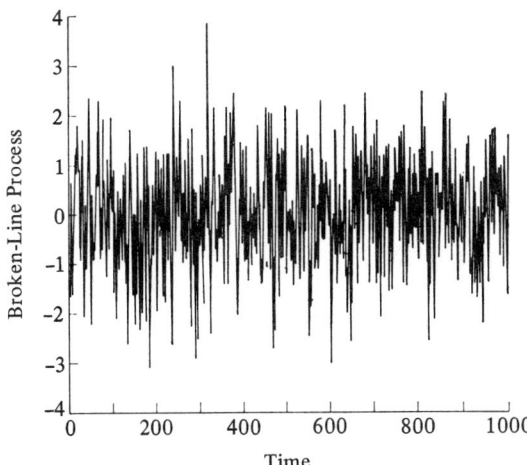

Figure 5.6 Sample realization of process 4, where $a_1 = 1$, $q = 1.1$, and $N = 10$ (from Mejía et al., 1972).

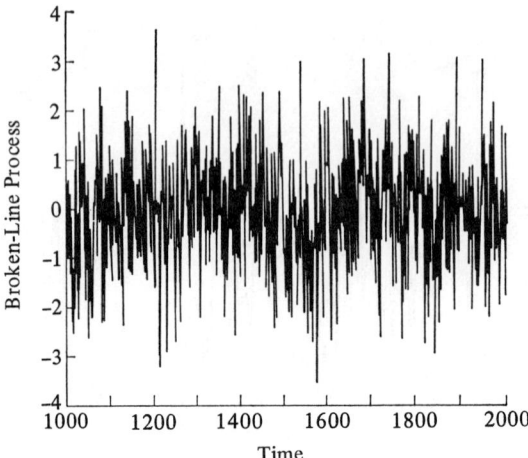

Figure 5.7 Sample realization of process 5, where $a_1 = 1$, $q = 5$, and $N = 5$ (from Mejía et al., 1972).

very short memory (process 4) to very large memory (process 2). Figure 5.11 shows the spectral densities for the five processes. It includes cases in which a narrow band of frequencies is present (process 1), as well as a case approaching white noise (process 4). Figures 5.12 to 5.14 show the growth of the Hurst range with time. These figures show that the transition region increases as the memory of the process increases. The transition region is the time period before the process yields a Hurst coefficient of 0.5. Process 4, which has the

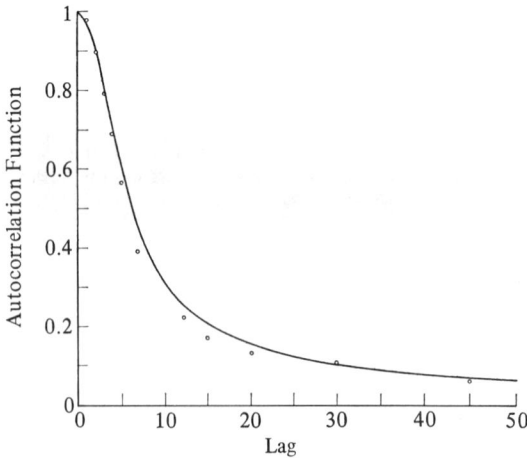

Figure 5.8 Autocorrelation function for process 1, where $a_1 = 5$, $q = 1.5$, and $N = 20$ (from Mejía et al., 1972).

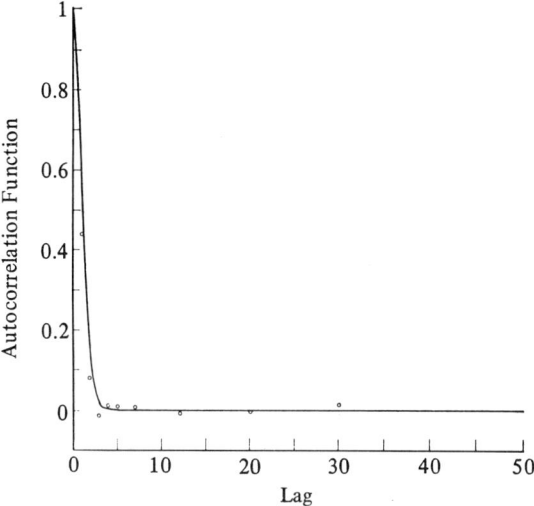

Figure 5.9 Autocorrelation function for process 4, where $a_1 = 1$, $q = 1.1$, and $N = 10$ (from Mejía et al. 1972).

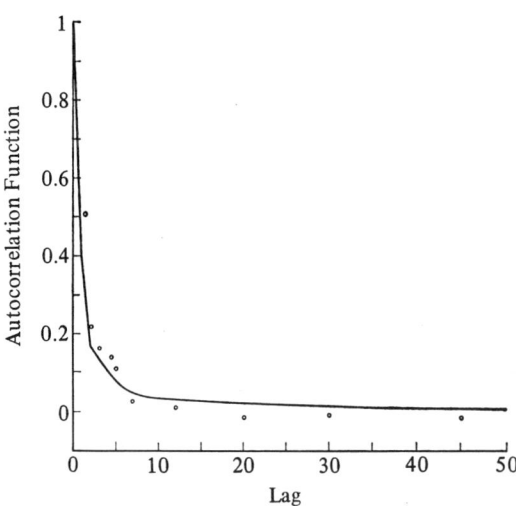

Figure 5.10 Autocorrelation function for process 5, where $a_1 = 1$, $q = 5$, and $N = 5$ (from Mejía et al. 1972).

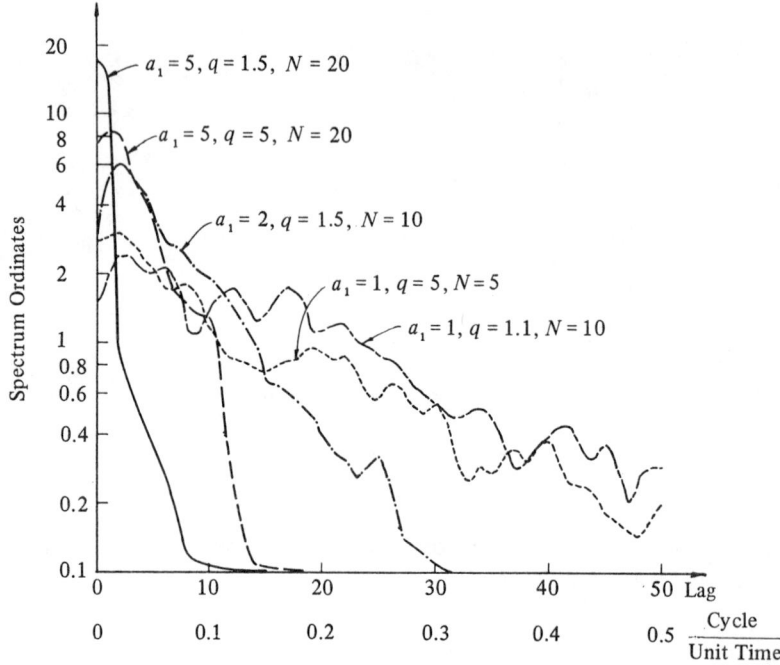

Figure 5.11 Spectral densities for the five processes analyzed (from Mejía et al., 1972).

shortest memory, will reach the asymptotic value of 0.5 for the Hurst coefficient in a short time (Fig. 5.13). However, process 1, which has an extremely long memory, still exhibits the Hurst phenomenon for a sample size of 10,000.

Another peculiarity of the broken-line model is its ability to preserve the second derivative at zero lags of the covariance function $\psi''(0)$. It is the only model that can do this explicitly, at the expense of not preserving the lag-one correlation. The details of how the fitting to a given second derivative at zero lag is achieved is discussed in the Appendix. The potential importance of this parameter is the role it plays in establishing extreme value and crossing properties of Gaussian processes. This is discussed in detail in Section 5.8. Section 5.8.3 will discuss the implications of using the broken-line model to preserve $\psi''(0)$ and will illustrate the performance of this model in simulating crossing properties.

5.8 CROSSING THEORY

Crossing theory deals with the properties of excursions of random processes above or below threshold values of the process. Since hydrologists are usually interested in extreme events, this translates to the study of floods and droughts.

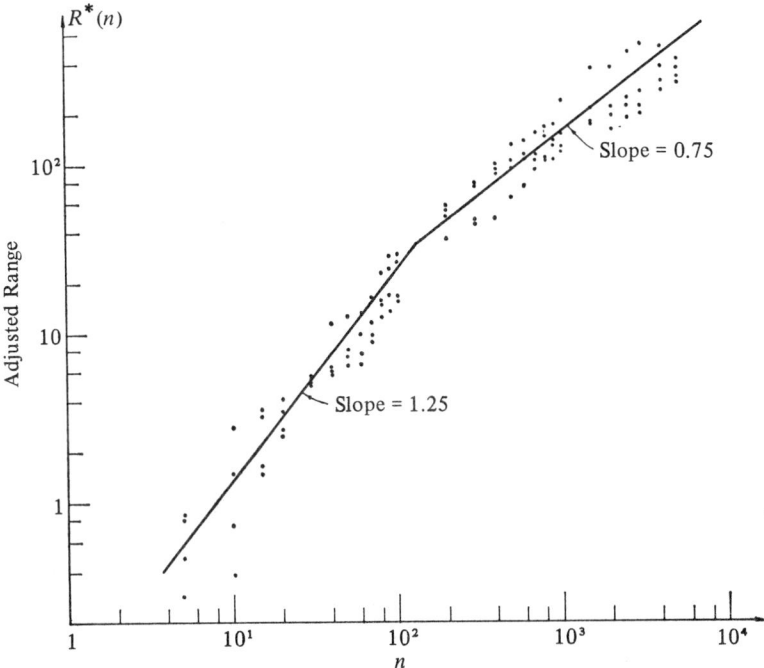

Figure 5.12 Growth of the adjusted range with time for process 1, where $a_1 = 5$, $q = 1.5$, and $N = 20$ (from Mejía et al., 1972).

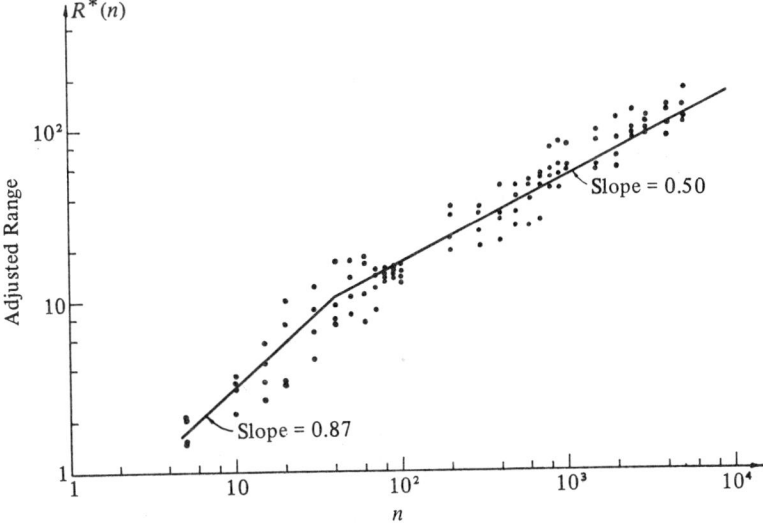

Figure 5.13 Growth of the adjusted range with time for process 4, where $a_1 = 1$, $q = 1.1$, and $N = 10$ (from Mejía et al., 1972).

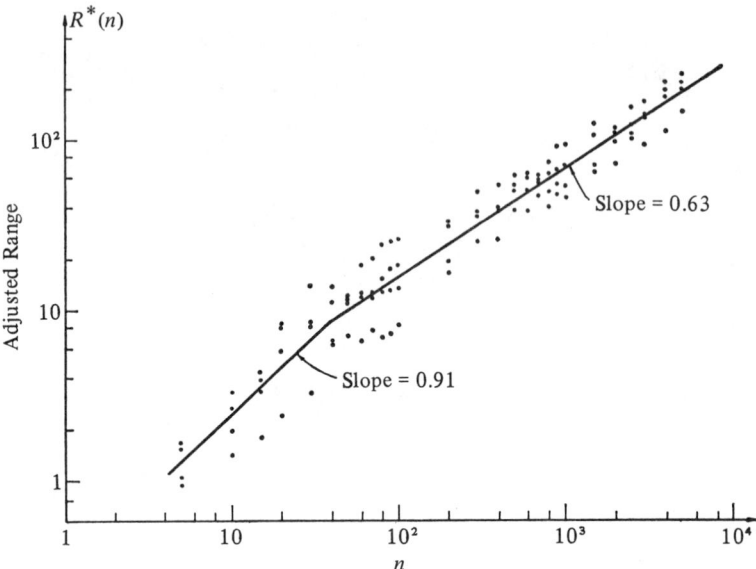

Figure 5.14 Growth of the adjusted range with time for process 5, where $a_1 = 1$, $q = 5$, and $N = 5$ (from Mejía et al., 1972).

Some of the quantities of interest are illustrated in Fig. 5.15 for a continuous random process $\xi(t)$ (normalized to have zero mean and unit variance). An upcrossing is defined as the point where the process cuts a specified level a with positive slope. A downcrossing is a cut of the threshold with a negative slope. Other quantities of interest are the time between upcrossings B_u^+ (time between floods), the duration of excursions above or below the threshold value (ℓ_u^+ or ℓ_{-u}^-), the duration of excursions above or below the mean ℓ_0, and the area (volume of flood) of an excursion above (below) the threshold level u ($-u$), A_u^+ (or A_u^-). A natural extension to discrete random processes can be made.

It is necessary to emphasize the importance of the above statistics in hydrology. If the process $\xi(t)$ represents river flow, the level u may be such that ℓ_u^+ represents the length of a flooding period and A_u^+ the excess volume above storage capacity. B_u^+ would be the time between successive floodings. Alternatively, $-u$ may be such that ℓ_{-u}^- represents the length of a drought period, A_u^- the deficit volume of water, and B_u^- the time between droughts.

For Gaussian random processes, the topic of excursions and intervals between crossings have been investigated by Rice (1954) and by Longuet-Higgins (1962), and more extensively by Cramer and Leadbetter (1967). Theoretical equations exist that relate the expected value of most of the above statistics to the correlation function. The crossing statistics of continuous processes depend on the second derivative at the origin of the correlation function, $\rho''(0)$. The behavior of the crossings for discrete processes depends mainly on the lag-one correlation of the process $\rho(1)$.

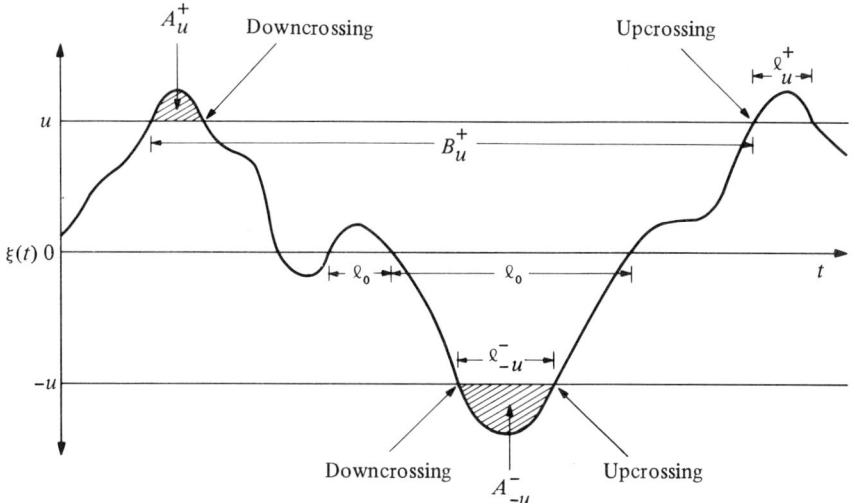

Figure 5.15 Illustration of crossing statistics for continuous random processes.

5.8.1 Continuous Processes

Let $\xi(t)$ denote a real-value, normal, continuous-parameter stationary process having, for convenience, zero mean. The $2i$th moment of the spectral function $\Phi(\omega)$ is denoted by λ_{2i}

$$\lambda_{2i} = \int_{-\infty}^{\infty} \omega^{2i}\Phi(\omega)\, d\omega, \qquad i = 0,1,2,\ldots \tag{5.81}$$

where ω is the angular frequency.

The mean number of crossings of the level u by $\xi(t)$ during the interval $(0, T)$ is given by Rice (1954) as

$$E[C_u(0,T)] = \frac{T}{\pi}\left(\frac{\lambda_2}{\lambda_0}\right)^{1/2} e^{-u^2/2\lambda_0}. \tag{5.82}$$

The total number of crossings C_u is made up of an equal number of upcrossings U_u and downcrossings D_u:

$$E[U_u] = E[D_u] = E\left[\frac{C_u}{2}\right].$$

Rice (1954) has shown that if the autocovariance function $\psi(\tau)$ has a finite $2i$th derivative at the origin (denoted by $\psi^{2i}(0)$), then it holds true that

$$\lambda_{2i} = (-1)^i \psi^{2i}(0) = (-1)^i \sigma^2 \rho^{2i}(0). \tag{5.83}$$

For this reason, the results in crossing theory for a continuous process can be written either in a function of λ_2 or of $\rho''(0)$.

Let the function $F_1(t)$ represent the conditional cumulative probability that there is at least one crossing in the interval $(0, t)$, given an upcrossing occurring at $t = 0$. That is, $F_1(t)$ may be regarded as the probability that the duration of an arbitrary upward excursion will not exceed t. Thus, it is possible to intuitively regard $F_1(t)$ as the distribution function of the length of an upward excursion. In a similar manner let $F_2(t)$ be the distribution function of the length of the interval between an arbitrarily chosen upcrossing and the next upcrossing.

To our knowledge the functions $F_1(t)$ and $F_2(t)$ have not yet been obtained in an explicit manner, but Cramer and Leadbetter (1967) gave the moments of these distributions as

$$\int_0^\infty t \, dF_1(t) = \mu^{-1}\left[P\{\xi(0) > u\} - v_0(\infty)\right] \tag{5.84}$$

$$\int_0^\infty t \, dF_2(t) = \mu^{-1}\left[1 - u_0(\infty)\right], \tag{5.85}$$

and for $n > 1$

$$\int_0^\infty t^n \, dF_1(t) = n(n-1)\mu^{-1}\int_0^\infty \left[v_0(t) - v_0(\infty)\right] t^{n-2} \, dt \tag{5.86}$$

$$\int_0^\infty t^n \, dF_2(t) = n(n-1)\mu^{-1}\int_0^\infty \left[u_0(t) - u_0(\infty)\right] t^{n-2} \, dt, \tag{5.87}$$

where

$$\mu = E\left[U_u(1)\right] = \frac{1}{2\pi}\left(\frac{\lambda_2}{\lambda_0}\right)^{1/2} e^{-u^2/2\lambda_0} \tag{5.88}$$

is the mean number of upcrossings of the level u per unit time and

$$u_0(t) = P\{U_u(0, t) = 0\}$$
$$v_0(t) = P\{\xi(0) > u, C_u(0, t) = 0\}$$
$$u_0(\infty) = \lim_{t \to \infty} u_0(t)$$
$$v_0(\infty) = \lim_{t \to \infty} v_0(t).$$

Since $u_0(t)$ is the probability of no upcrossing in the time interval $(0, t)$, it is clear that in many situations the limit $u_0(\infty)$ will be zero. This is not always the case, however, and there are many common processes that have $u_0(\infty)$ different from zero. Similar remarks apply to $v_0(\infty)$.

Equations (5.84) through (5.87) assume that $\xi(t)$ is a strictly stationary process. When $\xi(t)$ is stationary—meaning second-order stationary—and normal, it automatically becomes a special class of a strictly stationary process. In the latter case, and also when $\xi(t)$ is an ergodic process, Cramer and Leadbetter (1967) have shown that $u_0(\infty) = v_0(\infty) = 0$ and the means of $F_1(t)$

and $F_2(t)$ are given by

$$\int_0^\infty t \, dF_1(t) = \mu^{-1} P\{\xi(0) > u\} \tag{5.89}$$

$$\int_0^\infty t \, dF_2(t) = \mu \tag{5.90}$$

with corresponding simplifications for the higher moments.

In the case where $\xi(t)$ has a nonzero mean m, it is only necessary to replace u by $u - m$ in the above formulas, since $\xi(t)$ crosses the level u when $\xi(t) - m$ crosses $u - m$.

The total area cut off above an arbitrary level by this process has been investigated by Cramer and Leadbetter (1967) and represented by Z_n. Defining for any non-negative integer n

$$\begin{aligned} \eta_n(t) &= (\xi(t) - u)^n && \text{if} && \xi(t) > 0 \\ &= 0 && \text{otherwise} && \end{aligned} \tag{5.91}$$

and

$$Z_n(t) = \frac{1}{T} \int_0^T \eta_n(t) \, dt \qquad n = 0, 1, 2, \ldots . \tag{5.92}$$

Z_0 will be the proportion of time of $(0, T)$ that $\xi(t)$ spends above the level u. $T Z_1(T)$ is the area cut off by the process above the level u in $0 \le t \le T$. $E[Z_1(T)]$ is the expected value of the area bounded by $\xi(t)$ above the level u per unit time (Nordin and Rosbjerg, 1970):

$$E[Z_1(T)] = \int_u^\infty (x - u) \frac{1}{\sigma\sqrt{2\pi}} e^{-x^2/2\sigma^2} \, dx. \tag{5.93}$$

The area bounded by $\xi(t)$ above level u is denoted by A_u^+. The probability distribution of A_u^+ has not been obtained, but its mean value can be estimated by dividing the area cut off by the process $\xi(t)$ above the level u by the number of excursions of $\xi(t)$ above u. Nordin and Rosbjerg (1970) gave this result as

$$E[A_u^+] = 2 E[Z_1(T)] E[\ell_0] e^{u^2/2\sigma^2}. \tag{5.94}$$

The term $E[\ell_0]$ represents the expected interval between zero crossings:

$$E[\ell_0] = \pi \left[\frac{\psi(0)}{-\psi''(0)} \right]^{1/2} \tag{5.95}$$

There exist several methods for estimating λ_2 or $\rho''(0)$. The most common ones are

a. Estimation of the second spectral moment using the sample spectrum (Rodríguez-Iturbe, 1969, and Mejía, 1971);

b. Estimation of $\rho''(0)$ as a finite difference of the discrete-correlation function (Rodríguez-Iturbe, 1968), which leads to

$$\rho''(0) = 2\rho(1) - 2, \tag{5.96}$$

where $\rho(1)$ is the lag-one autocorrelation estimate; and

c. Estimation by fitting the number of crossings of the mean value of the sample.

The experiments of Bras et al. (1981) show that methods (b) and (c) are more adequate.

The explicit theoretical expressions are really valid only for Gaussian processes. The explicit non-Gaussian solution remains one of the unsolved problems of probability theory. Nevertheless, Nordin and Rosbjerg (1970) have analyzed streamflow series of 12 rivers throughout the world and concluded that the theoretical relationships were acceptable approximations even for clearly non-Gaussian streamflows. The agreement was better as normality conditions were approached. Rodríguez-Iturbe (1968, 1969) tested the theory on data from the Orinoco River (in Venezuela) and the Rhine River (in Switzerland) with similar favorable results. Bras et al. (1981) tested the theory on 31 rivers throughout the world. Their results corroborated some of the previously quoted conclusions but showed a stronger dependence on normality for good results. Some observed results were

1. Crossing theory, as expected, becomes very useful in predicting extreme-value statistics as the normality assumption is approached. Normality should be tested with at least three criteria:
   a. low skewness
   b. kurtosis near 3, and
   c. high level of acceptance based on a χ^2 test.

2. Mean area bounded by upward and downward excursions, and total area above and below a given level, is underestimated and overestimated, respectively, by the Gaussian theory when positive skewness is present. The reverse is true for negative skewness. The kurtosis affects the magnitude of the error in estimation.

3. Although no particular trend was detected on the behavior of other statistics (e.g., number of crossings, run length), it is clear they were still sensitive (although to a lesser degree) to the normality assumption.

4. The series analyzed were discrete, nevertheless, both continuous and discrete crossing theory results (based on $\rho''(0)$ and $\rho(1)$, respectively) proved to be valid.

Figure 5.16 shows good agreement of theory and reality for the nearly Gaussian behavior of the Rhine at Basel, Switzerland, as computed by Bras et al. (1981). Figure 5.17, from the same authors, shows considerable diver-

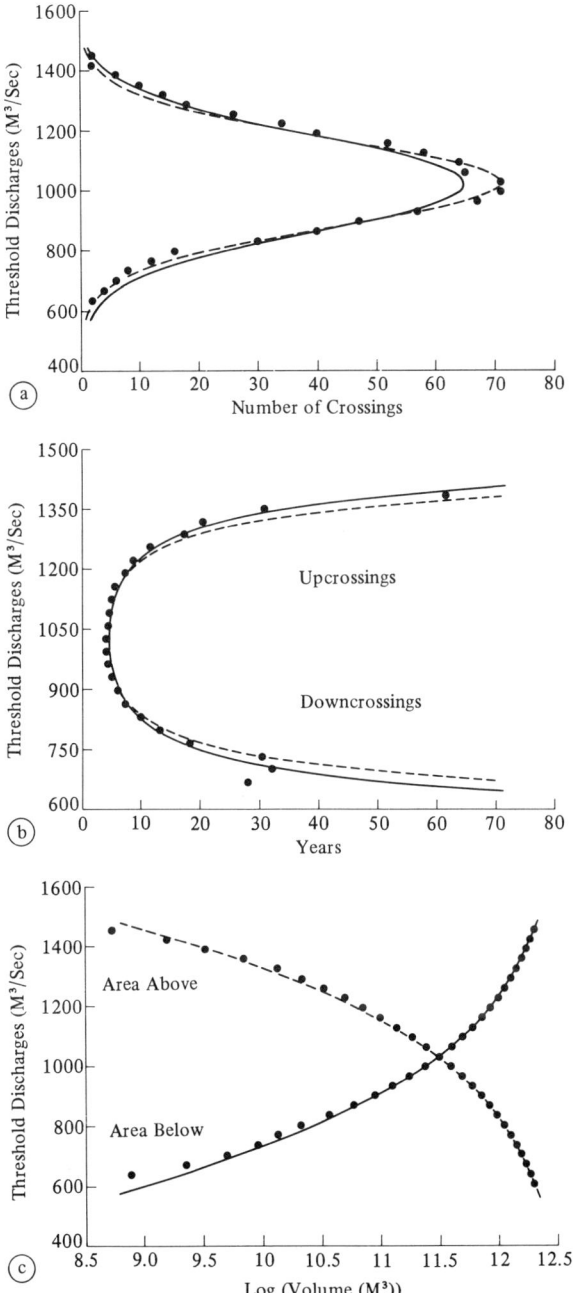

Figure 5.16 Threshold discharge values vs.: (a) mean number of crossings; (b) mean time between successive up- and downcrossings; and (c) mean total area above or below an excursion, as theoretically predicted and estimated from historical data of the Rhine River at Basel, Switzerland. See Table 5.8 for relevant data and explanation of symbols. (From Bras et al., 1981).

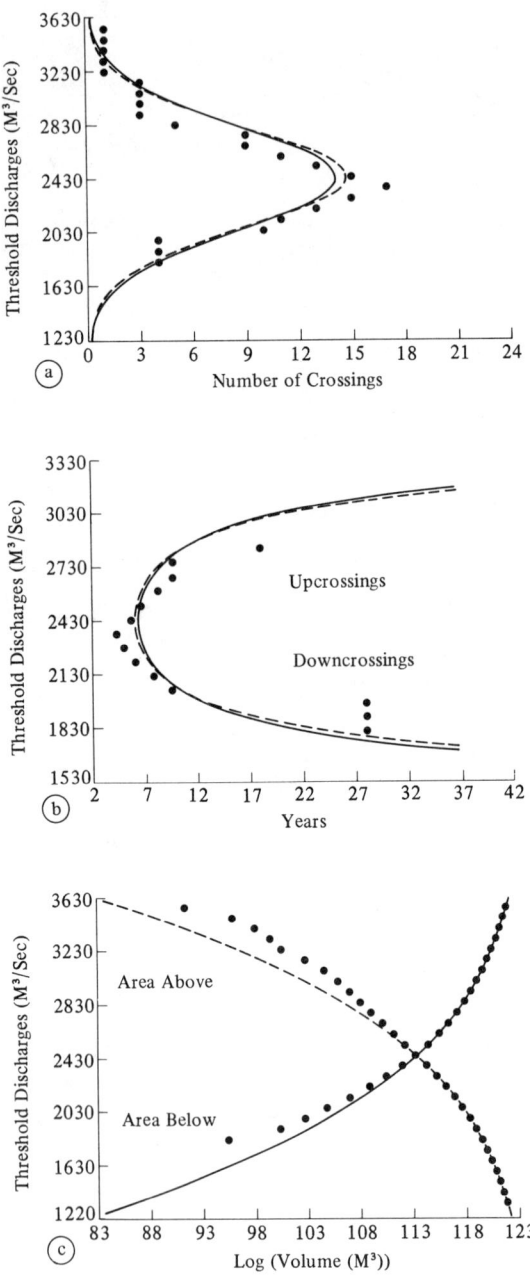

Figure 5.17 Threshold discharge values vs.: (a) mean number of crossings; (b) mean time between successive up- and downcrossings; and (c) mean total area above or below an excursion, as theoretically predicted and estimated from historical data of the Nile River at Hassanab, Sudan. See Table 5.9 for relevant data and explanation of symbols. (From Bras et al., 1981).

Table 5.8
Data and explanation of symbols in Fig. 5.16 for the Rhine River at Basel, Switzerland

Period of record:	1807–1957	Skewness coefficient	0.15
Mean flow (m^3/s)	1,025.75	Kurtosis coefficient	2.80
Standard deviation (m^3/s)	162.83	Lag-one correlation	0.08

Estimates of the second derivative at the origin:

spectral method	finite difference	crossings of the mean
−1.85	−2.94	−2.21

Symbols:

— = finite-difference estimator; ---- = discrete theory predictions; and ... = historical observations

Source: Bras et al. (1981).

gence of Gaussian results and data at Hassanab in the Nile River Basin. (Some of the divergence may be due to incorrect and limited data.) Tables 5.8 and 5.9 give data relevant to the above two figures.

5.8.2 Discrete Processes

Where observations are given as a series of mean values, each of which is assumed to apply over a given time interval Δt, the continuous curve of Fig. 5.15 should be replaced by the zero-mean, variance-1 step function shown in Fig. 5.18—ξ_k, $k = 1, 2, \ldots, n$. The crossing characteristics of ξ_k have been studied by Nordin and Rosbjerg (1970). This section follows their work closely.

Table 5.9
Data and explanation of symbols in Fig. 5.17 for the Nile River at Hassanab, Sudan

Period of record	1908–1952	Skewness coefficient	0.68
Mean flow (m^3/s)	2,436.68	Kurtosis coefficient	3.49
Standard deviation (m^3/s)	395.5	Lag-one correlation	0.50

Estimates of the second derivative at the origin:

spectral method	finite difference	crossings of the mean
−1.00	−1.64	−1.15

See Table 5.8 for explanation of symbols.
Source: Bras et al. (1981).

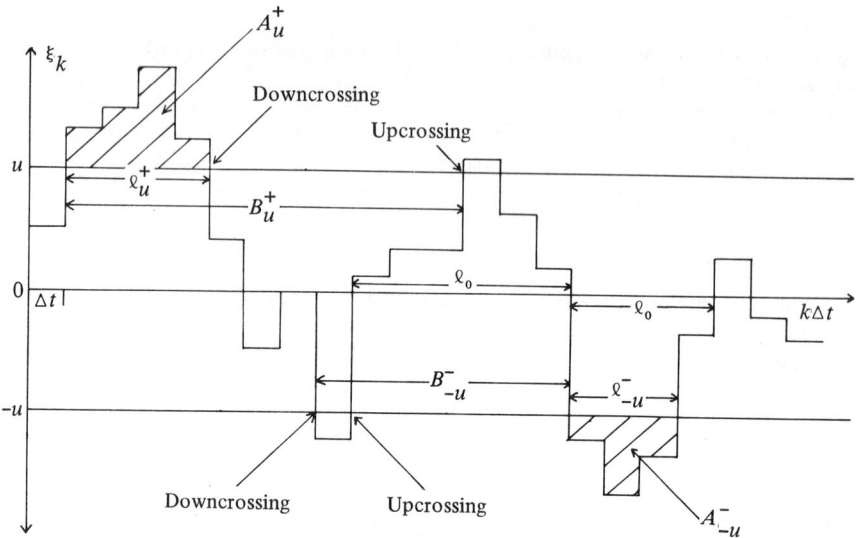

Figure 5.18 Definition sketch for a discrete time series where each value ξ_k is assumed to apply for the time interval Δt.

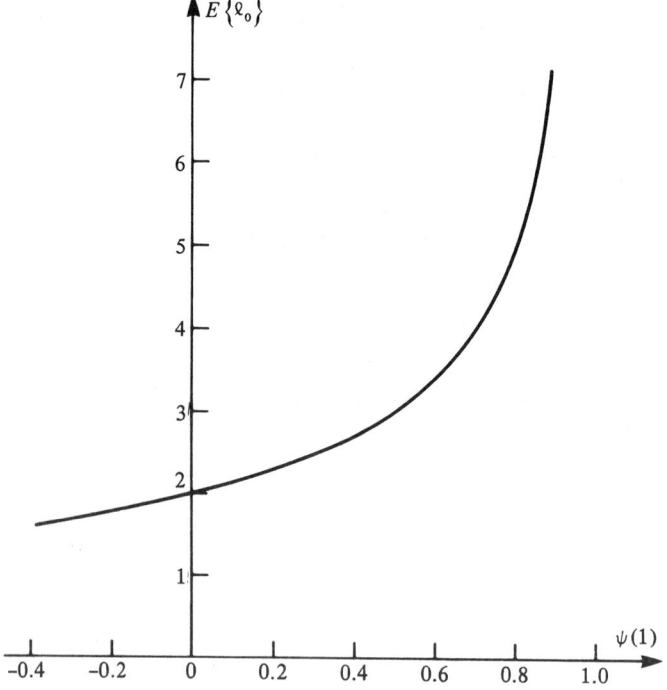

Figure 5.19 Average interval between zero crossings, $E\{\ell_0\}$, for the discrete time series (from Nordin and Rosbjerg, 1970).

The expected interval between zero crossings $E\{\ell_0\}$, is the reciprocal of the probability that two successive values (ξ_k, ξ_{k+1}), denoted by ξ_1 and ξ_2, have opposite signs. For a Gaussian process with zero mean and unit variance and with $\Delta t = 1$, this probability is given by Tick and Shaman (1966) as

$$P(\xi_1 < 0, \xi_2 > 0) + P(\xi_1 > 0, \xi_2 < 0) = \frac{1}{2} - \frac{1}{\pi} \arcsin \psi(1). \qquad (5.97)$$

Hence, $E\{\ell_0\}$, plotted in Fig. 5.19, is given by

$$E\{\ell_0\} = \left(\frac{1}{2} - \frac{1}{\pi} \arcsin \psi(1)\right)^{-1}, \qquad (5.98)$$

where $\psi(1)$ is the lag-one autocovariance or first serial correlation coefficient if the process has unit variance.

The average number of upcrossings per unit time of the level u by the step function ξ_k is

$$\begin{aligned}E[U_u(0,1)] &= \tfrac{1}{2}\{P[\xi_1 > u, \xi_2 < u] + P[\xi_1 < u, \xi_2 > u]\} \\ &= P[\xi_1 > u] - P[\xi_1 > u, \xi_2 > u].\end{aligned} \qquad (5.99)$$

The values of $E[U_u(0,1)]$ are most readily determined from tabulated values of the bivariate normal distributions. Curves showing values of $E[U_u(0,1)]$ for selected values of $\psi(1)$ are given in Fig. 5.20.

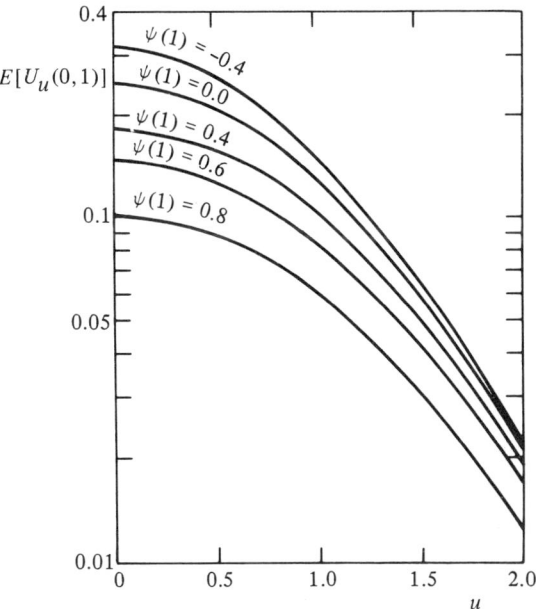

Figure 5.20 Expected number of upcrossings per unit time $E[U_u(0,1)]$ for the discrete time series (from Nordin and Rosbjerg, 1970).

The average length of the excursions of the function ξ_k above the level u is

$$E[\ell_u^+] = \frac{P[\xi > u]}{P[\xi > u] - P[\xi_1 > u, \xi_2 > u]}. \tag{5.100}$$

The joint distribution of ξ_1 and ξ_2 is bivariate normal with covariance $\psi(1)$, so the second term in the denominator is

$$P[\xi_1 > u, \xi_2 > u]$$
$$= \frac{1}{2\pi[1 - \psi^2(1)]^{1/2}} \int_u^\infty \int_u^\infty \exp\left\{-\frac{x_1^2 - 2x_1x_2 + x_2^2}{2[1 - \psi^2(1)]}\right\} dx_1\, dx_2.$$

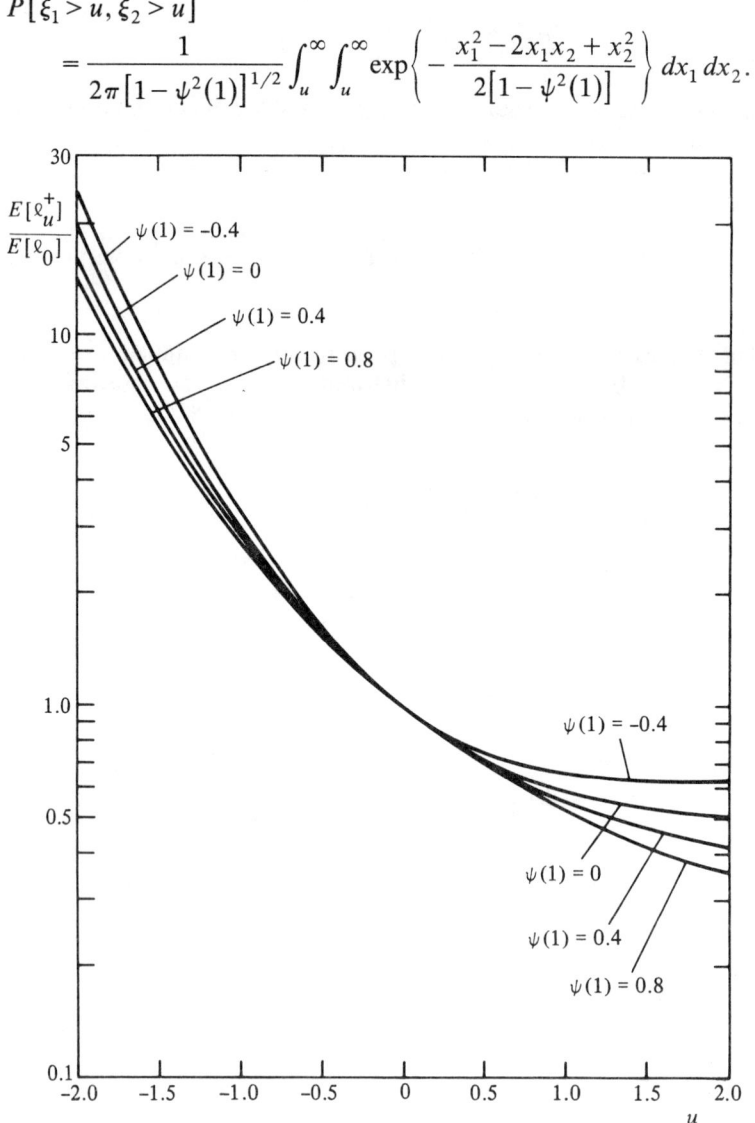

Figure 5.21 The ratio $E[\ell_u^+]/E[\ell_0]$ for the discrete time series (from Nordin and Rosbjerg, 1970).

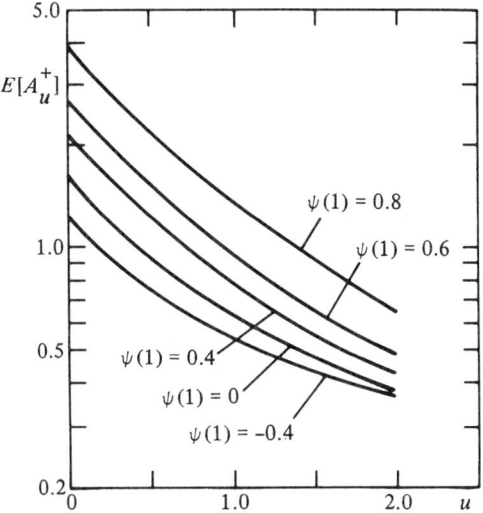

Figure 5.22 Expected area bounded by ξ_k and u for a single excursion of ξ_k above the level u (from Nordin and Rosbjerg, 1970).

The ratio $E[\ell_u^+]/E[\ell_0]$ for the discrete-time series is shown in Fig. 5.21 as a function of both u and $\psi(1)$.

The expected area per unit time bounded by the process $\xi(t)$ above the level u given in Eq. (5.93) is independent of time and applicable to both continuous and discrete series. The mean area per excursion of ξ_k above u is

$$E[A_u^+] = \frac{E[Z_1(T)]}{E[U_u(0,1)]},\qquad (5.101)$$

where the numerator is given by Eq. (5.93) and the denominator by Eq. (5.99). Figure 5.22 shows the behavior of $E[A_u^+]$ as a function of level u and $\psi(1)$.

5.8.3 Simulation of Crossing Characteristics

This section illustrates the properties of the broken-line streamflow generator with special emphasis on the crossing characteristics of the historical and simulated sequences. The experiments and results are those reported by Bras et al. (1981).

The historical streamflow records for four Nile River stations (Wadi Halfa, Mongalla, Roseires, and Malakal) were analyzed. A broken-line model was

Table 5.10
Common statistics for four Nile River stations

Data set		Mean (10⁶ m³/yr)	Standard deviation	Skewness	Lag-one
Wadi Halfa	historical (1905–1967)	85,627	12,741	0.025	0.184
	synthetic (100 yr)*	85,958	12,161	0.212	0.147
Mongalla	historical (1905–1967)	31,363	12,524	1.118	0.887
	synthetic (100 yr)*	31,151	11,154	0.797	0.858
Roseires	historical (1905–1967)	49,514	9,405	−0.182	0.162
	synthetic (100 yr)*	49,750	9,111	−0.037	0.191
Malakal	historical (1905–1967)	29,315	5,709	1.651	0.812
	synthetic (100 yr)*	29,306	5,145	0.628	0.762

*Average value over 200 runs.

Source: Bras et al. (1981).

then used to generate 200 multivariate sequences of 100 years each. Tables 5.10 and 5.11 show the historical statistics of the four stations, as well as the mean value of each statistic for the 200 synthetically generated sequences. In each run, the model was calibrated to preserve the lag-one correlation. It is clear that there is good agreement between the historical and synthetic sequences.

No adjustment for sample bias was made in either historical or generated statistics. This may be particularly important in regard to the skewness (see Bobée and Robitaille, 1975).

Figure 5.23 shows 106 years of annual historical discharges at Aswan. Of interest in this figure is the persistence of flows above or below the mean in groups of several years. In the 106 years shown, the hydrograph crosses the mean line 27 times. A series of 100 sequences of 100 years each were generated for Aswan (preserving $\rho(1)$, not $\rho''(0)$), and the number of crossings for each series was calculated. The average number of crossings for the 100 runs was 26.6, with the lowest and highest number of crossings being 16 and 36,

Table 5.11
Spatial correlation matrices for four Nile River stations

	Historical				Synthetic*			
Wadi Halfa	1.00	0.10	0.88	0.26	1.00	0.05	0.86	0.26
Mongalla	0.10	1.00	−0.04	0.88	0.05	1.00	−0.11	0.84
Roseires	0.88	−0.04	1.00	0.05	0.86	−0.11	1.00	0.12
Malakal	0.26	0.88	0.05	1.00	0.26	0.84	0.12	1.00

*Average value over 200 runs.

Source: Bras et al. (1981).

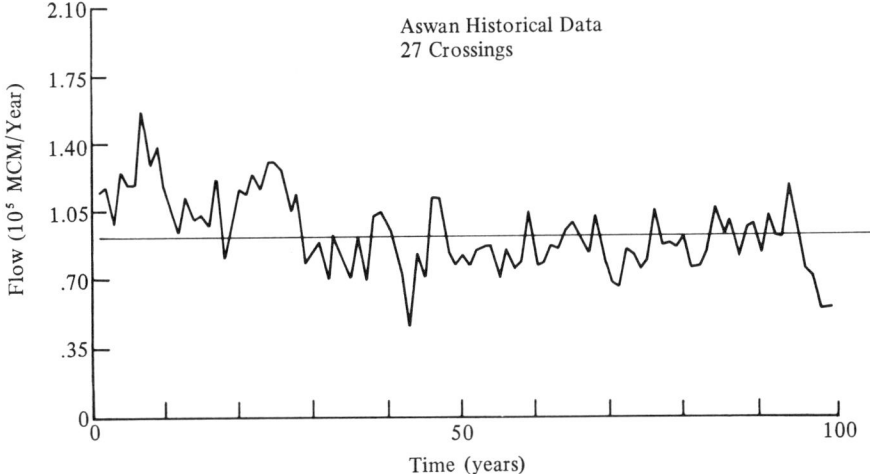

Figure 5.23 Historical streamflow sequences at Aswan, Nile River, Egypt (from Bras et al., 1981).

respectively. Figure 5.24 gives typical 100-year sequences of streamflows generated by the broken-line model for Aswan (assuming a Hurst coefficient of 0.72). Shown are runs with the average, medium, and maximum number of crossings, respectively, obtained from the 100-run experiment.

For comparison, 100 sequences of 100 years of annual streamflow for Aswan were generated with a multivariate autoregressive model (see Curry and Bras, 1978). This model does not attempt explicitly to preserve the Hurst coefficient. Figure 5.25 shows one particular sequence with 47 crossings, this being the average number of crossings over the 100 runs. Also shown are runs with the minimum and maximum number of crossings obtained from the 100-run experiment. Although it is hard to make conclusive remarks in comparing 100-year sequences, clearly, the Markov-generated sequences oscillate above and below the mean faster and with less persistency than the broken-line model.

The analyses reported in the literature cited throughout this section show that

1. Theory agrees well when the actual series is nearly Gaussian.
2. Both continuous and discrete theory do well when applicable. Furthermore, estimating $\rho''(0)$ with the finite difference approximation,

$$\rho''(0) = 2\rho(1) - 2, \qquad (5.102)$$

appears to be one of the better procedures.

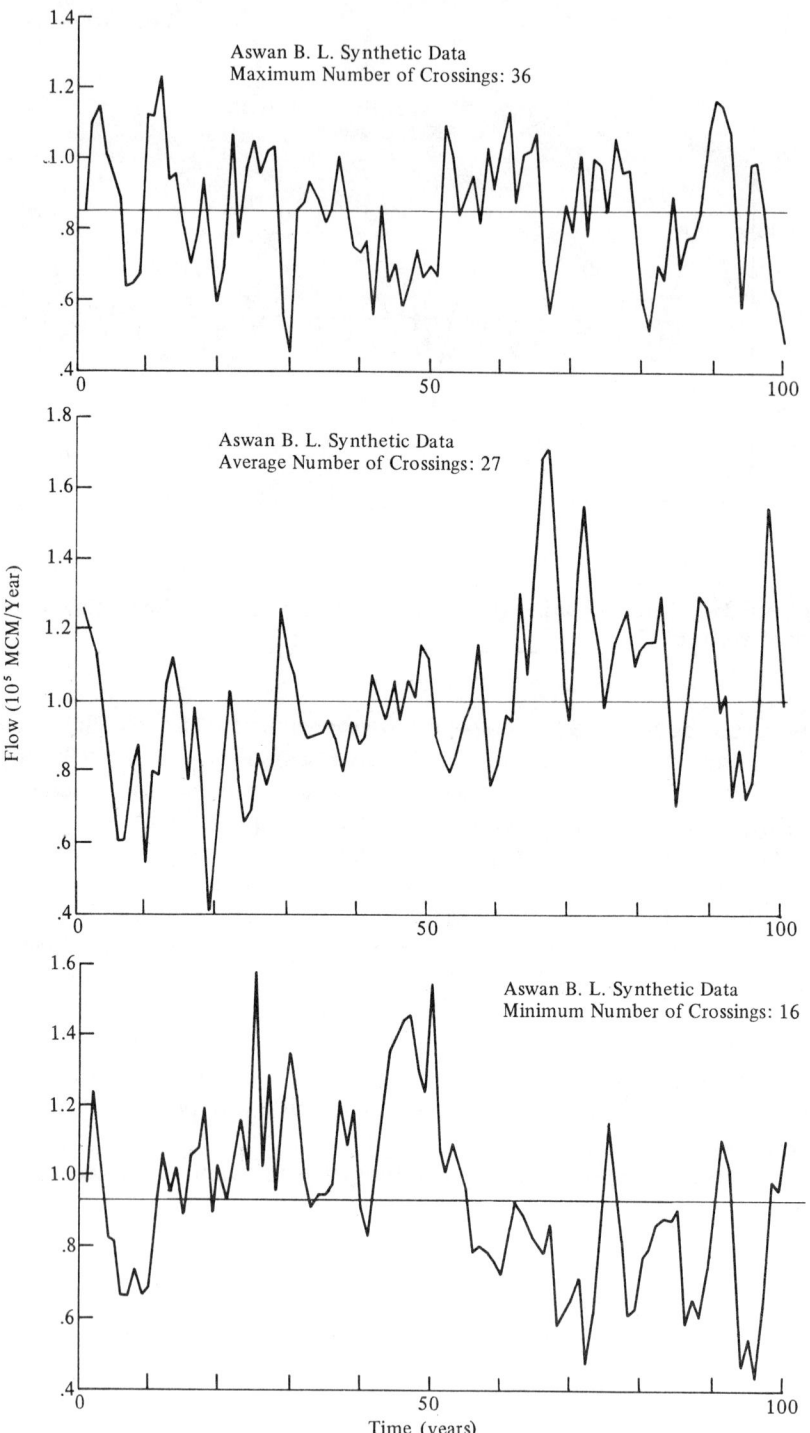

Figure 5.24 Broken-line-generated streamflow sequences at Aswan, Nile River, Egypt (from Bras et al., 1981).

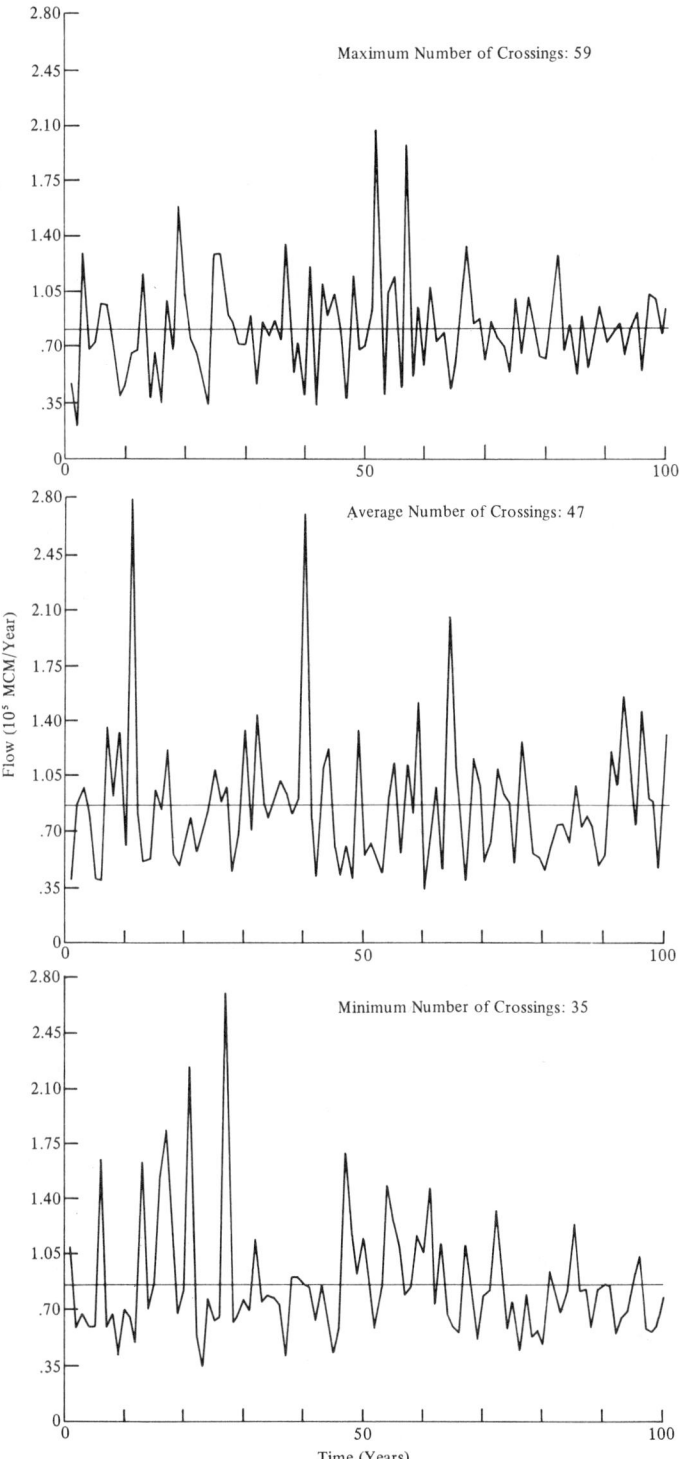

Figure 5.25 Typical streamflows at Aswan, Nile River, Egypt, generated by a multivariate autoregressive model (from Bras et al., 1981).

The above two conclusions imply that the broken-line model has no reason to preserve extreme value statistics except maybe for Gaussian or nearly Gaussian processes and that the preservation of $\rho(1)$ should be sufficient to do well in such cases. In other words, there seems to be no reason for preserving a value of $\rho''(0)$ independent of $\rho(1)$.

In fact, the broken-line model is unable to preserve arbitrary values of $\rho''(0)$ and $\rho(1)$. Figure 5.26 shows the region (shaded) of $\rho''(0)$ and $\rho(1)$ that the broken line can simultaneously preserve. Equation (5.102) is plotted as the straight line shown in the figure. Note that values of negative $\rho''(0)$ are limited between 0 and 2.

Bras et al. (1981) also used the broken-line model to simulate 1000 years of Gaussian discharge in the Rhine. As expected, all common statistics, including an assumed 0.72 Hurst coefficient, were very well preserved. Figure 5.27 shows the comparison of crossing-theory results to extreme value statistics computed from the 1000 years of synthetic streamflows. It is apparent that agreement is relatively good. Some disagreement exists in the curves corresponding to the mean area bounded per crossing and the total area bounded. The disagreement can be traced to an error in the mean area bounded above a threshold per unit of time. The results imply that the theory underestimates this quantity for the

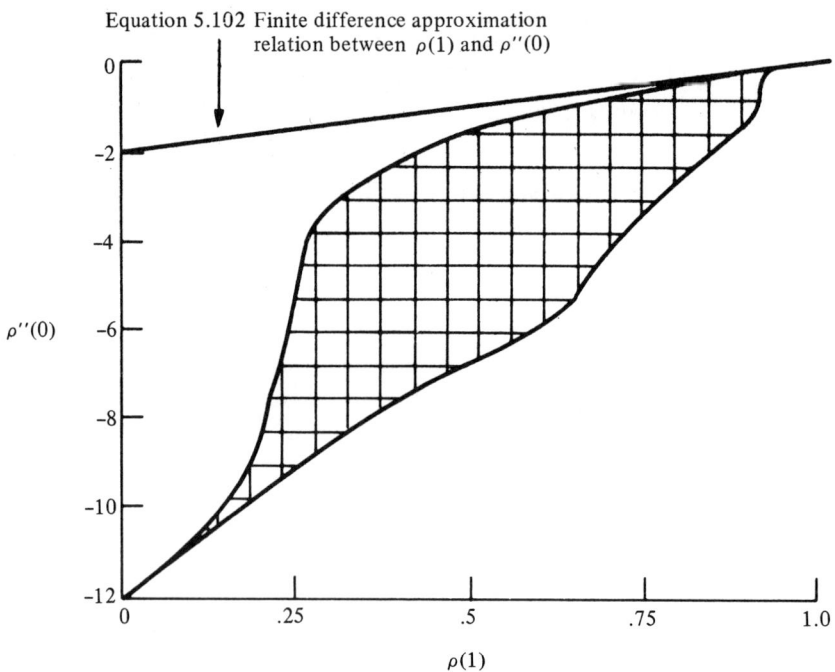

Figure 5.26 Region of $\rho(1)$ and $\rho''(0)$ that the broken line can preserve (from Bras et al., 1981).

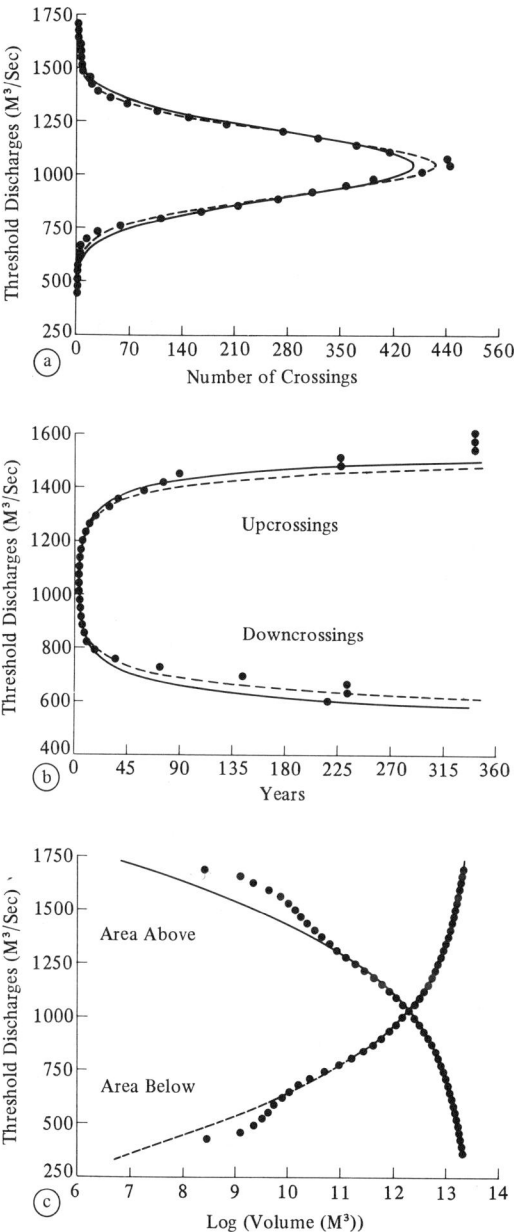

Figure 5.27   Threshold discharge values vs.: (a) mean number of crossings; (b) mean time between successive up- and downcrossing; and (c) mean total area above or below an excursion, as theoretically predicted and as computed from 1000 year of synthetic streamflows generated for the Rhine River at Basel, Switzerland, by the broken-line model. See Table 5.8 for relevant data and explanation of symbols. (From Bras et al., 1981).

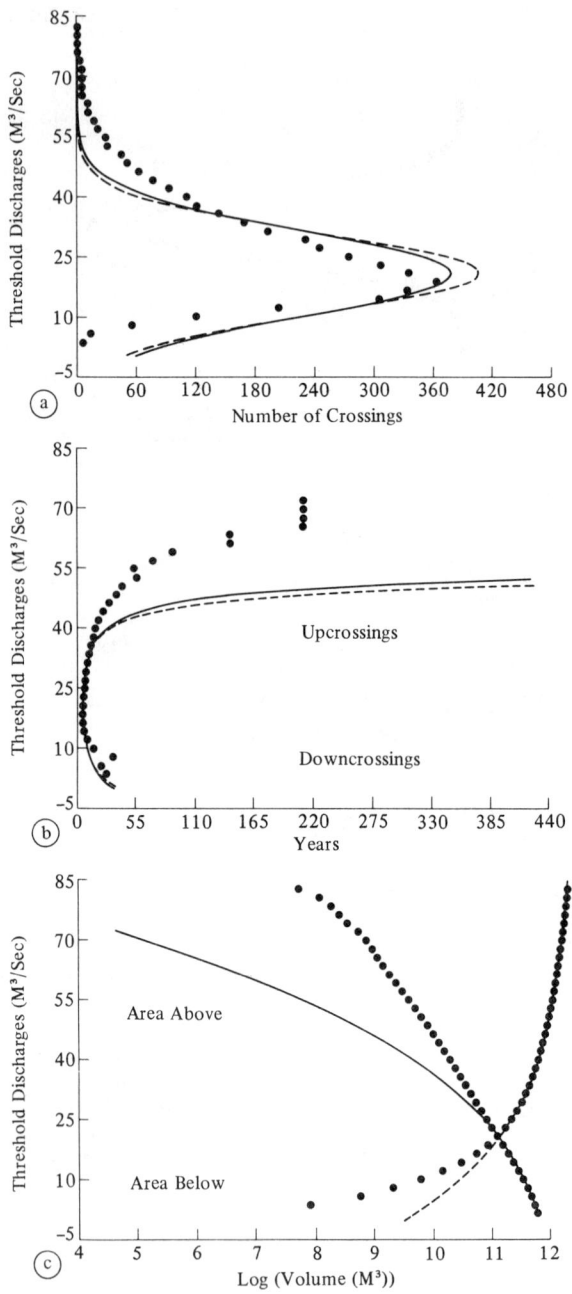

Figure 5.28 Threshold discharge values vs.: (a) mean number of crossings; (b) mean time between successive up- and downcrossing; and (c) mean total area above or below an excursion, as theoretically predicted and as computed from 1000 year of synthetic streamflows generated for Kiewa at Kiewa, Victoria, by the broken-line model. See Table 5.12 for relevant data and explanation of symbols. (From Bras et al., 1981).

Table 5.12
Data and explanation of symbols in Fig. 5.28 for the Kiewa River at Kiewa, Victoria

Period of record	1885–1957	Skewness coefficient	1.55
Mean flow (m^3/s)	20.64	Kurtosis coefficient	6.84
Standard deviation (m^3/s)	10.19	Lag-one correlation	0.29

Estimates of the second derivative at the origin:

spectral method	finite difference	crossings of the mean
-1.42	-2.09	-1.95

See Table 5.8 for explanation of symbols.
Source: Bras et al. (1981).

extremes. As expected, this quantity is strongly dependent on the Hurst coefficient. Since an assumed value was used in the experiment, there is no reason to doubt that the cause of the disagreement lies here.

As previously discussed, records strongly deviating from the normal assumption, particularly with high skewness, do not adjust well to crossing-theory results. Such a record (Kiewa at Kiewa, Victoria) was simulated with the broken-line model by Bras et al. (1981) and 1000 years of synthetic data were produced. A comparison of extreme-value statistics of the generated record with the theoretical predictions showed considerable divergence (Fig. 5.28, Table 5.12). The high positive skewness leads to a lesser number of crossings relative to theoretical results when threshold values are below the mean; and to more crossings relative to theoretical results when threshold values are above the mean. Mean negative run lengths are underestimated with the theory while the prediction of positive run lengths with discrete crossing theory is good. The mean time between successive upcrossings is overestimated by the theory and the mean time between downcrossings is underestimated. All of the above observations are logical results of high skewness. Skewness affects the mean area bounded per crossing and the total area above and below a threshold level in the same manner as previously discussed.

5.9 LONG-TERM PERSISTENCE AND RESERVOIR ANALYSIS

Streamflow-generating schemes belonging to the Brownian domain are generally short-memory models that cannot reproduce the high values of the Hurst coefficient (around $H = 0.7$) observed in long-duration–flow sequences. It is then natural to suspect that the use of such models can lead to underdesign of storage capacities. Nevertheless, it is well known that water-dependent systems tend to have a high degree of resiliency and can tolerate substantial temporal deficits without incurring the justifiably expected losses (Klemeš et al., 1981). The answer regarding the effect of long-term persistence in reservoir analysis lies mainly in the nature of the reliability measures used to assess the appropriateness of the design. Given the different socioeconomic considerations, conflicting interests, and changing priorities and standards, which vary from one case to another, there is no definite answer to the above question. Nevertheless it is important for the hydrologist to have some terms of reference that may be helpful in guiding his assessment of the importance of long-term persistence modeling in the design of reservoirs.

Lettenmaier and Burges (1977) compared three different models regarding the probability distributions of storage obtained through their use at a single site. Through simulation, they estimated the cumulative probability distributions of storage required to satisfy a given (constant) demand (expressed as demand $\div$ mean flow, $D^* = d/\mu_x$) over the operating life. Individual cumulative distributions were determined as follows: for each model, 1000 independent traces of annual flows, each having a length equal to the operating life of a hypothetical reservoir, were generated. Each trace was then processed, through the sequent peak algorithm (SPA) presented by Fiering (1967) to yield a value of required storage, such that if the reservoir were full at the onset of the worst critical period of the trace then the stored water plus inflow would be just sufficient to fully satisfy the demand. The ordered required storages for all traces then form the empirical probability distribution of storage.

Lettenmaier and Burges (1977) used three models in their simulations: fast fractional Gaussian noise, ARMA(1,1), and ARMA–Markov. The last of these models is a combination of the high-frequency lag-one Markov and low frequency-ARIMA(1,0,1). Our interest here is not in the differences among the models—which in any case were small and not very important—but in the changes of the cumulative distributions of storage for the different values of the parameter and specifically for different values of H.

Figures 5.29 and 5.30 show examples of the results. The skewness coefficient is denoted by G, CV is the coefficient of variation, ρ is the lag-one correlation, and N is the useful life of the storage. Even for a small difference of 0.75 to 0.80 in the Hurst coefficient, relatively large differences are obtained in the storage necessary to maintain a given degree of reliability when D^* is taken as 0.9. The study of Lettenmaier and Burges also shows that the storage requirement under the SPA is quite sensitive to the skewness, to the correlation ρ, and the useful life of the project. Additional information covering the sensitivity of reservoir size to variation in CV is shown by Burges and Lettenmaier (1975).

The SPA is a conservative algorithm that generates storage capacities that in many of the situations studied by Lettenmaier and Burges (1977) were impracticably large.

Klemeš et al. (1981) have shown conclusively how poor the nonfailure storage capacity can be as an indicator of reservoir performance. From this they go on to judge the adequacy of hydrologic models for reservoir analysis by the differences in hydrologic aspects of reservoir performance. Indeed the SPA does not distinguish between a case where a given storage would result in

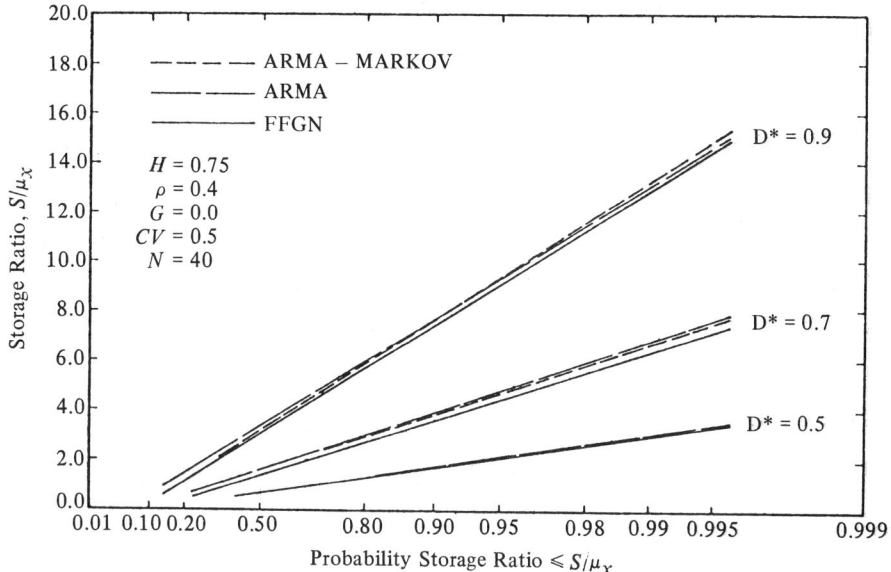

Figure 5.29 Probability storage diagram for $H = 0.75$, $G = 0$, and $N = 40$ years (from Lettenmaier and Burges, 1977).

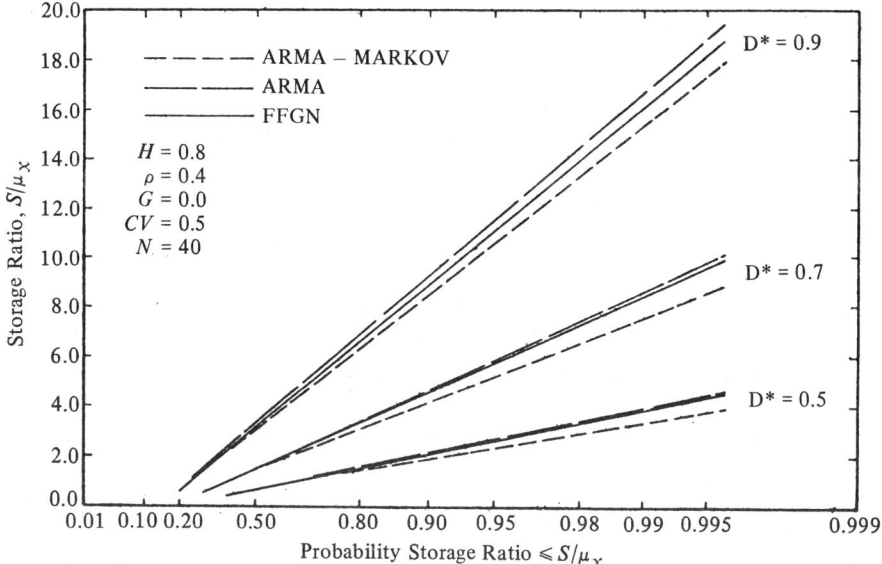

Figure 5.30 Probability storage diagram for $H = 0.80$, $G = 0$, and $N = 40$ years (from Lettenmaier and Burges, 1977).

a failure in water supply lasting, say, a year with supply falling to 10% of the target, and a case where a failure consists of a couple of days in which the supply is reduced to 95% of the target.

The most complete investigation of whether the use of longer-memory models is imperative to prevent serious underdesign has been carried out by Klemeš et al. (1981). The inflows to the reservoir were series of monthly flows obtained by disaggregation of annual flows generated by short- and long-memory models. The short-memory model was a lag-one Markov scheme and the long-memory process was simulated with a broken-line model. Klemeš et al. (1981) characterize the reliability of reservoir operation in three different ways:

a. Occurrence-based (annual) reliability R_a is the number of nonfailure years expressed as a percentage of the total number of years in the given period (equivalent to the probability of a nonfailure year).

b. Time-based reliability R_t is the total duration time of all nonfailure intervals expressed as a percentage of the total length of the given period. One may assume that the inflows are continuous within each month and are equal to the monthly mean flow.

c. Quantity-based reliability R_v is the actual amount of water supplied to the consumer expressed as a percentage of the total amount desired during the given period.

The following conclusions were obtained by Klemeš et al. (1981) regarding the difference in reliability resulting from the use of a short-memory inflow model instead of a long-memory model:

1. The short-memory inflow model leads to overestimation of reservoir performance reliability as compared to the long-memory model.

2. The overestimation is high for annual reliability (underestimation of the failure-year probability), lower for time-based reliability, and lowest for quantity-based reliability.

3. The overestimation of all three reliability characteristics is very small for reservoirs with storages up to $S/\mu_x = 1$ (mean annual flow volume) for any draft ratio. For reservoirs with a storage ratio larger than 1, the overestimation is very small for draft ratios up to about 0.6 to 0.8 and over 1.1, depending on the inflow parameters CV and ρ. This means that the length of memory in the annual inflow series is irrelevant if the regulation itself has no memory over one year. If the draft is low, the reservoir fills up every year; if it is high, the reservoir empties every year: in both cases, only seasonal regulation is involved.

4. Overestimation is maximal for draft ratios close to 1 and increases with storage ratios up to about 3.

5. The overestimation slightly increases with the coefficient of variation of the annual flows and decreases with the increase of the lag-one correlation.

The uncertainties existing in the statistical parameters of the streamflow simulation models and in the socioeconomic considerations influencing the design of a reservoir may have more weight in the final analysis than the choice of a short-memory model versus a long-memory one. Nevertheless, this does not justify ignoring the well-established concepts of long-memory analysis, especially for draft ratios close to 1.

Appendix to Chapter 5

The Broken-Line
Model — Details

A.1   PRESERVATION OF HURST COEFFICIENT WITH THE BROKEN-LINE MODEL: THE UNIVARIATE CASE

The most complete description and set of experiments related to the implementation of the broken-line processes as a fully operational model for multivariate hydrologic simulation is that of Curry and Bras (1978). This Appendix closely follows their work.

As pointed out in Section 5.3, fractional Gaussian noise (Mandelbrot and Van Ness, 1968) is a stationary process, with infinite memory, which reproduces the Hurst phenomena. The correlation function of the process is given by

$$C(\tau, H) = \tfrac{1}{2}\big[|\tau+1|^{2H} - 2|\tau|^{2H} + |\tau-1|^{2H}\big], \qquad (A.1)$$

which for relatively large τ, can be approximated as

$$C_1(\tau, H) \approx H(2H-1)\tau^{2H-2}. \qquad (A.2)$$

The error involved in this approximation tends to zero very rapidly as $\tau \to \infty$.

This correlation function can be reproduced by adding an infinite number of weighted simple broken-line processes of the same variance, σ_ξ^2, in which the parameter a varies continuously from 0 to ∞. Using the continuous equivalent of Eq. (5.77) and following the presentation of Curry and Bras, we have

$$C_1(\tau, H) = \frac{\int_0^\infty \sigma_\xi^2 f(a) \rho(\tau, a)\, da}{\int_0^\infty \sigma_\xi^2 f(a)\, da} = \frac{\int_0^\infty f(a) \rho(\tau, a)\, da}{\int_0^\infty f(a)\, da}, \quad (A.3)$$

where $\rho(\tau, a)$ is the correlation function of the simple broken-line process with parameter a at lag τ and $f(a)$ is the square of the weight given to each one of the simple broken lines with parameter a.

The correlation function in Eq. (5.73) can be rewritten as

$$\rho(\tau, a) = \begin{cases} 1 - \dfrac{3}{4}\left(\dfrac{\tau}{a}\right)^2\left(2 - \dfrac{\tau}{a}\right) & a \geq \tau \\[2mm] \dfrac{1}{4}\left(2 - \dfrac{\tau}{a}\right)^3 & \dfrac{\tau}{2} < a < \tau \\[2mm] 0 & a \leq \dfrac{\tau}{2}. \end{cases} \quad (A.4)$$

Assuming that

$$\int_0^\infty f(a)\, da = 1, \quad (A.5)$$

Eq. (A.3) becomes

$$\begin{aligned} C_1(\tau, H) &= \int_{\tau/2}^{\tau} \frac{f(a)}{4}\left(2 - \frac{\tau}{a}\right)^3 da \\ &\quad + \int_{\tau}^{\infty} f(a)\left[1 - \frac{3}{4}\left(\frac{\tau}{a}\right)^2\left(2 - \frac{\tau}{a}\right)\right] da \\ &= H(2H - 1)\tau^{2H-2}. \end{aligned} \quad (A.6)$$

A solution of this integral equation is of the form

$$f(a) = ba^{2H-3} \quad (A.7)$$

(Mejia, 1971) and thus Eq. (A.6) may be written as

$$\begin{aligned} C_1(\tau, H) &= \int_{\tau/2}^{\tau} \frac{ba^{2H-3}}{4}\left(2 - \frac{\tau}{a}\right)^3 da \\ &\quad + \int_{\tau}^{\infty} ba^{2H-3}\left[1 - \frac{3}{4}\left(\frac{\tau}{a}\right)^2\left(2 - \frac{\tau}{a}\right)\right] da \\ &= H(2H - 1)\tau^{2H-2}, \end{aligned} \quad (A.8)$$

which upon integration yields

$$\frac{6b\tau^{2H-2}(2^{3-2H}-1)}{(2H-2)(2H-3)(2H-4)(2H-5)} = H(2H-1)\tau^{2H-2}.$$

Now we can solve for b, which is given by

$$b = \frac{H(2H-1)(2H-2)(2H-3)(2H-4)(2H-5)}{6(2^{3-2H}-1)}. \qquad (A.9)$$

It will soon be apparent that the change of variables

$$a = a_1 B^v \qquad (A.10)$$

will be convenient. Thus once again, assuming that $\int_0^\infty f(a)\, da = 1$, Eq. (A.3) becomes

$$C_1(\tau, H) = \int_0^\infty \rho(\tau, a_1 B^v) f(a_1 B^v)\, d(a_1 B^v),$$

which can be expressed as

$$\begin{aligned}
C_1(\tau, H) &= \int_0^\infty \rho(\tau, a_1 B^v) b(a_1 B^v)^{2H-3} a_1\, d(e^{v \log_e B}) \\
&= \int_{-\infty}^\infty \rho(\tau, a_1 B^v) b(a_1 B^v)^{2H-3} a_1 B^v (\log_e B)\, dv. \qquad (A.11) \\
&= \int_{-\infty}^\infty \rho(\tau, a_1 B^v) a_1^{2H-2} b(\log_e B) B^{(2H-2)v}\, dv.
\end{aligned}$$

Because digital computers cannot evaluate an infinite sum of broken lines, a finite-sum approximation to Eq. (A.11) must be found. This can be accomplished as in fast fractional Gaussian noise, by dropping the very-high-frequency and very-low-frequency simple broken lines. Very-high-frequency broken lines have a very rapidly decaying correlation function, and hence do not contribute to the Hurst phenomena. From the viewpoint of simulation, the sum of such broken lines can be approximated by a single high-frequency component to be added in later. Very-low-frequency components vary slowly and may be ignored below a threshold value of frequency because they add little overall variance.

Therefore, dividing the variable v into N-unit length spans and neglecting very-high-frequency as well as very-low-frequency terms, Eq. (A.11) may be written as

$$C_1(\tau, H) \approx C_2(\tau, H) = a_1^{2H-2} b \log_e B \sum_{n=0}^{N-1} \rho(\tau, a_1 B^n) \int_{n-0.5}^{n+0.5} B^{2(H-1)v} \, dv$$

$$= a_1^{2H-2} b \log_e B \sum_{n=0}^{N-1} \rho(\tau, a_1 B^n) \left(\frac{1}{2(H-1)\log_e B} e^{2(H-1)v \log_e B} \Big|_{n-0.5}^{n+0.5} \right)$$

$$= \frac{a_1^{2H-2}}{2(H-1)} b \sum_{n=0}^{N-1} \rho(\tau, a_1 B^n)(B^{H-1} - B^{1-H}) B^{2(H-1)n}. \qquad (A.12)$$

Clearly, N and B affect the number and the frequency of the broken lines summed and thus the goodness of fit. Considerations leading to the choice of N and B were given in Section 5.5.

The variance of the process whose correlation is defined in Eq. (A.12) is approximately given by

$$\sigma_\xi^2 \int_{a_1 B^{-0.5}}^{\infty} f(a) \, da = \sigma_\xi^2 \int_{a_1 B^{-0.5}}^{\infty} ba^{2H-3} \, da$$

$$= \sigma_\xi^2 \left(\frac{ba_1^{2H-2} B^{1-H}}{2 - 2H} \right). \qquad (A.13)$$

Therefore, it is necessary to introduce another simple broken-line process such that the total variance is σ_ξ^2, as required by the assumption of Eq. (A.5). A high-frequency simple broken line, with parameter $a_0 (\le a_1)$ and variance σ_ξ^2 is added to minimize the effect on the Hurst phenomena (caused by low-frequency terms: high as). The square of the weight for the high-frequency term will be approximately

$$1 - \frac{ba_1^{2H-2} B^{1-H}}{2 - 2H}. \qquad (A.14)$$

The parameter a_1 is used to fit the correlation function of the new broken-line process to a given second derivative at the origin, $\rho''(0)$, or to a given lag-one correlation coefficient, $\rho(1)$. The interest in the possible preservation of $\rho''(0)$ lies in its role quantifying crossing properties and extreme-value characteristics of a process as discussed in Section 5.8.

The second derivative at the origin of the correlation function for a single broken line can be computed from Eq. (5.73) and is

$$\rho_i''(0, a_i) = \frac{-3}{a_i^2}. \qquad (A.15)$$

The fractional Gaussian noise correlation function approximation, Eq. (A.12), plus the high-frequency broken line with parameter $a_0 = a_1$, has a $\rho''(0)$

given by

$$\rho''(0) = \frac{a_1^{2H-2}}{2(H-1)} b(B^{H-1} - B^{1-H}) \sum_{n=0}^{N-1} \rho''_{n+1}(0, a_1 B^n) B^{2(H-1)n}$$

$$+ \left(1 - \frac{ba_1^{2H-2}B^{1-H}}{2-2H}\right) \rho''_0(0, a_1), \tag{A.16}$$

where $\rho''_0(0, a_1)$ is the second derivative of the correlation function at the origin associated with the high-frequency simple broken line with approximate square weight given by Eq. (A.14) and $\rho''_{n+1}(0, a_1 B^n)$ is the function associated with the simple broken line with square weight equal to

$$f(a_1 B^n) = \frac{a_1^{2H-2}}{2(H-1)} b(B^{H-1} - B^{1-H}) B^{2(H-1)n}. \tag{A.17}$$

Equation (A.16) may be rewritten as

$$\rho''(0) = -\frac{3}{a_1^2} \left(1 - \frac{ba_1^{2H-2}B^{1-H}}{2-2H}\right)$$

$$- \frac{3a_1^{2(H-2)} b(B^{H-1} - B^{1-H})}{2(H-1)} \sum_{n=0}^{N-1} B^{2(H-2)n}. \tag{A.18}$$

Letting

$$C = -\frac{3b(B^{H-1} - B^{1-H})}{2(H-1)} \sum_{n=0}^{N-1} B^{2(H-2)n} + \frac{3bB^{1-H}}{2-2H}, \tag{A.19}$$

Eq. (A.19) may be simplified to

$$\rho''(0) = -\frac{3}{a_1^2} + Ca_1^{2(H-2)} \tag{A.20}$$

thus

$$a_1 = \sqrt{\frac{-3 + Ca_1^{2(H-1)}}{\rho''(0)}}. \tag{A.21}$$

By an iterative trial-and-error procedure, the value of a_1 can be found that satisfies Eq. (A.21).

The lag-one correlation function for a single broken line with parameter a_i is given by

$$\rho_i(1, a_i) = \begin{cases} 1 - \frac{3}{4}\frac{1}{a_i^2}\left(2 - \frac{1}{a_i}\right) & a_i \geq 1 \\ \frac{1}{4}\left(2 - \frac{1}{a_i}\right)^3 & 0.5 < a_i < 1 \\ 0 & a_i \leq 0.5. \end{cases} \tag{A.22}$$

Hence, after the high-frequency broken line with parameter a_0 is added to the approximation given in Eq. (A.12), the $\rho(1)$ of the new process is given by

$$
\rho(1) = \frac{a_1^{2H-2}}{2(H-1)} b\left(B^{H-1} - B^{1-H}\right) \sum_{n=0}^{N-1} \rho_{n+1}\left(1, a_1 B^n\right) B^{2(H-1)n}
$$
$$
+ \left(1 - \frac{b a_1^{2H-2} B^{1-H}}{2-2H}\right) \rho_0\left(1, a_0\right). \tag{A.23}
$$

The characteristics of the above equation have been studied in detail by Curry and Bras (1978). The value of a_0 will determine what range of $\rho(1)$ can be satisfied by varying the parameter a_1. The behavior of Eq. (A.23) for $a_0 \leq 0.5$ is shown in Fig. A.1. Although the lag-one correlation of a single broken line increases with an increase in parameter a_1, the overall behavior of $\rho(1)$, which is constructed from weighted simple broken lines according to Eq. (A.23) with $a_0 \leq 0.5$, is to decrease as a_1 increases. The behavior is caused by rapidly decreasing weighting of the simple broken lines with parameters $a_1, a_1 B^1, \ldots, a_1 B^{N-1}$ and the increased weighting on the high-frequency broken line, which contributes nothing to the overall $\rho(1)$ since $a_0 \leq 0.5$ and $\rho_0(1, a_0) = 0$ (see Eq. A.22). The $\rho(1)$ approaches zero since the weighting on the high-frequency broken line tends to one as a_1 increases as shown in Fig. A.2. The behavior of the weight on the high-frequency broken line as a_1 is

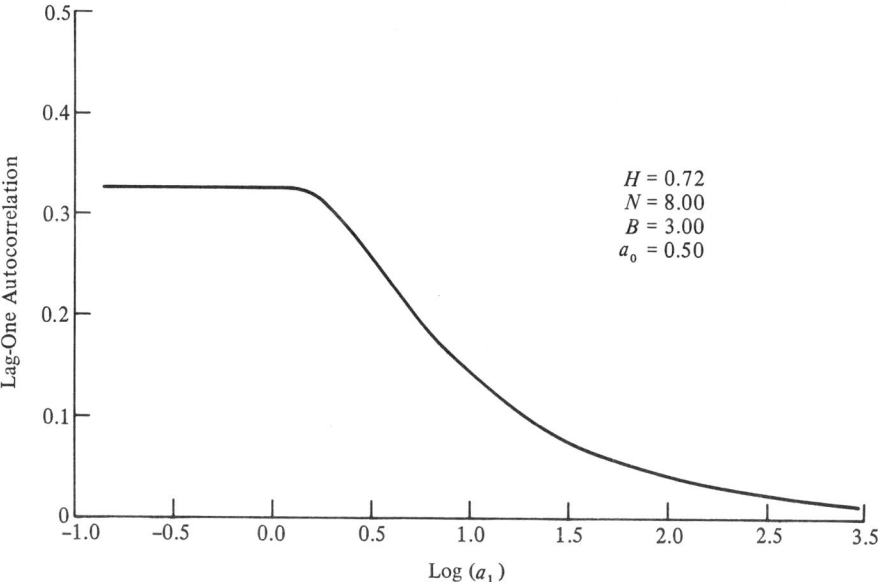

Figure A.1 Behavior of $\rho(1)$ for the broken-line process as given by Eq. (A.23) (from Curry and Bras, 1978).

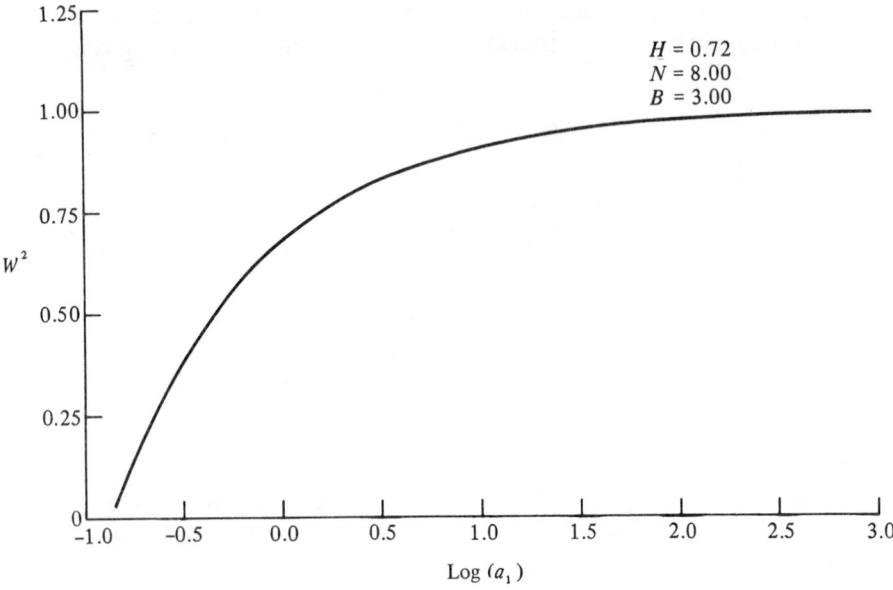

Figure A.2 Behavior of the square of the weight for the high-frequency term in the broken-line process (Eq. A.14) (from Curry and Bras, 1978).

increased, combined with the behavior illustrated in Fig. A.2 implies that Eq. (A.23) will approach $\rho_0(1, a_0)$ as a_1 increases for any value of a_0.

Thus, to ensure that an arbitrarily small $\rho(1)$ can be preserved, a_0 must be less but close to 0.5 (Curry and Bras [1978] recommended 0.49) so that $\rho_0(1, a_0) = 0$. In this case, using Eq. (A.22), Eq. (A.23) may be written as

$$\rho(1) = \frac{a_1^{2H-2}}{2(H-1)} b \sum_{n=0}^{N-1} \left[\left(\left[1 - \frac{3}{4} \frac{1}{(a_1 B^n)^2} \left(2 - \frac{1}{a_1 B^n} \right) \right] I_1(a_1 B^n) \right. \right.$$

$$\left. \left. + \frac{1}{4} \left(2 - \frac{1}{a_1 B^n} \right)^3 I_2(a_1 B^n) \right) (B^{H-1} - B^{1-H}) B^{2(H-1)n} \right], \quad \text{(A.24)}$$

where

$$I_1(x) = \begin{cases} 1 & \text{if } x \geq 1 \\ 0 & \text{otherwise} \end{cases}$$

$$I_2(x) = \begin{cases} 1 & 0.5 < x < 1 \\ 0 & \text{otherwise.} \end{cases} \quad \text{(A.25)}$$

So long as the historical lag-one correlation of the process being modeled is less than the maximum of Eq. (A.24) (see Fig. A.1), the value of a_1 that satisfies the relationship can be found by a trail-and-error iterative procedure. When the historical lag-one correlation is larger than the maximum, the solution procedure yields successive a_1s which rapidly converge toward zero. In this case a_0 must be set to a value greater than 0.5. However, a_0 should be kept as small as possible, thus forming a high-frequency broken line so that Eq. (A.12) remains valid and the Hurst coefficient will not be biased. This can be accomplished by letting $a_0 = a_1$. Equation (A.23) may then be written:

$$\rho(1) = \frac{a_1^{2H-2}}{2(H-1)} b \sum_{n=0}^{N-1} \left[\left[\left[1 - \frac{3}{4} \frac{1}{(a_1 B^n)^2} \left(2 - \frac{1}{a_1 B^n} \right) \right] I_1(a_1 B^n) \right. \right.$$

$$+ \frac{1}{4} \left(2 - \frac{1}{a_1 B^n} \right)^3 I_2(a_1 B^n) \right] (B^{H-1} - B^{1-H}) B^{2(H-1)n} \right]$$

$$+ \left(1 - \frac{ba_1^{2H-2} B^{1-H}}{2 - 2H} \right) \left[\left(1 - \frac{3}{4} \frac{1}{a_1^2} \left(2 - \frac{1}{a_1} \right) \right) I_1(a_1) \right.$$

$$\left. + \frac{1}{4} \left(2 - \frac{1}{a_1} \right)^3 I_2(a_1) \right], \tag{A.26}$$

where $I_1(x)$ and $I_2(x)$ are as defined in Eq. (A.25).

Given a historical lag-one coefficient greater than the maximum of Eq. (A.24), Eq. (A.26) can be solved for a_1 by a trial-and-error iterative procedure.

Figure A.3 shows the value of a_1 required to satisfy a lag-one correlation coefficient in the range [0.01, 0.99] using Eq. (A.24) or (A.26).

To summarize the results derived thus far, a process $\beta(t)$, exhibiting the Hurst phenomena and a given lag-one autocorrelation coefficient or $\rho''(0)$ may be constructed from the summation of $N+1$ simple broken lines,

$$\beta(t) = \sum_{i=0}^{N} W_i \xi_i(t), \tag{A.27}$$

where

$$W_i = \sqrt{\frac{a_1^{2(H-1)}}{2(H-1)} b (B^{H-1} - B^{1-H}) B^{2(H-1)(i-1)}} \qquad i = 1, 2, 3, \dots, N \tag{A.28}$$

$$W_0 = \sqrt{1 - \sum_{i=1}^{N} W_i^2} \tag{A.29}$$

$$b = \frac{H(2H-1)(2H-2)(2H-3)(2H-4)(2H-5)}{6(2^{3-2H}-1)}. \tag{A.30}$$

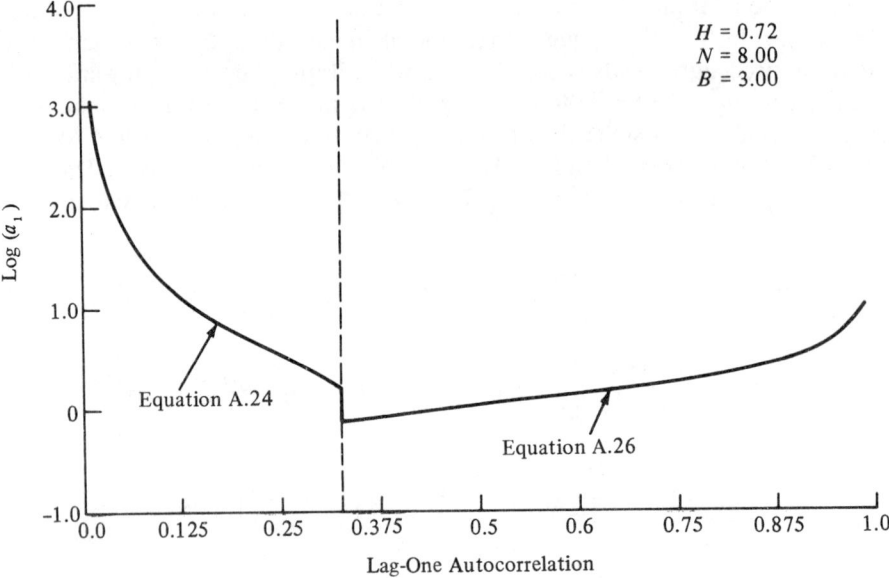

Figure A.3 Value of a_1 required to satisfy a lag-one correlation coefficient using Eqs. (A.24) and (A.26) (from Curry and Bras, 1978).

H is the Hurst coefficient; ξ_i the simple broken-line process with variance σ_ξ^2 and parameter $a_i = a_1 B^{(i-1)}$ constructed from the random variables $\eta(0, \sigma_\eta^2)$; ξ_0 the simple broken-line process with variance σ_η^2 and parameter $a_0 = a_1$ if $\rho''(0)$ is being preserved and $a_0 = 0.49$ or $a_0 = a_1$ [see Eqs. A.24 and A.26] if $\rho(1)$ is being preserved, constructed from the random variables $\eta(0, \sigma_\eta^2)$

$$N = \left\| \frac{\log QT}{\log B} \right\|;$$

Q, B the quality parameters; T the total number of years of simulation required; and a_1 defined to fit $\rho(1)$ or $\rho''(0)$ of the process being modeled. For this process

$$E[\beta(t)] = \sum_{i=0}^{N} W_i E[\xi_i(t)] = 0 \qquad (A.31)$$

$$\text{var}[\beta(t)] = E[\beta(t)^2] = \sum_{i=0}^{N} W_i^2 E[\xi_i(t)^2]$$

$$= \sum_{i=0}^{N} W_i^2 E[\xi_i(0)^2] = \frac{2}{3}\sigma_\eta^2 \qquad (A.32)$$

$$\gamma(\beta) = \text{skewness coefficient} = \frac{E\left[\beta(t) - \overline{\beta(t)}\right]^3}{\sigma_\beta^3}$$

$$= \left(\frac{3}{2}\right)^{3/2} \frac{\gamma(\eta)}{2} \sum_{i=0}^{N} W_i^3.$$

(A.33)

The process $\beta(t)$ can be given a particular skewness by specifying $\gamma(\eta)$. Therefore, it can be used to simulate a phenomenon of a given mean, standard deviation, skewness, and $\rho(1)$ or $\rho''(0)$ that exhibits the Hurst phenomena.

For example, the process, $\beta'(t)$, constructed from $\beta(t)$ of Eq. (A.27)

$$\beta'(t) = \mu_{\beta'} + \sigma_{\beta'} \left(\frac{3}{2\sigma_\eta^2}\right)^{1/2} \beta(t)$$

(A.34)

has a mean of $\mu_{\beta'}$, variance $\sigma_{\beta'}^2$, Hurst coefficient H, and $\rho(1)$ or $\rho''(0)$ defined by the choice of a_1. If the sequence of ηs used for each simple broken line has skewness coefficient

$$\gamma(\eta) = \frac{2}{\sum_{i=0}^{N} W_i^3} \left(\frac{2}{3}\right)^{3/2} \gamma(\beta')$$

(A.35)

then the skewness of $\beta'(t)$ is $\gamma(\beta')$.

The broken-line model is the only time-series–analysis tool that explicitly preserves the Hurst phenomena and exists as an operational multivariate version. The derivation and details of the multivariate case is given in Section A.2.

A.2   THE MULTIVARIATE BROKEN-LINE MODEL

Consider the generating process

$$\beta(t) = \mu + \left(\frac{3}{2\sigma_\eta^2}\right)^{1/2} \sum_{i=0}^{N} \sigma W_i(a_i) \xi_{a_i}(t),$$

(A.36)

where $\beta(t)$ represents a realization of the multivariate process at time t,

$$\beta(t) = \begin{bmatrix} \beta_1(t) \\ \vdots \\ \beta_n(t) \end{bmatrix},$$

(A.37)

and n is the number of stations that are being considered, μ is a vector formed by the mean of each one of the n processes,

$$\mu = \begin{bmatrix} \mu_{\beta_1} \\ \vdots \\ \mu_{\beta_n} \end{bmatrix}, \tag{A.38}$$

σ is a diagonal matrix formed by the standard deviation of each one of the processes,

$$\sigma = \begin{bmatrix} \sigma_{\beta_1} & \cdot & \cdot & \cdot & \cdot & 0 \\ \cdot & \sigma_{\beta_2} & \cdot & \cdot & \cdot & \cdot \\ \cdot & \cdot & & & \ddots & \cdot \\ \cdot & \cdot & & & & \cdot \\ 0 & \cdot & \cdot & \cdot & \cdot & \sigma_{\beta_n} \end{bmatrix}, \tag{A.39}$$

and $\mathbf{W}_i(a_i)$ is a diagonal matrix formed by the weights given to the simple broken lines with the common parameter a_i in each process,

$$\mathbf{W}_i(a_i) = \begin{bmatrix} W_i^1(a_i) & \cdots & & 0 \\ \vdots & W_i^2(a_i) & & \vdots \\ & & \ddots & \\ 0 & \cdots & & W_i^n(a_i) \end{bmatrix}. \tag{A.40}$$

Assume these weights have been normalized in such a way that

$$\sum_{i=0}^{N} \mathbf{W}_i(a_i)\mathbf{W}_i^T(a_i) = \mathbf{I}, \tag{A.41}$$

where $\mathbf{I}$ is the identity matrix. $\xi_{a_i}(t_i)$ is a vector where each element represents a simple broken-line process with parameters a_i and σ_η^2:

$$\xi_{a_i}(t) = \begin{bmatrix} \xi_{a_i}^1(t) \\ \vdots \\ \xi_{a_i}^n(t) \end{bmatrix}. \tag{A.42}$$

An individual element of $\beta(t)$ may be written as

$$\beta_j(t) = \mu_{\beta_j} + \left(\frac{3}{2\sigma_\eta^2} \right)^{1/2} \sigma_{\beta_j} \sum_{i=0}^{N} W_i^j(a_i)\xi_{a_i}^j(t). \tag{A.43}$$

Note that this is the generation process presented in the univariate example, Eq. (A.34).

The cross-correlation among processes in $\boldsymbol{\beta}$ can be chosen by generating $\xi_{a_i}(t)$s that have a determined cross-correlation. Consider the covariance matrix

$$
E\left[\{\boldsymbol{\beta}(t)-\boldsymbol{\mu}\}\{\boldsymbol{\beta}(t)-\boldsymbol{\mu}\}^T\right] = E\left[\frac{3}{2\sigma_\eta^2}\sum_{i=0}^{N}\boldsymbol{\sigma}\mathbf{W}_i(a_i)\xi_{a_i}(t)\xi_{a_i}^T(t)\mathbf{W}_i^T(a_i)\boldsymbol{\sigma}^T\right].
$$
(A.44)

Postmultiplying Eq. (A.44) by $[\boldsymbol{\sigma}^T]^{-1}$ and premultiplying it by $\boldsymbol{\sigma}^{-1}$ and taking into account that $\boldsymbol{\sigma}^T = \boldsymbol{\sigma}$ and $\mathbf{W}_i(a_i) = \mathbf{W}_i(a_i)^T$ yields

$$
E\left[\boldsymbol{\sigma}^{-1}\{\boldsymbol{\beta}(t)-\boldsymbol{\mu}\}\{\boldsymbol{\beta}(t)-\boldsymbol{\mu}\}^T\boldsymbol{\sigma}^{-1}\right] = \frac{3}{2\sigma_\eta^2}\sum_{i=0}^{N}\mathbf{W}_i(a_i)E\left[\xi_{a_i}(t)\xi_{a_i}^T(t)\right]\mathbf{W}_i(a_i)
$$
(A.45)

and hence, from Eq. (5.75),

$$
\begin{bmatrix} 1 & \cdots & \rho_{1n} \\ \vdots & \ddots & \vdots \\ \rho_{n1} & \cdots & 1 \end{bmatrix} = \frac{3}{2\sigma_\eta^2}\sum_{i=0}^{N}\left[\mathbf{W}_i(a_i)\frac{2}{3}\begin{bmatrix} \sigma_\eta^2 & \lambda_{12} & \cdots & \lambda_{1n} \\ \vdots & \vdots & \cdots & \vdots \\ \lambda_{n1} & \cdot & \cdots & \sigma_\eta^2 \end{bmatrix}\mathbf{W}_i(a_i)\right],
$$
(A.46)

where ρ_{ij} is the cross-correlation among broken-line sequences i and j and λ_{ij} is the cross-covariance between ηs used to simulate processes i and j (see Eqs. 5.74 and 5.75).

A typical element for Eq. (A.46) will be

$$
\rho_{ij} = \frac{\lambda_{ij}}{\sigma_\eta^2}\sum_{k=0}^{N}W_k^i(a_k)W_k^j(a_k),
$$
(A.47)

from which

$$
\lambda_{ij} = \frac{\rho_{ij}\sigma_\eta^2}{\displaystyle\sum_{k=0}^{N}W_k^i(a_k)W_k^j(a_k)}.
$$
(A.48)

Therefore, using the historical cross-correlation between the process, the values of the cross-covariance among the ηs in the broken-line processes can be evaluated. The problem reduces to the question of how to generate ηs with these cross-covariances. This can be accomplished as follows

Consider the vector

$$\eta_k = \begin{bmatrix} \eta_k^1 \\ \vdots \\ \eta_k^n \end{bmatrix}, \tag{A.49}$$

where the ηs fulfill the conditions given by Eq. (5.74). Consider another vector ε_k composed of independent zero mean, variance σ_η^2, random variables with skewness $\gamma(\xi^j)$, and related to η_k by

$$\eta_k = \mathbf{D} \begin{bmatrix} \varepsilon_k^1 \\ \varepsilon_k^2 \\ \vdots \\ \varepsilon_k^n \end{bmatrix}, \tag{A.50}$$

where $\mathbf{D}$ is an $n \times n$ matrix. To satisfy Eq. (A.50), $\mathbf{D}$ must have the property

$$E\left[\eta_k \eta_k^T\right] = \begin{bmatrix} \sigma_\eta^2 & \cdots & \lambda_{1n} \\ \vdots & \ddots & \vdots \\ \lambda_{n1} & \cdots & \sigma_\eta^2 \end{bmatrix} = E\left[\mathbf{D}\varepsilon_k \varepsilon_k^T \mathbf{D}^T\right] = \mathbf{D}\mathbf{D}^T. \tag{A.51}$$

$\mathbf{D}$ can be solved for by an decomposition method as seen in Chapter 3.

The required skewness for each of the ε^j can now be considered. For the preservation of $\gamma(\beta_j)$, Eq. (A.33) requires

$$\gamma\left(\eta_k^j\right) = \frac{2}{\sum\limits_{i=0}^{N} \left(W_i^j\right)^3} \left(\frac{2}{3}\right)^{3/2} \gamma(\beta_j) \qquad j=1,2,\ldots,n \tag{A.52}$$

and that

$$\gamma\left(\eta_k^j\right) = \gamma\left(\eta_\ell^j\right) \qquad \forall\, k,\ell. \tag{A.53}$$

By Eq. (A.50)

$$\gamma(\eta^j) = E\left[\sum_{i=1}^{j} \left(d_{ji}\xi^i\right)^3\right] \Big/ \sigma_\eta^3$$

$$= \sum_{i=1}^{j} d_{ji}^3 \gamma(\varepsilon^i)/\sigma_\eta^3 \qquad j=1,2,\ldots,n. \tag{A.54}$$

Beginning with $j = 1$, Eq. (A.54) can be solved sequentially for $\gamma(\varepsilon^i)$:

$$\gamma(\eta^1) = d_{11}^3 \gamma(\varepsilon^1)/\sigma_\eta^3$$

$$\gamma(\eta^2) = \left(d_{21}^3 \gamma(\varepsilon^1) + d_{22}^3 \gamma(\varepsilon^2)\right)/\sigma_\eta^3 \qquad (A.55)$$

$$\vdots \qquad\qquad \vdots$$

$$\gamma(\eta^n) = \left(d_{n1}^3 \gamma(\varepsilon^1) + \cdots + d_{nn}^3 \gamma(\varepsilon^n)\right)/\sigma_\eta^3.$$

The last problem to be solved for the multivariate broken-line–processes simulation is how to simultaneously fit each one of the broken-line processes to its respective $\rho''(0)$ or $\rho(1)$.

The fitting of $\rho''(0)$ is done as follows. For each separate process $\beta_j(t)$, a_1 is found according to Eq. (A.21). The largest of these values is used for all processes. A second additional broken line with parameter $a = 1$ is added to each process (Curry and Bras recommend values of 0.985 or 1.015 instead of 1) so that now each process is composed of $N + 2$ simple broken lines having parameters $1, a_1, a_1, a_1 B, a_1 B^2, \ldots$. Weights are computed as indicated before, except for the weights for the broken line with parameter 1 and the first one with parameter a_1. As given by Eq. (A.14), the approximate weight of the high-frequency terms should be such that

$$1 - \frac{b a_1^{2H-2} B^{1-H}}{2 - 2H} = W_1^2 + W_{a_1}^2. \qquad (A.56)$$

Simultaneously, the equation for the second derivative at the origin of the correlation function must be satisfied and thus, similarly to Eq. (A.16):

$$\rho''(0) = -\frac{3}{a_1^2} W_{a_1}^2 - 3W_1^2 - \frac{3a_1^{2(H-2)} b\left(B^{H-1} - B^{1-H}\right)}{2(H-1)} \sum_{n=0}^{N-1} B^{2(H-2)n}. \qquad (A.57)$$

Equations (A.56) and (A.57) can be solved easily for W_1 and W_{a_1} for each one of the processes. This method works because the process with the least negative $\rho''(0)$ will have the largest a_1. By successively putting more weight on the high-frequency term with parameter 1 and less on the extra broken line with parameter a_1, a more negative $\rho''(0)$ for a process can be realized since the smaller the a the more negative $\rho''(0)$ is for a single broken line.

The simultaneous fitting of each process to its respective $\rho(1)$ is done much the same as for the fitting of $\rho''(0)$, although more complications arise. Again, for each separate process $\beta_j(t)$, a_1 is found according to Eq. (A.24) or Eq. (A.26) as appropriate, and the largest of these values is used for all processes.

From Fig. A.3 it can be seen that the process from which a_1 will be chosen for all processes is the one with either the lowest or highest lag-one autocorrelation coefficient. If a_1 is chosen from the process with the lowest lag-one autocorrelation, implying $a_0 = 0.49$, a second high-frequency simple broken line with parameter $a_1' > 0.5$ ($\rho(1, a_1') > 0$) must be added to each of the processes so that their higher lag-one autocorrelation coefficients can be met. Each of the processes' correlation function is tailored to its respective $\rho(1)$ by varying the weighting of the two high-frequency broken lines with parameters 0.49 and a_1'. Thus, to minimize the error introduced in Eq. (A.12), the biasing of the Hurst coefficient, a_1' should be as small as possible and hence used to construct the highest-frequency broken line possible. The lowest value of a_1' feasible is that which will satisfy the highest lag-one coefficient if zero weight is put on the high-frequency broken line with parameter $a_0 = 0.49$. This value can be calculated as follows:

$$
\begin{aligned}
\rho^*(1) = \frac{a_1^{2H-2}}{2(H-1)} b \sum_{n=0}^{N-1} & \left[\left(\left[1 - \frac{3}{4} \frac{1}{(a_1 B^n)^2} \left(2 - \frac{1}{a_1 B^n} \right) \right] I_1(a_1 B^n) \right. \right. \\
& \left. + \frac{1}{4} \left(2 - \frac{1}{a_1 B^n} \right)^3 I_2(a_1 B^n) \right) (B^{H-1} - B^{1-H}) B^{2(H-1)n} \right] \\
& + W^* \left(\left[1 - \frac{3}{4} \frac{1}{(a_1')^2} \left(2 - \frac{1}{a_1'} \right) \right] I_1(a_1') + \frac{1}{4} \left(2 - \frac{1}{a_1'} \right)^3 I_2(a_1') \right),
\end{aligned}
$$

(A.58)

where $\rho^*(1)$ is the largest lag-one coefficient of the processes being modeled, W^* is the sum of the squared weights of the high-frequency broken lines (Eq. A.14), and $I_1(x)$ and $I_2(x)$ are as defined in Eq. (A.25). However, often this value of a_1' will create a weighting matrix that will cause the matrix defined by Eq. (A.48) (a transformation of the correlation matrix) to be inconsistent. It is not possible to a priori calculate the value of a_1' that will make the matrix consistent. In case of inconsistency, the value of a_1' computed from Eq. (A.58) is incrementally increased until the matrix of Eq. (A.48) is consistent.

If the largest a_1 found according to Eq. (A.14) or (A.16) is from the process with the highest lag-one autocorrelation coefficient, implying $a_0 = a_1$ (calculated according to Eq. A.16), a second high-frequency simple broken line with parameter $a_1' = 0.49$ ($\rho(1, 0.49) = 0$) must be added to each of the processes so that the lower lag-one autocorrelation coefficients can be met. As before, each of the processes' correlation functions is tailored to its respective $\rho(1)$ by varying the weighting of the two high-frequency broken lines with parameters $a_1' = 0.49$ and $a_0 = a_1$ while constraining the sum of weights equal to the value defined in Eq. (A.14). The process with the highest lag-one will have zero weight on the added high-frequency broken line with parameter $a_1' = 0.49$. For the other processes, the lower their lag-one coefficient the higher the weight placed on the broken line with parameter $a_1' = 0.49$.

CHAPTER 6

Multidimensional Hydrologic Processes

6.1 INTRODUCTION

Spatial patterns of variation of geophysical phenomena can be studied by means of stochastic processes representing these variations in a continuous or discrete way over the space considered. The stochastic process associated with the discrete representations is called a multivariate process. It represents a family of random vectors $\mathbf{X}(t) = \{X_1(t), \ldots, X_n(t)\}^T$ depending on the parameter t, where each one of the components is the value assumed by the process at specified points on the area at time t. In contrast, the continuous representation of such processes requires the theory of random fields or multidimensional processes, in which the stochastic process $X(\mathbf{u}, t)$ represents a random variable depending on the parameter $\mathbf{u} = \{u_1, \ldots, u_n\}^T$, representing the coordinates of a point in an n-dimensional space. The applications of the theory of multivariate processes to the description of geophysical phenomena have increased significantly over the last few years, whereas random fields have not been used frequently by the applied hydrostatistician. However, the study of random fields is justified, since the effect of coarse discrete representations of the phenomenon under study may not adequately capture the physical reality.

Furthermore, the use of large multivariate techniques with a sizable number of points involves operations with matrices whose size may render the problem impractical.

This chapter explores the representation of multidimensional hydrologic processes, the synthesis of these processes, and the sampling schemes necessary to adequately represent them.

6.2 CORRELATION AND SPECTRUM

As in the case of unidimensional processes, correlation and spectrum play a crucial role in the theory and application of random fields. Mejía and Rodríguez-Iturbe (1974) and Rhenals-Figueredo et al. (1974) describe the theory incorporated in this section.

Let $\xi(\mathbf{x}) = \xi(x_1, \ldots, x_n)$ be a real random function of n real variables, i.e., a real random function of position in the n-dimensional space $\mathcal{R}^n$; such functions are called random fields in $\mathcal{R}^n$. The mean and covariance of the random field $\xi(\mathbf{x})$, $\mathbf{x} \in \mathcal{R}^n$, are

$$m(\mathbf{x}) = E\{\xi(\mathbf{x})\} \tag{6.1}$$

$$C(\mathbf{x}_1, \mathbf{x}_2) = E\{\xi(\mathbf{x}_1)\xi(\mathbf{x}_2)\} - m(\mathbf{x}_1) \cdot m(\mathbf{x}_2). \tag{6.2}$$

The correlation function is defined as

$$\rho(\mathbf{x}_1, \mathbf{x}_2) = \frac{C(\mathbf{x}_1, \mathbf{x}_2)}{\left[C(\mathbf{x}_1, \mathbf{x}_1) \cdot C(\mathbf{x}_2, \mathbf{x}_2)\right]^{1/2}}. \tag{6.3}$$

The random field is said to be "wide-sense" homogeneous if

$$m(\mathbf{x}) = \text{constant}$$

and

$$C(\mathbf{x}_1, \mathbf{x}_2) = C(\mathbf{x}_2 - \mathbf{x}_1) = C(\mathbf{v}), \tag{6.4}$$

where

$$\mathbf{v} = \mathbf{x}_2 - \mathbf{x}_1.$$

It is clear that the concept of a homogeneous random field in $\mathcal{R}^1$ coincides with the concept of a stationary random process, so that stationary processes are included in the class of homogeneous random fields.

The homogeneous correlation function $\rho(\mathbf{v})$ can be expressed (Yaglom, 1962, p. 82) in the form of the Fourier–Stieltjes integral

$$\rho(\mathbf{v}) = \int_{\mathcal{R}^n} e^{i(\boldsymbol{\omega} \cdot \mathbf{v})} \, dF(\boldsymbol{\omega}), \tag{6.5}$$

where ω is a vector, $(\omega \cdot v)$ indicates the scalar product of the vectors, and $F(\omega)$ is called the spectral distribution function of the field.

From Eq. (6.5) we see that if the correlation function $\rho(v)$ is continuous everywhere, there exists a random variable ω that has $\rho(v)$ as a characteristic function.

Thus,

$$\rho(v) = E\{e^{i(\omega \cdot v)}\} = \int_{\mathscr{R}^n} e^{i(\omega \cdot v)} dF(\omega),$$

$F(\omega)$ being the distribution function on the n-dimensional random vector $\omega = (\omega_1, \ldots, \omega_n)$.

Equation (6.5) can be written as

$$\rho(v) = \int_{\mathscr{R}^n} e^{i(\omega \cdot v)} f(\omega) \, d\omega \qquad d\omega = (d\omega_1, \ldots, d\omega_n). \qquad (6.6)$$

The function $f(\omega)$ is called the spectral density of the random field.

Since for real fields $\rho(v) = \rho(-v)$, then $f(\omega) = f(-\omega)$, and Eq. (6.6) becomes

$$\rho(v) = \int_{\mathscr{R}^n} f(\omega) \cos(\omega \cdot v) \, d\omega. \qquad (6.7)$$

The field $\xi(x)$ is called homogeneous and isotropic if

$$\rho(v) = \rho(|v|). \qquad (6.8)$$

That is, the autocorrelation function depends only on the length of the vector v and not on its direction. To deal with this case, it is convenient to introduce the n-dimensional spherical coordinate system. In the case of $n = 2$, we use polar coordinates.

Let $G(\omega)$ be the distribution function of $|\omega|$, and call it the radial spectral distribution function of the field. If its derivative exists, it will be denoted by $G'(\omega)$, and referred to as the radial spectral density function.

The relation between the distribution function of $|\omega|$ and that of ω can be established as follows:

$$G'(\omega) = \int_{|\omega| = \omega} f(\omega) \, d\omega. \qquad (6.9)$$

Because $f(\omega)$ has a constant value on $|\omega| = \omega$ due to the isotropy condition, if we denote this constant value as $f(\omega)$, Eq. (6.9) is equivalent to

$$G'(\omega) = f(\omega) \int_{|\omega| = \omega} d\omega.$$

The integral corresponds to the surface area of the n-dimensional sphere of radius ω, which can be expressed as

$$\int_{|\omega|} d\omega = \frac{2\pi^{n/2}\omega^{n-1}}{\Gamma(n/2)}$$

where $\Gamma(\cdot)$ denotes the gamma function. Therefore $G'(\omega)$ and $f(\omega)$ are related by

$$G'(\omega) = \frac{f(\omega)2\pi^{n/2}\omega^{n-1}}{\Gamma(n/2)}, \tag{6.10}$$

where $\omega = |\omega|$. For the case of a two-dimensional random field ($n = 2$) we get

$$G'(\omega) = 2\pi\omega f(\omega). \tag{6.11}$$

From now on, the results will be given for $n = 2$; that is, for a homogeneous and isotropic random field in the plane. Note that for this case $\rho(v) = \rho(x_1, x_2)$ is a surface whose contours on the x_1, x_2-plane are circles on a plane defined by orthogonal axes x and y.

Introducing polar coordinates in Eq. (6.7), we get

$$\rho(v) = \int_0^\infty \left[\int_0^{2\pi} \cos(\omega v \cos\theta) \, d\theta \right] f(\omega) \omega \, d\omega$$

$$\rho(v) = 2\pi \int_0^\infty J_0(\omega v) f(\omega) \omega \, d\omega, \tag{6.12}$$

where $J_0(x)$ is the Bessel function of order zero, $\omega = |\omega| = (\omega_1^2 + \omega_2^2)^{1/2}$, $v = |v| = (v_1^2 + v_2^2)^{1/2}$, and v_1, v_2 are elements of the distance vector v. From Eq. (6.12):

$$f(\omega) = \frac{1}{2\pi} \int_0^\infty \rho(v) J_0(\omega v) v \, dv. \tag{6.13}$$

This result shows that the spectral density function of a bidimensional homogeneous and isotropic random field is the Fourier–Bessel transform (or Hankel transform) of the correlation function $\rho(v)$ multiplied by a constant.

Functional forms, satisfying certain conditions that will be considered later, can be correlation functions of a stationary one-dimensional process, as well as correlation functions of a homogeneous and isotropic random field.

This is the case, for example, for the exponential function

$$\rho(v) = e^{-\alpha|v|}$$

where bars around a scalar imply the absolute value. However, the spectral density of the one-dimensional process is the Fourier transform of $\rho(v)$ multiplied by a constant, but the spectral density of the bidimensional random field is the Fourier–Bessel transform of $\rho(v)$ multiplied by a constant. It is worthwhile to study this example in some detail.

One-Dimensional Process

Consider the stationary stochastic process $\{X(t), -\infty < t < \infty\}$ with zero mean, unit variance, and correlation function given by

$$\rho(v) = e^{-\alpha|v|}; \qquad -\infty < v < \infty.$$

The spectral density is

$$f_1(\omega) = \frac{1}{2\pi} \int_{-\infty}^{\infty} e^{-i\omega v} e^{-\alpha|v|} \, dv$$

$$= \frac{1}{2\pi} \left[\int_0^{\infty} e^{-(\alpha+i\omega)v} \, dv + \int_{-\infty}^0 e^{(\alpha-i\omega)v} \, dv \right]$$

$$= \frac{1}{2\pi} \left[\frac{1}{\alpha + i\omega} + \frac{1}{\alpha - i\omega} \right].$$

Thus, $f_1(\omega) = \alpha(\alpha^2 + \omega^2)^{-1}/\pi; \quad -\infty < \omega < \infty.$

Two-Dimensional Random Field

Let $\xi(\mathbf{x})$ be a homogeneous and isotropic random field in $\mathscr{R}^2$ with zero mean, unit variance, and isotropic correlation function given by

$$\rho(v) = e^{-\alpha v} \qquad v \geq 0.$$

From Eq. (6.13), the spectral density of the field is

$$f_2(\omega) = \frac{1}{2\pi} \int_0^{\infty} e^{-\alpha v} J_0(\omega v) v \, dv.$$

From Gradshteyn and Ryzhik (1965), we get

$$f_2(\omega) = \frac{\alpha}{2\pi} (\alpha^2 + \omega^2)^{-3/2} \qquad 0 \leq \omega < \infty.$$

Matern (1960) presents results that relate the correlation function of a random field and the correlation function of a one-dimensional stochastic process. Any isotropic correlation function in $\mathcal{R}^2$ can be a correlation function in $\mathcal{R}^1$. However, it does not follow that every correlation function in $\mathcal{R}^1$ can be an isotropic correlation function in $\mathcal{R}^2$.

Yaglom (1957) proved a general theorem for random fields in $\mathcal{R}^n$, which we rephrase here only for the case of homogeneous and isotropic random fields: A necessary and sufficient condition that the function $\rho(v)$ be the correlation function of a homogeneous and isotropic random field $\xi(\mathbf{x})$ in $\mathcal{R}^n$ is that it be representable in the form

$$\rho(v) = \int_0^\infty Y_n(\omega v) \, d\Phi(\omega), \tag{6.14}$$

where

$$Y_n(t) = \Gamma\left(\frac{n}{2}\right)\left(\frac{2}{t}\right)^{(n-2)/2} J_{(n-2)/2}(t)$$

$$= 1 - \frac{t^2}{2 \cdot n} + \frac{t^4}{2 \cdot 4 \cdot n(n+2)} - \cdots . \tag{6.15}$$

$J_{(n-2)/2}(t)$ is the Bessel function of order $(n-2)/2$ and $\Phi(\omega)$ is a real nondecreasing function on $[0, \infty]$ such that

$$\int_0^\infty \frac{d\Phi(\omega)}{(1+\omega^2)^p} < \infty$$

for some non-negative p.

Also, it can be shown that

$$\Phi(\omega) = \frac{1}{2^{(n-2)/2}\Gamma(n/2)} \int_0^\infty Y_{n/2}(v\omega)(v\omega)^{n/2} \frac{\rho(v)}{v} \, dv.$$

For $n = 2$, Eq. (6.15) becomes:

$$Y_2(t) = J_0(t) \tag{6.16}$$

and

$$\rho(v) = \int_0^\infty J_0(\omega v) \, d\Phi(\omega). \tag{6.17}$$

Substitution of Eq. (6.11) in Eq. (6.12) yields

$$\rho(v) = \int_0^\infty J_0(\omega v) G'(\omega) \, d\omega; \tag{6.18}$$

therefore

$$\rho(v) = \int_0^\infty J_0(\omega v)\, dG(\omega). \tag{6.19}$$

Comparing Eqs. (6.17) and (6.19) we see that

$$\Phi(\omega) \equiv G(\omega), \tag{6.20}$$

i.e., $\Phi(\omega)$ is the radial spectral distribution function. The condition that $\Phi(\omega)$ is a nondecreasing function implies $G'(\omega) \geq 0$. Of course, this assumes that $G(\omega)$ is differentiable. Since,

$$G'(\omega) = 2\pi\omega f(\omega) \qquad 0 \leq \omega < \infty,$$

the above condition is equivalent to

$$f(\omega) \geq 0. \tag{6.21}$$

A very general family of isotropic random fields is characterized by a correlation function of the type

$$\rho(v) = 2\left(\frac{bv}{2}\right)^{s-(n/2)} K_{s-(n/2)}(bv)\left[\Gamma\left(s - \frac{n}{2}\right)\right]^{-1}, \tag{6.22}$$

where K is the modified Bessel function of the second kind and n is the dimension of the process. The spectral density of such a process is given by

$$f(\omega) = \text{const}\left(1 + \frac{\omega^2}{b^2}\right)^{-s}. \tag{6.23}$$

The radial density function can be found according to Eq. (6.10):

$$G'(\omega) = \text{const}\,\frac{\omega^{n-1}}{\left[1 + (\omega^2/b^2)\right]^s}. \tag{6.24}$$

Of special importance in this family are the random fields for which $s = (n+1)/2$ and those for which $s = 1 + n/2$.

In the first case, when $s = (n+1)/2$, we have

$$\rho(v) = \exp(-bv), \qquad v \geq 0 \tag{6.25}$$

which is the classic exponentially decaying correlation function. For $n = 2$ it represents processes varying over an area for which

$$G'(\omega) = \frac{\omega}{b^2\left[1 + (\omega^2/b^2)\right]^{3/2}} \qquad 0 \leq \omega < \infty \tag{6.26}$$

and

$$G(\omega) = 1 - \frac{1}{\left[1 + \left(\omega^2/b^2\right)\right]^{1/2}} \qquad 0 \le \omega < \infty. \qquad (6.27)$$

For $s = 1 + n/2$,

$$\rho(v) = vbK_1\{vb\}, \qquad v \ge 0 \qquad (6.28)$$

which represents the correlation function for a physically meaningful process in which a point is related symmetrically to those around it. It has been used to describe rainfall correlation structure in hydrology (Rodríguez-Iturbe and Mejía, 1974a). In the case of $n = 2$ we have, for a process varying over an area,

$$G'(\omega) = \frac{2\omega}{b^2\left[1 + \left(\omega^2/b^2\right)\right]^2} \qquad 0 \le \omega < \infty \qquad (6.29)$$

and

$$G(\omega) = 1 - \frac{b^2}{\left(\omega^2 + b^2\right)} \qquad 0 \le \omega < \infty. \qquad (6.30)$$

6.3 SYNTHESIS OF SAMPLE FUNCTIONS OF A RANDOM FIELD: SAMPLING FROM THE SPECTRUM

The spectral representation of a homogeneous and isotropic random field provides a means of generating sample functions of that field when its spectral density is assumed to be known. This method has been described by Mejía and Rodríguez-Iturbe (1974).

Consider the process

$$\xi(t) = \cos(wt + \theta),$$

where w is a constant, t represents the time parameter of the process, and θ is a random variable uniformly distributed from 0 to 2π. This is the well-known random-phase model, which is both stationary and ergodic, since averages both in time and across the ensemble are

$$m = 0$$
$$\sigma^2 = 1/2$$
$$\rho(v) = \cos wv.$$

A modification of this model can be obtained as

$$\xi(t) = \cos(\omega t + \theta), \qquad (6.31)$$

where ω is a random variable obtained from a distribution function $F(\omega)$ that represents the spectral distribution function corresponding to the isotropic correlation function $\rho(v)$, and θ is again a random variable uniformly distributed from 0 to 2π. The process described by Eq. (6.31) has the following ensemble moments:

$$m_1 = 0$$

and

$$\rho_1(v) = \int_{-\infty}^{\infty} \cos(\omega v)\, dF(\omega) = E\{\cos \omega v\}.$$

According to Eq. (6.5), we have

$$\rho_1(v) = \rho(v),$$

which shows the stationarity of the process.

However, averages in time will give

$$m_2 = 0$$
$$\rho_2(v) = \cos \omega_1 v,$$

where ω_1 is an outcome of a random experiment from the distribution function $F(\omega)$. Therefore the process is not ergodic, since every realization will have a different correlation function. This difficulty can be bypassed by considering the process

$$\xi(t) = \left(\frac{2}{N}\right)^{1/2} \sum_{i=1}^{N} \cos(\omega_i t + \theta_i), \tag{6.32}$$

where the ω_i and the θ_i are mutually independent random variables distributed as described for Eq. (6.31).

Averages across the ensemble will produce

$$m_1 = 0$$
$$\sigma_1^2 = 1$$
$$\rho_1(v) = \rho(v).$$

Averages in time are

$$m_2 = 0$$
$$\sigma_2^2 = 1$$
$$\rho_2(v) = \frac{1}{N} \sum_{i=1}^{N} \cos \omega_i v.$$

By taking the limit as $N \to \infty$,

$$\rho_2(v) = \int_{-\infty}^{\infty} \cos \omega v \, dF(\omega) = \rho(v),$$

which proves the asymptotic ergodicity of the process. In addition, the process is asymptotically Gaussian, since the first two moments of the distribution of $\cos(\omega t + \theta)$ are finite and the central-limit theorem applies.

For finite N,

$$E\{\rho_2(v)\} = \int_{-\infty}^{\infty} \cos(\omega v) \, dF(\omega) = \rho(v)$$

and

$$\mathrm{var}\{\rho_2(v)\} = \frac{\mathrm{var}\{\cos \omega v\}}{N}.$$

We can write

$$\mathrm{var}\{\cos \omega v\} = E\{\cos^2 \omega v\} - \rho^2(v)$$

and

$$E\{\cos^2 \omega v\} = \int_{-\infty}^{\infty} \frac{\cos 2\omega v + 1}{2} \, dF(\omega) = \frac{\rho(2v)+1}{2}.$$

Therefore

$$\mathrm{var}\{\rho_2(v)\} = \frac{\rho(2v) - 2\rho^2(v) + 1}{2N}, \tag{6.33}$$

which tends to zero as $N \to \infty$.

Equation (6.32) suggests a procedure for the generation of stationary and isotropic sequences that are asymptotically Gaussian and ergodic. The method lies in the repeated sampling from the distribution function $F(\omega)$ for ω between 0 and 2π, followed by the use of these values in Eq. (6.32). The number of harmonics that must be generated depends on the desired agreement between the theoretical correlation function and the correlation function obtained by an infinite time average of the process. This agreement is measured by the variance expressed in Eq. (6.33).

As an example, Mejía and Rodríguez-Iturbe (1974) consider the generation of a discrete first-order autoregressive process. Although it is not practical, the generation of such a process by using the suggested procedure will illustrate the methodology and will shed light on the synthesis of more complicated processes. The sampling for the spectral distribution function was done by using the inverse method, which is based on the fact that the solutions $\{y\}$ to the equation $\{u = F(y)\}$ (where u is a sequence of independent and uniformly distributed random variables on the interval $0,1$) form a sequence of independent variables with cumulative distribution function $F(y)$. According to such a procedure it is necessary to find the inverse function, $y = F^{-1}(u)$.

In the case of a first-order discrete autoregressive process,

$$\rho(v) = \rho(1)^{|v|} \; ;$$

its spectral density in the continuous case is given by

$$f(\omega) = \frac{T}{\pi(1 + \omega^2 T^2)} \qquad -\infty \leq \omega < \infty,$$

where

$$\rho(1) = e^{-(1/T)}.$$

Therefore the spectral distribution function is given by

$$F(\omega) = \frac{1}{\pi}\left\{\tan^{-1}\omega T + \frac{\pi}{2}\right\},$$

where $\tan^{-1}(\cdot)$ is defined from $-\pi/2$ to $\pi/2$.

In the discrete case, with one unit of time as the sampling interval, the spectral density is given by

$$f(\omega) = \frac{1 - \rho^2(1)}{2\pi(1 + \rho^2(1) - 2\rho(1)\cos\omega)} \qquad -\pi \leq \omega \leq \pi$$

and

$$F(\omega) = \frac{1}{\pi}\left\{\tan^{-1}\left(\frac{1 - \rho(1)}{1 + \rho(1)}\tan\frac{\omega}{2}\right) + \frac{\pi}{2}\right\} \qquad -\pi \leq \omega \leq \pi. \quad (6.34)$$

From Eq. (6.34) one gets

$$\omega = 2\tan^{-1}\left\{\frac{1 + \rho(1)}{1 - \rho(1)}\tan\left(\pi u - \frac{\pi}{2}\right)\right\}; \qquad u = \{0,1\},$$

which presents a simple procedure for the generation of random variables with $F(\omega)$ as the cumulative distribution function.

It is worth noting that in some cases $u = F(y)$ cannot be solved explicitly for y. In such cases, numerical procedures such as the Newton–Raphson method have been used successfully. After having generated ω according to the above procedure and θ from a uniform distribution between $(0, 2\pi)$, they are

Table 6.1
Theoretical and sample correlation function for first-order autoregressive processes generated using 200 harmonics and 1500 sample points

	$\rho = 0.2$		$\rho = 0.5$	
	Theoretical	*Generated*	*Theoretical*	*Generated*
$\rho(1)$	0.2	0.1971	0.5	0.4893
$\rho(2)$	0.04	0.0408	0.25	0.2478
$\rho(3)$	0.008	-0.1551	0.125	0.1160
$\rho(4)$	0.0016	-0.1070	0.0625	-0.0146
$\rho(5)$	0.00032	-0.0125	0.03125	-0.0848
$\rho(6)$	0.000064	0.0201	0.015625	-0.2213
$\rho(7)$	0.0000128	0.0243	0.0078125	-0.1334
$\rho(8)$	0.00000256	-0.0406	0.00390625	-0.1284
$\rho(9)$	0.000000512	-0.0135	0.001953125	-0.1555
$\rho(10)$	0.0000001024	-0.2162	0.0009765625	-0.1214

Source: Mejía and Rodríguez-Iturbe (1974).

used in Eq. (6.32). Table 6.1 illustrates the theoretical and sample correlation functions obtained for two first-order autoregressive processes for which $\rho(1) = 0.2$ and 0.5. The number of harmonics added is equal to 200, and the number of data generated was 1500. The lack of conformance of the sampling correlation function to the theoretical one is due mainly to the effect of the finite number of harmonics. However, the correlation function of the generated data reproduces, within a reasonable margin, the theoretical one.

The synthesis of random fields follows a similar approach. Let us consider now the process given by

$$\xi(\mathbf{x}) = \cos\{\mathbf{x}^T \mathbf{Y}\omega + \theta\},$$

where $\mathbf{x}$ represents a vector of coordinates $\{x_1, \ldots, x_n\}^T$ in n-dimensional Euclidean space; $\mathbf{Y}$ is an n-dimensional random variable equidistributed on the surface of the sphere of unit radius in $\mathcal{R}_n$; ω is a random variable equivalent to $|\omega|$, whose radial distribution function is $G(\omega)$, corresponding to the isotropic correlation function $\rho(v)$; and θ is uniformly distributed over the range 0 to 2π, all the random variables (ω, θ, and $\mathbf{Y}$) being mutually independent. As we did in the unidimensional case, let us consider the ensemble averages of the process given by

$$m(\mathbf{x}) = 0$$

and

$$\rho_1(\mathbf{x}_1, \mathbf{x}_2) = E\left\{\cos\left[(\mathbf{x}_1 - \mathbf{x}_2)^T \mathbf{Y}\omega\right]\right\}.$$

Averages over an arbitrary straight line on the space are such that the correlation function will depend on the value $Y\omega$ taken by such a random variable in each one of the realizations, showing the nonergodicity of the process. Consider, as before, the process

$$\xi(\mathbf{x}) = \left(\frac{2}{N}\right)^{1/2} \sum_{i=1}^{N} \cos(\mathbf{x}^T \mathbf{Y}_i \omega_i + \theta_i), \tag{6.35}$$

where the meaning of each one of the variables is the same as it was in the previous process and all are mutually independent. The average over a straight line on the space for such a process will have a correlation function

$$\rho_2(v) = \frac{1}{N} \sum_{i=1}^{N} \cos(\mathbf{v}^T \mathbf{Y}_i \omega_i),$$

where $\mathbf{v} = \mathbf{x}_1 - \mathbf{x}_2$ and $v = |\mathbf{v}|$.

The expected value and variance of the correlation are given by

$$E\{\rho_2(v)\} = \rho(v)$$

and

$$\mathrm{var}\{\rho_2(v)\} = \frac{\mathrm{var}\{\cos(\mathbf{v}^T \mathbf{Y}_i \omega_i)\}}{N},$$

but

$$E\{\cos^2(\mathbf{v}^T \mathbf{Y}_i \omega_i)\} = E\left\{\frac{\cos(2\mathbf{v}^T \mathbf{Y}_i \omega_i) + 1}{2}\right\} = \frac{\rho(2v) + 1}{2}.$$

Therefore

$$\mathrm{var}\{\rho_2(v)\} = \frac{\rho(2v) - 2\rho^2(v) + 1}{2N},$$

which shows the asymptotic ergodicity of the process.

For a random field in two dimensions, Eq. (6.35) becomes

$$\xi(\mathbf{x}) = \sqrt{\frac{2}{N}} \sum_{i=1}^{N} \cos[\mathbf{x}^T \cdot \mathbf{Y}_i \omega_i + \theta_i], \tag{6.36}$$

where $\mathbf{x}$ represents a vector of coordinates (x_1, x_2) in $\mathscr{R}^2$; $\mathbf{Y}_i$ is a two-dimensional random variable (Y_{i1}, Y_{i2}) equidistributed on the circle of unit radius; ω_i is a random variable whose distribution is the radial distribution function

$G(\omega)$ corresponding to an isotropic correlation function; and θ_i is uniformly distributed between 0 and 2π. All the above random variables are mutually independent; $(\mathbf{x}^T \cdot \mathbf{Y}_i)$ denotes the vector inner product.

Since $\mathbf{Y}_i$ is equidistributed on the unit circle, we can write $\mathbf{Y}_i = (\cos\alpha_i, \sin\alpha_i)$, where α_i is uniformly distributed between 0 and 2π.

Thus Eq. (6.36) becomes

$$\xi(x_1, x_2) = \sqrt{\frac{2}{N}} \sum_{i=1}^{N} \cos\left[\omega_i(x_1\cos\alpha_i + x_2\sin\alpha_i) + \theta_i\right], \qquad (6.37)$$

which represents a process that is homogeneous, isotropic, and asymptotically Gaussian and ergodic with zero mean and unit variance. Fixing the number N for terms of the summation, and sampling independently ω_i, α_i, and θ_i from the corresponding distributions, we can get a sample function of the random field given by Eq. (6.37).

Equation (6.37), developed by Mejía and Rodríguez-Iturbe (1974), has been used extensively in hydrology for the generation of spatial processes. The next two sections will highlight multidimensional hydrologic synthesis.

6.4 A MULTIDIMENSIONAL MODEL FOR RAINFALL GENERATION

The synthesis of rainfall data plays many important roles in studies of water resources. Hydrologists are commonly confronted with the fact that historical records are insufficient for analysis and decision making. Under these conditions the ability to synthesize and simulate historical series becomes extremely important.

Rhenals-Figueredo et al. (1974) suggest that existing rainfall models can be classified as follows:

1. *Point rainfall models* generate time sequences of rainfall depth at a single point.
2. *Multivariate rainfall models* consider several rain gages simultaneously and are intended to preserve the covariance structure of the historical rainfall data existing in those points.
3. *Multidimensional rainfall models* attempt to characterize the rainfall phenomenon at every point over the area of interest.

The above classifications may be subdivided into rainfall exterior models, which generate storm exterior characteristics such as total depth, duration of event, and time between events, and rainfall interior models, which generate the time distribution of the total rainfall depth within each event.

This section presents the multidimensional nonstationary model of rainfall exteriors and interiors developed by Bras and Rodríguez-Iturbe (1976).

The development of the model was based on the following basic knowledge describing the behavior of storms.

1. Each storm moves with an average velocity U over the area of interest and follows a certain trajectory. Individual disturbances within the storm move at about the same velocity (Zawadzki, 1973b; Houze, 1969; Grayman and Eagleson, 1971).

2. Water falling at any instant is statistically correlated to what happened at previous times (Zawadzki, 1973a; Leclerc and Schaake, 1973).

3. Correlation in space of rainfall at any time is observed (Huff, 1970; Zawadzki, 1973b; Rhenals-Figueredo et al., 1974; Rodríguez-Iturbe and Mejía, 1974; Eagleson, 1967).

4. Spatial and time correlations are neither separable nor independent (Zawadzki, 1973b, Lenton and Rodríguez-Iturbe, 1974).

5. Rainfall is a nonstationary process. The mean and variance differ over time at all points in space (Zawadzki, 1973b; Leclerc and Schaake, 1973; Pilgrim et al., 1969).

At this point it is assumed that the storm interior of an event with given depth and duration can be modeled as

$$i(\mathbf{x}_i, t) = i_\mu(\mathbf{x}_i, t) + \eta(\mathbf{x}_i, t), \qquad (6.38)$$

where $i(\mathbf{x}_i, t)$ is the rainfall intensity at a point with coordinate vector $\mathbf{x}_i$ at time t, $i_\mu(\mathbf{x}_i, t)$ is the mean intensity at $(\mathbf{x}_i, t)$, where the mean value is taken over the ensemble of all possible storms of the same characteristics, and $\eta(\mathbf{x}_i, t)$ is the noisy zero-mean residual obeying a certain covariance function in time and space.

It is generally accepted that storms of a given type (e.g., frontal storms) in an area can be represented in a non-dimensional form of the type shown in Fig. 6.1 (Eagleson, 1970; Pilgrim et al., 1969). That figure gives an example of the average temporal rainfall distribution in terms of percentage of precipitation versus percentage of duration. Usually, the above information is obtained from data on mass accumulation of rainfall in one or more rain gages in the region. Here it is assumed that every point in space will have the same average mass distribution. By multiplying the nondimensional mass curve of Fig. 6.1a by a given depth and duration, the mean temporal behavior $i_a(t)$ of the storm at all points relative to the storm's starting time at each point is obtained. It is assumed that storm duration is the same everywhere. In order to represent all stations at the same time, an absolute time scale is defined within the area. Starting time is the moment the moving storm hits the first point in the area.

Thus data of the form of Fig. 6.1b are, for every point, translated to absolute time by the distance from the origin to the point in the direction of the storm movement. If it is assumed for simplicity that the storm moves

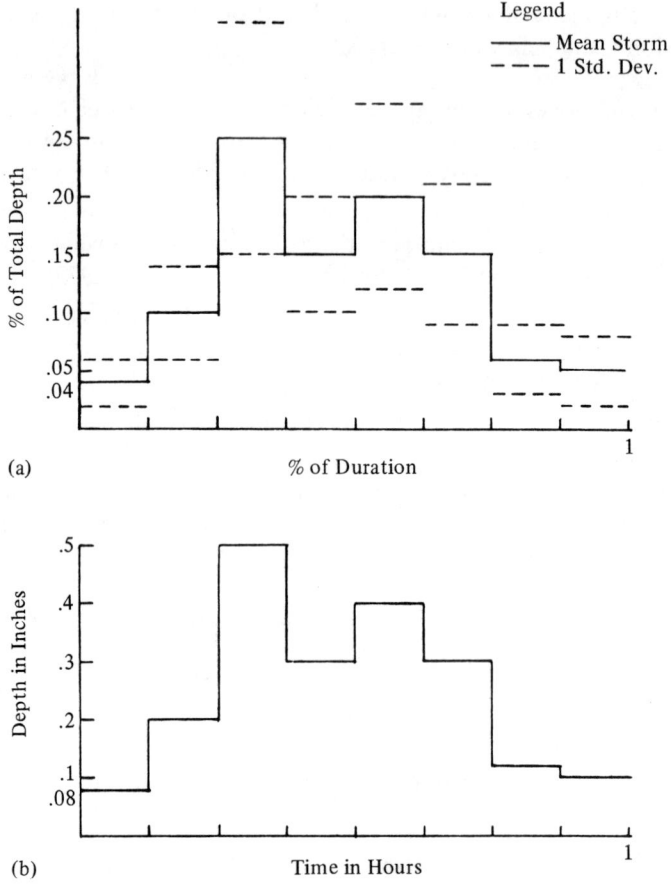

Figure 6.1 (a) Nondimensional storm representation. (b) Mean temporal behavior $i_a(t)$ resulting for a 2-in. storm of one-hour duration (from Bras and Rodríguez-Iturbe, 1976).

parallel to the x-axis (see Fig. 6.2) it is clear that

$$i_\mu(\mathbf{x}_i, t) = i_a[t - (x_i/U)], \qquad (6.39)$$

where $i_\mu(\mathbf{x}_i, t)$ is the mean intensity at $\mathbf{x}_i$ and time t after the storm reaches the boundaries of the area of interest; $i_a(t)$ the average precipitation at time t, where t is now the actual time during which it has been raining at a given point (see Fig. 6.1b); x_i the x coordinate of point i; and U the storm average velocity in direction x.

Note that because of the hyetograph form of the input data (see Fig. 6.1), a discretization in time is unavoidable. Rain accumulation is lumped within finite time intervals.

The next step is to hypothesize the form of the covariance function of the noisy residuals. In its most general form,

$$E\left[\eta(x_i, t')\eta(x_j, t'')\right] = C(x_i, t'; x_j, t'').$$

Here we write the above as

$$E\left[\eta(x_i, t')\eta(x_j, t'')\right] = \sigma(x_i, t')\sigma(x_j, t'')\rho(x_i, t'; x_j, t''),$$

where E is the expectation operator; $C(\cdot)$ the functional covariance; $\sigma(x_i, t)$ the standard deviation of rainfall intensity at point x_i and time t; $\rho(x_i, t'; x_j, t'')$ the general functional form of normalized covariance.

Standard deviation $\sigma(x_i, t)$ corresponds to the variation around $i_\mu(x_i, t)$ and is similarly obtained from the data (see Fig. 6.1). The same time translation is applicable, so

$$\sigma(x_i, t) = \sigma_a\left[t - (x_i/U)\right].$$

Equation (6.38) can then be expressed as

$$i(x_i, t) = i_a\left[t - (x_i/U)\right] + R(x_i, t)\sigma_a\left[t - (x_i/U)\right], \qquad (6.40)$$

where $R(x_i, t)$ is the standardized residual at point x_i and time t with zero mean and unit variance, and $E[R(x_i, t')R(x_j, t'')] = \rho(x_i, t'; x_j, t'')$, as was defined previously.

The statistical behavior of the residual $R(x_i, t)$ embodies the spatial and time correlation of the rainfall process. The function $\rho(x_i, t'; x_j, t'')$ must then be defined.

At this point we introduce a basic assumption about the behavior of rainfall. It is assumed that Taylor's hypothesis (Taylor, 1935) of turbulence is valid within a storm. This hypothesis implies that the correlation in time is equivalent to that in space if time is transformed to space in the mean direction of storm movement. If the storm moves in direction x, Taylor's hypothesis implies

$$\rho(x_i, t'; x_j, t'') = \rho(y_i, x_i, t'; y_j, x_j, t'')$$

$$= \rho(y_i, x_i - Ut'; y_j, x_j - Ut'').$$

Taylor's hypothesis is applicable in translating processes with relatively weak time dependence within their moving-coordinate system such that time dependence in a fixed-coordinate system is dominated by the average motion. In other words it is assumed that "noise" or "turbulence" is connected along the mean flow velocity without evolving appreciably in a "reasonable" distance. Such reasoning has been found to be applicable in general fluid turbulence (Hinze, 1959), wind and gust studies (Harris, 1971), and large-scale atmospheric processes (Gandin, 1965; Panchev, 1971). It has been tested and corroborated once for particular rainstorms by Zawadzki (1973b). Note again that in the present model it is the unit-variance residuals that obey Taylor's hypothesis. Zawadzki's work justifies Taylor's hypothesis for intervals of time of less than one hour, and this fact should be taken into consideration in using this assumption in rainfall generation.

The second assumption used is that rainfall intensities (or depths per time interval) have isotropic spatial correlation functions at any point in time. Such an assumption is justified by the works of Huff (1970) and Zawadzki (1973b). Therefore

$$\rho\left(\mathbf{x}_i, t'; \mathbf{x}_j, t''\right) = \rho\left\{\sqrt{(y_j - y_i)^2 + \left[(x_j - Ut'') - (x_i - Ut')\right]^2}\right\}. \tag{6.41}$$

Note that the above equation implies that the standardized residuals (mean zero and unit variance) are stationary and isotropic in the variables $x_1 = y$ and $x_2 = x - Ut$. The above does not imply stationarity of the "noise" element $\eta(\mathbf{x}_i, t)$ in Eq. (6.38), nor does it imply isotropy with respect to x, y, and t, separately.

It is necessary now to have a scheme for synthesizing random fields of a prescribed correlation structure. This scheme will then be used to generate rainfall events over an area. Equation (6.37) can be used to generate a random process in the $\mathcal{R}^2$ space defined by $x_1 = y$ and $x_2 = x - Ut$. To do so, the random variable ω_i must be sampled from the radial spectral distribution function corresponding to Eq. (6.41).

Values of $\xi(x_1, x_2)$ generated through Eq. (6.37) are samples of a stationary and isotropic two-dimensional process in the variables x_1 and x_2. When Eq. (6.41) describes the correlation function to be used in the generation, the generated values will be stationary and isotropic in the variables $x_1 = y$ and $x_2 = x - Ut$, but again this does not imply stationarity or represent isotropy of the rainfall residuals $\eta(x, y, t)$. The spatial and time correlations of the generated values approach Eq. (6.41) as the number of harmonics N goes to infinity.

The functional form of the correlation function, Eq. (6.41), would naturally depend on the hydrologic conditions of the area at hand. A set of commonly used isotropic correlation functions are:

$$\rho(v) = e^{-\alpha|v|} \qquad \text{(single exponential)} \qquad (6.42)$$

$$\rho(v) = e^{-\alpha^2 v^2} \qquad \text{(quadratic exponential)} \qquad (6.43)$$

$$\rho(v) = |v| b K_1(|v| b) \qquad \text{(Bessel form)}, \qquad (6.44)$$

where $K_1(\quad)$ is the modified Bessel function of the second kind and order 1, b, and α are parameters, and $|v|$ is the distance between points.

The corresponding radial spectral densities and distributions of Eqs. (6.42) and (6.44) were given in Eqs. (6.26) and (6.29).

The quadratic exponential has the following radial spectral density and distribution:

$$G'(\omega) = \left(\omega/2\alpha^2\right) \cdot \exp\left(-\omega^2/4\alpha^2\right) \qquad 0 \le \omega < \infty \qquad (6.45)$$

$$G(\omega) = 1 - \exp\left(-\omega^2/4\alpha^2\right) \qquad 0 \le \omega < \infty. \qquad (6.46)$$

The works of Huff (1970) and Zawadzki (1973b) suggest the use of the single exponential function to represent the spatial correlation structure of rainfall intensities.

Bras and Rodríguez-Iturbe's (1976) rainfall model consists of an algorithm uniting existing techniques of rainfall exterior generation and the techniques of rainfall interior simulation just presented. Required inputs are basically the following:

1. Marginal distribution (form and parameters) of time between events t_b.

2. Marginal distribution of storm duration t_d (it is assumed that rainfall accumulates during the same amount of time at each point).

3. Conditional distribution of area average total depth D given the storm duration.

4. Distribution of storm average velocities and directions.

5. Probability of occurrence of various types of events.

6. Nondimensional time distribution of rainfall (holding over all points in space) and corresponding standard deviations of rainfall accumulation in each time step (Fig. 6.1) for each event type.

7. Form of time and space correlations given in Eq. (6.41).

The steps of the generation algorithm are then:

1. Set total time of generation desired.

2. Sample t_b, t_d, and D from corresponding distributions.

3. Sample storm velocity, direction, and type.

4. Construct average time distribution $i_a(t)$ of generated storm by scaling the nondimensional hyetograph by the generated depth and duration.

5. Construct corresponding standard deviation $\sigma_a(t)$.

6. Specify points in space where generation is desired.

7. Generate zero-mean and unit-variance residuals with Eq. (6.37).

8. Create storm interior at desired points with Eq. (6.40).

9. Go back to Step 2 to generate a consecutive event and repeat for desired period of generation.

The generation of storm interiors (using Eq. 6.40) extends for the storm duration plus the time the storm takes to traverse the area. The storm duration at each point, though, is t_d, since $i_a[t - (x_i/U)]$ and $\sigma_a[t - (x_i/U)]$ are zero or near zero before the storm reaches the point x_i and become zero again after the storm passes the point in question. To illustrate, a simple numerical example of rainfall generation is given. It is assumed that only one type of storm is generated (no seasonal variations or different storm classifications).

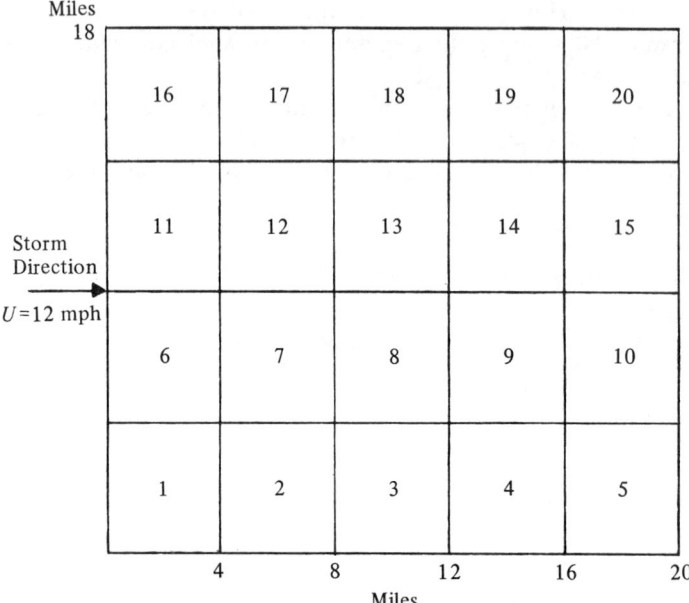

Figure 6.2 Diagram of 360-mi² area showing discretization, storm velocity, and direction (from Bras and Rodríguez-Iturbe, 1976).

Storm speed is given as a deterministic input. For convenience, storm direction is fixed so as to coincide with the definition of the x-coordinate axis.

The area of generation is a 360-mi² rectangular area shown in Fig. 6.2. Its length is 20 mi in the x direction and 18 mi in the y direction.

The residual correlation function is of the single exponential type, Eq. (6.42), with parameter 0.15. It takes the form

$$\rho\left(x_j, t''; x_i, t'\right) = \exp\left\{-0.15\left[\left(y_j - y_i\right)^2 + \left(\left(x_j - Ut''\right) - \left(x_i - Ut'\right)\right)^2\right]^{1/2}\right\}.$$

(6.47)

Rainfall exterior parameters are assumed to obey simple exponential distributions. Similar distributions have been used previously by Leclerc and Schaake (1972) and Eagleson (1971).

The nondimensional (in time) rainfall distribution was shown in Fig. 6.1b together with values of one standard deviation. The hyetograph was discretized in eight intervals, as shown in that figure.

Note that the correlation-function parameter was kept constant at 0.15, even though as the generator stands different storms would have different time intervals having different correlation parameters. The value of 0.15 is one that is reasonable for some types of storms in the light of Huff's (1970) work.

Table 6.2
Point-intensity values of a generated storm with the multidimensional rainfall model

Point	Depth (in.)										Total
	0.000 hr	1.791 hr	3.582 hr	5.372 hr	7.163 hr	8.954 hr	10.745 hr	12.536 hr	14.327 hr	16.117 hr	
1	0.0000	0.0439	0.1140	0.2402	0.1422	0.2295	0.1547	0.0575	0.0519	0.0000	1.034
2	0.0001	0.0445	0.1048	0.2384	0.1435	0.2318	0.1578	0.0616	0.0501	0.0000	1.033
3	0.0600	0.0476	0.1068	0.2499	0.1626	0.2193	0.1557	0.0627	0.0432	0.0000	1.048
4	0.0000	0.0482	0.0987	0.2540	0.1575	0.2162	0.1664	0.0582	0.0511	0.0000	1.050
5	0.0000	0.0455	0.0959	0.2629	0.1611	0.2213	0.1594	0.0559	0.0561	0.0000	1.058
6	0.0000	0.0427	0.1127	0.2350	0.1489	0.2355	0.1536	0.0555	0.0503	0.0000	1.034
7	0.0000	0.0452	0.1111	0.2354	0.1592	0.2361	0.1570	0.0636	0.0514	0.0000	1.059
8	0.0001	0.0451	0.1046	0.2597	0.1599	0.2356	0.1592	0.0590	0.0525	0.0000	1.076
9	0.0001	0.0456	0.0945	0.2572	0.1576	0.2208	0.1573	0.0636	0.0543	0.0000	1.051
10	0.0000	0.0446	0.0913	0.2498	0.1622	0.2201	0.1459	0.0621	0.0518	0.0000	1.028
11	0.0000	0.0462	0.0961	0.2304	0.1704	0.2062	0.1474	0.0627	0.0579	0.0001	1.017
12	0.0000	0.0416	0.0954	0.2330	0.1718	0.2105	0.1555	0.0609	0.0585	0.0001	1.027
13	0.0000	0.0426	0.1000	0.2712	0.1640	0.2076	0.1577	0.0540	0.0613	0.0001	1.059
14	0.0000	0.0386	0.1016	0.2610	0.1621	0.2219	0.1487	0.0618	0.0607	0.0000	1.056
15	0.0000	0.0371	0.1052	0.2683	0.1653	0.2125	0.1442	0.0669	0.0574	0.0000	1.057
16	0.0000	0.0455	0.0940	0.2438	0.1741	0.2048	0.1433	0.0611	0.0566	0.0001	1.023
17	0.0000	0.0447	0.0941	0.2644	0.1746	0.2093	0.1596	0.0622	0.0584	0.0000	1.067
18	0.0000	0.0415	0.1039	0.2659	0.1742	0.2122	0.1503	0.0633	0.0579	0.0001	1.069
19	0.0000	0.0365	0.1029	0.2612	0.1650	0.2097	0.1595	0.0651	0.0623	0.0001	1.062
20	0.0000	0.0349	0.0999	0.2704	0.1645	0.1945	0.1565	0.0626	0.0583	0.0001	1.042

Velocity is 12.0 mi/hr, duration is 14.33 hr, total mean area depth is 1.079 in., and date of occurrence from initial time is 19.498 days.

Source: Bras and Rodriguez-Iturbe (1976).

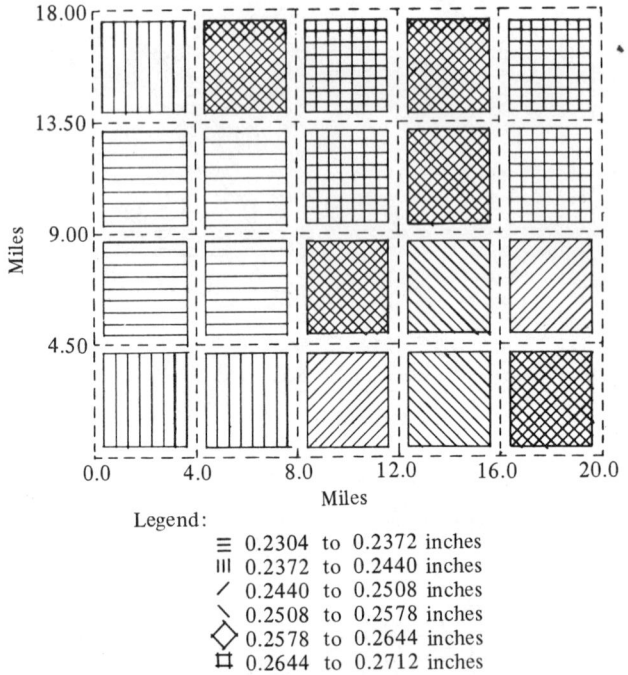

Legend:

≡ 0.2304 to 0.2372 inches
III 0.2372 to 0.2440 inches
╱ 0.2440 to 0.2508 inches
╲ 0.2508 to 0.2578 inches
◇ 0.2578 to 0.2644 inches
⌂ 0.2644 to 0.2712 inches

Figure 6.3 Distribution of point rainfall intensities over an area at 5.372 hours into the storm (from Bras and Rodríguez-Iturbe, 1976).

By using a storm velocity of 12.0 mi/hr and the data described in previous paragraphs, several storms were generated. In these examples, only 50 harmonics were used in the generation procedure.

Table 6.2 gives the point-intensity values of a generated storm. The generation points are the centers of the individual subareas. Figure 6.3 shows the spatial distribution of point intensities at 5.372 hours into the storm occurring 19.5 days after the beginning of the generation.

The little variation in total depth among the different points at any given time during the storm is to be expected in this example, where the intensity, total duration, and correlation decay (0.15) belong to a frontal type of precipitation generally affecting a large region, of which the area under study is only a small fraction.

In many basins it is doubtful that a certain nondimensional storm representation in time, such as that shown in Fig. 6.1, will be applicable to all points in the region. Orographic effects are a clear case where both the mean and standard deviation of such a representation will change from point to point in the area. In order to incorporate this kind of behavior in the model, the average temporal character $i_a(t)$ is described by a set of hyetographs, which correspond to different subregions in the area, instead of being solely described by a hyetograph such as Fig. 6.1. This set of temporal-rainfall characteristics will be then multiplied by different amounts of total expected depth of rainfall at different regions of the area under consideration.

6.5 DEPENDENCE OF STORM RUNOFF ON THE MULTIDIMENSIONAL STRUCTURE OF RAINFALL AND OTHER PARAMETERS

This section first describes the importance and effects of the spatial characteristics of the precipitation input to a runoff model of small catchments as presented by Wilson et al. (1979). Two models are used: a deterministic rainfall–runoff model (Wilson, 1976) and the multidimensional nonstationary time-varying rainfall-generation model described in Section 6.4. The rainfall–runoff model is based on the solution of the conservation of mass equation and the kinematic simplification of the momentum equation. Infiltration is represented by using the U.S. Soil Conservation Service (1971) method, and appropriate curve numbers were used in each segment.

The effect of the spatial variability of rainfall is studied here by sampling each storm imposed upon the rainfall–runoff model of a basin in two different manners. In the first case, each storm generated multidimensionally by the

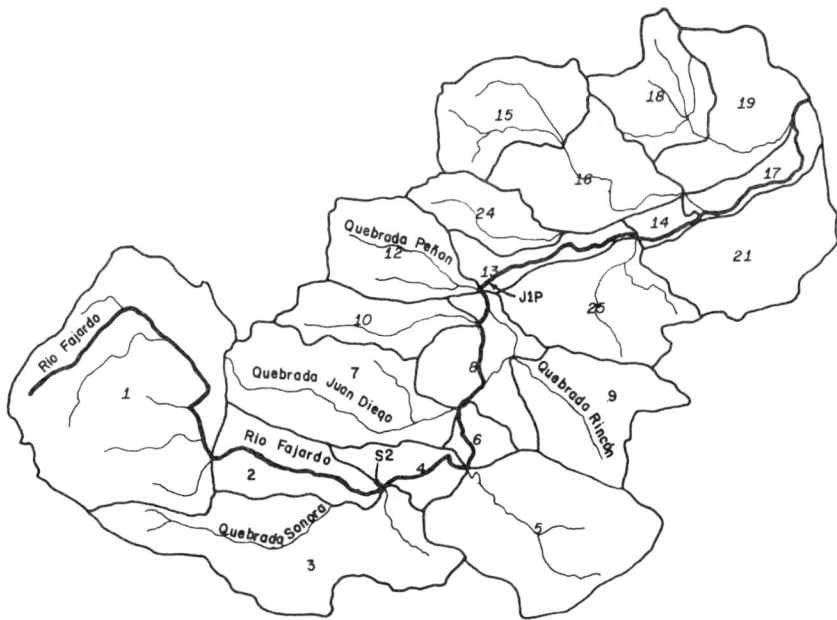

Figure 6.4 Map of Río Fajardo Basin showing breakdown into subcatchments (from Rodríguez-Iturbe, 1975).

precipitation model is recorded at ground level by a dense network of rain gages, which preserve the temporal and spatial characteristics of the input into the basin. In the second case, only one gage was assumed to be sampling the rainfall event; the precipitation recorded in time by this gage is assumed to be taking place over all the physical space of the basin. The runoff response is then compared for both of the above cases.

The catchment selected in the study by Wilson et al. (1979) is the Río Fajardo in Puerto Rico. Located in the northeastern part of Puerto Rico, the Río Fajardo drains an area of 26.5 mi^2. A detailed representation was made of the basin by means of the rainfall–runoff model with 21 individual subcatchments. The basin and the breakdown in subcatchments are shown in Fig. 6.4.

It is important to emphasize that the objective here is not to model the Río Fajardo basin for any particular design purpose. The Río Fajardo physiographic characteristics were used in order to give realistic values and magnitudes to the different parameters of the rainfall–runoff model. For the purposes of the example, any well-documented basin would have accomplished the same purpose.

Rainfall exteriors were again characterized with simple exponential distributions. Only events with mean area depths above 2 in. were used for storm-interior generation and in the experiments. Smaller mean depths were either completely infiltrated or produced an insignificant hydrograph with the

Table 6.3
Rainfall characteristics for the experiments relating multidimensional precipitation patterns and hydrograph characteristics

Storm	Duration (hr)	Mean area depth (in.)	Correlation parameter α
1	20.55	2.11	2.0
2	11.29	2.25	2.0
3	22.24	5.34	2.0
4	12.63	4.10	2.0
5	8.37	2.93	2.0
6	20.55	2.11	0.6
7	11.29	2.25	0.6
8	22.24	5.34	0.6
9	12.63	4.10	0.6
10	8.37	2.93	0.6
11	20.55	2.11	0.15
12	11.29	2.25	0.15
13	22.24	5.34	0.15
14	12.63	4.10	0.15
15	8.37	2.93	0.15

Source: Wilson et al. (1979).

parameters used in the basin representation. In all cases the storms are assumed to have the dimensionless representation shown in Fig. 6.1a and to travel with a velocity of 12 mi/hr in the positive x direction. The exponential correlation function, Eq. (6.42), was used by Wilson et al. (1979).

Fifteen storms with the characteristics described in Table 6.3 were imposed upon the rainfall–runoff model. The storms are generated over a 50-mi^2 rectangular area, 7.9 mi long and 6.32 mi wide, which completely contains the 26.5-mi^2 Fajardo basin. These are controlled experiments in which the generated storms are not representative of the Fajardo basin, which is only playing the role of a real-world watershed. The storm area is broken down into 20 blocks, with each block 2.5 mi^2 in area, as shown in Fig. 6.5. The storms are recorded in two ways. First, a rain gage is assumed to be located in the center of each block; second, only one rain gage, in Block 20, is assumed to be in operation.

The isohyetal maps show that the storm surface gets progressively less uniform as the correlation parameter α goes from 0.15 to 2.0. Examples of this are shown in Figs. 6.6 and 6.7, which reflect the fact that a large α implies a fast decrease in correlation with distance.

The peak discharge, time to peak discharge, and total runoff volume were determined from the hydrographs resulting from the storm recorded by 20

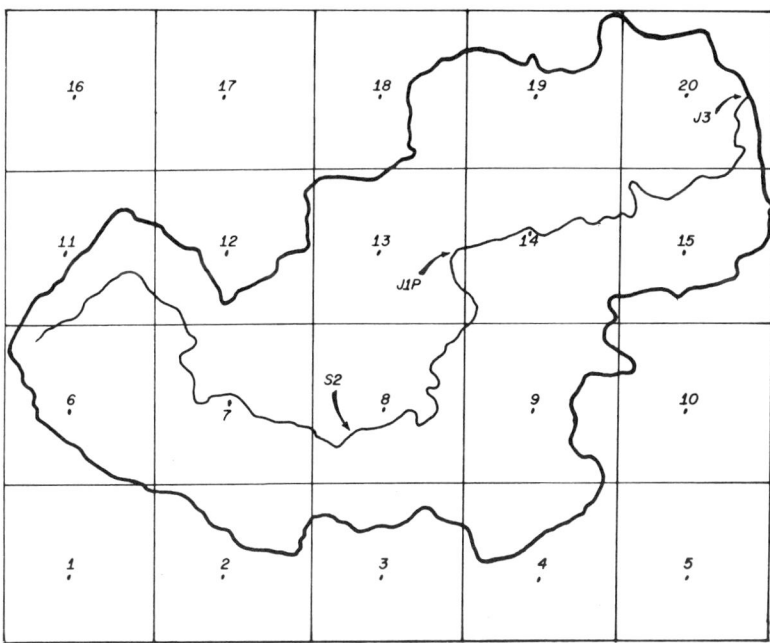

Figure 6.5 Grid over which rainfall is generated in the Río Fajardo Basin example (from Wilson, et al., 1979).

Figure 6.6 Isohyetal map for Storm 2, with $\alpha = 2.0$ (from Wilson et al., 1979).

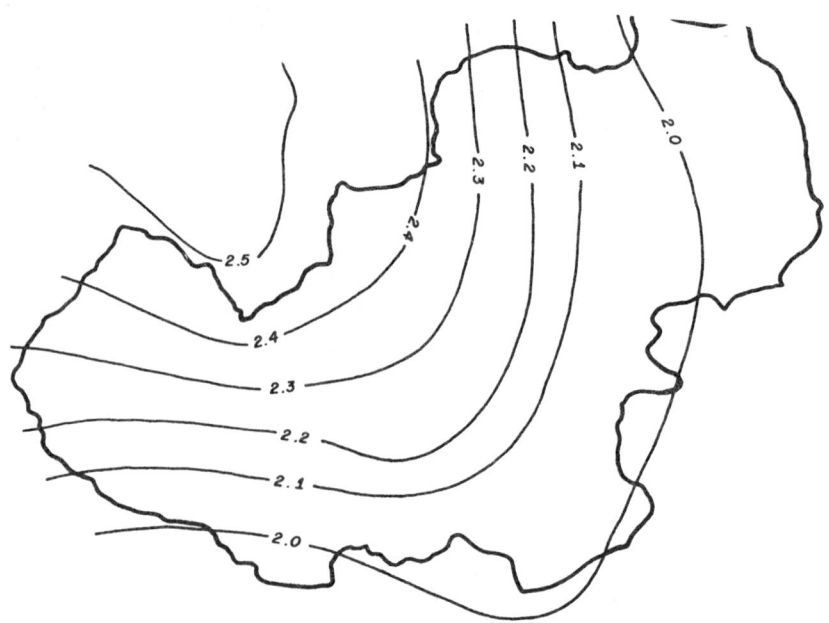

Figure 6.7 Isohyetal map for Storm 12, with $\alpha = 0.15$ (from Wilson et al., 1979).

Table 6.4
Values of hydrograph parameters for the hydrograph at catchment outlet

Storm	Correlation parameter α	Rain gages	t_p (min)	Peak (cfs)	Volume $(10^6 \times ft^3)$	Mean depth (in.)
\multicolumn{7}{c}{Storm Duration, 20.55 hr; Storm Depth, 2.10 in.}
1	2.0	20	800	3,489.04	132.96	1.98
	2.0	1	930	3,557.97	127.20	1.91
6	0.6	20	795	4,362.77	144.00	2.06
	0.6	1	810	3,974.50	133.20	1.97
11	0.15	20	640	3,886.97	136.80	1.98
	0.15	1	650	3,622.79	97.32	1.61
\multicolumn{7}{c}{Storm Duration, 11.29 hr; Storm Depth, 2.25 in.}
2	2.0	20	525	2,718.00	59.67	2.21
	2.0	1	550	3,263.20	62.54	2.27
7	0.6	20	515	3,187.07	69.18	2.39
	0.6	1	475	3,999.69	75.50	2.51
12	0.15	20	505	3,099.89	60.94	2.16
	0.15	1	550	2,180.54	46.39	1.90
\multicolumn{7}{c}{Storm Duration, 22.24 hr; Storm Depth, 5.34 in.}
3	2.0	20	835	13,360.21	537.15	5.44
	2.0	1	510	21,002.47	703.93	6.66
8	0.6	20	510	13,918.60	532.75	5.41
	0.6	1	510	21,220.70	611.94	5.91
13	0.15	20	865	15,735.51	616.74	6.00
	0.15	1	545	14,013.34	596.74	5.80
\multicolumn{7}{c}{Storm Duration, 12.63 hr; Storm Depth, 4.09 in.}
4	2.0	20	500	8,630.00	195.95	4.26
	2.0	1	515	8,559.37	170.84	3.87
9	0.6	20	565	7,313.94	171.48	4.12
	0.6	1	350	6,798.96	159.64	3.98
14	0.15	20	495	9,288.62	188.43	4.14
	0.15	1	495	9,352.54	176.76	3.95
\multicolumn{7}{c}{Storm Duration, 8.37 hr; Storm Depth, 2.93 in.}
5	2.0	20	395	3,570.52	56.25	2.85
	2.0	1	405	3,899.90	64.05	3.10
10	0.6	20	405	3,791.98	62.55	3.03
	0.6	1	400	3,529.14	62.48	3.07
15	0.15	20	395	2,577.20	47.40	2.72
	0.15	1	385	3,420.08	61.95	3.00

Source: Wilson et al. (1979).

gages and from the storm recorded with only one gage. The results are shown in Table 6.4. Note that the mean storm depth is different from the mean rainfall depth of the 20-gage case because the basin occupies only a little more than 50% of the storm area.

Comparing the mean storm depths for the one-gage and the 20-gage cases, they are not very different at all; the mean absolute difference is 8% and the maximum 22%. This is expected from the size of the basin and the nature of the parameters of the storm model, which in this case are representative of frontal storms of long duration and low intensity. Since the purpose of the experiments of Wilson et al. (1979) was to study the influence of the spatial pattern of rainfall on the storm runoff, it is convenient that the total volume of rainfall input does not differ much in each case.

Consider now the differences in volume of the runoff hydrograph at the outlet of the catchment. With the exception of one storm, the volume is underpredicted if the depth is underpredicted and overpredicted if the depth is overpredicted. This is somewhat expected, since the infiltration and interception remain the same for both the one-gage and the 20-gage cases. Nevertheless, an interesting point emerges here, namely, the percentage of error in the depths is in most cases amplified within the volume. Take, for example, Storm 12, for which the mean depth is underestimated by 12% when using the one-gage network but the total runoff volume produced when using the one-gage rainfall input is 24% under the total runoff volume coming from the 20-gage rainfall input. The mean absolute error in the runoff volumes of the one-gage case with respect to the 20-gage case is 13%.

No one disputes that the spatial distribution of rainfall has a direct effect on the amount of storm runoff, but the argument has been made in the past that errors in the rainfall input will be lessened when routed through a basin. Without trying to make a general conclusion in the opposite direction, the results from the experiments of Wilson et al. do not support the above argument and point to the fact that even in cases where the total depth of rainfall is not in serious error, the distribution of the input, when not preserved, may lead to discrepancies in the volume of the runoff output.

Wilson et al. (1979) also analyzed the time to peak and the peak of the runoff events. Again, the spatial character of the rainfall plays an important role, and it is not possible to speak of lessening of the error estimation when going from rainfall depth to hydrograph peak. Sometimes, as in Storm 8, the rainfall depth is overestimated by 9% and the peak is overestimated by 52%. In others, such as Storm 11, the total depth is underestimated by 18% and the hydrograph peak is underestimated by 7%.

Freeze (1980) has developed a stochastic–conceptual mathematical model of hydrologic processes in a hillslope. The model was used to investigate the influence of the spatial stochastic properties of the hillslope parameters on the statistical properties of the resulting runoff events. The model involves spatially distributed calculations of infiltration rates, moisture contents, and water-table positions.

The approach followed by Freeze is an event-based analysis. One complete experiment involves the simulation of N hydrologic events of M time steps each. The output is in the form of N hydrographs, each of the form Q^m, $m = 1, 2, \ldots, M$. The outflow is calculated on the basis of a distributed set of physically based equations that represent the hydrologic process on the hillslope.

The model incorporates time-independent and time-dependent hillslope variables. The time-independent variables such as the saturated hydraulic conductivity, K_{ij}^s, are defined on a two-dimensional grid in space. These variables remain constant throughout time in each of the simulated rainfall events.

The time-dependent hillslope variables such as the infiltration at a certain point (x_i, y_j), f_{ij}^m, are calculated at each time step over time intervals corresponding to the superscript $m = 1, 2, \ldots, M$. Other superscripts refer to specific times; thus p_{ij}^p represents the precipitation p_{ij} at the time of ponding, θ_{ij}^0 stands for the initial moisture content, and K_{ij}^s is the hydraulic conductivity at saturation.

For any given event Freeze presents a six-step procedure to operate his conceptual model.

1. Generate the time-independent hillslope parameters. These include the topographic elevation z_{ij}, the surface runoff travel times T_{ij}, and three soil properties: saturated hydraulic conductivity K_{ij}^s; porosity n_{ij}; and a soil-storage parameter B_{ij}.

2. Generate the external storm properties for each rainfall event. These include the time since the previous storm t_b, the storm duration t_D, the storm velocity U, and the storm's total rainfall depth D.

3. Generate the initial hillslope conditions for each event. These include the water-table elevation z_{ij}', the unsaturated soil depth z_{ij}'', the initial moisture content θ_{ij}^0, and the initial moisture deficit S_{ij}^0.

4. Generate the internal rainfall-intensity pattern p_{ij}^m for each time step of each event.

5. Calculate the infiltration rate f_{ij}^m and the rainfall excess r_{ij}^m for each time step of each event.

6. Calculate the streamflow hydrograph Q^m, $m = 1, 2, \ldots, M$ for each event.

Following the simulation of N events, Freeze carries out a statistical analysis of the main characteristics of the outflow hydrographs. All the generation of random fields followed the sampling from the spectrum technique. The model by Bras and Rodríguez-Iturbe (1976) was used for rainfall generation.

Among the conclusions obtained by Freeze (1980) there are several of special importance in the context of this chapter; two of them will be listed here.

1. The mean, variance, and correlation structure of the representation of hydraulic conductivity as a spatial stochastic process exert an important influence on the statistical properties of runoff events arising from a hillslope under a given climatic regime.

2. If one tries to simulate a heterogeneous hillslope on which the actual log-transformed hydraulic conductivities have a given spatial correlation, mean and variance, with a representative homogeneous hillslope on which the hydraulic conductivity is the mean of the field at all points, the statistical properties of the predicted runoffs generated from a given stochastic sequence of storm events may be seriously in error.

6.6 SIMULATION OF RANDOM FIELDS WITH THE TURNING-BANDS METHOD

This section presents the turning-bands method (TBM) for simulation of random fields. This method was first presented by Matheron (1973) and applied by the Ecole des Mines de Paris (e.g., Journel, 1974; Delhomme, 1979).

With a significant number of points in the field, the TBM is more efficient than the process of sampling from the spectrum. Here, we closely follow the very clear exposition contained in Mantoglou and Wilson (1982).

Instead of synthesizing the multidimensional field directly, the TBM performs simulations along lines in space. Then a weighted sum of values of the line process are assigned to each point in the region $\mathcal{R}^n$.

Let us designate the multidimensional stationary and isotropic process that we want to synthesize by $Z(\cdot)$. We will assume the random field has zero mean and is normally distributed. If this is not the case, a transformation must be applied first. A two-dimensional example is shown in Fig. 6.8 where random lines are generated from an arbitrary origin. The angle θ of each line with the horizontal axis is a random variable uniformly distributed between 0 and 2π. Along each line i, a unidimensional process with the mean and covariance function $C_1(\zeta)$ is then generated. The points along line i in which the unidimensional process $Z_i(\zeta)$ will be generated, are the projections along that line of the points on the two-dimensional region where we want to generate values of the random field. Figure 6.8 shows one point in space with location vector $\mathbf{x}_N$; its projected counterpoint on line i is ζ_{Ni} and the value of the unidimensional process at that point is $Z_i(\zeta_{Ni})$. We can write $Z_i(\zeta_{Ni}) = Z_i(\mathbf{x}_N \cdot \mathbf{u}_i)$, where $\mathbf{u}_i$ is the unit vector on line i and the parentheses contain the inner product of the vectors $\mathbf{x}_N$ and $\mathbf{u}_i$.

For each of L lines an independent realization of the unidimensional process with covariance $C_1(\zeta)$ is then generated and the point N of the two-dimensional region is assigned the simulated value

$$Z_S(\mathbf{x}_N) = \frac{1}{\sqrt{L}} \sum_{i=1}^{L} Z_i(\mathbf{x}_N \cdot \mathbf{u}_i). \tag{6.48}$$

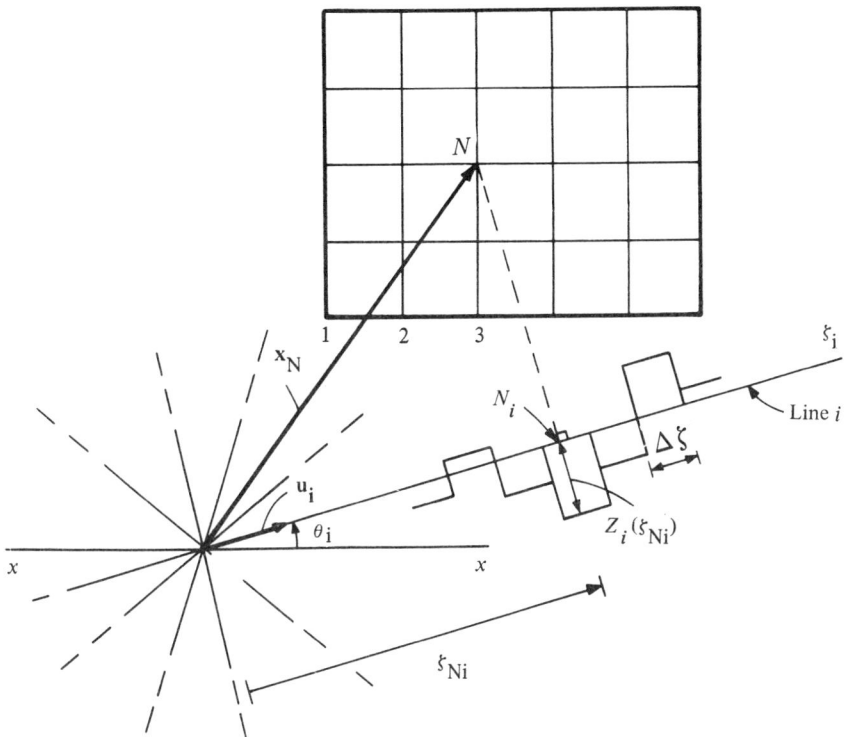

Figure 6.8 Schematic representation of the field and the turning-bands lines (from Mantoglou and Wilson, 1982).

The Turning Bands Method (TBM) gets its name from the discretized segments—or bands—appearing on each random line because of the discrete character of the unidimensional generation (see Fig. 6.8). Since the lines "turn" between 0 and 2π, the bands also "turn," and this is the reason for the name coined by Matheron (1973).

The key question now is the relation between the covariance $C_1(\zeta)$ of the unidimensional process and the covariance $C_S(v)$ of the random field simulated in two or three dimensions.

Following Mantoglou and Wilson (1981), take two points of the field with location vectors x_1 and x_2 as shown in Fig. 6.9. The simulated values in these points are given by

$$Z_S(x_1) = \frac{1}{\sqrt{L}} \sum_{i=1}^{L} Z_i(x_1 \cdot u_i) \tag{6.49}$$

and

$$Z_S(x_2) = \frac{1}{\sqrt{L}} \sum_{j=1}^{L} Z_j(x_2 \cdot u_j). \tag{6.50}$$

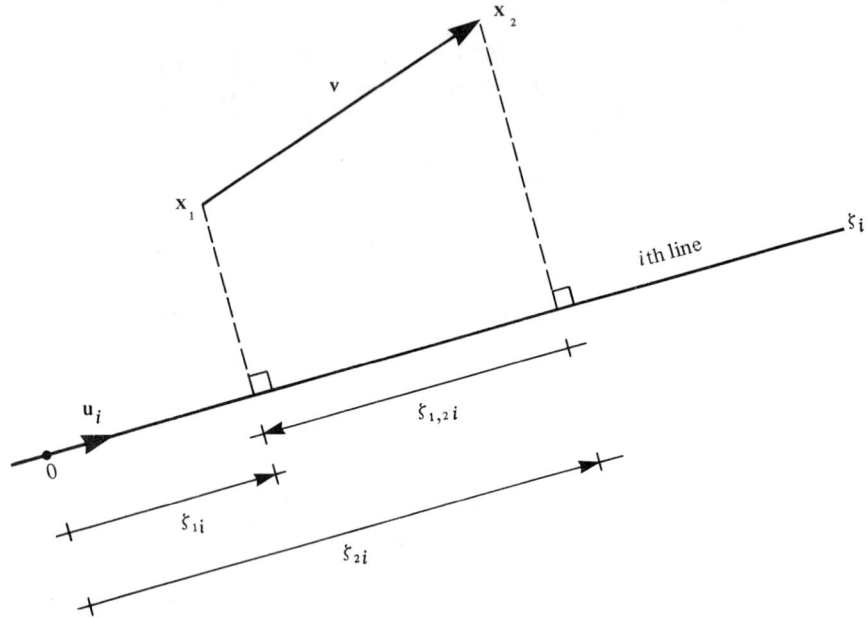

Figure 6.9 Projection of the vector **v** on a turning-bands line (from Mantoglou and Wilson, 1981).

Since the unidimensional process has zero mean, the field Z_S also has zero mean and the covariance between the two points is

$$C_S(\mathbf{x}_1, \mathbf{x}_2) = E\left[\left(\frac{1}{\sqrt{L}} \sum_{i=1}^{L} Z_i(\mathbf{x}_1 \cdot \mathbf{u}_i)\right)\left(\frac{1}{\sqrt{L}} \sum_{j=1}^{L} Z_j(\mathbf{x}_2 \cdot \mathbf{u}_j)\right)\right]$$

$$= \frac{1}{L} \sum_{i=1}^{L} \sum_{j=1}^{L} E\left[Z_i(\mathbf{x}_1 \cdot \mathbf{u}_i) Z_j(\mathbf{x}_2 \cdot \mathbf{u}_j)\right].$$

Because of the independence of the unidimensional processes along the different lines, the expected value in the above summation will be zero unless $i = j$. Therefore

$$C_S(\mathbf{x}_1, \mathbf{x}_2) = \frac{1}{L} \sum_{i=1}^{L} E\left[Z_i(\mathbf{x}_1 \cdot \mathbf{u}_i) Z_i(\mathbf{x}_2 \cdot \mathbf{u}_i)\right]. \tag{6.51}$$

We now write

$$E\left[Z_i(\mathbf{x}_1 \cdot \mathbf{u}_i) Z_i(\mathbf{x}_2 \cdot \mathbf{u}_i)\right] = C_1(\mathbf{v} \cdot \mathbf{u}_i), \tag{6.52}$$

where $C_1(\zeta)$ is the covariance of the unidimensional process assumed to be stationary and $\mathbf{v} = \mathbf{x}_2 - \mathbf{x}_1$. The covariances $C_S(v)$ and $C_1(\zeta)$ are then related through

$$C_S(\mathbf{x}_1, \mathbf{x}_2) = \frac{1}{L} \sum_{i=1}^{L} C_1(\mathbf{v} \cdot \mathbf{u}_i). \tag{6.53}$$

The vector $\mathbf{u}_i$ is uniformly distributed over the unit circle or the unit sphere depending on if we work in two or three dimensions.

When $L \to \infty$, we have

$$C_S(v) = \lim_{L \to \infty} \frac{1}{L} \sum_{i=1}^{L} C_1(\mathbf{v} \cdot \mathbf{u}_i) = E[C_1(\mathbf{v} \cdot \mathbf{u})], \tag{6.54}$$

where $v = |\mathbf{v}|$.

The expectation of Eq. (6.54) can be written as

$$E[C_1(\mathbf{v} \cdot \mathbf{u})] = \int_{\mathscr{C}} C_1(\mathbf{v} \cdot \mathbf{u}) f(\mathbf{u}) \, d\mathbf{u}, \tag{6.55}$$

where $f(\mathbf{u})$ denotes the probability density function of $\mathbf{u}$ and $\mathscr{C}$ the unit circle or unit sphere. The term $d\mathbf{u}$ denotes the differential length or area on $\mathscr{C}$ at the end of vector $\mathbf{u}$.

In the two-dimensional case, we have

$$f(\mathbf{u}) = \frac{1}{2\pi}. \tag{6.56}$$

When working in three dimensions

$$f(\mathbf{u}) = \frac{1}{4\pi}. \tag{6.57}$$

Using Eq. (6.55) and the above relations, we then have for the two-dimensional and three-dimensional cases

$$C_S(v) = \frac{1}{2\pi} \int_{\substack{\text{unit} \\ \text{circle}}} C_1(\mathbf{v} \cdot \mathbf{u}) \, d\mathbf{u} \tag{6.58}$$

and

$$C_S(v) = \frac{1}{4\pi} \int_{\substack{\text{unit} \\ \text{sphere}}} C_1(\mathbf{v} \cdot \mathbf{u}) \, d\mathbf{u}. \tag{6.59}$$

Although the two-dimensional field is more common in surface hydrology, we will include here the three-dimensional case because of its use mainly in the simulation of aquifer characteristics and groundwater conditions. In practice you know or assume to know the covariance function $C(v)$ to be preserved during the simulation. So $C_S(v)$ in Eqs. (6.58) and (6.59) is set equal to $C(v)$ and the problem now is to obtain the one-dimensional covariance function $C_1(\zeta)$ in an explicit manner from the previous equations. Again we follow Mantoglou and Wilson's (1981) derivation.

6.6.1 Three-Dimensional Fields

Define (orthogonal) axes (x, y, z) with origin at the point $\mathbf{x}_1$ and with the z-axis in the direction of the vector $\mathbf{v} = \mathbf{x}_2 - \mathbf{x}_1$, as shown in Fig. 6.10. In spherical coordinates

$$\mathbf{v} \cdot \mathbf{u} = v \cos \phi,$$

where $v = |\mathbf{v}|$ and $d\mathbf{u} = \sin \phi \, d\phi \, d\theta$. Equation (6.59) is then written as

$$C_S(v) = \frac{1}{4\pi} \int_0^{2\pi} d\theta \int_0^{\pi} C_1(v \cos \phi) \sin \phi \, d\phi$$

$$= \frac{1}{2} \int_0^{\pi} C_1(v \cos \phi) \sin \phi \, d\phi. \tag{6.60}$$

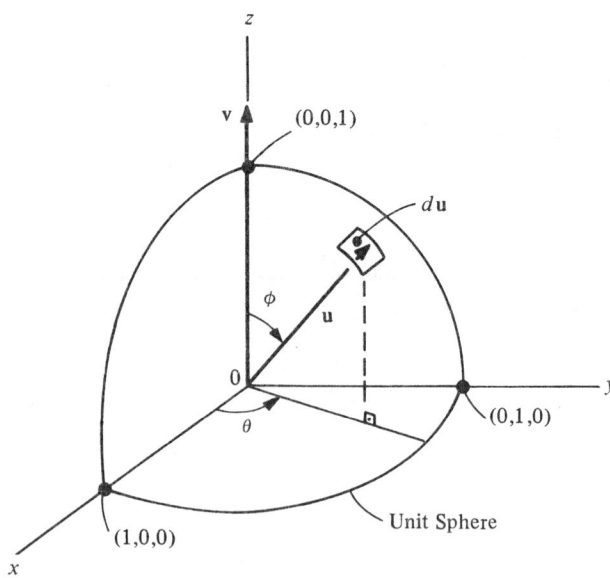

Figure 6.10 Definition sketch for the turning-bands method in the three-dimensional case (from Mantoglou and Wilson, 1982).

The projection of v along the line is

$$\zeta = v \cos \phi; \qquad d\zeta = -v \sin \phi \, d\phi.$$

Substituting into Eq. (6.60) and because of the symmetry of the unidimensional covariance function, one obtains

$$C_S(v) = \frac{1}{v} \int_0^v C_1(\zeta) \, d\zeta. \qquad (6.61)$$

Using the Leibnitz rule of differentiation and changing the notation, you get

$$C_1(v) = \frac{d}{d\zeta}(\zeta C_S(\zeta)), \qquad (6.62)$$

which relates the three-dimensional covariance function to the unidimensional one.

Mantoglou and Wilson (1981) give several examples of unidimensional equivalents to typical three-dimensional covariance functions, as shown below.

Exponential Model

For a three-dimensional exponential covariance function of the form

$$C(v) = \sigma^2 e^{-\alpha v} \qquad 0 \le v < \infty,$$

the corresponding one-dimensional covariance function is given by Eq. (6.62) as

$$C_1(\zeta) = \sigma^2 \frac{d}{d\zeta}(\zeta e^{-\alpha\zeta}) = \sigma^2(1 - \alpha\zeta)e^{-\alpha\zeta} \qquad 0 \le \zeta < \infty. \qquad (6.63)$$

Double Exponential Model

Expressed by the formula

$$C(v) = \sigma^2 e^{-\alpha^2 v^2} \qquad 0 \le v \le \infty,$$

the corresponding unidimensional covariance function is given by

$$C_1(\zeta) = \sigma^2(1 - 2\alpha^2\zeta^2)e^{-\alpha^2\zeta^2} \qquad 0 \le \zeta < \infty. \qquad (6.64)$$

Spherical Model

If the spherical model is given by

$$C(v) = \begin{cases} \sigma^2\left[1 - \frac{3}{2}\frac{v}{a} + \frac{1}{2}\frac{v^3}{a^3}\right] & 0 \le v \le a \\ 0 & a < v, \end{cases}$$

the corresponding one-dimensional covariance function is

$$C_1(\zeta) = \begin{cases} \sigma^2 \left[1 - \dfrac{3\zeta}{a} + \dfrac{2\zeta^3}{a^3} \right] & 0 \le \zeta \le a \\ 0 & a < \zeta. \end{cases} \quad (6.65)$$

Using these results for the unidimensional covariance function, any method based on the generation of a line process directly from the covariance function can be used to synthesize three-dimensional fields through Eq. (6.48). An effective method to accomplish the unidimensional generation is through a moving-average process (Journel and Huijbregts, 1978). This method is now described as presented by Mantoglou and Wilson (1981).

Generation Along the Turning-Bands Lines Using a Moving-Average Process

Let the segment D in Fig. 6.11 represent one of the random lines on which we want to generate the unidimensional process with covariance $C_1(\zeta)$. The process will be generated at discrete points located in the middle of discrete bands or intervals of width $\Delta\zeta$. Now take a segment T parallel to D and on it generate an independent process having zero mean, unit variance, and a

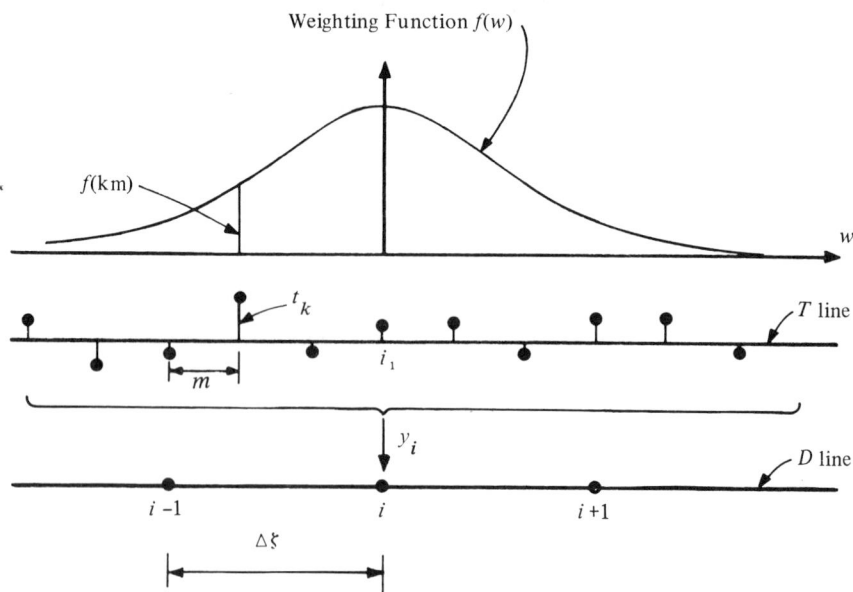

Figure 6.11 Schematic representation of the moving-average process (from Mantoglou and Wilson, 1981).

uniform distribution. Let the discretization length of this process be $m = \Delta\zeta/\mu$, where μ is an integer, so that each point i on D corresponds to a point i_1 on T.

Let us define a moving-average process y_i on line D,

$$y_i = \sum_{k=-\infty}^{\infty} f(km) t_{i_1+k}, \tag{6.66}$$

where t_{i_1+k} are realizations of the random process T generated on the segment parallel to D. The weighting function $f(km)$ must be specified in such a manner that y_i has the desired covariance function $C_1(\zeta)$.

The discrete covariance C_1^d of the process y_i is defined by

$$C_1^d(\zeta) = E[y_i y_{i+\zeta}]$$

$$= E\left\{ \sum_{k=-\infty}^{\infty} T_{i_1+k} f(km) \sum_{k'=-\infty}^{\infty} T_{\zeta_1+k'} f(k'm)\right\},$$

where $\zeta_1 = (i_1 + \mu\zeta)$. The discrete covariance $C_1^d(\zeta)$ can now be written as

$$C_1^d(\zeta) = E\left\{ \sum_{k=-\infty}^{\infty} \sum_{\ell=-\infty}^{\infty} T_{i_1+k} T_{i_1+\ell} f(km) f((\ell - \mu\zeta)m)\right\}, \tag{6.67}$$

where $\ell = (\mu\zeta + k')$. The above expectation is

$$C_1^d(\zeta) = \sum_{k=-\infty}^{\infty} f(km) f(km - \mu\zeta m), \tag{6.68}$$

because T is an independent unit-variance process. With $m \to 0$, Eq. (6.68) becomes

$$C_1(\zeta) = \int_{-\infty}^{\infty} f(w) f(w - \zeta) \, dw = \int_{-\infty}^{\infty} f(w) \tilde{f}(\zeta - w) \, dw, \tag{6.69}$$

where $\tilde{f}(w) = f(-w)$. Equation (6.69) specifies the structure of the weighting function to be used in the moving-average process of Eq. (6.66). The function $f(w)$ for some common models goes to zero when $w \to \infty$; thus in Eq. (6.66) it is necessary only to use a finite number, k_{max}, of summation terms.

All univariate covariances corresponding to the generally used covariance functions of three-dimensional fields can be expressed as a convolution product of the type given by Eq. (6.69). The corresponding weighting functions $f(w)$ have been obtained by Journel and Huijbregts (1978) and are given below.

Exponential Model

The three-dimensional covariance function and the corresponding one-dimensional covariance function were already derived (Eq. 6.63). $C_1(\zeta)$ can be expressed as a convolution product of the type given by Eq. (6.69) with weighting function

$$f(w) = \begin{cases} 2\sigma\sqrt{\alpha}\,(1-\alpha w)e^{-\alpha w} & w \geq 0 \\ 0 & w < 0. \end{cases} \qquad (6.70)$$

In practice the function $f(w)$ is significantly different from zero only in the interval $[0, 4/\alpha]$. The number of summation terms in Eq. (6.66) is approximately given by

$$k_{\max} = \frac{4}{\alpha m}.$$

The discretization length m should be small in order that C_1^d of Eq. (6.68) approximates well the covariance function C_1 of Eq. (6.69). This means that the (αm) chosen should be a small value.

Double Exponential Model

The weighting function is given by

$$f(w) = \frac{16\sigma^2\alpha^3}{\sqrt{\pi}}we^{-2\alpha^2w^2} \qquad -\infty \leq w < \infty. \qquad (6.71)$$

In practice this function is significantly nonzero only in the interval $[-2/\alpha, 2/\alpha]$.

Spherical Model

The weighting function is in this case

$$f(w) = \begin{cases} \sigma\sqrt{12/a^3}\,w & -a/2 \leq w \leq a/2 \\ 0 & \text{otherwise.} \end{cases} \qquad (6.72)$$

6.6.2 Two-Dimensional Fields

Equation (6.58) gives the relation between the covariance function of the two-dimensional field and the one-dimensional covariance of the line process in the TBM. Define now the origin of the orthogonal axes (x, y) in the plane of the field, with the origin at point x_1 and the y-axis in the direction of the

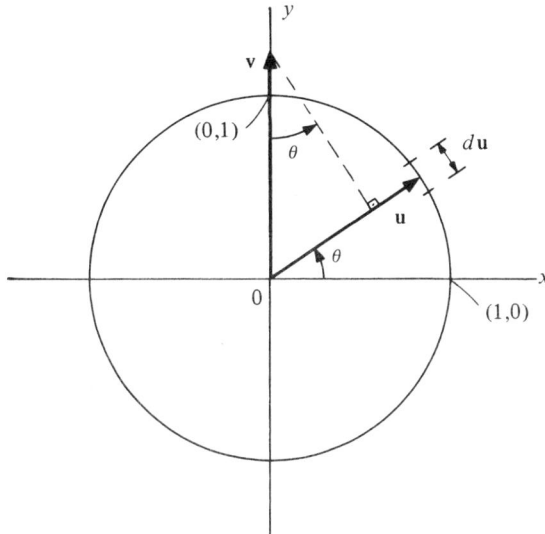

Figure 6.12 Definition sketch for the turning-bands method in the two-dimensional case (from Mantoglou and Wilson, 1982).

vector $\mathbf{v} = x_2 - x_1$ as shown in Fig. 6.12. In polar coordinates we have $\mathbf{v} \cdot \mathbf{u} = v \sin \theta$ and $d\mathbf{u} = d\theta$. Equation (6.58) then becomes

$$C_S(v) = \frac{1}{2\pi} \int_0^{2\pi} C_1(v \sin \theta) \, d\theta,$$

which simplifies to

$$C_S(v) = \frac{2}{\pi} \int_0^{\pi/2} C_1(v \sin \theta) \, d\theta. \tag{6.73}$$

Defining $\zeta = v \sin \theta$, $d\zeta = v \cos \theta \, d\theta$, we may write Eq. (6.73) as

$$C_S(v) = \frac{2}{\pi} \int_0^v \frac{C_1(\zeta)}{\sqrt{v^2 - \zeta^2}} \, d\zeta. \tag{6.74}$$

Equation (6.74) relates the two-dimensional covariance function $C(v) = C_S(v)$ to the unidimensional one $C_1(v)$ along the turning-bands lines. It is an integral equation in which we cannot directly express $C_1(v)$ as a function of $C(v)$ as we can in the three-dimensional case. There are particular solutions for some common cases; specifically, Mantoglou and Wilson (1981) show through the spectral density function that $C_1(\zeta)$ can be obtained for the exponential covariance two-dimensional field and for the Bessel type of two-dimensional covariance.

Exponential Model

The two-dimensional model has a covariance function

$$C(v) = \sigma^2 e^{-\alpha v} \tag{6.75}$$

and the unidimensional one is given by

$$C_1(\zeta) = \sigma^2 \left\{ 1 - \frac{\pi}{2} \alpha\zeta \left[I_0(\alpha\zeta) - L_0(\alpha\zeta) \right] \right\}, \tag{6.76}$$

where I_0 is a Bessel function of order zero, and L_0 is a modified Struve function of order zero (Abramowitz and Stegun, 1965).

Bessel Model

The two-dimensional model has a covariance function

$$C(v) = \sigma^2 bv K_1(bv) \tag{6.77}$$

and the unidimensional one is given by

$$C_1(\zeta) = \sigma^2 \left\{ 1 - \frac{b\zeta}{2} \left[e^{-b\zeta} Ei(b\zeta) - e^{b\zeta} Ei(-b\zeta) \right] \right\}, \tag{6.78}$$

where Ei is the exponential integral function.

The problem now is the generation of the unidimensional process with covariance functions as given by Eqs. (6.76) and (6.78). Mantoglou and Wilson (1981) show that Eqs. (6.76) and (6.78) are very similar in shape to the so-called hole covariance function described by Eq. (6.63). Moreover, they also show that the two-dimensional covariance function corresponding in the turning-bands method to the hole-type unidimensional covariance is given by

$$C(v) = \sigma^2 \left\{ I_0(\alpha v) - L_0(\alpha v) + \alpha v \left[I_1(\alpha v) - L_1(\alpha v) - \frac{2}{\pi} \right] \right\}, \tag{6.79}$$

where I_1 and L_1 are Bessel and Struve functions of order 1 and α is the parameter of $C_1(\zeta)$ in Eq. (6.63). Mantoglou and Wilson (1981) demonstrate that Eq. (6.79) is very similar in shape to Eqs. (6.75) and (6.77) ($\alpha = b$). Thus, one can fit a model of the type of Eq. (6.79) to the exponential and Bessel processes and proceed to the generation of the unidimensional process along the turning-bands lines using a moving-average process with the weighting function defined in Eq. (6.70).

In the TBM, Mantoglou and Wilson (1981) found that for two-dimensional fields, the use of 4 to 16 lines gave sufficient accuracy. In three dimensions, Journel and Huijbregts (1978) have shown that 15 lines joining the midpoints of the opposite edges of a regular isocahedron are adequate. The TBM can also be used to generate nonisotropic as well as integrated processes. It is also useful to obtain a particular class of nonstationary field, represented by so-called generalized covariances, to be discussed in Chapter 7.

6.7 TRANSFORMATION OF POINT RAINFALL TO AREA RAINFALL

Point-rainfall depths are usually transformed to area averaged precipitation by using a correction factor of the kind developed by the U.S. Weather Bureau (1957). This factor was developed to transform point-rainfall frequency curves into area frequency curves and it is shown in Fig. 6.13. Unfortunately, in many regions of the world Fig. 6.13 has been used without consideration of local characteristics. The U.S. Weather Bureau reduction factor has been expressed by Eagleson (1972) in the following terms:

$$K = \frac{d_A}{d} = 1 - \exp\left(-1.1 t_r^{1/4}\right) + \exp\left(-1.1 t_r^{1/4} - 0.01 A\right), \qquad (6.80)$$

where the duration t_r is in hours and the area A is in square miles. For example, if the average value of a 10-year, six-hour rainfall at a certain place is

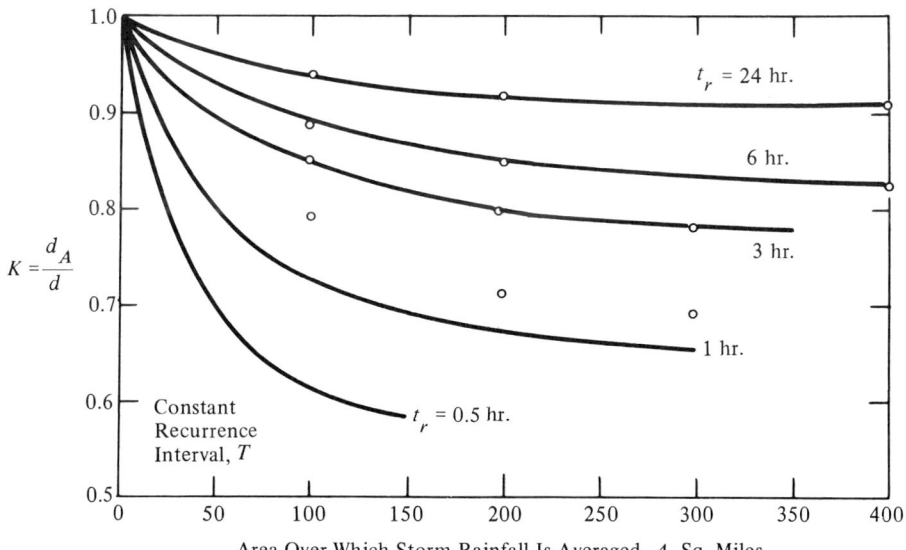

Figure 6.13 Spatial reduction of point-rainfall intensity according to the U.S. Weather Bureau relationship. Points represent fitted relationship $d_A/d = 1 - \exp(-1.1 t_r^{1/4}) + \exp(-1.1 t_r^{1/4} - 0.01 A)$ (from Eagleson, 1972).

3.45 in., the average six-hour depth over a drainage area of 200 mi^2 will be 85% of the point value, or 2.9 in.

Without going into extensive review of Fig. 6.13, it is worthwhile to mention some of its aspects. The curves were originally developed from networks covering areas from 100 to 400 square miles located mostly east of the Mississippi River. The average number of recorders in each network was six, and the length of record varied between 7 and 15 years (U.S. Weather Bureau, 1957). A year later, data from 13 other networks were judged in agreement with the original curves (U.S. Weather Bureau, 1958). The curves of Fig. 6.13 represent the ratio of maximum annual mean depth over the area to the mean maximum annual point rainfall. The numerator of the ratio is obtained from the annual series of values, each of which is the maximum average depth for a given area during the year, the times of beginning and ending of the 24-hour duration, for example, being the same for each station in the area. The denominator of the ratio is the mean of the individual station values. The area covered by n gages was assumed equal to n circles having diameters equal to the average station spacing; this definition is subjective owing to the uncertainty in regard to the area covered by the network. None of the networks had records of sufficient length to evaluate the effect of the return period on the point–area relationship. Storm magnitude (or return period) was tentatively accepted as a nonimportant parameter for this problem on the basis of an analysis (not shown in the references) of storm-rainfall data.

6.7.1 M. Roche Methodology

A general method with a theoretical base for the transformation of point rainfall to area rainfall was developed by Roche (1963). We will follow here the review of this method presented by Rodríguez-Iturbe and Mejía (1974b). The problem to be solved was framed as follows.

Given the point rainfall for a certain level of probability at an arbitrary point on a surface S, what is the average rainfall over the surface for the same level of probability? For practical purposes the question could be restated as follows. Given information for the point-rainfall process, what inferences can be made with respect to area-rainfall probabilities that in the final analysis are the ones of interest for the hydrologist?

The process will be assumed to be homogeneous, meaning that the rainfall at any point on the surface in consideration follows the same probability law. Define a surface S with a homogeneous distribution of points $1, 2, \ldots, n$ which are associated with the rainfall depths $h_1, h_2, \ldots, h_n$. Those depths are random variables assumed to follow the probability density $p(h_1, h_2, \ldots, h_n)$. The probability that the average rainfall depth over the area S is larger than a value h is given by

$$\int_1 \int_2 \cdots \int_n p(h_1, h_2, \ldots, h_n) \, dh_1, \ldots, dh_n = p \qquad (6.81)$$

with the condition

$$(h_1 + h_2 + \cdots + h_n)/n \geq h, \qquad (6.82)$$

where the exact expression for the probability is given by the limit of the previous equations when $n \to \infty$.

Roche (1963) starts the analysis by considering two points, 1 and 2, on the surface S separated by a distance x_{12}. Let $p(h_1, h_2)$ be the density function for the couple (h_1, h_2) and let $z = (h_1 + h_2)/2$ be the average rainfall over the two points in consideration. We may now construct a graph such as the one shown in Fig. 6.14 in which the curves represent the lines of constant probability density $p_1, p_2, \ldots$. Owing to the isotropic assumption these curves are symmetric with respect to the bisector of the quadrant $[h_1, h_2]$. Clearly, all the curves are contained in the first quadrant, since negative rainfall depths are impossible. We may now draw the line $h_1 + h_2 = 2z$, and the area above this line enclosed by the curves previously constructed will represent the probability that the average rainfall over points 1 and 2 is larger than z:

$$\int\int p(h_1, h_2)\, dh_1\, dh_2 \qquad h_1 + h_2 \geq 2z, \qquad (6.83)$$

the integration being extended to the domain of variation of h_1 and h_2. From Fig. 6.14, Roche then repeats the analysis for several values of z and obtains the variation of z as a function of the probability as shown in Fig. 6.15. One may follow the same reasoning, now varying the distance among the points x_{12}, and construct Fig. 6.16, which shows the variation of z as a function of the distance between the two points in consideration and the level of probability.

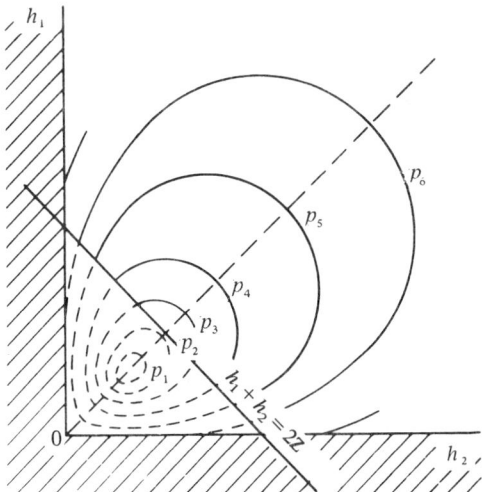

Figure 6.14   Curves of equal probability density $p(h_1, h_2)$ for the pair (h_1, h_2) (from Roche, 1963).

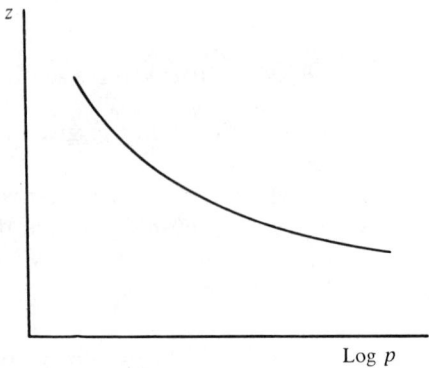

Figure 6.15 Average rainfall depth over two points at a fixed distance as a function of the probability level (from Roche, 1963).

The transformation of point rainfall to area rainfall can now be done in the following manner. Consider a rectangle of length L and width ℓ. The average rainfall over its surface is

$$P = \frac{1}{\ell L} \int_0^L dx \int_x^{x+\ell} z(y)\, dy, \tag{6.84}$$

where $z(y)$ is the average rainfall over two points separated by a distance y. Equation (6.84) can be written

$$P = \frac{1}{\ell L} \int_0^L \left[\int_0^{x+\ell} z(y)\, dy - \int_0^x z(y)\, dy \right] dx, \tag{6.85}$$

which can be integrated graphically from the curves shown in Fig. 6.16 for different levels of probability. Thus, for example, one is able to find the

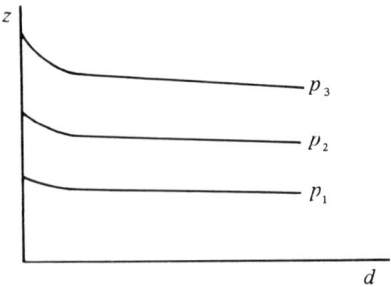

Figure 6.16 Average rainfall depth over two points as a function of the distance between them and the probability level (from Roche, 1963).

average rainfall over an area P given the point rainfall with the same return period. In his examples, Roche finds that the reduction factors are nevertheless independent of the considered frequency of occurrence.

Although the above method is correct in its analysis, it is extremely cumbersome because for each case it will be necessary to construct laborious graphs such as those in Figs. 6.14 to 6.16 and then perform the graphical integrations. Moreover, Fig. 6.14 is really a set of figures because there will be one for each spatial correlation coefficient that we want to test between points 1 and 2. Later in this chapter an example will be worked out with all the different methods, and the practical difficulties of this approach will be evident.

6.7.2 Multidimensional Methodology

Rodríguez-Iturbe and Mejía (1974b) developed a multidimensional method that leads to more general results than those of Roche (1963).

Let $f(\mathbf{x}, t)$ represent the point-precipitation intensity at the point of spatial coordinates $\mathbf{x}$ during the time t. We assume that the process is stationary and isotropic and also that the correlation function can be factored in a spatial and temporal part

$$\rho(v, \tau) = r(v)\rho(\tau), \tag{6.86}$$

where v represents the distance between two points in the space and τ is their time difference. Equation (6.86) does not mean that the same spatial correlation or temporal correlation is used for different storms but that for every event or type of storm the correlation structure can be factored in a spatial part and in a temporal part. Mathematically this is an extremely convenient property whose physical realism needs more study.

Rainfall is a multidimensional process occurring in time and space. For a particular type of rainfall event, it is possible to conceive different realizations occurring along the time scale; in the problem we are dealing with, we are interested in the mean value over an area of all possible realizations of the event in consideration.

Without loss of generality we can consider the process $f(\mathbf{x}, t)$ as having zero mean and σ_p^2 variance (where the notation σ_p^2 stands for point variance). The rainfall volume for a fixed period centered around time t over a given area A can be expressed as

$$f'(t) = \int_A f(\mathbf{x}, t)\, d\mathbf{x}, \tag{6.87}$$

which has zero mean and whose covariance structure in the time domain is given by

$$E\left[\int_A f(\mathbf{x}_1, t_1)\, d\mathbf{x}_1 \int_A f(\mathbf{x}_2, t_2)\, d\mathbf{x}_2\right] = \int_A \int_A E[f(\mathbf{x}_1, t_1)f(\mathbf{x}_2, t_1 + \tau)]\, d\mathbf{x}_1\, d\mathbf{x}_2$$

$$= \sigma_p^2 \rho_2(\tau) \int_A \int_A r(\mathbf{x}_1 - \mathbf{x}_2)\, d\mathbf{x}_1\, d\mathbf{x}_2$$

$$= A^2 \sigma_p^2 \rho(\tau) \bar{r}(\mathbf{x}_1 - \mathbf{x}_2|A), \qquad (6.88)$$

where $\bar{r}(\mathbf{x}_1 - \mathbf{x}_2|A)$ represents the expected correlation coefficient between two points randomly chosen in the area A.

It is worthwhile to remark that the expected value in the previous integration applies over all possible realizations of the event in consideration.

We may then state that the total rainfall process $f'(t)$ has a covariance structure given by Eq. (6.88), and its variance is equal to $A^2 \sigma_p^2 \bar{r}(\mathbf{x}_1 - \mathbf{x}_2|A)$. Thus the point-rainfall process is distributed with zero mean and variance σ_p^2, whereas the total-rainfall process has zero mean and variance $A^2 \sigma_p^2 \bar{r}(\mathbf{x}_1 - \mathbf{x}_2|A)$. Therefore a point-rainfall value with a given return period can be transformed into a total-rainfall value with the same return period when it is multiplied by the ratio of the standard deviations $A[\bar{r}(\mathbf{x}_1 - \mathbf{x}_2)|A]^{1/2}$. If one is interested in the correction factor relating rainfall depths over the area, the correction factor is then

$$K = \left[\bar{r}(\mathbf{x}_1 - \mathbf{x}_2|A)\right]^{1/2}, \qquad (6.89)$$

which is a geographically fixed area reduction ratio in the sense that for a given spatial correlation structure, this factor depends only on the characteristics of the area in question.

When the point process is Gaussian, the area process will also be Gaussian, and the application of the correction factor K of Eq. (6.89) will relate identical frequencies or return periods. On the other hand, when the point process is non-Gaussian, the area process will still tend toward normal, and there will be no exact correspondence between the frequencies of both processes.

Rodríguez-Iturbe and Mejía (1974b) studied the cases of the exponential and the Bessel correlation structures:

$$r(v) = e^{-\alpha|v|}$$

and

$$r(v) = vbK_1(bv).$$

The expected value of the correlation coefficient $[\bar{r}(\mathbf{x}_1 - \mathbf{x}_2|A)]$ over a square area is easily obtained from the well-known density function for the distance between two randomly chosen points in the square $G(v)$ used in the integral

$$\int_0^d r_1(v)G(v)\, dv,$$

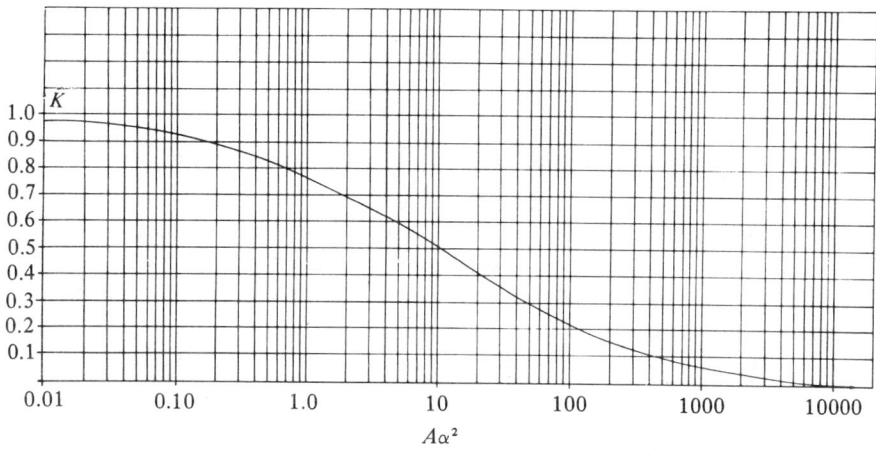

Figure 6.17 Spatial reduction of point-rainfall intensity for the case of exponential correlation structure (from Rodríguez-Iturbe and Mejía, 1974b).

which can be numerically evaluated, as described by Rodríguez-Iturbe and Mejía (1974a). Here, d represents the longest distance in the area (in the case of a square, the diagonal). For the presentation of the results, it is important to note that the value of the expected correlation coefficient $\bar{r}(x_1 - x_2|A)$ is the same when A and α or b vary as long as the product Ab^2 or $A\alpha^2$ remains constant (Rodríguez-Iturbe and Mejía, 1974a).

Figures 6.17 and 6.18 present the values of $[\bar{r}(x_1 - x_2|A)]^{1/2}$ as a function of $A\alpha^2$ and Ab^2 for the exponential and Bessel-type correlation functions, respectively. It is convenient to point out that the geometrical shape of the area

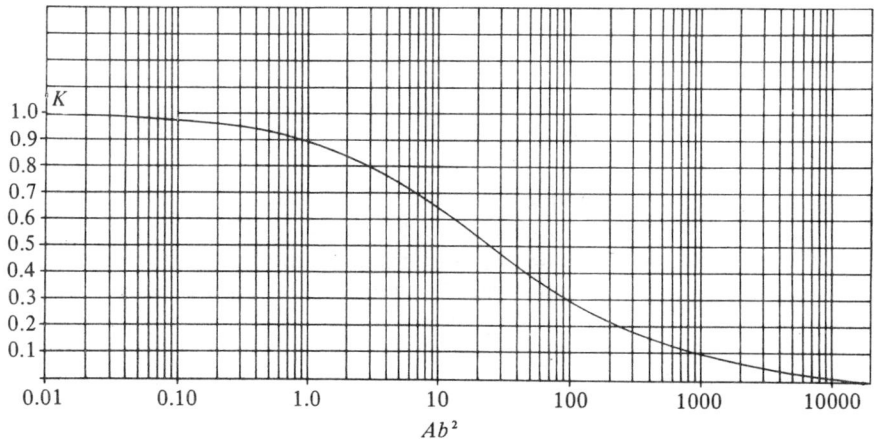

Figure 6.18 Spatial reduction of point-rainfall intensity for the case of Bessel-type correlation structure (from Rodríguez-Iturbe and Mejía, 1974b).

of study has a small influence on $\bar{r}(\mathbf{x}_1 - \mathbf{x}_2 | A)$ except for exceptionally elongated shapes, where one is changing from a two-dimensional to a one-dimensional space (Rodríguez-Iturbe and Mejía, 1974a).

6.7.3 Spatial Reduction of Point Rainfall for the Flakoho Basin

The example given by Roche (1963) was used by Rodríguez-Iturbe and Mejía (1974b) for comparison of the different methods for spatial reduction of point rainfall. The area is located on the Ivory Coast in the Flakoho basin near Ferkessedougou, clearly a tropical region. It is a rectangle of about 9 km by 5.5 km (50 km^2) with nine rain gages on it; the problem is to estimate the reduction factor for transforming the frequencies of daily point rainfall into frequencies of daily rainfall over the whole area. The estimated mean value and standard deviation for the point process are $h_m = 17.48$ mm and $\sigma_h = 2.02$ mm.

Roche Methodology

Roche (1963) starts his analysis by fitting a truncated normal law to the natural logarithms of the point values

$$F(h) = F_0 + (1 - F_0) \frac{1}{(2\pi\sigma_y)^{1/2}} \cdot \int_{-\infty}^{y} \exp\left[-\frac{1}{2}\left(\frac{y - \bar{y}}{\sigma_y}\right)^2\right] dy, \quad (6.90)$$

with $y = \log h$, $\bar{y} = 2.86$, $\sigma_y = 0.704$, and $F_0 = 0.85$. He then makes a study of the correlation decay with distance, obtaining $r(1 \text{ km}) = 0.90$, $r(3 \text{ km}) = 0.73$, $r(8 \text{ km}) = 0.57$, and $r(14 \text{ km}) = 0.50$.

From Eq. (6.90) it is possible to deduce that the curves of equal density needed for the construction of a graph such as that in Fig. 6.14 are given by

$$-2(1 - r^2)\sigma_y^2 \log\left[2\pi\sigma_y^2(1 - r^2)^{1/2} p\right],$$

where p is the level of probability and r is the correlation between the two points.

Figure 6.19 shows examples of the sets of graphs that are involved in the construction of Fig. 6.14. From this set of graphs it is possible to construct Fig. 6.20 from which it is possible, after graphical integration, to estimate the reduction factor that is independent of frequency and that Roche estimates to be 0.86.

Multidimensional Methodology

It is necessary first to estimate the parameter of the correlation structure. A method presented in Section 6.8 estimates the correlation parameter by fitting the correlation function so that it preserves the estimated correlation coefficient

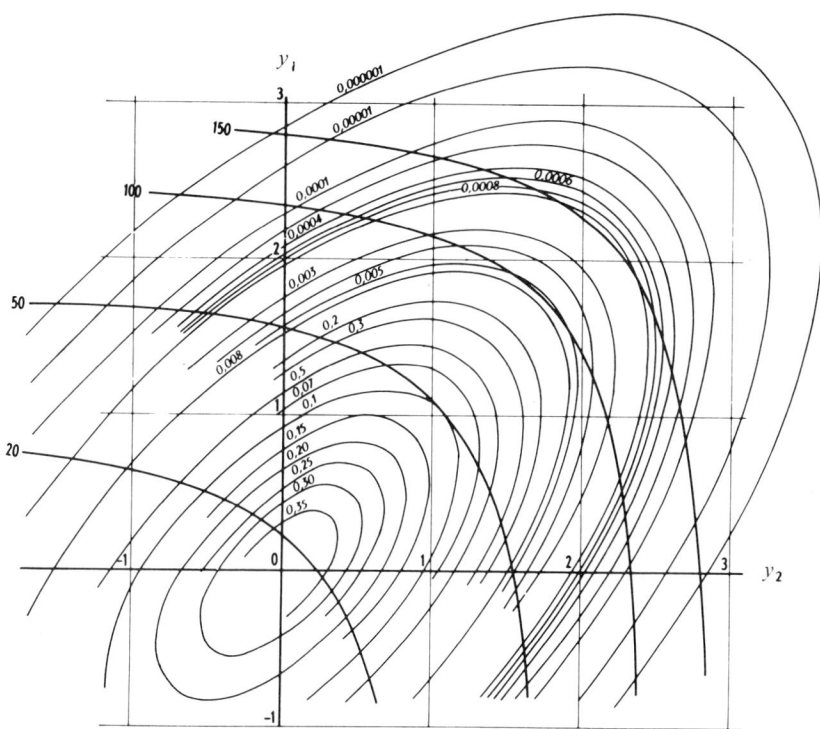

Figure 6.19 Curves of equal probability density for a couple of points in the Flakoho basin with a correlation coefficient of 0.60 (from Roche, 1963).

at a typical distance in the space under consideration. This distance is called the "characteristic correlation distance" and measures the mean separation between two points randomly chosen in the area. Its method of estimation will be described in the next section of this chapter and for a rectangle of 9 by 5.5 km it is approximately equal to 3.80 km. Thus the estimation of the correlation function will be made by preserving the value $r(4 \text{ km}) = 0.68$ given by Roche for the historical data,

$$r(4 \text{ km}) = 0.68 = e^{-4\alpha}$$
$$r(4 \text{ km}) = 0.68 = 4bK_1(4b),$$

which yield $\alpha = 0.096$ and $b = 0.202$, for the two types of correlation, respectively.

By using $A\alpha^2 = 0.460$ and $Ab^2 = 2.040$, Figs. 6.17 and 6.18 give correction factors of 0.86 for transforming point rainfall to area rainfall for both types of correlation functions. The obtained value is identical to the one found by Roche (1963) and moreover confirms that the type of correlation function to be

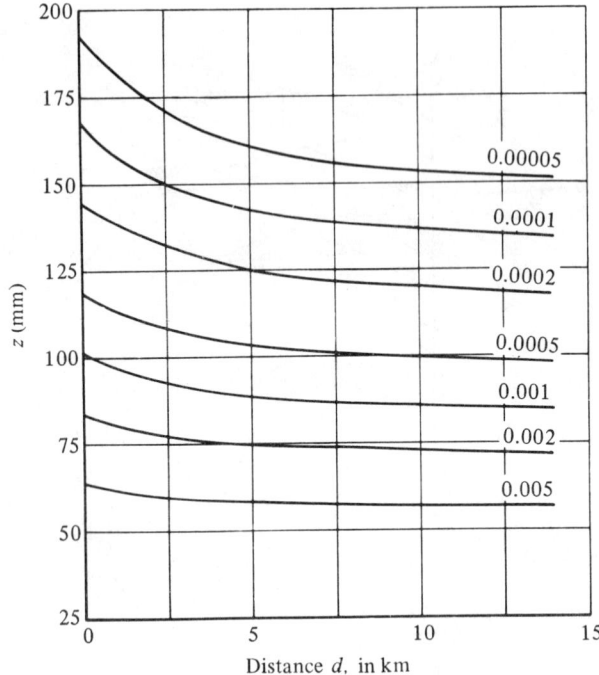

Figure 6.20 Average rainfall depth over two points as a function of their separation and the probability level for the Flakoho basin (from Roche, 1963).

used is of minor importance as long as its parameters are evaluated in the neighborhood of the characteristic correlation distance.

U.S. Weather Bureau Relationship

The U.S. Weather Bureau graphs are not directly applicable in this case because daily rainfall is made up of different kinds of events with many durations. Nevertheless, for purposes of illustration and not of comparison, Rodríguez-Iturbe and Mejía (1974b) have evaluated Eq. (6.80) with $A = 19.3$ mi^2 and $t_r = 24$ hours to obtain a reduction factor of 0.985.

6.8 SAMPLING OF HYDROLOGIC RANDOM FIELDS

The planning, development, and management of water resources require information about a large quantity of physical, economic, and social factors. There is always uncertainty associated with the data that engineers and planners must use for water-resource problems, and it is this uncertainty that causes the questions of how much information is enough and what kind of data is needed to deal with real-life problems. The answer will always depend on the particular objectives being pursued; this is why it is so difficult to provide guidelines for the design of data-collection programs. Data-sampling activities should never be divorced from a preplanned program of interpretation and modeling of the system at hand. The choice of the variables to be measured and their sampling locations and sampling rates all depend upon the objectives of the programs, the type of models to be used to synthesize or to represent the system, and the sensitivity of the decision-making process to errors in the input information. Moreover, the validity of the information itself is determined by the conditions of sampling, the accuracy of the measurements, and the time–space variability of the phenomena being sampled.

The U.S. Office of Water Data Coordination has defined three levels of information concerning data-network design. Level 1 provides a base level of information for wide regional or national planning to be used for resource inventory and as background information for the design of more intensive and specific network systems. Level 2 concerns networks called to provide general water-resource–planning data, and Level 3 is restricted to data-collection programs for specific planning and managing activities (Rodda, 1969).

Levels 1 and 2 can be classified in systems to provide regional estimation–type of data, because they will provide the planner with spatial rainfall averages and a primary type of streamflow-gaging stations from which a regionalization approach can be undertaken to hedge against the development of unanticipated needs.

Level 3 is oriented toward the collection of accurate information regarding hydrologic characteristics influencing a preplanned program of economic development for the area. At this level of data collection we are interested not only in the regional estimation–type of problems but also in the local estimation of engineering variables at specific sites and subregions.

There exists a clear difference in the design process for data-collection systems corresponding to Levels 1 and 2 and those under Level 3. When work is done in Level 3–type networks, we have an available forecast of the economic development and relations between the errors of estimate of the hydrologic characteristics and the losses incurred in the developments envisaged. On the basis of this information it is possible to use mathematical programming techniques to obtain the optimal variation of the network characteristics during the planning period.

It is much more difficult to set up an economic optimization objective for the natural-regime network; in this case the information is only loosely related to economic factors, and the marriage between the sampling activity and the preplanned program of interpretation may be a difficult one to accomplish.

Two basic questions are involved in the design of data-collection systems.

1. How much effort should be expended in the network, and how should it be allocated to return the most information?

2. What inferences should be drawn from data collected by the network, and what are the uncertainties in those inferences?

When a network is designed to collect Level 3 information, it is not very difficult to set up economic optimization objectives, which usually come to the fact that marginal benefits produced by additional data should never be less than the cost to collect the data itself. There is a fair amount of work related to this topic in the hydrologic literature (Tschannerl, 1970; Dawdy et al., 1970; Moss, 1970; Duckstein and Kisiel, 1971; Davis and Dvoranchik, 1971). One appropriate line of attack in these cases is based on Bayesian decision theory. This approach is described in detail by Davis and Dvoranchik (1971), and is based on the premise that information is valuable only if its possession may cause a change of decision or action, and its value is measured by the economic gain associated with the change of action.

A much less-defined problem is the network design for areas in which economic development is not foreseen (at least, not with enough clarity) within the planning horizon: Level 1 and part of Level 2 information. When we try to pursue lines of attack similar to those used for Level 3 information, a dilemma arises: in most cases we know neither a decision space (for design) nor a utility function. The many uses to which Level 1 hydrologic information will be applied cannot be foreseen at the time that the data are being collected, and even if they could be, there would be so many uses so varied in character that it would be infeasible to construct a decision formulation.

Thus optimizing allocations in a well-defined Bayesian framework is not possible in designing networks for the base level of information. The planner is then left with two basic principles for Level 1 data-collection programs:

1. A network operated on a fixed budget should be designed to minimize the error of estimation of the hydrologic variables involved.
2. A network operated on a criteria of minimum acceptable accuracy should be designed for a minimum cost.

Any decision formulation that leads to optimizing allocations in this type of network will have to include one of two parameters: an estimate of the error made by the data-collection systems, or a measure of the amount of information collected by the system. The approach to study those two parameters is different if we wish to set up a network that will allow the engineer to construct point estimates of a hydrologic characteristic at an ungaged site, or will allow the engineer to estimate mean values of a hydrologic variable over a whole region.

The first case is that of stream gaging, in which for ungaged sites the selection of those that are to be gaged can be approached by means of a regional analysis of the information at the stations in the existing network or in similar systems. This type of analysis has been set forth by Matalas (1969) and Matalas and Gilroy (1968) and is based on a regression scheme that relates the means of the existing stations to physiographic and meteorologic parameters. By using the same regression, it is possible to obtain estimates of the mean values at ungaged sites. Other statistical parameters may also be regionalized as in Benson (1964) and Benson and Matalas (1967).

The second case is that in which the engineer needs a network to estimate mean values of a hydrologic variable over a certain region. This is typically the problem in precipitation networks.

Precipitation is normally the most variable hydrologic element over a territory, and its characterization is most commonly needed for water-balance studies and for flood forecasting.

In water-balance studies, what is needed is the long-term mean area precipitation during a certain period of time. The period depends mainly on the variability of rainfall, and it may be monthly, seasonal, or annual depending on how representative the measures in the time scale are; in other words, it depends on the time scale of stationarity of the phenomenon and the use that is going to be made of the data.

In flood-forecasting studies, precipitation data are commonly used for the construction of area–depth duration curves and as input to rainfall–runoff types of models. In these cases, what is needed is an estimation of the contribution of a particular type of storm to the area under consideration; this will be called mean area rainfall for an event.

Rodríguez-Iturbe and Mejía (1974a) formulated a method for the design of precipitation networks that incorporates the multidimensional correlation structure of the rainfall process. They developed a general framework to estimate the variance of the sample long-term mean area precipitation and mean area rainfall of a storm event. The variance is expressed as a function of correlation in time, correlation in space, length of operation of the network, and geometry of the gaging array. This method is presented here essentially as given by Rodríguez-Iturbe and Mejía (1974a).

6.8.1 Long-Term Mean Area Rainfall

The rainfall process is considered as a multidimensional random-field $f(\mathbf{x}_i, t)$ function of the spatial coordinates $\mathbf{x}_i$ and the time t. For the determination of the mean value of this process, the correlation structure of the field in both space and time is of great importance. For this particular problem we will assume that the process is stationary and, furthermore, that its correlation function is separable in terms of its spatial and temporal structure. The second assumption means that the covariance structure of $f(\mathbf{x}_i, t)$ can be written in the form

$$\operatorname{cov}\left[f(\mathbf{x}_i, t), f(\mathbf{x}_{i'}, t')\right] = \sigma_p^2 r(\mathbf{x}_i - \mathbf{x}_{i'}) \rho^*(t - t'), \qquad (6.91)$$

where σ_p^2 is the point variance of $f(\mathbf{x}_i, t)$; $r(\mathbf{x}_i - \mathbf{x}_{i'})$ is the spatial correlation structure; and $\rho^*(t - t')$ is the temporal correlation structure. This assumption seems logical when one observes that long-term area mean values are estimated by first forming a spatial average during each period of time considered (years, months, etc.) and then adding them up in a discrete fashion.

The assumption of weak stationarity means that the expectation is a constant and the covariance function exists and is only a function of the difference between the spatial and temporal coordinates and not of the position or time itself. This limits the applicability of the technique to not-very-large areas and furthermore to regions that can be considered fairly uniform with respect to hydrologic behavior, important factors such as orographic effects thus being ruled out or neglected.

The temporal correlation structure of rainfall in terms of years, months, or weeks appears to be quite weak and can be approximated by a simple Markovian scheme:

$$\rho^*(t - t') = \rho^{|t'-t|}, \tag{6.92}$$

where ρ denotes the first autocorrelation coefficient, which is practically always less than 0.250.

The spatial correlation structure $r(\mathbf{x}_i - \mathbf{x}_{i'})$ in Eq. (6.91) poses a different problem. The available data are not sufficient for statistical discrimination between different kinds of functions that could represent the spatial correlation structure. Two problems may be distinguished in this aspect.

1. What kind of function do we use to represent the spatial correlation?
2. How do we fit the parameters of that function with unevenly spaced data?

With respect to the first question, it has been common to look for correlation functions that decay as a function of distance; one of this type is the exponentially decaying function

$$r(x, y) = \exp\left[- \alpha \left(x^2 + y^2\right)^{1/2}\right],$$

which for an isotropic process becomes the familiar

$$r(v) = e^{-\alpha v}, \tag{6.93}$$

where v represents the distance between any two points. A different correlation function is the Bessel type used several times in this chapter

$$r(v) = bv K_1(bv), \tag{6.94}$$

which relates rainfall at a point in space $f(x, y)$ symmetrically to rainfall at all points around (x, y). All the developments of this section will be made for both correlation functions in Eqs. (6.93) and (6.94).

It is important in the design of rainfall networks to develop information regarding the range of variation existing in the parameters of the correlation function. Eagleson (1967) presents the correlation structure as a function of distance for a 1250-mi^2 catchment in Australia. Using annual data with 13 rain gages and 17 years of data, he finds a correlation of 0.5 for a distance of 19 mi. This will correspond to values $\alpha = 0.0365$ mi^{-1} and $b = 0.0684$ mi^{-1} in Eqs. (6.93) and (6.94), respectively.

Stol (1972), working with monthly data in the Netherlands, shows correlation decays that are much faster for July than for January: α(January) = 0.0010 km^{-1} and α(July) = 0.0096 km^{-1}. Hendrick and Comer (1970), using daily precipitation data during the months of June to August (1961 to 1966) in northern Vermont, show a strong decay in correlation that produces $r = 0.51$ at a distance of 5 mi. This corresponds to $\alpha = 0.135$ mi^{-1} and $b = 0.26$ mi^{-1}. Rodríguez-Iturbe and Mejía (1974a) chose a homogeneous area of 30,000 km^2 with 26 rain gages in central Venezuela in order to investigate correlation decay as a function of distance. The region is shown in Fig. 6.21. They found that although it is true that smooth correlation structures can be "fitted" to the data, it makes a great difference from the point of view of network design how the actual fitting was performed. By using a sophisticated scheme that takes into account both the relative position of the stations and the length of the records, it was estimated that r(10 km) = 0.942 and r(50 km) = 0.533. These values will render exponentially decaying factors of α(10 km) = 0.0060 km^{-1} and α(50 km) = 0.012 km^{-1}, which are quite different and as will be seen later would require different network densities to maintain the required precision and the length of operation of the network constant.

Now the question is which α or b to use? The answer intrinsically depends on the size of the area being analyzed. Thus if the area is small, the correlation in space should be fitted with the criterion of preserving estimated correlation coefficients for a short distance. The opposite will be true when we deal with a large region. A typical distance that characterizes the size and shape of the area being analyzed is the mean distance between two randomly chosen points in the region; this is defined as the "characteristic correlation distance." Clearly, other methods could also be used for the estimation of the correlation parameters.

Ghosh (1951) derives the distribution of the distance between two points chosen at random in a plane convex region. The region is denoted by S, and its area and perimeter by A and P, respectively. When S is a rectangle with sides A_1 and A_2, $\lambda = A_1/A_2$, the frequency function can be written as

$$1/(A)^{1/2} G\left[v/(A)^{1/2}, (A_1/A_2)^{1/2}\right], \tag{6.95}$$

where v is the distance among the points and

$$G(w, a) = 2w\left[f_1(w, a) + f_2(wa, a) + f_2(w/a, 1/a)\right]$$

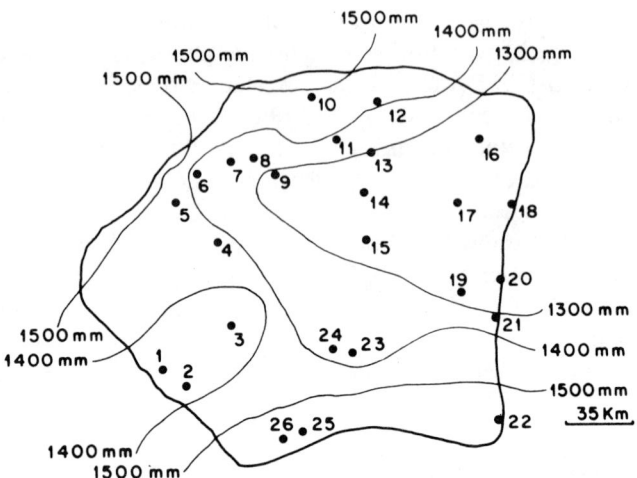

Figure 6.21 Central Venezuela region (Portuguesa state) used as case study for rainfall-network design (from Rodríguez-Iturbe and Mejía, 1974a).

with

$$f_1(w, a) = \pi + w^2 - 2w(a + 1/a) \qquad 0 < w < (a^2 + a^{-2})^{1/2}$$
$$f_1(w, a) = 0 \qquad \text{otherwise}$$
$$f_2(w, a) = 2(w^2 - 1)^{1/2} - 2\cos^{-1}(1/w) - a^{-2}(w - 1)^2$$
$$\qquad 1 < w < (1 + a^4)^{1/2}$$
$$f_2(w, a) = 0 \qquad \text{otherwise.}$$

Matern (1960) has used Eq. (6.95) and equations applicable to other geometries to compute the mean distance between two randomly chosen points in seven regions of area 1: circle, 0.5108; hexagon, 0.5126; square, 0.5214; equilateral triangle, 0.5544; rectangle $\lambda = 2$, 0.5691; rectangle $\lambda = 4$, 0.7137; and rectangle $\lambda = 16$, 1.3426.

In working with a region of area A, it is only necessary to adjust those factors with proportionality coefficients made up of the ratio of two corresponding distances in the figure of area A and the figure of unit area. Thus in the region of central Venezuela, the zone can be approximated by a rectangle with a side ratio equal to 2; with an area of 30,000 km² this gives a diagonal of 268 km. A unit-area rectangle of the same shape has a diagonal of 1.58, and thus the characteristic correlation distance is in this case $0.5691 \times 268/1.58 = 97$ km.

We need then to fit the correlation structure in space, preserving estimated correlation coefficients for distances on the order of 100 km, which were found to be equal to 0.21. This in turn produces $\alpha = 0.0156$ km^{-1} and $b = 0.0234$ km^{-1}, which will be the correlation parameters to be used later in this section to design a network for this particular region. The watershed analyzed by Eagleson (1967) has an area of 1250 mi^2 and can also be approximated by a rectangle with $\lambda = 2$. This gives a diagonal of 56 mi and a characteristic correlation distance of 20 mi, which in this case agrees with Eagleson's correlation radius r_0 of 19 mi, defined to be the distance at which the correlation function drops to 0.5.

With the previous concepts regarding the correlation structure of long-term area rainfall, we proceed now to present the framework for estimating the sampling requirements of such a process. The development is taken from Rodríguez-Iturbe and Mejía (1974a).

Let us establish the following notation: $f(\mathbf{x}_i, t)$ is the difference between rainfall depth at the point of spatial coordinates $\mathbf{x}_i$, during year, month, or season t, and the mean of the process; N is the number of stations in the network; T is the number of years, months, or seasons that the network is in operation; and A is the area under consideration. The hydrologist wants to estimate

$$P = \lim_{T \to \infty} \frac{1}{AT} \sum_{t=1}^{T} \int_A f(\mathbf{x}_i, t)\, d\mathbf{x}_i \tag{6.96}$$

by means of

$$\hat{P} = \frac{1}{NT} \sum_{i=1}^{N} \sum_{t=1}^{T} f(\mathbf{x}_i, t). \tag{6.97}$$

The precision of the estimation is measured by the variance of $\hat{P}$:

$$\text{var}[\hat{P}] = E\left[\frac{1}{NT} \sum_{i=1}^{N} \sum_{t=1}^{T} f(\mathbf{x}_i, t) - \lim_{T \to \infty} \frac{1}{AT} \sum_{t=1}^{T} \int_A f(\mathbf{x}_i, t)\, d\mathbf{x}_i \right]^2. \tag{6.98}$$

We will now prove that the mean value given by Eq. (6.96) has zero variance and therefore can be considered as a constant.

$$E\left[\lim_{T \to \infty} \frac{1}{AT} \sum_{t=1}^{T} \int_A f(\mathbf{x}_i, t)\, d\mathbf{x}_i \right]^2$$

$$= E\left[\lim_{T \to \infty} \frac{1}{A^2 T^2} \sum_{t=1}^{T} \sum_{t'=1}^{T} \int_A \int_A f(\mathbf{x}_i, t) \cdot f(\mathbf{x}_{i'}, t')\, d\mathbf{x}_i\, d\mathbf{x}_{i'} \right]$$

$$= \lim_{T \to \infty} \frac{1}{A^2 T^2} \sigma_p^2 \cdot \sum_{t=1}^{T} \sum_{t'=1}^{T} \cdot \int_A \int_A r(\mathbf{x}_i - \mathbf{x}_{i'}) \rho^{|t - t'|}\, d\mathbf{x}_i\, d\mathbf{x}_{i'},$$

where use has been made of Eqs. (6.91) and (6.92). We now write the previous expression as

$$\lim_{T \to \infty} \frac{1}{A^2 T^2} \sigma_p^2 \left[\sum_{t=1}^{T} \rho^{|0|} + 2 \sum_{t=1}^{T-1} \sum_{t'=t+1}^{T} \rho^{|t'-t|} \right] \int_A \int_A r(\mathbf{x}_i - \mathbf{x}_{i'}) \, d\mathbf{x}_i \, d\mathbf{x}_{i'}.$$

(6.99)

Now we have

$$\sum_{t=1}^{T} \rho^{|0|} = T$$

and

$$\sum_{t=1}^{T-1} \sum_{t'=t+1}^{T} \rho^{|t'-t|} = \sum_{t=1}^{T-1} \rho + \rho^2 + \cdots + \rho^{T-t}.$$

(6.100)

Calling

$$S = 1 + \rho + \rho^2 + \cdots + \rho^{T-t-1}$$

and subtracting ρS from both sides, we can write

$$S = (1 - \rho^{T-t})/(1 - \rho)$$

and Eq. (6.100) is equal to

$$\sum_{t=1}^{T-1} \frac{\rho(1 - \rho^{T-t})}{1 - \rho} = \frac{\rho}{1 - \rho} \left[(T - 1) - \frac{\rho}{1 - \rho}(1 - \rho^{T-1}) \right].$$

(6.101)

Substituting Eq. (6.101) into (6.99), we obtain

$$\lim_{T \to \infty} \frac{1}{A^2 T^2} \sigma_p^2 \left\{ T + 2 \frac{\rho}{1 - \rho} \left[(T - 1) - \frac{\rho}{1 - \rho}(1 - \rho^{T-1}) \right] \right\}$$
$$\cdot \int_A \int_A r(\mathbf{x}_i - \mathbf{x}_{i'}) \, d\mathbf{x}_i \, d\mathbf{x}_{i'} = 0.$$

In this manner without loss of generality we can consider

$$E[f(\mathbf{x}_i, t)] = 0$$

and

$$E[f^2(\mathbf{x}_i, t)] = \sigma_p^2.$$

The variance of the regional mean $\hat{P}$ (Eq. 6.97) is then given by

$$\text{var}[\hat{P}] = \frac{1}{N^2T^2} E\left[\sum_{t=1}^{T}\sum_{i=1}^{N} f(\mathbf{x}_i, t)\right]^2. \tag{6.102}$$

The problem that we face now is to evaluate $\text{var}[\hat{P}]$ as a function of the correlation structure of the process in both space and time, the number of stations in the network, the sampling geometry of the network, and the length of operation of the stations. To this end we write

$$\text{var}[\hat{P}] = \frac{1}{N^2T^2} E\left[\sum_{t=1}^{T}\sum_{i=1}^{N}\sum_{i'=1}^{N} f(\mathbf{x}_i, t)f(\mathbf{x}_{i'}, t)\right.$$

$$\left. + 2\sum_{t=1}^{T-1}\sum_{t'=t+1}^{T}\sum_{i=1}^{N}\sum_{i'=1}^{N} f(\mathbf{x}_i, t)f(\mathbf{x}_{i'}, t')\right] \tag{6.103}$$

$$= \frac{1}{N^2T^2}\sigma_p^2\left\{\left[\sum_{i=1}^{N}\sum_{i'=1}^{N} r(\mathbf{x}_i - \mathbf{x}_{i'})\right]\right.$$

$$\left. \cdot \left[\sum_{t=1}^{T} 1 + 2\sum_{t=1}^{T-1}\sum_{t'=t+1}^{T} \rho^{t'-t}\right]\right\}.$$

Let us now call

$$F_2(N) = \frac{\displaystyle\sum_{i=1}^{N}\sum_{i'=1}^{N} r(\mathbf{x}_i - \mathbf{x}_{i'})}{N^2} \tag{6.104}$$

and

$$F_1(T) = \frac{1}{T^2}\left[\sum_{t=1}^{T} 1 + 2\sum_{t=1}^{T-1}\sum_{t'=t+1}^{T} \rho^{t'-t}\right]$$

$$= \left[T + 2\sum_{t=1}^{T-1} \frac{\rho}{1-\rho}(1 - \rho^{T-t})\right]T^{-2} \tag{6.105}$$

$$= \left\{T + 2\frac{\rho}{1-\rho}\left[(T-1) - \frac{\rho}{1-\rho}(1 - \rho^{T-1})\right]\right\}T^{-2}.$$

Equation (6.103) can then be written as

$$\text{var}[\hat{P}] = \sigma_p^2[F_1(T)][F_2(N)], \tag{6.106}$$

where the variance of the regional mean is expressed as a function of the point

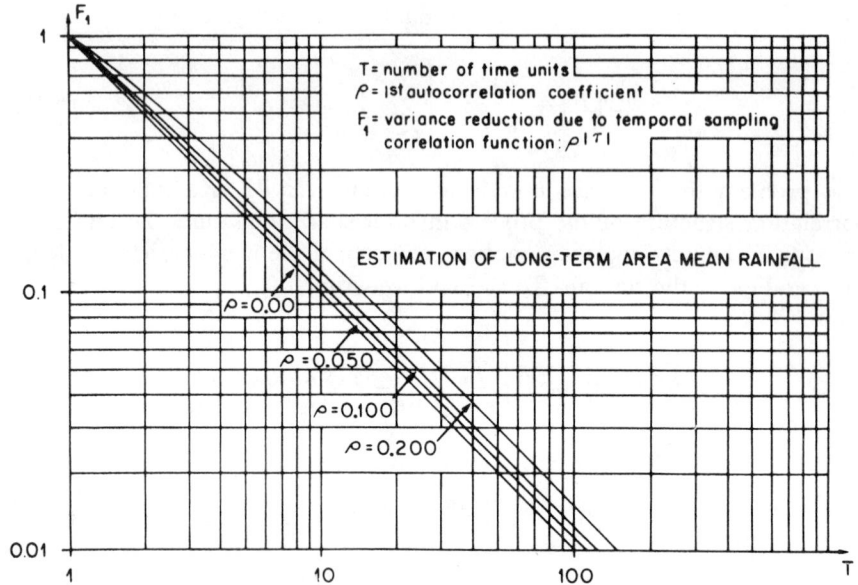

Figure 6.22 Variance reduction factor due to temporal sampling used in the estimation of long-term mean area rainfall (from Rodríguez-Iturbe and Mejía, 1974a).

variance of the process multiplied by two reduction factors, one of them $F_2(N)$ due to sampling in space and the other $F_1(T)$ due to sampling in time.

$F_1(T)$ is independent of the number of stations and the spatial properties of the process; it is only a function of the correlation in time and the length of time that the network has been in operation. Figure 6.22 shows $F_1(T)$ as a function of ρ and T; when $T =$ one year, month, or season, the variance reduction due to temporal sampling is equal to 1, an indication that there is no reduction at all in the variance of the long-term area mean with respect to the point variance of the process.

The variance reduction due to spatial sampling $F_2(N)$ depends on the correlation structure in space, the sampling geometry, and the number of stations. Three types of sampling schemes can be considered.

1. In the simple random-sampling type of network each station is located with a uniform probability distribution over the whole space A independently of the other stations.
2. In the stratified random-sampling case we assume that our space A is divided into a number of non-overlapping congruent strata s_i. From each stratum, k points are chosen randomly where the rain gages will be located.
3. In the systematic-sampling scheme the cluster of sampling units forms some regular geometric pattern.

Simple random sampling and stratified sampling can be realistic schemes for practical hydrologic purposes in which we either distribute the stations more or less randomly or divide the region in several subareas of similar sizes where the position of the stations in each subarea is determined by conditions such as accessibility and nearby population centers.

Variance Reduction Factor Due to Spatial Random Sampling

We need to evaluate

$$F_2(N) = \frac{1}{N^2} E\left[\sum_{i=1}^{N} \sum_{i'=1}^{N} r(\mathbf{x}_i - \mathbf{x}_{i'}) \right]$$

$$= \frac{1}{N^2} E\left[\sum_{i=1}^{N} r(0) + 2 \sum_{i=1}^{N-1} \sum_{i'=i+1}^{N} r(\mathbf{x}_i - \mathbf{x}_{i'}) \right].$$

Since the sampling scheme makes the terms random, we work with expected values. The above further reduces to

$$F_2(N) = \frac{1}{N^2} \{ N + N(N-1) E[r(\mathbf{x}_i - \mathbf{x}_{i'})|A] \}, \tag{6.107}$$

where $E[r(\mathbf{x}_i - \mathbf{x}_{i'})|A]$ represents the expected value of the correlation between two points randomly located on an area of size A. It depends on both the shape of the area and the type of spatial correlation structure that characterizes the process.

$$E[r(\mathbf{x}_i - \mathbf{x}_{i'})|A] = \int_0^R r(v)G(v)\,dv, \tag{6.108}$$

where $r(v)$ represents the spatial correlation assumed to be isotropic and $G(v)$ is the frequency function of the distance v between two randomly chosen points in the area A. Here, R represents the largest distance existing in A. It was previously seen that $G(v)$ is a function of the shape of the area under study (see Eq. 6.95), but fortunately it varies little for the shapes normally found in nature. Only for the rectangle with $\lambda = 16$ can we notice a sizable difference in values because in this case we are moving from a two-dimensional case to a transect or one-dimensional space. Because of this we will perform all our computations for a square region, which is the simplest to evaluate. Furthermore, we need a scheme to generalize the computations for different combinations of areas and correlation parameters.

Let us consider an area of size A on which we superimpose a process with correlation parameter α (or b) equal to 1. Equation (6.108) then takes the form

$$E[r(\mathbf{x}_i - \mathbf{x}_{i'})|A] = \int_0^d r(v)G(v)\,dv, \tag{6.109}$$

where d is the length of the diagonal of the square region.

Let us analyze a similar region with area A/α^2, where α is the magnitude of the parameter of the correlation structure. Equation (6.108) now has the form

$$E\left[r(\mathbf{x}_i - \mathbf{x}_{i'})|A\right] = \int_0^{d/\alpha} r_1(v)G'(v)\, dv, \qquad (6.110)$$

where d/α is the length of the new diagonal, and we must analyze the form of $r_1(v)$ and $G'(v)$. Here $r_1(v)$ is simply given by

$$r_1(v) = r(\alpha v)$$

and $G'(v)$ will be of the same form as $G(v)$, but it is affected by a factor of proportionality equal to $1/\alpha$ that reflects the change made in going from area A to A/α^2:

$$(1/\alpha)G'(v/\alpha) = G(v)$$

or

$$G'(v) = \alpha G(v\alpha).$$

Equation (6.110) can then be written as

$$E\left[r(\mathbf{x}_i - \mathbf{x}_{i'})|A\right] = \int_0^{d/\alpha} r(\alpha v)\alpha G(\alpha v)\, dv$$

and making $\alpha v = v'$, we obtain

$$E\left[r(\mathbf{x}_i - \mathbf{x}_{i'})|A\right] = \int_0^d r(v')G(v')\, dv', \qquad (6.111)$$

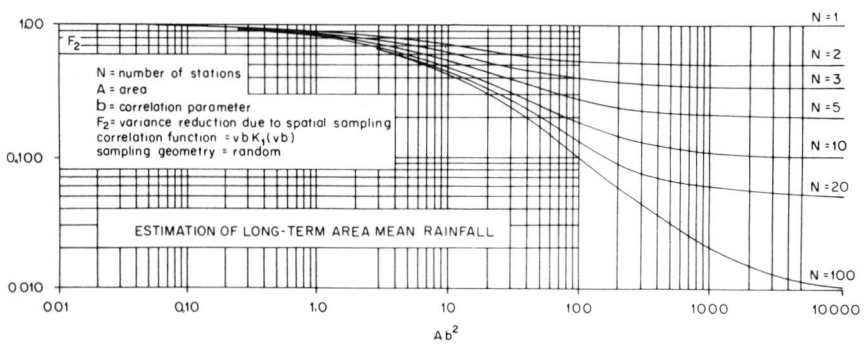

Figure 6.23 Variance reduction factor due to spatial sampling with random design used in the estimation of long-term mean area rainfall with $r(v) = vbK_1(vb)$ (from Rodríguez-Iturbe and Mejía, 1974a).

which is identically equal to Eq. (6.109). In this manner we have shown that an area of size A with a process with a correlation parameter equal to 1 has the same expected value of the correlation between two randomly chosen points as a homologous area of size A/α^2 over which is acting a process with correlation parameter α. Thus what remains constant is the product $A\alpha^2$ if we want to obtain the same value of $E[r(\mathbf{x}_i - \mathbf{x}_{i'})|A]$.

Equation (6.107) was evaluated by Rodríguez-Iturbe and Mejía (1974a) by calculating the integral given by Eq. (6.108) for a large range of values of $A\alpha^2$, maintaining a fixed N, and then varying N and repeating the procedure. The evaluation of Eq. (6.108) was performed numerically with the use of the expression for $G(v)$ given by Eq. (6.95). Two sets of values were obtained, one for each of the correlation structures given by Eqs. (6.93) and (6.94).

The results are presented in Figs. 6.23 and 6.24, which will be studied in detail later in their application to practical cases.

Variance Reduction Factor Due to Spatial Stratified Sampling

We need to evaluate

$$F_2(N) = \frac{1}{N^2} E\left[\sum_{i=1}^{N}\sum_{i'=1}^{N} r(\mathbf{x}_i - \mathbf{x}_{i'})\right]$$

$$= \frac{1}{N^2} E\left\{\sum_{i=1}^{N}\left[r(0) + \sum_{\substack{i'=1 \\ i'\neq i}}^{N} r(\mathbf{x}_i - \mathbf{x}_{i'})\right]\right\}. \tag{6.112}$$

We will now call $\mathbf{x}_i$ and $\mathbf{y}_i$ the two different points randomly located in the same stratum and $\mathbf{x}_i$ and $\mathbf{y}_{i'}$ the two different points randomly located in

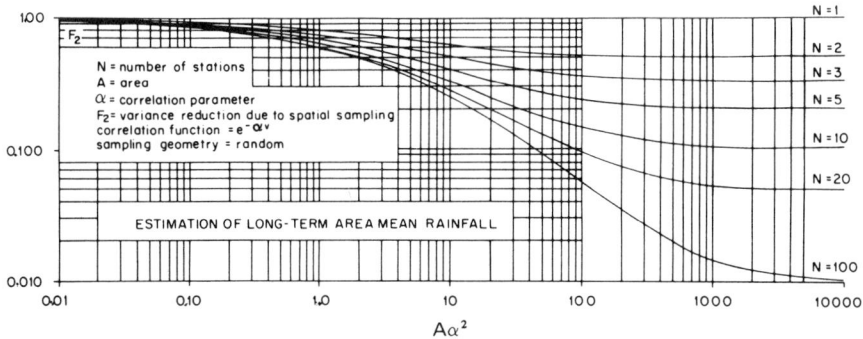

Figure 6.24   Variance reduction factor due to spatial sampling with random design used in the estimation of long-term mean area rainfall with $r(v) = e^{-\alpha v}$ (from Rodríguez-Iturbe and Mejía, 1974a).

different strata. Equation (6.112) becomes

$$F_2(N) = \frac{1}{N^2} E\left\{ N + \sum_{i=1}^{N}\left[\sum_{i'=1}^{N} r(\mathbf{x}_i - \mathbf{y}_{i'}) - r(\mathbf{x}_i - \mathbf{y}_i) \right] \right\}, \qquad (6.113)$$

where the points in the second summation are now divided into points in different strata (first term) and points in the same stratum (second term). We can further write

$$F_2(N) = \frac{1}{N^2}(N + W_1 - W_2),$$

where by taking expected values as done before,

$$W_1 = E\left[\sum_{i=1}^{N} \sum_{i'=1}^{N} r(\mathbf{x}_i - \mathbf{y}_{i'}) \right].$$

Since $\mathbf{x}_i$ and $\mathbf{y}_{i'}$ are random variables with uniform distributions $1/a$ and $1/a'$,

$$W_1 = \sum_{i=1}^{N} \sum_{i'=1}^{N} \int_{a_i}\int_{a_{i'}} \frac{r(\mathbf{x}_i - \mathbf{y}_{i'})}{a^2}\, d\mathbf{x}_i\, d\mathbf{y}_{i'}. \qquad (6.114)$$

Here a_i and $a_{i'}$ represent the area of the strata i and i' assumed to be equal in size. Equation (6.114) is easily simplified when we note that the double summation in i and i' of the integrals over a_i and $a_{i'}$ gives in effect two integrals over the whole area A:

$$\begin{aligned} W_1 &= \int_A\int_A \frac{r(\mathbf{x}_i - \mathbf{y}_{i'})\, d\mathbf{x}_i\, d\mathbf{y}_{i'}}{a^2} \\ &= \frac{A^2 E\left[r(\mathbf{x}_i - \mathbf{y}_{i'})|A \right]}{a^2}. \end{aligned} \qquad (6.115)$$

We will now assume one station per stratum, and the area of the stratum will be adjusted according to the number of stations in order to meet this assumption. In this manner,

$$W_1 = N^2 E\left[r(\mathbf{x}_i - \mathbf{y}_{i'})|A \right]. \qquad (6.116)$$

Equation (6.116) can now be evaluated with the same procedure as for Eq. (6.108), previously described in detail. The term W_2 in Eq. (6.113) is equal to

$$W_2 = E \sum_{i=1}^{N} r(\mathbf{x}_i - \mathbf{y}_i).$$

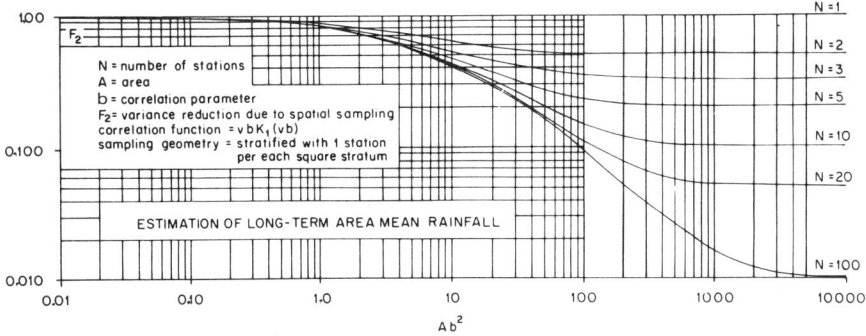

Figure 6.25 Variance reduction factor due to spatial sampling with stratified design used in the estimation of long-term mean area rainfall with $r(v) = vbK_1(vb)$ (from Rodríguez-Iturbe and Mejía, 1974a).

By taking expected values as done before,

$$W_2 = NE\left[r(\mathbf{x}_i - \mathbf{x}_{i'})|(A/N)\right], \tag{6.117}$$

where $E[r(\mathbf{x}_i - \mathbf{x}_{i'})|(A/N)]$ represents the expected value of the correlation between two points randomly chosen in the area (A/N). Equation (6.117) can also be evaluated with the procedure already described for Eq. (6.108).

As in the case of spatial random sampling the study of stratified sampling was made for a square region, and moreover, the strata were assumed to be square in order to avoid boundary or frontier problems.

The results of this part of the analysis showing the variance reduction factor due to spatial stratified sampling (Eq. 6.113) are shown graphically in Figs. 6.25 and 6.26.

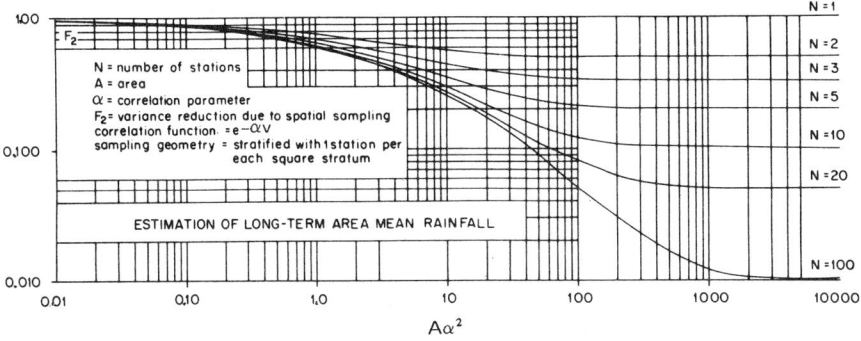

Figure 6.26 Variance reduction factor due to spatial sampling with stratified design used in the estimation of long-term mean area rainfall with $r(v) = e^{-\alpha v}$ (from Rodríguez-Iturbe and Mejía, 1974a).

Figures 6.23 to 6.26 provide an analytical tool for trading time versus space in the estimation of long-term spatial averages of precipitation. Their use in network design will be illustrated with two examples, developed by Rodríguez-Iturbe and Mejía (1974a).

6.8.2   Trading Time versus Space

Central Venezuela Region

The area of approximately 30,000 km^2 is shown in Fig. 6.21. Its 26 rain gages, the stations with their mean annual precipitation values, the standard deviations, and the length of the records are given in Table 6.5. To start with, it is

Table 6.5
Description of rainfall data used in the central Venezuela example

Station	Years of record	Mean (mm)	Standard deviation (mm)
1	1958–1971	1445	193
2	1955–1971	1412	131
3	1955–1965	1269	234
4	1951–1971	1404	127
5	1956–1971	1500	156
6	1955–1971	1342	189
7	1953–1971	1328	234
8	1948–1971	1294	248
9	1954–1971	1144	195
10	1943–1971	1440	194
11	1956–1971	1341	223
12	1950–1971	1427	261
13	1944–1971	1318	200
14	1958–1971	1296	210
15	1953–1971	1308	225
16	1961–1971	1240	228
17	1961–1971	1255	215
18	1954–1971	1155	206
19	1952–1964	1325	221
20	1958–1971	1252	276
21	1952–1971	1269	246
22	1952–1971	1514	318
23	1961–1971	1462	283
24	1948–1965	1370	194
25	1952–1965	1429	191
26	1946–1965	1452	199

Source: Rodríguez-Iturbe and Mejía (1974a).

necessary to adjust a correlation structure in both time and space that will be representative of the whole region. The parameter b to use in the correlation function $vbK_1(vb)$ was estimated from the equation

$$\sum_{i=1}^{N}\sum_{j=1}^{N}\sum_{k=k_i(i,j)}^{k_f(i,j)} f_{i,k}f_{j,k} = \sum_{i=1}^{N}\sum_{j=1}^{N}\left[k_f(i,j)-k_i(i,j)+1\right]v_{i,j}bK_1(v_{i,j}b),$$

(6.118)

where $k_i(i, j)$ represents the first or initial year for which the records of both stations i and j exist, $k_f(i, j)$ represents the final year for which the records of both stations i and j exist, $f_{i,k}$ is the standardized amount of rainfall during year k at station i (it could be also a monthly or seasonal amount), $v_{i,j}$ is the distance between stations i and j, and N is the number of stations. Equation (6.118) will yield a value of b that weights all the stations according to the length of their record. In this example a value of b equal to 0.0234 km^{-1} was obtained. As discussed in the previous section, the characteristic correlation distance is in this case about 100 km, which yields $r(100) = 0.0234 \times 100 \times K_1(0.0234 \times 100) = 0.21$. Fitting now an exponential decay at this distance, we get $r(100) = e^{-100\alpha} = 0.21$; $\alpha = 0.0156$. Thus the two equations to be used for describing the spatial correlation structure are

$$r(v) = 0.0234vK_1(0.0234v)$$

(6.119)

and

$$r(v) = e^{-0.0156v},$$

(6.120)

where v is in kilometers.

The variance reduction factors due to spatial sampling $F_2(N)$ are given in Table 6.6. We can see that $F_2(N)$ decreases much more in going from one station to five than when the gages are increased from 5 to 100.

Table 6.6
Variance reduction factor due to spatial sampling, $F_2(N)$, with $Ab^2 = 16.43$ and $A\alpha^2 = 7.30$ in the central Venezuela example

	Bessel correlation		Exponential correlation	
N	*Random*	*Stratified*	*Random*	*Stratified*
1	1.00	1.00	1.00	1.00
2	0.66	0.60	0.65	0.60
3	0.55	0.48	0.54	0.48
5	0.46	0.40	0.43	0.39
10	0.40	0.36	0.37	0.34
20	0.37	0.34	0.33	0.31
100	0.34	0.32	0.31	0.29

Source: Rodríguez-Iturbe and Mejía (1974a).

Before reaching any practical conclusions we must estimate the variance reduction factor due to temporal sampling $F_1(T)$. For this we must first estimate the time autocorrelation coefficient representative for the whole area. Thus ρ can be estimated as the solution of

$$\sum_{i=1}^{N} \sum_{j=1}^{N} \sum_{k=k_i'(i,j)}^{k_f'(i,j)} f_{i,k} f_{j,k+1}$$

$$= \rho \sum_{i=1}^{N} \sum_{j=1}^{N} \left[k_f'(i,j) - k_i'(i,j) + 1 \right] v_{i,j} b K_1(v_{i,j} b), \qquad (6.121)$$

where $k_i'(i,j)$ and $k_f'(i,j)$ represent the initial and final year for which both the record of station i and the record of station j in the following year exist. The other terms are the same as those in Eq. (6.118) where b has been estimated. For the annual data in this example the obtained ρ was 0.00. For $\rho = 0.00$, the values of the variance reduction factor due to temporal sampling are given in Table 6.7.

Combining Tables 6.6 and 6.7, we can estimate the efficiency of different network schemes for the area considered. In the case of one station in operation during 20 years, we can expect a total variance reduction factor of

$$F_1(T) \cdot F_2(N) = 0.050 \times 1 = 0.050.$$

In other words, this network will produce an estimate of the long-term area mean with a variance on the order of 5% of the variance of the point-rainfall process (Eq. 6.106). If we wish for that type of precision in a lapse of 10 years, we will need

$$F_2(N) = 0.050 / F_1(10) = 0.50.$$

This corresponds to $N = 4$ stations in the case of random sampling for both

Table 6.7
Variance reduction factor due to temporal sampling with $\rho = 0.00$

T	$F_1(T)$	T	$F_1(T)$
1	1.000	15	0.067
2	0.500	20	0.050
3	0.333	30	0.033
5	0.200	50	0.020
7	0.140	75	0.013
10	0.100	100	0.010

Source: Rodríguez-Iturbe and Mejía (1974a).

correlation structures given by Eqs. (6.119) and (6.120) and to $N = 3$ stations when the network is stratified.

It is interesting to observe that the same precision of 0.050 cannot be obtained in a lapse of five years because it will be necessary for

$$F_2(N) = 0.050/0.20 = 0.25,$$

which is a value smaller than the asymptotic one of Eq. (6.107) when N goes to infinity. From the graphs it can be seen that with $A\alpha^2 = 7.30$ and $F_2(N) = 0.25$, the corresponding value of N is still larger than 100. We thus have the important conclusion that we can trade time versus space in hydrologic data collection when we do not reduce the time interval too much, but no miracles can be expected in short times even from the most dense of all possible networks.

Table 6.8 presents the combined factors $F_1(T) \times F_2(N)$ for the example under consideration. This product represents the total reduction in variance relative to variance of point rainfall when the long-term area mean with N stations during T years is estimated. Table 6.8 was constructed for the Bessel-type correlation function given by Eq. (6.119), but the use of Eq. (6.120) gives practically identical results. Even for quite a small number of years (2, 5, or 10), five stations will accomplish most of the possible reduction in variance, and there is little justification for going over this number. It can also be observed that $F_1(T)$ weights more than $F_2(N)$ in the reduction of the variance of the long-term area mean: when $T = 5$ yr, $F_1(T) = 0.200$, and a value $F_2(N) = 0.200$ cannot be obtained in this example. This shows again that trading time versus space, although it is possible and in some instances necessary, is an expensive proposition.

Another important conclusion is that the form of the correlation function in space does not affect the results provided that the estimation of the parameters is carried out with objective criteria.

Table 6.8
Total factor of variance reduction due to temporal and spatial sampling, $F_1(T) \times F_2(N)$, in the central Venezuela region constructed for the Bessel correlation with a randomly designed network

N	$T = 2$	$T = 5$	$T = 10$
1	0.500	0.200	0.100
2	0.320	0.132	0.066
3	0.275	0.110	0.055
5	0.230	0.092	0.046
10	0.200	0.080	0.040
20	0.185	0.074	0.037
100	0.170	0.068	0.034

Source: Rodríguez-Iturbe and Mejía (1974a).

Regarding the estimation of the point variance σ_p^2 in Eq. (6.106), it can be done through the point records at each station. Because the process has been assumed to be stationary, σ_p^2 must be the same in all stations, and thus we can put back-to-back all the individual records and compute the variance of that series of data. This variance will be an estimate of σ_p^2. For the region in central Venezuela, the obtained result was 5.44×10^4 mm^2 of rainfall, and thus the total reduction of $F_1(T) \cdot F_2(N)$ should be applied to this value in order to obtain the variance of our estimated long-term area mean. For the case in which $F_1(T) \cdot F_2(N) = 0.050$, which we discussed previously, we obtain an estimate of the long-term area mean that has a variance of

$$\text{var}[\hat{P}] = 0.050 \times 5.44 \times 10^4 \text{ mm}^2$$

or equivalently, the standard deviation of $(\hat{P})$ equals 52.16 mm.

New South Wales (Australia)

This second example considers a catchment near Lismore along the northern coast of New South Wales as described by Eagleson (1967). It was seen in Section 6.8.1 that this region has a characteristic correlation distance of 20 mi, which will be used now in the fitting of the spatial correlation structure.

From Eagleson's data we have that $r(20 \text{ mi}) = 0.47$, which in turn yields $\alpha = 0.037$ mi^{-1} and $b = 0.0684$ mi^{-1} in Eqs. (6.93) and (6.94), respectively. Thus

$$r(v) = e^{-0.037v} \qquad\qquad A\alpha^2 = 1.71$$
$$r(v) = 0.0684 v K_1(0.0684 v) \qquad\qquad Ab^2 = 5.85$$

will be used as spatial correlation functions in this example.

Table 6.9

Variance reduction factor due to spatial sampling, $F_2(N)$, with a random design for $Ab^2 = 5.85$ and $A\alpha^2 = 1.71$ in the New South Wales example

N	Bessel correlation	Exponential correlation
1	1.00	1.00
2	0.76	0.76
3	0.69	0.69
5	0.62	0.62
10	0.56	0.56
20	0.54	0.54
100	0.53	0.53

Source: Rodríguez-Iturbe and Mejía (1974a).

Table 6.9 gives the values of $F_2(N)$ for a randomly designed network. It is observed that after three stations (or even two), there is a very small decrease in the variance reduction factor due to spatial sampling. The relative importance of this decrease will be even slighter when $F_1(T)$ is brought into action. Thus for 10 years of data and assuming that $\rho(1)$ in time is equal to zero, we get $F_1 = 0.10$. By fixing $N = 5$, $F_1(10) \cdot F_2(5) = 0.062$. Similarly, $F_1(10) \cdot F_2(100) = 0.053$. There is a decrease of only 1% in the variance of the estimated long-term area mean when the number of stations is increased from 5 to 100, hardly an economical step.

6.8.3 Area Mean of Rainfall Event

In general, the difference between the mean rainfall P_a over an area and the storm center point value has the following characteristics (Eagleson, 1970):

1. Increases with decreasing total rainfall depth,
2. Decreases with increasing duration,
3. Is greater for convective and orographic precipitation than for cyclonic rainfall, and
4. Increases with increasing area.

For convective storms in Arizona, Woolhiser and Schwalen (1959) have fitted the average area rainfall with the function

$$P_a/P_t(0) = 1 - [0.14/P_t(0)] A_s^{0.6}, \tag{6.122}$$

where $P_t(0)$ is the total depth in inches at the storm center, P_a is the average depth over the circular area A_s (in square miles) surrounding the center, and radial symmetry is assumed. Since

$$A_s = \pi r^2, \qquad P_a = \frac{1}{\pi r^2} \int_0^r 2\pi r P_t(r)\, dr.$$

Eagleson (1967) then gives the following structure for precipitation as a function of radial distance from the storm center,

$$P_t(r)/P_t(0) = 1 - 0.72(r/r_0), \tag{6.123}$$

where $r_0 = 1.73 P_t(0)$ is the correlation radius in miles, already defined. The shape of the functions given by Eqs. (6.93) and (6.94) suggests that they may be appropriate for the description of the spatial correlation structure of a rainfall event. It is nevertheless necessary to get an idea of the range of variation of the parameters.

Fogel and Duckstein (1969) present data that show storm center depths for convective rainfall in Arizona varying from 0.75 to 5 in. This is equivalent to correlation radii from 1.30 mi for the weaker storms up to 9 mi for the more intensive ones. These values in turn produce correlation structures of the form

or

$$r(v) = e^{-0.533v}$$
$$r(v) = 0.93vK_1(0.93v)$$

$$(6.124)$$

for the storm with a center depth of 0.75 in. and

or

$$r(v) = e^{-0.080v}$$
$$r(v) = 0.13vK_1(0.13v)$$

$$(6.125)$$

for 5-in. storm centers.

It will be shown later in this section that Eqs. (6.124) and (6.125) lead to different types of networks; the question then posed is What type of storm should be the commanding one in the design of a network for the estimation of the mean area precipitation of an event? This question generally cannot be answered, and it is by necessity linked to the economic criteria involved with the problem at hand.

It is also necessary to get an idea of the area extension of convective storms when one is in the process of designing a network. This has been done by using the relationship presented by Fogel and Duckstein (1969)

$$P_t(r) = P_t(0)\exp(-\pi r^2 t),\qquad(6.126)$$

where t is a dispersion parameter given by

$$t = 0.27e^{-0.67}P_t(0).$$

We will arbitrarily fix the limit of the storm at a depth of 0.1 in., obtaining in this manner areas of $A = 12$ mi^2 for $P_t(0) = 0.75$ in. and $A = 435$ mi^2 for $P_t(0) = 5.0$ in.

For great cyclonic storms in the United States, Boyer (1957) fits the average area rainfall distribution with the storm-centered function

$$P_t(r)/P_t(0) = e^{-ar},\qquad(6.127)$$

where the parameter a is given by Eagleson (1967) as

$$a = 1.68/r_0.\qquad(6.128)$$

A typical isohyetal pattern for this type of storm is shown in Fig. 6.27, which shows the rainfall produced on August 12 and 13, 1955, in the Baltimore

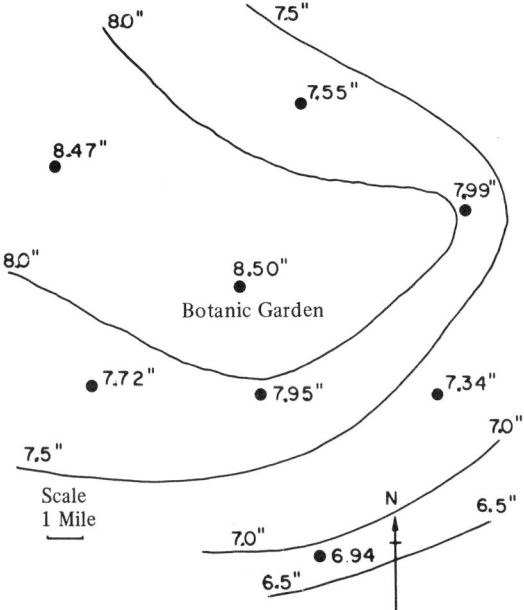

Figure 6.27 Isohyetal pattern of cyclonic storm of Hurricane Connie, August 12 and 13, 1955 (from Rodríguez-Iturbe and Mejía, 1974a).

area by Hurricane Connie. This particular example can be described with $P_t(0) = 8.50$ in. and $P_t(8.3 \text{ mi}) = 7.0$ in.

By using Eq. (6.127) to compute a and Eq. (6.128) to estimate the correlation radius, a value of 73 mi is obtained for r_0. This in turn corresponds to correlation structures of the type

$$r(v) = e^{-0.009v} \qquad r(v) = 0.016 v K_1(0.016v). \qquad (6.129)$$

The area extension of this type of storm is so large [$P_t(r) = 0.5$ in. corresponds to $A = 18,000$ mi^2] that it will cover any region that we may consider homogeneous for the purpose of network design.

We proceed now to introduce the framework for estimating the sampling requirements involved in the estimation of the mean area rainfall of an event. The development is taken from Rodríguez-Iturbe and Mejía (1974a).

In this case we want to estimate mean rainfall over a certain area A:

$$P = \frac{1}{A} \int_A f(\mathbf{x}_i) \, d\mathbf{x}_i, \qquad (6.130)$$

where $f(\mathbf{x}_i)$ denotes rainfall depth on the point with spatial coordinates $\mathbf{x}_i$ in the space A. In practice, the hydrologist estimates P by the arithmetic mean of

N point samples represented by the rain-gage stations

$$\hat{P} = \frac{1}{N} \sum_{i=1}^{N} f(\mathbf{x}_i). \qquad (6.131)$$

The performance of the network can be characterized by the variance

$$E\left[(P - \hat{P})^2\right] = \sigma_N^2, \qquad (6.132)$$

where it is important to understand that the expectation is taken over all possible outcomes that the rainfall $f(\mathbf{x}_i)$ may produce over the space A. Now $\hat{P}$ is a random variable, and this is an important difference from Eq. (6.96) in estimating long-term mean area rainfall. Also note that now the time element does not play an explicit role.

We can write Eq. (6.132) as

$$E\left[\frac{1}{N}\sum_{i=1}^{N} f(\mathbf{x}_i) - \frac{1}{A}\int_A f(\mathbf{x}_i)\,d\mathbf{x}_i\right]^2 = \frac{\sigma_p^2}{N^2}\left[\sum_{i=1}^{N} 1 + 2\sum_{i=1}^{N-1}\sum_{i'=i+1}^{N} r(\mathbf{x}_i - \mathbf{x}_{i'})\right]$$

$$- \frac{2}{NA}\sigma_p^2 \cdot \sum_{i=1}^{N}\int_A r(\mathbf{x}_i - \mathbf{x}_{i'})\,d\mathbf{x}_{i'}$$

$$+ \frac{\sigma_p^2}{A^2}\int_A\int_A r(\mathbf{x}_i - \mathbf{x}_{i'})\,d\mathbf{x}_i\,d\mathbf{x}_{i'}, \qquad (6.133)$$

where σ_p^2 represents the point variance of the process. Equation (6.133) will now be evaluated for two types of network designs: simple random sampling and stratified sampling.

Variance Reduction Factor Due to Random Sampling

In this case, Eq. (6.133) can be written as

$$\sigma_N^2 = \frac{\sigma_p^2}{N^2}\left\{N + E\left[r(\mathbf{x}_i - \mathbf{x}_{i'})|A\right] \cdot N \cdot (N-1)\right\}$$

$$- \frac{2}{NA} \cdot \sigma_p^2 \cdot \frac{N}{A} \cdot \int_A\int_A r(\mathbf{x}_i - \mathbf{x}_{i'})\,d\mathbf{x}_i\,d\mathbf{x}_{i'} + E\left[r(\mathbf{x}_i - \mathbf{x}_{i'})|A\right]\sigma_p^2$$

$$= \frac{\sigma_p^2}{N^2}\left\{N + N(N-1) \cdot E\left[r(\mathbf{x}_i - \mathbf{x}_{i'})|A\right]\right\} \qquad (6.134)$$

$$- 2\sigma_p^2 E\left[r(\mathbf{x}_i - \mathbf{x}_{i'})|A\right] + E\left[r(\mathbf{x}_i - \mathbf{x}_{i'})|A\right]\sigma_p^2$$

$$= \frac{\sigma_p^2}{N} - \frac{\sigma_p^2}{N} \cdot E\left[r(\mathbf{x}_i - \mathbf{x}_{i'})|A\right] = \sigma_p^2 F_2(N),$$

where

$$F_2(N) = \{1 - E[r(\mathbf{x}_i - \mathbf{x}_{i'})|A]\}/N.$$

Variance Reduction Factor Due to Stratified Sampling

With the same notation used for Eq. (6.113), Eq. (6.133) is now written as

$$\sigma_N^2 = \sigma_p^2 \left\{ \frac{1}{N^2} \left[N + \sum_{i=1}^{N} \sum_{i'=1}^{N} r(\mathbf{x}_i - \mathbf{y}_{i'}) - \sum_{i=1}^{N} r(\mathbf{x}_i - \mathbf{y}_i) \right] \right\}$$
$$- \sigma_p^2 \frac{2}{NA} \frac{N}{A} \sum_{i=1}^{N} \int_A \int_{a_i} r(\mathbf{x}_i - \mathbf{y}_{i'}) \, d\mathbf{x}_i \, d\mathbf{y}_{i'} + \frac{\sigma_p^2}{A^2} \int_A \int_A r(\mathbf{x}_i - \mathbf{y}_{i'}) \, d\mathbf{x}_i \, d\mathbf{y}_{i'},$$

which can be rewritten as

$$\sigma_N^2 = \frac{\sigma_p^2}{N^2} \{ N + N^2 E[r(\mathbf{x}_i - \mathbf{y}_{i'})|A] - NE[r(\mathbf{x}_i - \mathbf{y}_{i'})|(A/N)] \}$$
$$- 2\sigma_p^2 E[r(\mathbf{x}_i - \mathbf{y}_{i'})|A] + \sigma_p^2 E[r(\mathbf{x}_i - \mathbf{y}_{i'})|A] \qquad (6.135)$$
$$= \sigma_p^2 F_2(N),$$

where $F_2(N) = \{1 - E[r(\mathbf{x}_i - \mathbf{y}_{i'})|(A/N)]\}/N$ and the assumption has been made of one station per stratum.

Figures 6.28 to 6.31 show Eqs. (6.134) and (6.135) for the two types of correlation functions used in this section. It is observed that for very large values of $A\alpha^2$ and Ab^2 the curve corresponding to $N = 1$ goes to $F_2(1) = 1$, an indication that one gage alone will give an estimate of the area mean with variance equal to the point variance of the process σ_p^2. Nevertheless, for small values of $A\alpha^2$ and Ab^2, even one gage alone will produce an estimate of the area mean whose variance is considerably smaller than σ_p^2.

Equations (6.134) and (6.135) will now be used for the sampling in space of convective and cyclonic storms with the purpose of estimating the area mean of these types of events.

Analysis of Convective Storms

We studied reasonable values of the correlation parameter (α or b) depending on the intensity of the storm. The following tabulation gives the values already discussed.

$P_t(0)$	α	b
5	0.080	0.130
2	0.200	0.355
0.75	0.533	0.930

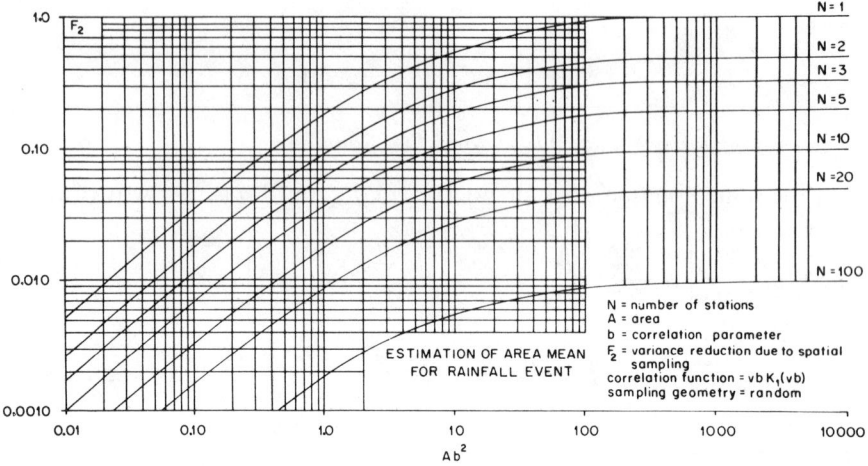

Figure 6.28 Variance reduction factor due to spatial sampling with random design used in the estimation of mean area rainfall of an event with $r(v) = vbK_1(vb)$ (from Rodríguez-Iturbe and Mejía, 1974a).

The values for $P_t(0)$ are given in inches. Tables 6.10 and 6.11 give the results of estimating the area mean values of these three storms over areas of 1500, 500, and 50 mi^2 when a network of three gages is used. It is seen that the variance reduction factor $F_2(N)$ is much smaller for the case of a heavy storm in all cases, an indication that the error made in these cases is of less importance. The larger the area, the larger $F_2(N)$, an indication that more stations are needed to maintain the same precision in the estimate. It is also

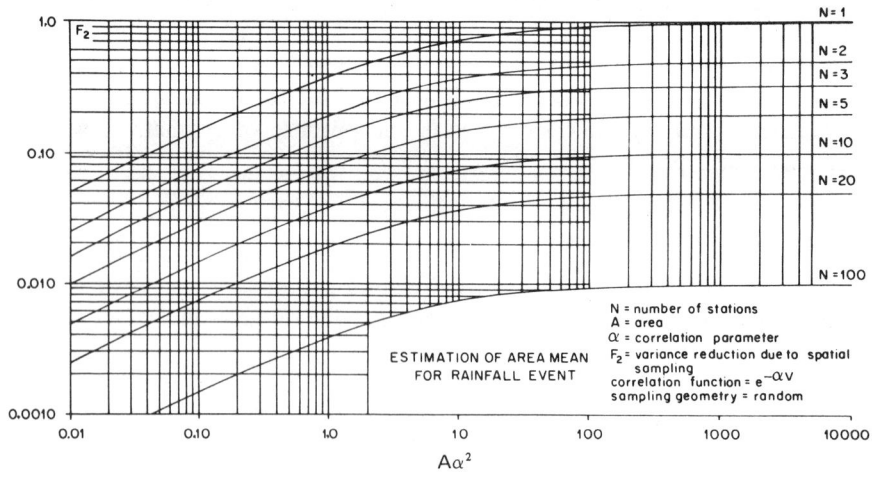

Figure 6.29 Variance reduction factor due to spatial sampling with random design used in the estimation of mean area rainfall of an event with $r(v) = e^{-\alpha v}$ (from Rodríguez-Iturbe and Mejía, 1974a).

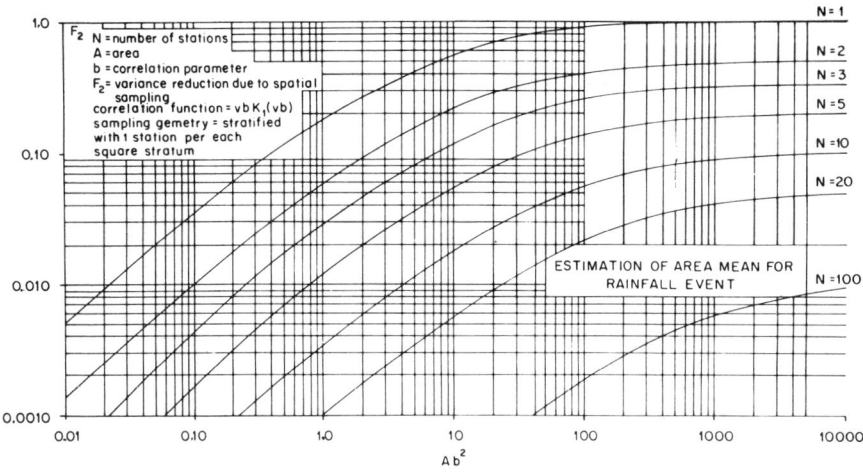

Figure 6.30 Variance reduction factor due to spatial sampling with stratified design used in the estimation of mean area rainfall of an event with $r(v) = vbK_1(vb)$ (from Rodríguez-Iturbe and Mejía, 1974a).

important to note that stratification can significantly reduce the value of $F_2(N)$; with only three stations, reduction largely depends on the size of the area under consideration because the larger the area, the more similar the random and the stratified schemes are.

From Figs. 6.28 to 6.31, it is seen that the number of stations now plays a very important role in the estimation process, somewhat differently from that in the estimation of long-term area means.

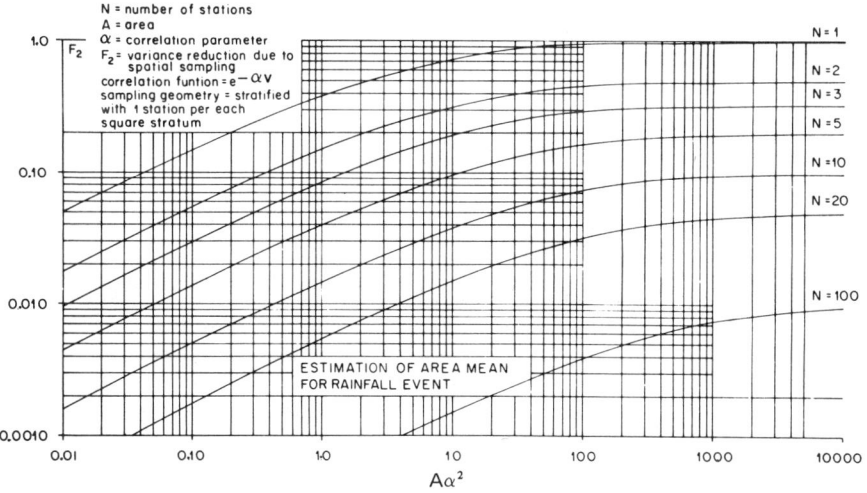

Figure 6.31 Variance reduction factor due to spatial sampling with stratified design used in the estimation of mean area rainfall of an event with $r(v) = e^{-\alpha v}$ (from Rodríguez-Iturbe and Mejía, 1974a).

Table 6.10
Variance reduction factors, $F_2(N)$, for Bessel correlation function for different combinations of areas and convective storms with a network of three stations

	Random design			Stratified design		
Area (mi²)	$P_t(0) = 5$	$P_t(0) = 2$	$P_t(0) = 0.75$	$P_t(0) = 5$	$P_t(0) = 2$	$P_t(0) = 0.75$
1500	0.240	0.320	0.320	0.180	0.280	0.320
500	0.175	0.280	0.320	0.110	0.230	0.310
50	0.055	0.160	0.260	0.025	0.094	0.210

The values for $P_t(0)$ are given in inches.
Source: Rodríguez-Iturbe and Mejía (1974a).

Analysis of Cyclonic Storms

Hurricane Connie in the Baltimore area was described with Eq. (6.127) using $\alpha = 0.009$ and $b = 0.016$. Assuming that we want to estimate the area mean depth of this event over an area of 1500 mi² with a network of three gages, we obtain the following:

Random
$$A\alpha^2 = 0.12 \rightarrow F_2(3) = 0.054 \qquad Ab^2 = 0.38 \rightarrow F_2(3) = 0.032$$

Stratified
$$F_2(3) = 0.032 \qquad F_2(3) = 0.014$$

Because of the area coverage of the storm and its intensity, there are large differences in $F_2(N)$ according to the type of correlation structure used in the analysis.

Table 6.11
Variance reduction factors, $F_2(N)$, for exponential correlation function for different combinations of areas and convective storms with a network of three stations

	Random design			Stratified design		
Area (mi²)	$P_t(0) = 5$	$P_t(0) = 2$	$P_t(0) = 0.75$	$P_t(0) = 5$	$P_t(0) = 2$	$P_t(0) = 0.75$
1500	0.250	0.310	0.320	0.190	0.280	0.320
500	0.190	0.280	0.320	0.140	0.230	0.310
50	0.082	0.165	0.260	0.050	0.110	0.210

The values for $P_t(0)$ are given in inches.
Source: Rodríguez-Iturbe and Mejía (1974a).

CHAPTER 7

Estimation of Static Linear Hydrologic Systems

7.1 INTRODUCTION

There are many problems in hydrology where the estimation of the state of the system by using available data and knowledge about measurement conditions is the point of interest. Estimation is the process of extracting information from data. The data itself may come from different sources and may contain errors; diverse measurements must then be blended to form estimates that reflect optimality with respect to a chosen criterion.

This chapter will concentrate on estimation problems for hydrologic systems that are both linear and static. The assumption of linearity is one of mathematical convenience, and is very appealing in hydrology, where many nonlinear transformations may be adequately approximated by a linear one. (Chapter 9 will deal with the estimation aspects of nonlinear hydrologic systems.) In addition to the linearity assumption, we will concentrate here on static systems or, in other words, on problems of hydrologic estimation where the dynamics of the process do not play an important role. (Chapter 8 will focus on the estimation of dynamic hydrologic systems.)

There are many hydrologic examples where the main issue is the estimation of an area average or the optimal contouring of a certain parameter. The existence of a correlation structure and the fact that the data contain errors in measurement and almost always are collected at unequally spaced intervals, makes the estimation problem a challenging and important one, which may be tackled under the general framework of a linear and static system.

The importance of an optimal weighting of the different pieces of data for the above kind of problems is easily seen following an example given by Schweppe (1973). Define an unknown but constant variable X, which is measured twice with some error,

$$Z_1 = X + V_1$$
$$Z_2 = X + V_2,$$

where Z_i is an observation (measurement) and the V_is are random errors with the following first- and second-moment properties:

$$E[V_1] = E[V_2] = E[V_1V_2] = 0$$
$$E[V_1^2] = 1$$
$$E[V_2^2] = 4.$$

Ignoring the variance of the observation errors, an obvious linear estimator of X, $\hat{X}$, would be

$$\hat{X} = \mathbf{WZ},$$

where the weighting vector $\mathbf{W}$ is

$$\mathbf{W} = [1/2, 1/2]$$

and $\mathbf{Z}$ is a vector

$$\mathbf{Z}^T = [Z_1, Z_2].$$

So,

$$\hat{X} = \frac{Z_1 + Z_2}{2}.$$

The mean square error of the estimator is given by

$$E[(X - \hat{X})^2] = E\left[\left(X - \frac{Z_1 + Z_2}{2}\right)^2\right]$$

$$= E\left[\left(X - \frac{(X + V_1) + (X + V_2)}{2}\right)^2\right]$$

$$= E\left[\left(-\frac{V_1 + V_2}{2}\right)^2\right] = \frac{1}{4}E[V_1^2 + 2V_1V_2 + V_2^2]$$

$$= \frac{1}{4}[1 + 4] = \frac{5}{4}.$$

Had the weighting vector been $\mathbf{W} = [4/5, 1/5]$, leading to

$$\hat{X} = \frac{4}{5}Z_1 + \frac{1}{5}Z_2,$$

the mean square error of estimation would be

$$
\begin{aligned}
E\left[(X - \hat{X})^2\right] &= E\left[\left(X - \frac{4}{5}Z_1 - \frac{1}{5}Z_2\right)^2\right] \\
&= E\left[\left(X - \frac{4}{5}(X + V_1) - \frac{1}{5}(X + V_2)\right)^2\right] \\
&= E\left[\left(-\frac{4}{5}V_1 - \frac{1}{5}V_2\right)^2\right] = \frac{1}{25}E\left[16V_1^2 + 8V_1V_2 + V_2^2\right] \\
&= \frac{1}{25}[16 + 4] = \frac{20}{25} = \frac{4}{5}.
\end{aligned}
$$

Clearly, the new weighting system yields a smaller mean square error. It seems logical then to weigh the observations in terms of their relative accuracy. The next few sections will discuss the general expressions for linear filters that will guarantee the minimum mean square error in estimating a static linear system.

7.2 OPTIMAL STATIC LINEAR ESTIMATOR OF A RANDOM VARIABLE WITH KNOWN MEAN AND COVARIANCE

Define a static linear observation of a random vector $\mathbf{X}$.

$$\mathbf{Z} = \mathbf{HX} + \mathbf{V}, \tag{7.1}$$

where $\mathbf{Z}$ is an $(m \times 1)$ vector of observations, $\mathbf{X}$ is a random $(n \times 1)$ vector, and $\mathbf{V}$ is an $(m \times 1)$ vector of measurement noise. The $(m \times n)$ matrix $\mathbf{H}$ defines the experiment. For continuity of notation with later chapters, $\mathbf{X}$ will be called the state.

The following probabilistic information is available,

$$
\begin{aligned}
E[\mathbf{X}] &= \mathbf{M} \\
E\left[(\mathbf{X} - \mathbf{M})(\mathbf{X} - \mathbf{M})^T\right] &= \psi \\
E[\mathbf{V}] &= \mathbf{0} \\
E[\mathbf{V}\mathbf{V}^T] &= \mathbf{R} \\
E[\mathbf{H}\mathbf{X}\mathbf{V}^T] &= \mathbf{0}.
\end{aligned}
\tag{7.2}
$$

The optimal (in the sense of a minimum mean square error) linear estimator of $\mathbf{X}$ is then

$$\hat{\mathbf{X}} = \Sigma \mathbf{H}^T \mathbf{R}^{-1} \mathbf{Z} + \Sigma \psi^{-1} \mathbf{M}, \tag{7.3a}$$

where the mean-square-error-of-estimation matrix is

$$\Sigma = E\left[(\mathbf{X} - \hat{\mathbf{X}})(\mathbf{X} - \hat{\mathbf{X}})^T\right]$$
$$= (\mathbf{H}^T \mathbf{R}^{-1} \mathbf{H} + \psi^{-1})^{-1}. \tag{7.3b}$$

The estimator is unbiased:

$$E[\hat{\mathbf{X}}] = \mathbf{M}.$$

Using the matrix-inversion lemma, the optimal estimator can also be expressed as

$$\hat{\mathbf{X}} = \mathbf{M} + \psi \mathbf{H}^T \left[\mathbf{H} \psi \mathbf{H}^T + \mathbf{R}\right]^{-1} (\mathbf{Z} - \mathbf{H} \mathbf{M})$$
$$\Sigma = \psi - \psi \mathbf{H}^T \left[\mathbf{H} \psi \mathbf{H}^T + \mathbf{R}\right]^{-1} \mathbf{H} \psi. \tag{7.4}$$

Minimization of Σ implies that any other estimator with the mean-square-error matrix $\tilde{\Sigma}$ will be such that $\tilde{\Sigma} - \Sigma$ is positive semidefinite. All elements of the optimal estimator $\hat{\mathbf{X}}$ are optimal estimates of the corresponding elements of $\mathbf{X}$. Therefore, no one element estimate can be improved at the expense of others.

The best estimate of any linear operation on $\mathbf{X}$,

$$\mathbf{Y} = \mathbf{C} \mathbf{X}, \tag{7.5}$$

is the same linear operation on $\hat{\mathbf{X}}$

$$\hat{\mathbf{Y}} = \mathbf{C} \hat{\mathbf{X}} \tag{7.6}$$

with the mean-square-error-of-estimation matrix

$$\Sigma_Y = \mathbf{C} \Sigma_X \mathbf{C}^T. \tag{7.7}$$

The derivation of the result in Eqs. (7.3) and (7.4) can follow several paths. Following Sage and Melsa (1971), the minimum-variance Bayes estimator, $\hat{\mathbf{X}}$, minimizes the square cost function

$$C(\mathbf{X}, \hat{\mathbf{X}}) = (\mathbf{X} - \hat{\mathbf{X}}(\mathbf{Z}))^T \mathbf{S} (\mathbf{X} - \hat{\mathbf{X}}(\mathbf{Z})) \tag{7.8}$$

over all possible values of $\hat{\mathbf{X}}$ and $\mathbf{Z}$. The matrix $\mathbf{S}$ must be symmetric and non-negative definite but otherwise arbitrary. The Bayes risk is given by

$$B = \int_{-\infty}^{\infty} \int_{-\infty}^{\infty} (\mathbf{X} - \hat{\mathbf{X}}(\mathbf{Z}))^T \mathbf{S}(\mathbf{X} - \hat{\mathbf{X}}(\mathbf{Z})) f(\mathbf{X}, \mathbf{Z}) \, d\mathbf{X} \, d\mathbf{Z}, \qquad (7.9)$$

where $f(\mathbf{X}, \mathbf{Z})$ is the joint distribution of random vectors $\mathbf{X}$ and $\mathbf{Z}$. Equation (7.9) can be rewritten as

$$B = \int_{-\infty}^{\infty} \left\{ \int_{-\infty}^{\infty} (\mathbf{X} - \hat{\mathbf{X}}(\mathbf{Z}))^T \mathbf{S}(\mathbf{X} - \hat{\mathbf{X}}(\mathbf{Z})) f(\mathbf{X}|\mathbf{Z}) \, d\mathbf{X} \right\} f(\mathbf{Z}) \, d\mathbf{Z}, \quad (7.10)$$

where $f(\mathbf{X}|\mathbf{Z})$ is the conditional distribution of $\mathbf{X}$ given $\mathbf{Z}$. Minimization of Eq. (7.10) is the same as minimization of the inner integral with respect to $\mathbf{X}$. Thus, minimization of the expected value of the cost of an estimation error (Bayes risk, Eq. 7.9) is the same as minimizing the expected value of the cost given the observations $\mathbf{Z}$. The conditional Bayes risk to minimize is then

$$B(\hat{\mathbf{X}}|\mathbf{Z}) = E\left[C(\mathbf{X}, \hat{\mathbf{X}})|\mathbf{Z} \right] = \int_{-\infty}^{\infty} (\mathbf{X} - \hat{\mathbf{X}})^T \mathbf{S}(\mathbf{X} - \hat{\mathbf{X}}) f(\mathbf{X}|\mathbf{Z}) \, d\mathbf{X}, \quad (7.11)$$

where the dependence of $\hat{\mathbf{X}}$ on $\mathbf{Z}$ is not shown for the sake of brevity. Taking the first derivative of the above yields

$$\frac{\partial B(\hat{\mathbf{X}}|\mathbf{Z})}{\partial \hat{\mathbf{X}}} = 0 = 2 \int_{-\infty}^{\infty} \mathbf{S}(\hat{\mathbf{X}} - \mathbf{X}) f(\mathbf{X}|\mathbf{Z}) \, d\mathbf{X}, \qquad (7.12)$$

which is satisfied if

$$\hat{\mathbf{X}} \int_{-\infty}^{\infty} f(\mathbf{X}|\mathbf{Z}) \, d\mathbf{X} = \int_{-\infty}^{\infty} \mathbf{X} f(\mathbf{X}|\mathbf{Z}) \, d\mathbf{X} = \hat{\mathbf{X}}. \qquad (7.13)$$

So, the minimum-variance estimator is the mean of the conditional distribution of $\mathbf{X}$ given $\mathbf{Z}$. The above estimator is valid for a variety of symmetric loss functions (Sage and Melsa, 1971). If in Eq. (7.1) the random variables $\mathbf{X}$ and $\mathbf{V}$ are assumed Gaussian with the first two moments as given by Eq. (7.2), the conditional probability density becomes

$$f(\mathbf{X}|\mathbf{Z}) = \frac{f(\mathbf{Z}|\mathbf{X}) f(\mathbf{X})}{f(\mathbf{Z})}, \qquad (7.14)$$

where all the above densities are also Gaussian with parameters

$$E[\mathbf{Z}|\mathbf{X}] = \mathbf{H}\mathbf{X}$$
$$\mathrm{var}[\mathbf{Z}|\mathbf{X}] = \mathbf{R}$$
$$E[\mathbf{X}] = \mathbf{M}$$
$$\mathrm{var}[\mathbf{X}] = \psi$$
$$E[\mathbf{Z}] = \mathbf{H}\mathbf{M}$$
$$\mathrm{var}[\mathbf{Z}] = \mathbf{H}\psi\mathbf{H}^T + \mathbf{R}.$$

The probability density functions are then

$$f(\mathbf{Z}|\mathbf{X}) = \frac{1}{(2\pi)^{m/2}|\mathbf{R}|^{1/2}} \exp\left[-\frac{1}{2}(\mathbf{Z}-\mathbf{HX})^T\mathbf{R}^{-1}(\mathbf{Z}-\mathbf{HX})\right]$$

$$f(\mathbf{X}) = \frac{1}{(2\pi)^{n/2}|\psi|^{1/2}} \exp\left[-\frac{1}{2}(\mathbf{X}-\mathbf{M})^T\psi^{-1}(\mathbf{X}-\mathbf{M})\right]$$

$$f(\mathbf{Z}) = \frac{1}{(2\pi)^{m/2}\left|(\mathbf{H}\psi\mathbf{H}^T+\mathbf{R})\right|^{1/2}}$$
$$\cdot\exp\left[-\frac{1}{2}(\mathbf{Z}-\mathbf{HM})^T(\mathbf{H}\psi\mathbf{H}^T+\mathbf{R})^{-1}(\mathbf{Z}-\mathbf{HM})\right],$$

where $|\cdot|$ represents the determinant. Substitution in Eq. (7.14) yields

$$f(\mathbf{X}|\mathbf{Z}) = \frac{|\mathbf{H}\psi\mathbf{H}^T+\mathbf{R}|^{1/2}}{(2\pi)^{n/2}|\psi|^{1/2}|\mathbf{R}|^{1/2}} \exp\left[-\frac{1}{2}(\mathbf{X}-\hat{\mathbf{X}})^T\mathbf{\Sigma}^{-1}(\mathbf{X}-\hat{\mathbf{X}})\right],$$

where

$$\hat{\mathbf{X}} = \left(\psi^{-1}+\mathbf{H}^T\mathbf{R}^{-1}\mathbf{H}\right)^{-1}(\mathbf{H}^T\mathbf{R}^{-1}\mathbf{Z}+\psi^{-1}\mathbf{M})$$

$$= \mathbf{\Sigma}(\mathbf{H}^T\mathbf{R}^{-1}\mathbf{Z}+\psi^{-1}\mathbf{M}) \tag{7.15}$$

$$\mathbf{\Sigma}^{-1} = \psi^{-1}+\mathbf{H}^T\mathbf{R}^{-1}\mathbf{H}.$$

Equations (7.15) are the results given in Eqs. (7.3) and (7.4).

The above results were derived using the Gaussian assumption; nevertheless, the estimators remain optimal in comparison to all other linear forms (Schweppe, 1973).

Some other important properties of the estimator are

$$E\left[(\mathbf{X}-\hat{\mathbf{X}})\mathbf{Z}^T\right] = \mathbf{0}$$
$$E\left[(\mathbf{X}-\hat{\mathbf{X}})\hat{\mathbf{X}}^T\right] = \mathbf{0} \tag{7.16}$$

implying that the residual error is orthogonal to the observations and any linear function of the observations. Other results are

$$E[\mathbf{X}\hat{\mathbf{X}}^T] = E[\hat{\mathbf{X}}\hat{\mathbf{X}}^T]$$
$$\mathbf{\Sigma} = E[\mathbf{X}\mathbf{X}^T] - E[\hat{\mathbf{X}}\hat{\mathbf{X}}^T] \tag{7.17}$$
$$E[\mathbf{X}\mathbf{X}^T] \geq E[\hat{\mathbf{X}}\hat{\mathbf{X}}^T].$$

7.3 OPTIMAL ESTIMATION OF UNKNOWN CONSTANTS

Schweppe (1973) refers to the estimator discussed in the previous section as Bayesian. It utilizes prior knowledge to estimate a realization of a random variable. Of interest to the engineer-scientist may be the observation and estimation of an unknown constant. Define the observation vector,

$$\mathbf{Z} = \mathbf{HX} + \mathbf{V}, \tag{7.18}$$

where $\mathbf{X}$ is a completely unknown constant, and $\mathbf{V}$ is a random error of observation with properties,

$$E[\mathbf{V}] = \mathbf{0}$$
$$E[\mathbf{VV}^T] = \mathbf{R}. \tag{7.19}$$

Although the problem of estimating $\mathbf{X}$ from Eq. (7.18) is different from that of the Bayesian formulation, it is heuristically possible to obtain its solution from Eq. (7.4). A completely unknown constant can be interpreted as a random variable with infinite prior covariance, $\psi \rightarrow \infty \mathbf{I}$. Using this limit, Eq. (7.3) becomes

$$\hat{\mathbf{X}} = \Sigma \mathbf{H}^T \mathbf{R}^{-1} \mathbf{Z}$$
$$\Sigma = (\mathbf{H}^T \mathbf{R}^{-1} \mathbf{H})^{-1} = E\left[(\mathbf{X} - \hat{\mathbf{X}})(\mathbf{X} - \hat{\mathbf{X}})^T\right]. \tag{7.20}$$

Equation (7.20) is the solution to Eq. (7.18) or, in Schweppe's (1973) terminology, the Fisher problem. There are several differences between the estimator of Eq. (7.20) and that of Eq. (7.3). In particular, the unbiased condition now implies

$$E[\hat{\mathbf{X}}] = \mathbf{X}.$$

No orthogonality exists between errors and observations.

$$E\left[(\mathbf{X} - \hat{\mathbf{X}})\mathbf{Z}^T\right] = -[\mathbf{H}^T \mathbf{R}^{-1} \mathbf{H}]^{-1} \mathbf{H}^T \neq \mathbf{0}.$$

Furthermore, the mean-square-error matrix exists only if the order of the observation vector, m, is greater than or equal to that of the state vector $\mathbf{X}$, n. If this condition is not satisfied, $\mathbf{X}$ remains undefined and Σ does not exist since $[\mathbf{H}^T \mathbf{R}^{-1} \mathbf{H}]$ cannot be inverted for $m < n$ (it will be singular).

Assuming normally distributed measurement error, $\mathbf{V}$, the estimator of Eq. (7.20) becomes a maximum-likelihood estimate. It is obtained by maximizing the likelihood function, $f(\mathbf{Z}|\mathbf{X})$. Solving,

$$\frac{\partial f(\mathbf{Z}|\mathbf{X})}{\partial \mathbf{X}} = 0$$

leads to

$$\hat{\mathbf{X}} = (\mathbf{H}^T \mathbf{R}^{-1} \mathbf{H})^{-1} \mathbf{H}^T \mathbf{R}^{-1} \mathbf{Z}.$$

The estimator given in Eq. (7.20) remains the best linear estimator of the Fisher problem (in a minimum-mean-square-error sense) even for non-Gaussian conditions (see Schweppe, 1973, for proof).

7.4 GENERAL COMMENTS ON THE OPTIMAL STATIC ESTIMATOR

It is valuable to study briefly the form of the linear estimators presented in the previous two sections. Particularly interesting is the structure of the Bayesian result—Eqs. (7.3) and (7.4). The optimal estimator of a random variable is a linear combination of the prior mean $\mathbf{M}$ and the deviation of the observation $\mathbf{Z}$ from the prior mean value. The weight given to the observations is

$$\mathbf{W} = (\mathbf{H}^T \mathbf{R}^{-1} \mathbf{H} + \psi^{-1})^{-1} \mathbf{H}^T \mathbf{R}^{-1}.$$

If the observations are very uncertain (large $\mathbf{R}$) the weighting matrix will tend to

$$\mathbf{W} \approx \psi \mathbf{H}^T \mathbf{R}^{-1},$$

which will ultimately go to zero as $\mathbf{R} \to \infty$. In such cases the observations have little value and the best estimate of $\mathbf{X}$ remains simply the prior mean vector (the model).

At the other extreme, a completely unknown vector (large ψ) leads to the Fisher formulation where all the weight is given to the observations and prior model knowledge plays no role.

It is also important to understand that the algorithm is a function of $\mathbf{R}$, ψ, and $\mathbf{H}$. For example, in a Fisher formulation, the information content of the observations is measured by $\mathbf{H}^T \mathbf{R}^{-1} \mathbf{H}$ and not by $\mathbf{R}$ alone. It is possible for $\mathbf{H}$ to be such that large weight is given to a relatively uncertain observation.

One of the major advantages of linear estimators is the independence of the mean-square-error matrix from the observations. The implication is that a measure of the accuracy of the data-collection system is available before a single observation is made. The mean square error of estimation then becomes a magnificent tool in the design of data-collection networks. The use of this concept in hydrology will be discussed in Section 7.6.

The mean-square-error matrix Σ is a function of $\mathbf{H}$, $\mathbf{R}$, and ψ. Increased prior information (low ψ) or increased measurement accuracy (low $\mathbf{R}$) results in lower mean-square-estimation errors. Ordinarily, though, data-collection–network design has the objective of varying the matrix $\mathbf{H}$, or essentially altering the network configuration, until minimum error is achieved.

The form of Σ in Eq. (7.4) is particularly interesting,

$$\Sigma = \psi - \psi H^T \left[H \psi H^T + R \right]^{-1} H \psi.$$

The mean-square-error matrix is the prior model covariance ψ minus a positive definite quantity that measures the amount of information gained by the observation. The matrix Σ can then be interpreted as the posterior covariance matrix in a Bayesian formulation. Repeated observations and applications of the static filter would yield smaller and smaller posterior covariances as information is incorporated in the analysis. This concept will be exploited later in the study of dynamic systems (Chapter 8).

The two alternative formulations for the mean-square-error matrix in the Bayesian problem offer various computational advantages. To obtain Σ in Eq. (7.3), two inversions of $n \times n$ matrices are required. For large dimensions, this could be expensive and inaccurate. On the other hand, the expression is simple and compact. Equation (7.4) requires more multiplications but only one inversion of an $m \times m$ matrix. Since m is usually smaller than n, this is generally a considerable computational advantage.

Equations (7.3) and (7.20) were derived from a probabilistic framework. Similar equations result from other approaches. Of interest is the popular weighted least-squares formulation, a purely statistical method. An equation identical to Eq. (7.20) results from minimizing a function

$$J(X) = (Z - HX)^T R^{-1} (Z - HX). \tag{7.21}$$

Now, though, R is any positive definite weighting matrix.

A result similar to the Bayesian equation (Eq. 7.3) comes from minimizing

$$J(X) = (Z - HX)^T R^{-1} (Z - HX) + (X - M)^T \psi^{-1} (X - M).$$

Again, R plays the role of an arbitrary weighting matrix selected by the user.

Schweppe (1973) discusses a formulation where X and V are unknown variables, but bounded within a given region in n and m space, respectively. A possible solution to this problem leads to results similar to Eq. (7.3).

7.5   NONOPTIMAL ESTIMATORS

In practice, it is fairly common to use nonoptimal estimators. Uncertainty in model parameters is usually the cause of nonoptimal performance. Many times, model structure in terms of H, ψ, and R is unknown. Nonoptimal linear filters can still be very useful as long as their performance can be evaluated. Schweppe (1973) gives the resulting bias and mean-square-error matrix from a nonoptimal linear filter of the Bayesian type,

$$\hat{X} = WZ + W_0$$

as

$$\mathbf{b} = E[\mathbf{X}] - E[\hat{\mathbf{X}}]$$
$$= [\mathbf{I} - \mathbf{WH}]\mathbf{M} - \mathbf{W}_0 \qquad (7.22)$$

and

$$\Sigma = E\left[(\mathbf{X} - \hat{\mathbf{X}} - \mathbf{b})(\mathbf{X} - \hat{\mathbf{X}} - \mathbf{b})^T\right]$$
$$= [\mathbf{I} - \mathbf{WH}]\psi[\mathbf{I} - \mathbf{WH}]^T + \mathbf{WRW}^T.$$

For the unknown constant case,

$$\mathbf{X} - E[\hat{\mathbf{X}}] = \mathbf{b} = [\mathbf{I} - \mathbf{WH}]\mathbf{X} - \mathbf{W}_0$$
$$\Sigma = \mathbf{WRW}^T. \qquad (7.23)$$

Note that in the Fisher formulation the bias is a function of $\mathbf{X}$, which implies the possibility of unbounded error.

7.6 DESIGN OF RAINFALL DATA-COLLECTION NETWORKS: AN EXAMPLE

The independence of the mean square error of estimation from observations makes the static linear problem very well suited for studying and designing data-collection networks. This section is adapted from Bras and Rodríguez-Iturbe (1976b) and deals with designing a data-collection network for obtaining the mean area precipitation from a rainfall event. The work illustrates the use of static linear estimation to evaluate the accuracy of possible rain-gage configurations.

The area mean of an event has innumerable uses in hydrology. It is a traditional parameter in runoff and water-yield studies. Given a set of rain gages and data points, hydrologists have devised numerous techniques to obtain estimates of the mean of the process over a given area. The well-known methods of Thiessen polygon and isohyetal analyses are among the available estimators of the area mean of an event. These data-analysis techniques acknowledge uncertainty, due to spatial variability, of the rainfall process. Rainfall is a random function in time and space. Total rainfall depth of an event is a random function in space.

Data collection is viewed as an estimation problem. A stochastic event, total rainfall, which is continuous in space, must be estimated from discrete incomplete noisy observations in space. Designing a network is then the definition of the system that gives the best estimate of the true event (total rainfall in space). "Best" must be defined in terms of some function of accuracy and cost.

Consider a region A, as shown in Fig. 7.1. The hydrologist is interested in the estimation of

$$P = \frac{1}{A} \int_A f(x, y)\, dx\, dy, \qquad (7.24)$$

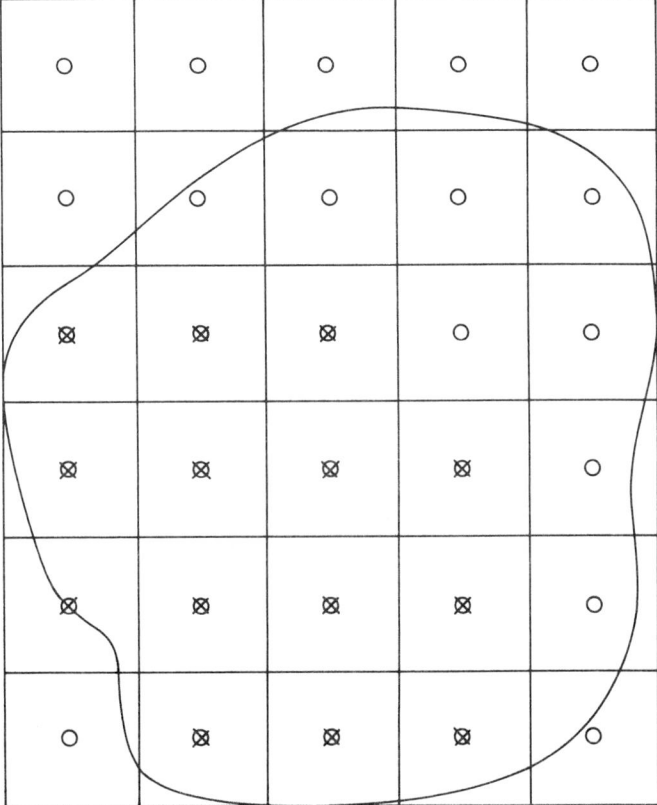

○ Discretized Points in Area A, $n = 30$

✕ Possible Station Locations

Figure 7.1 Area A and superimposed discretizing grid (from Bras and Rodríguez-Iturbe, 1976b).

where A is the area of region A, $f(x, y)$ is the function describing depth of a storm event over area A, and P is the area average of the total depth of the storm $f(x, y)$ over area A.

Under usual conditions, Eq. (7.24) is approximated by discretizing area A (see Fig. 7.1) and then evaluating

$$\hat{P} = \sum_{i=1}^{n} \rho_i f(\mathbf{x}_i) = \gamma^T \mathbf{f}, \tag{7.25}$$

where $\hat{P}$ is the approximation of true area depth P; $f(\mathbf{x}_i)$ the discrete value of storm $f(x, y)$ at point i, defined by coordinate vector $\mathbf{x}_i$; ρ_i weights on discrete storm-depth values; n the number of discrete points where $f(\mathbf{x}_i)$ is defined; γ^T the transpose of $n \times 1$ vector of weights ρ_i; and $\mathbf{f}$ the $n \times 1$ vector of $f(\mathbf{x}_i)$ values.

In a uniform discretization the weights ρ_i would be $A_i/A = 1/n$, where A_i is the subarea corresponding to each grid subdivision. As a matter of convenience, assume uniform discretization. The theory, though, is applicable with any values of ρ_i.

Equation (7.25) implies perfect knowledge of rainfall depth at all the discrete points defined in the grid over area A (see Fig. 7.1). In fact, observations are generally far from perfect and available only in a limited number $m < n$ of grid points (see Fig. 7.1). These m points of noisy observations do not necessarily form any recognizable geometric pattern. The hydrologist must filter and extrapolate these noisy observations to obtain an approximation of $\hat{P}$, which is in turn an estimator of P.

This approximation may be written

$$\tilde{P} = \frac{1}{n} \sum_{i=1}^{n} \hat{f}(\mathbf{x}_i)$$

or

$$\tilde{P} = \gamma^T \hat{\mathbf{f}}, \tag{7.26}$$

where $\hat{f}(\mathbf{x}_i)$ is the estimated value of $f(\mathbf{x}_i)$ obtained from noisy observations at a limited number of points m and extrapolated to the whole grid of n points, $\hat{\mathbf{f}}$ is the $(n \times 1)$ vector of estimated-rainfall-depth values $\hat{f}(\mathbf{x}_i)$ ($i = 1, \ldots, n$), and $\tilde{P}$ is the approximated value of $\hat{P}$ and final estimation of P.

The network-design problem then translates into obtaining the number m and the location of rain gages that will give us the best estimate $\tilde{P}$ of the true mean area precipitation P. "Best estimate" and "best design" are key terms that must be defined with respect to some design criteria.

Network design for area mean estimation should be done in terms of accuracy and cost objectives. Since generally it is not possible to define a utility function, the design is achieved by minimizing an objective function of the form

$$\min \left\{ 0\left[\delta(m, \mathbf{x}_i); C(m, \mathbf{x}_i) \right] = \delta(m, \mathbf{x}_i) + C_\Delta \cdot C(m, \mathbf{x}_i) \right\} \tag{7.27}$$

over all m and $\mathbf{x}_i$, $i = 1, \ldots, m$, where $\delta(m, \mathbf{x}_i)$ is the measure that decreases as accuracy increases. It is a function of m, the number of stations, and $\mathbf{x}_i$ ($i = 1, \ldots, m$), the location of stations. $C(m, \mathbf{x}_i)$ is the cost of m stations at locations $\mathbf{x}_i$ ($i = 1, \ldots, m$), and C_Δ is the measure of accuracy equivalent to a unitary change in cost.

Equation (7.27) assumes that the measure of accuracy is such that an inverse relation between cost and accuracy exists. That is, as the cost increases $\delta(m, \mathbf{x}_i)$ decreases.

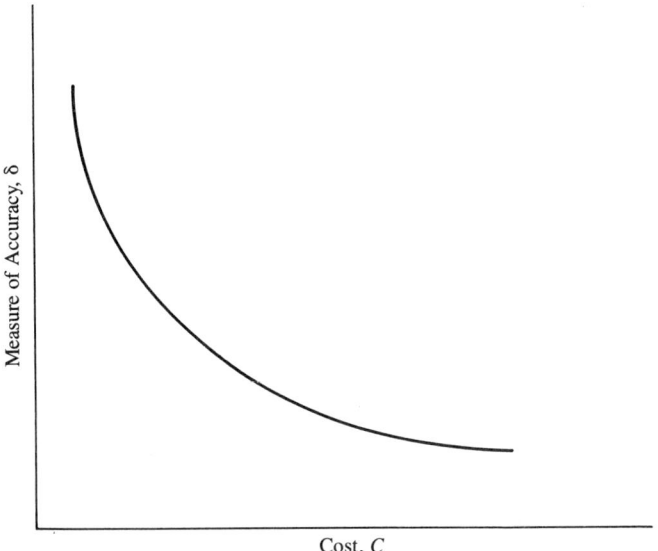

Figure 7.2 Plot of optimal combinations of cost and measure of accuracy, defining the transformation curve (from Bras and Rodríguez-Iturbe, 1976b).

For a given trade-off coefficient C_Δ the design process consists of finding the optimal combination of m and $\mathbf{x}_i$, which in turn defines the desired data-collection network.

Solving the design problem for different values of C_Δ allows the construction of a "transformation curve" of the type sketched in Fig. 7.2. Every point in the curve represents the values of the cost and accuracy associated with a network that minimizes the objective function for a particular value of C_Δ.

The cost term in Eq. (7.27) is taken as a linear summation of costs of particular locations

$$C(m,\mathbf{x}_i) = \sum_{i=1}^{m} C(\mathbf{x}_i) \qquad i = 1,\ldots, m, \qquad (7.28)$$

where $C(\mathbf{x}_i)$ is the cost at location $\mathbf{x}_i$.

The accuracy term $\delta(m,\mathbf{x}_i)$ is defined as the mean square error of estimating P by $\tilde{P}$. The mean square error (MSE) is defined as

$$\text{MSE} = \delta(m, \mathbf{x}_i) = E[(P - \tilde{P})^2], \qquad (7.29)$$

where E is the expectation operator.

By adding and subtracting $\hat{P}$ (Eq. 7.25), the MSE becomes

$$\text{MSE} = E\left[[(P - \hat{P}) + (\hat{P} - \tilde{P})]^2\right]$$

$$= E\left[(P - \hat{P})^2\right] + E\left[(\hat{P} - \tilde{P})^2\right] + 2E\left[(P - \hat{P})(\hat{P} - \tilde{P})\right] \quad (7.30)$$

$$= \sigma_m^2 + \sigma_e^2 + 2\sigma_{me}^2.$$

Term $\sigma_m^2 = E[(P - \hat{P})^2]$ is the mean square value of the "model" error $P - \hat{P}$. It represents the error resulting from approximating a continuous integral (Eq. 7.24) by a discrete summation (Eq. 7.25). Term $\sigma_e^2 = E[(\hat{P} - \tilde{P})^2]$ is the mean square value of the "estimation error" $\hat{P} - \tilde{P}$. It represents the error involved in estimating the summation of discrete values $f(x_i)$, $i = 1, \ldots, n$ (Eq. 7.25), by a summation of estimated values $\hat{f}(x_i)$, $i = 1, \ldots, n$, obtained from noisy incomplete observations. Term $\sigma_{me}^2 = E[(P - \hat{P})(\hat{P} - \tilde{P})]$ represents the dependence between the model error and the estimation error.

7.6.1 Accuracy Expressions

The terms of Eq. (7.30) can all be expressed in terms of the covariance function of the process $f(\mathbf{x})$, $\text{cov}(\mathbf{x}_i, \mathbf{x}_j)$. To simplify the notation, assume in computing the relevant expressions that rainfall is a zero-mean process; this does not diminish the generality of the results. The estimation of mean square errors will not be affected by this assumption.

The mean square model error term can be expanded as

$$\sigma_m^2 = E\left[(P - \hat{P})^2\right] = E[PP] + E[\hat{P}\hat{P}] - 2E[P\hat{P}]. \quad (7.31)$$

The evaluation of Eq. (7.31) depends on the sampling technique used. Sampling is done here with a fixed geometric grid pattern (starting point fixed and known). For that case and by substituting Eqs. (7.24) and (7.25), the following expression is obtained

$$\sigma_m^2 = \sum_{i=1}^{n} \sum_{j=1}^{n} \rho_i \rho_j \text{cov}(\mathbf{x}_i, \mathbf{x}_j) + \frac{1}{A^2} \int_A \int_A \text{cov}(\mathbf{x}_1, \mathbf{x}_2) \, d\mathbf{x}_1 \, d\mathbf{x}_2$$

$$- \frac{2}{nA} \sum_{i=1}^{n} \int_A \text{cov}(\mathbf{x}, \mathbf{x}_i) \, d\mathbf{x}, \quad (7.32)$$

where $\text{cov}(\mathbf{x}_i, \mathbf{x}_j) = E[f(\mathbf{x}_i)f(\mathbf{x}_j)]$.

The two integral expressions of Eq. (7.32) require a functional definition of the two-dimensional covariance structure of the process. Isotropic and homogeneous covariance functions are two other conditions that make the two integral terms manageable. This limitation of the proposed technique can be

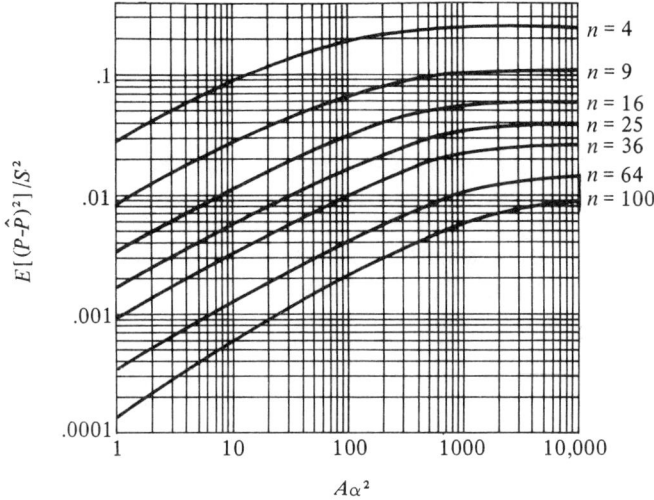

Figure 7.3 Mean square model error for single exponential covariance as a function of $A\alpha^2$ and n (from Bras and Rodríguez-Iturbe, 1976a).

somewhat relaxed, as will be shown later. Furthermore, the integral expressions have no closed solutions for most valid covariance functions. Fortunately, they can be reduced to one-dimensional expectation operations, which lend themselves to relatively easy and accurate numerical integrations. The reader is referred to the work of Bras and Rodríguez-Iturbe (1975, 1976a) to obtain details of the integration procedure and the evaluation of Eq. (7.32) for different covariance function forms. Figure 7.3 shows the result of Eq. (7.32) when $\mathrm{cov}(x_i, x_j) = \sigma^2 e^{-\alpha|x_i - x_j|}$. The correlation decay parameter is α and A is the area of interest, so the product $A\alpha^2$ is nondimensional. The number of potential station locations or grid size is n. The process point variance, assumed homogeneous throughout the area, is σ^2.

Now define the data-collection network by an equation of the form

$$\mathbf{Z} = \mathbf{Hf} + \mathbf{V}, \tag{7.33}$$

where $\mathbf{Z}$ is the $m \times 1$ vector of noisy observations at each discrete point in space $m \le n$; $\mathbf{f}$ the $n \times 1$ vector of true values of rainfall $f(\mathbf{x}_i)$ at the n discrete grid points; $\mathbf{V}$ the $m \times 1$ vector of white noise representing instrument error; and $\mathbf{H}$ the $m \times n$ matrix defining the data-collection network. The matrix $\mathbf{H}$ is of the form $\mathbf{H} = \{H_{ij}\} = h_{ij}$, if $\mathbf{x}_j$ is the location of an observation, and $\mathbf{H} = \{H_{ij}\} = 0$, otherwise. It is clear that $\mathbf{H}$ is a very sparse matrix of m rows; nonzero elements will appear only where there is an observation.

The following terms are also required in the analysis: $\psi = E\{\mathbf{f}\mathbf{f}^T\}$, the $n \times n$ covariance matrix of the true process; $\mathbf{R} = E[\mathbf{V}\mathbf{V}^T]$, the $m \times m$ covariance matrix of white-noise vector $\mathbf{V}$. Vectors $\mathbf{f}$ and $\mathbf{V}$ are assumed to be uncorrelated. They are also zero-mean random vectors.

Define $\hat{\mathbf{f}}$ as the best linear estimate of $\mathbf{f}$, where "best" means the linear estimate yielding the minimum-mean-square-error matrix. The resulting expression for $\hat{\mathbf{f}}$, as a function of $\mathbf{Z}$ was discussed in Section 7.4 and is repeated here:

$$\hat{\mathbf{f}} = \Sigma\mathbf{H}^T\mathbf{R}^{-1}(\mathbf{Z}-\mathbf{Z}_0)+\Sigma\psi^{-1}\mathbf{M}, \tag{7.34}$$

where $\Sigma = E[(\mathbf{f}-\hat{\mathbf{f}})(\mathbf{f}-\hat{\mathbf{f}})^T]$ is the mean square error of estimation. The mean vector $\mathbf{M}$ is assumed to be zero in this example with no loss of generality. The vector of observation bias $\mathbf{Z}_0$ is also assumed to be zero.

The mean-square-error-of-estimation matrix $\Sigma = E[(\mathbf{f}-\hat{\mathbf{f}})(\mathbf{f}-\hat{\mathbf{f}})^T]$ is given as

$$\Sigma = \left[\psi^{-1}+\mathbf{H}^T\mathbf{R}^{-1}\mathbf{H}\right]^{-1} = \psi - \psi\mathbf{H}^T\left[\mathbf{H}\psi\mathbf{H}^T+\mathbf{R}\right]^{-1}\mathbf{H}\psi. \tag{7.35}$$

By substituting Eqs. (7.25) and (7.26) in the expression of the mean square error of estimation, $\sigma_e^2 = E[(\hat{P} - \tilde{P})^2]$,

$$\sigma_e^2 = E\left[(\gamma^T\mathbf{f}-\gamma^T\hat{\mathbf{f}})^2\right] = E\left[(\gamma^T\mathbf{f}-(\hat{\mathbf{f}}^T\gamma^T))^2\right]$$
$$= \gamma^T E\left[(\mathbf{f}-\hat{\mathbf{f}})(\mathbf{f}-\hat{\mathbf{f}})^T\right]\gamma = \gamma^T\Sigma\gamma. \tag{7.36}$$

By using Eq. (7.35), the following equation is derived:

$$\sigma_e^2 = E\left[(\hat{P} - \tilde{P})^2\right] = \gamma^T\left[\psi^{-1}+\mathbf{H}^T\mathbf{R}^{-1}\mathbf{H}\right]^{-1}\gamma$$
$$= \gamma^T\left[\psi - \psi\mathbf{H}^T(\mathbf{H}\psi\mathbf{H}^T+\mathbf{R})^{-1}\mathbf{H}\psi\right]\gamma. \tag{7.37}$$

Note that Eq. (7.37) is not limited by any functional covariance expressions. Parameter σ_e^2 is a function of $\mathbf{H}$, $\mathbf{R}$, γ, and ψ. The covariance matrix ψ can be of any form and is not limited to isotropy or homogeneity conditions.

Left to be defined is the cross term σ_{me}^2 in Eq. (7.30). This term can be expanded as

$$\sigma_{me} = E\left[(P - \hat{P})(\hat{P} - \tilde{P})\right] = E[P\hat{P}]- E[P\tilde{P}]$$
$$- E[\hat{P}\hat{P}]+ E[\hat{P}\tilde{P}]. \tag{7.38}$$

By using Eqs. (7.24) and (7.25), the first term becomes

$$E[P\hat{P}] = E\left[\left(\frac{1}{A}\int_A f(\mathbf{x})\,d\mathbf{x}\right)\frac{1}{n}\sum_{i=1}^{n} f(x_i)\right]$$
$$= \frac{1}{nA}\sum_{i=1}^{n}\int_A \text{cov}(\mathbf{x},\mathbf{x}_i)\,d\mathbf{x}, \tag{7.39}$$

which is a summation of integrals similar to that encountered in Eq. (7.32).

Substitution of Eqs. (7.24) and (7.26) together with Eqs. (7.33) and (7.34) (with $\mathbf{M}$ and $\mathbf{Z}_0$ equal to $\mathbf{0}$) gives the second term in Eq. (7.38) as

$$E[P\tilde{P}] = \gamma^T \psi \mathbf{H}^T (\mathbf{H}\psi \mathbf{H}^T + \mathbf{R})^{-1} E[PZ]. \tag{7.40}$$

Using the expression for $\mathbf{Z}$ (Eq. 7.33) and the fact that $\hat{\mathbf{f}}$ and $\mathbf{Z}$ are uncorrelated gives

$$
\begin{aligned}
E[P\tilde{P}] &= \gamma^T \psi \mathbf{H}^T (\mathbf{H}\psi \mathbf{H}^T + \mathbf{R})^{-1} \{ \mathbf{H} E[Pf] + E[PV] \} \\
&= \gamma^T \psi \mathbf{H}^T (\mathbf{H}\psi \mathbf{H}^T + \mathbf{R})^{-1} \mathbf{H} E[Pf],
\end{aligned}
\tag{7.41}
$$

where

$$E[Pf] = \begin{bmatrix} a_1 \\ \vdots \\ a_n \end{bmatrix}$$

$$a_1, a_2, \ldots, a_n = \frac{1}{A} \int_A \mathrm{cov}(\mathbf{x}, \mathbf{x}_i) \, d\mathbf{x} \qquad i = 1, \ldots, n.$$

For an evaluation of integrals such as the above, the reader is referred to Bras and Rodríguez-Iturbe (1975).

The third term in (7.38) is simply

$$E[\hat{P}\hat{P}] = E[\gamma \mathbf{f} \gamma^T \mathbf{f}] = \gamma^T E[\mathbf{f}\mathbf{f}^T] \gamma = \gamma^T \psi \gamma. \tag{7.42}$$

The last term $E[\hat{P}\tilde{P}]$ can be written as

$$
\begin{aligned}
E[\hat{P}\tilde{P}] &= \gamma^T E[\mathbf{f}\mathbf{f}^T] \gamma = \gamma^T E\left[\mathbf{f} \left\{ \psi \mathbf{H}^T (\mathbf{H}\psi \mathbf{H}^T + \mathbf{R})^{-1} \mathbf{Z} \right\} \right] \gamma \\
&= \gamma^T E\left[\mathbf{f}\mathbf{Z}^T \left\{ \psi \mathbf{H}^T (\mathbf{H}\psi \mathbf{H}^T + \mathbf{R})^{-1} \right\}^T \right] \gamma.
\end{aligned}
$$

Substituting Eq. (7.33) and again using independence between $\mathbf{f}$ and $\mathbf{V}$ gives

$$
\begin{aligned}
E[\hat{P}\tilde{P}] &= \gamma^T (\psi \mathbf{H})^T \left\{ \psi \mathbf{H}^T (\mathbf{H}\psi \mathbf{H}^T + \mathbf{R})^{-1} \right\}^T \gamma \\
&= \gamma^T \psi \mathbf{H}^T (\mathbf{H}\psi \mathbf{H}^T + \mathbf{R})^{-1} \left[\psi \mathbf{H}^T \right]^T \gamma \\
&= \gamma^T \left[\psi \mathbf{H}^T (\mathbf{H}\psi \mathbf{H}^T + \mathbf{R})^{-1} \mathbf{H}\psi^T \right] \gamma
\end{aligned}
$$

and because of the symmetry of the covariance matrices

$$E[\hat{P}\tilde{P}] = \gamma^T \left[\psi \mathbf{H}^T (\mathbf{H}\psi \mathbf{H}^T + \mathbf{R})^{-1} \mathbf{H}\psi \right] \gamma. \tag{7.43}$$

In summary, the cross term σ_{me}^2 becomes (by using Eqs. 7.39, 7.41, 7.42, and 7.43)

$$\sigma_{me}^2 = \frac{1}{nA} \sum_{i=1}^{n} \int_A \text{cov}(\mathbf{x}, \mathbf{x}_i) \, d\mathbf{x} - \boldsymbol{\gamma}^T \boldsymbol{\psi} \mathbf{H}^T (\mathbf{H} \boldsymbol{\psi} \mathbf{H} + \mathbf{R})^{-1} \mathbf{H} E[Pf]$$

$$- \boldsymbol{\gamma}^T \boldsymbol{\psi} \boldsymbol{\gamma} + \boldsymbol{\gamma}^T \left[\boldsymbol{\psi} \mathbf{H}^T (\mathbf{H} \boldsymbol{\psi} \mathbf{H}^T + \mathbf{R})^{-1} \mathbf{H} \boldsymbol{\psi} \right] \boldsymbol{\gamma}, \qquad (7.44)$$

where $E[Pf]$ is as defined in Eq. (7.41).

With Eqs. (7.32), (7.37), and (7.44) all the terms of the chosen measure of accuracy, $\delta(m, \mathbf{x}_i)$, (Eq. 7.30) have been defined. The objective function defined by Eq. (7.26) can be minimized by varying the nature of matrix $\mathbf{H}$, which in turn affects $\delta(m, \mathbf{x}_i)$ and the total cost.

Minimization of the objective function was done with a numerically efficient algorithm that searched alternative network configurations, always moving in the direction of largest partial reductions in the objective function values. Bras and Rodríguez-Iturbe (1976b) discuss the algorithm in some detail.

7.6.2 Case Study

The above theory was tested in a realistic but imaginary sample problem. The objective was to design a rain-gage network for estimating the area mean of a certain type of convective storm over a 360-mi^2 area. The area is a rectangle with sides of 24 mi and 15 mi.

The area was divided into a 3-by-3 grid pattern, as shown in Fig. 7.4 (corresponding identification numbers for each grid are given in the figure). All

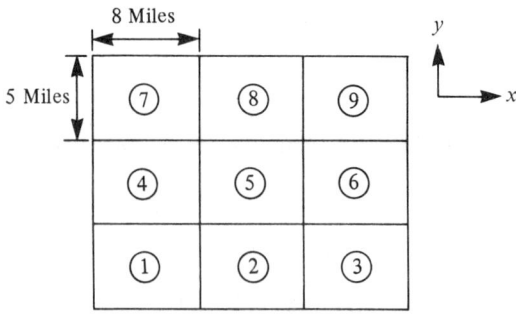

Total Area = 360 Sq. Miles

$\bigcirc$ = Identification number corresponding to discrete point defined as the center of the grid.

Figure 7.4 Diagram of test area showing discretization and other relevant information for a nine-point grid pattern (from Bras and Rodríguez-Iturbe, 1976b).

nine discrete locations are considered possible observation sites. The network optimization is performed with respect to a loosely defined objective function of the form given in Eq. (7.27). The trade-off coefficient C_Δ was not known accurately, so the generation of a transformation curve was necessary.

The process correlation function was assumed to be homogeneous and isotropic and given by

$$r(v) = bvK_1(bv), \qquad (7.45)$$

where v is the distance between any two points in space, K_1 the modified Bessel function of the second kind, and b the correlation function parameter.

The Bessel parameter b was given a value of 0.13. This value corresponds to that calculated by Rodríguez-Iturbe and Mejía (1974) using data from Fogel and Duckstein (1969) corresponding to convecting storms in Arizona with a maximum center depth of 5 in. and adequate for an area on the order of 400 mi^2. The point variance corresponding to the above data was 1.32 in.^{2} (Rodríguez-Iturbe and Mejía, 1974). This correlation structure and the parameter values were discussed in detail in Chapter 6. A diagonal error-covariance matrix is used.

The measurement error variance and data-collection costs associated with each discrete point are given in Table 7.1. The error variances were assigned so as to resemble variances obtained from data given by Morgan and Lourence (1969) for errors observed in readings of Fisher and Porter, Standard 8-in., and USSR-GG13000 rain gages, in comparison to lysimeter measurements. In doing these calculations to obtain the desired "ball-park" figures, it was assumed that the lysimeter readings represented the true mean value, that rainfall processes were stationary, and that rain gages behave similarly for different storms. The absolute values actually assigned to each point (variations around calculated variances) were chosen to simulate the fact that similar rain gages behave differently (have different errors) depending on location, altitude, wind conditions, etc. (Grayman and Eagleson, 1971; Larson and Peck, 1974; Sevruk, 1974).

The costs given vary around the values given by Grayman and Eagleson (1971) for telemetric rain-gage installations (around $2,000) and maintenance

Table 7.1
Measurement error variance and cost assigned to each grid in a nine-point discretization

	Grid point								
	1	*2*	*3*	*4*	*5*	*6*	*7*	*8*	*9*
Error variance $(10^{-3}, \text{in.}^2)$	5.0	5.0	7.0	4.0	5.5	7.0	4.0	5.0	5.5
Cost($)	4,000	2,500	1,500	4,000	4,000	2,500	4,100	2,500	4,000

(from \$1,000 to \$2,000 per year). Different cost values were assigned to represent different location conditions.

The first results to be presented were obtained by using the full expression for the measure of accuracy in the objective function. Table 7.2 summarizes the obtained results for values of C_Δ (the trade-off coefficient) of 10^{-5} and 10^{-6}. The table shows the breakdown of the measure of accuracy in the three types of mean square errors, namely, estimation, model, and cross-term errors.

From Table 7.2 you can observe differences in design for the two different trade-off coefficients. Different optimal designs are obtained in locating three stations and in locating seven stations. With $C_\Delta = 10^{-6}$, costs are weighted less, and the solution favors minimizing mean square error with the more expensive alternatives. Using the optimal designs given in Table 7.2, together with Fig. 7.4, it is clear the solutions are not obvious, but they are consistent and logical. For example, the solution for three stations with $C_\Delta = 10^{-6}$ is to place stations in grid points 4, 5, and 6. With $C_\Delta = 10^{-6}$, as was mentioned before, accuracy is given more attention and the above solution is the one that minimizes the distance between observed and unobserved locations, a solution which is consistent with the idea of maximizing accuracy (minimizing mean square error). Arguments such as the above are dangerous, though, in that they ignore instrument noise and costs that the solution takes into account.

Figures 7.5 and 7.6 show graphically part of the results in Table 7.2. Figure 7.5 is a plot of objective function value versus number of stations (it also gives optimal station location) for $C_\Delta = 10^{-5}$. For that trade-off coefficient a minimum objective function was achieved with four stations at locations 2, 4, 6, and 8.

In Figure 7.6, for $C_\Delta = 10^{-6}$, no minimum was achieved, since the plot of objective function versus number of stations was seemingly still slowly going down at the last point for eight stations. Such results imply that the user's discretization (in this case nine points) is not adequate. A finer discretization would allow more stations and would lead to an optimal solution in a plot of objective function versus number of stations.

Following the above idea, the area of interest was divided into an 18-point rectangular grid as shown in Fig. 7.7. Again dimensions and identification numbers are shown. The data of this new model correspond spatially to that of the nine-point grid as shown in Table 7.3.

Table 7.4 shows the network-design results with $C_\Delta = 10^{-6}$ and the 18-point–grid model. Again the total mean square error is broken down into its different components.

First note the similarity of solutions (same general pattern) between the 9-point and 18-point grid models. Variations occur only in the solutions for three and seven stations. Note too the similarity of the objective function values.

Figure 7.8 is a plot of the results given in Table 7.4, again objective function value versus number of stations. This time the curve corresponding to $C_\Delta = 10^{-6}$ achieves a minimum at seven stations.

Table 7.2
Full design results for a nine-point grid model

$C_\Delta = 10^{-5}$

	Number of stations							
	1	2	3	4	5	6	7	8
Objective function	0.210088	0.1648847	0.138452	0.137553	0.148998	0.184069	0.220427	0.256327
Cost ($)	4,000	6,500	7,500	11,000	12,500	16,500	20,500	24,600
Optimal design identification	5	4,6	4,3,8	2,4,6,8	2,3,4,6,8	2,3,4, 6,8,9	1,2,3,4, 6,8,9	1,2,3,5, 6,7,8,9
MSEE	0.1838550	0.0969951	0.0602530	0.0220466	0.0183342	0.0131858	0.0093826	0.0042084
Model	0.0065720	0.0065720	0.0065720	0.0065720	0.0065720	0.0065720	0.0065720	0.0065720
Cross term	−0.0103395	−0.0036824	−0.0033733	−0.0010658	−0.0009079	−0.0006890	−0.0005243	−0.0004530
Total*	0.1799880	0.0998847	0.0634516	0.0275528	0.0239982	0.0190687	0.0154266	0.0103274

$C_\Delta = 10^{-6}$

	1	2	3	4	5	6	7	8
Objective function	0.1740880	0.1063847	0.0592322	0.0385528	0.0364982	0.0355687	0.0354568	0.0349274
Cost ($)	4,000	6,500	10,500	11,000	12,500	16,500	20,600	24,600
Optimal design identification	5	4,6	4,5,6	2,4,6,8	2,3,4,6,8	2,3,4, 6,8,9	1,2,3,6, 7,8,9	1,2,3,5, 6,7,8,9
MSEE	0.1738550	0.0969951	0.0435945	0.0220466	0.0183342	0.0131858	0.0091497	0.0042084
Model	0.0065720	0.0065720	0.0065720	0.0065720	0.0065720	0.0065720	0.0065720	0.0065720
Cross term	−0.0103395	−0.0036824	−0.0014342	−0.0001066	−0.0009079	−0.0006891	−0.0008649	−0.0004530
Total*	0.1700880	0.0998847	0.0487322	0.0275528	0.0239982	0.0190687	0.0148568	0.0103274

* Total is the sum of the mean square error of estimation (*MSEE*), model, and cross-term values.

Source: Bras and Rodríguez-Iturbe (1976b).

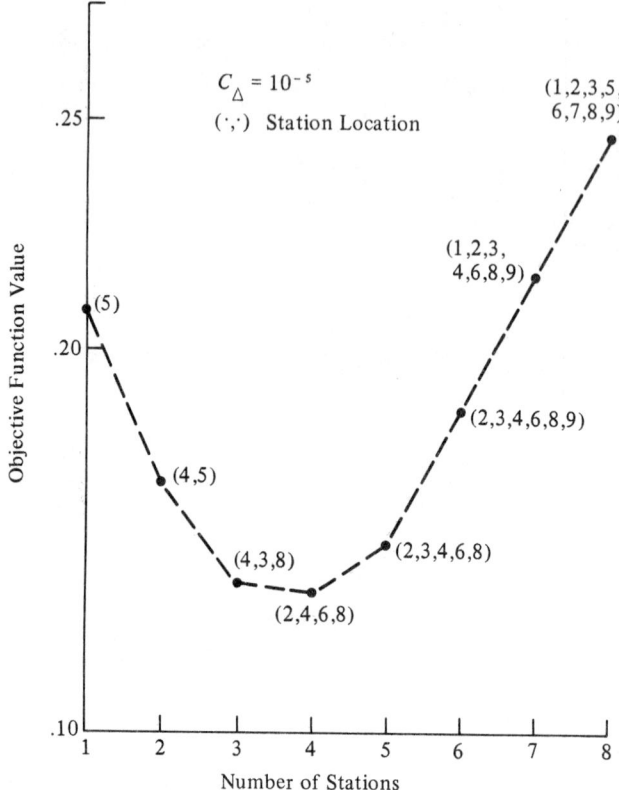

Figure 7.5 Optimal solutions for $C_\Delta = 10^{-5}$ using full measure of accuracy (from Bras and Rodríguez-Iturbe, 1976b).

The results given in Tables 7.2 and 7.4 give considerable insight with respect to the relative importance of the various terms in the total mean square error. For a covariance function of the Bessel type, the cross-term component of the mean square error is always negative. This implies a reduction in the sum of estimation and model mean square errors, which, in turn, implies that a straight sum of the latter two will produce some double counting of error. As is apparent from Eq. (7.44), the cross term depends strongly on the degree of discretization and the magnitudes of the event covariance matrix ψ and the error covariance matrix **R**. The results obtained show that indeed the cross term of the 18-point grid sample is consistently smaller than that of the 9-point grid sample. It can also be concluded that for the problem at hand this term is smaller than the estimation-error term, the difference being close to an order of magnitude.

The mean square model error behaves as expected; it is smaller for finer discretization and it is constant once the discretization is fixed. It remains

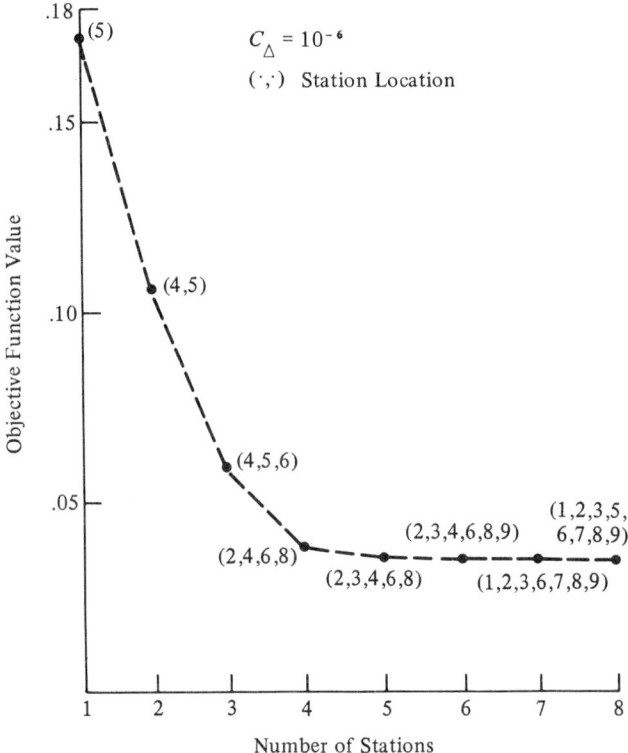

Figure 7.6 Optimal solutions for $C_\Delta = 10^{-6}$, using full measure of accuracy (from Bras and Rodríguez-Iturbe, 1976b).

smaller than the estimation-error term except for large numbers of stations when they become of the same order. In relation to the cross term, the model-error term is larger (in absolute value) except in the case of a very small number of stations. Actually, there is a region where the cross term and model term are of the same order and tend to cancel their effect on the total mean square error.

From the above comments, it can be concluded that the estimation term of the total mean square error dominates the design procedure for most cases of interest, i.e., cases where the number of stations to be allotted is small in relation to the area to be monitored. For a large number of stations, the cross term remains small, but the model term approaches or may exceed the value of the estimation term. Even in this case designing with only the estimation-error criteria may be sufficient if the grid pattern over the area is fixed (then the model error term is constant). The validity of this assumption not only reduces computation effort but eliminates any restrictions of isotropy or homogeneity

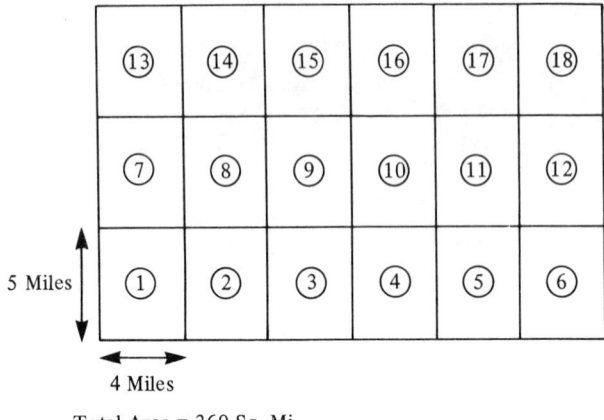

Total Area = 360 Sq. Mi.

◯ Identification number corresponding to discrete point defined as the center of the grid.

Figure 7.7 Diagram of test area showing discretization and other relevant information for an 18-point grid pattern (from Bras and Rodríguez-Iturbe, 1976b).

in the process covariance. As was said previously, the estimation error-term accepts any matrix-covariance formulation.

An alternative presentation of the results (versus Figs. 7.5 and 7.6) is a plot of mean square error of estimation versus cost. This corresponds to the previously mentioned transformation curve. Figure 7.9 is such a curve for the example problem at hand. It is constructed by plotting the optimal solutions obtained for various trade-off coefficients and each number of stations. Thus this curve eliminates nonoptimal points (points to the northeast in relation to others). It must be kept in mind that even though Figure 7.9 is referred to as a curve, it consists of discrete points. The decision maker with completely undefined utility function can obtain an approximate solution by entering Figure 7.9 with a budget or accuracy constraint. The solution will be approximate unless the entry corresponds to a plotted cost or accuracy measure. This is because the points in Fig. 7.9 are results from an unconstrained optimization procedure. If the user goes in with a cost or accuracy not corresponding to the given points, he can be sure that the true value of his unconstrained variable (accuracy or cost, respectively) is bounded within the values corresponding to the two closest points in the curve.

7.6.3 Conclusion

The network-design method previously presented depends on two basic inputs: the process-covariance matrix and the costs of observations. In normal situations, most input uncertainties are in the definition of the covariance structure.

Table 7.3
Measurement error variance and cost assigned to each grid in an 18-point discretization

	Grid point																	
	1	2	3	4	5	6	7	8	9	10	11	12	13	14	15	16	17	18
Error variance $(10^{-3}, \text{in.}^2)$	5.0	5.0	5.0	5.0	7.0	7.0	4.0	4.0	5.5	5.5	7.0	7.0	4.0	4.0	5.0	5.0	5.5	5.5
Cost ($)	4,000	4,000	2,500	2,500	1,500	1,500	4,000	4,000	4,000	4,000	2,500	2,500	4,100	4,100	2,500	2,500	4,000	4,000

Source: Bras and Rodriguez-Iturbe (1976).

Table 7.4
Full design results for an 18-point grid model

	Number of stations										
	1	2	3	4	7	8	9	10	12	14	16
Objective function	0.2193950	0.0773306	0.0590817	0.0429734	0.0335077	0.0343790	0.0348293	0.0347238	0.0388833	0.0447665	0.0518020
Cost ($)	4,000	6,500	10,600	11,000	20,500	24,600	26,100	28,100	33,100	41,100	49,200
Optimal design	9	8,11	2,11,14	3,7,11,15	3,5,7,9, 12,15,17	1,3,5,6,8, 13,15,17	1,3,5,6,8, 11,13,15,17	1,3,4,6,8,11, 13,15,16,18	1,3,4,5,6,8, 11,12,13,15, 16,18	1,3,4,5,6,8, 10,11,12,13, 15,16,17,18	1,2,3,4,5,6, 7,9,11,12,13, 14,15,16,17,18
MSEE	0.1816210	0.0699075	0.0477573	0.0309776	0.0117859	0.0069176	0.0074766	0.0053941	0.0040311	0.0023201	0.0012462
Model	0.0014417	0.0014417	0.0014417	0.0014417	0.0014417	0.0014417	0.0014417	0.0014417	0.0014417	0.0014417	0.0014417
Cross term	-0.0036677	-0.0005186	-0.0007173	-0.0004459	-0.0002199	-0.0003545	-0.0001890	-0.0002120	-0.0001895	-0.0000953	-0.0000859
Total*	0.1793950	0.0708306	0.0484817	0.0319734	0.0130077	0.0097790	0.0087293	0.0066238	0.0052833	0.0036655	0.0026020

$C_\Delta = 10^{-6}$

*Total is the sum of the mean square error of estimation (MSEE), model, and cross-term values.
Source: Bras and Rodriguez-Iturbe (1976).

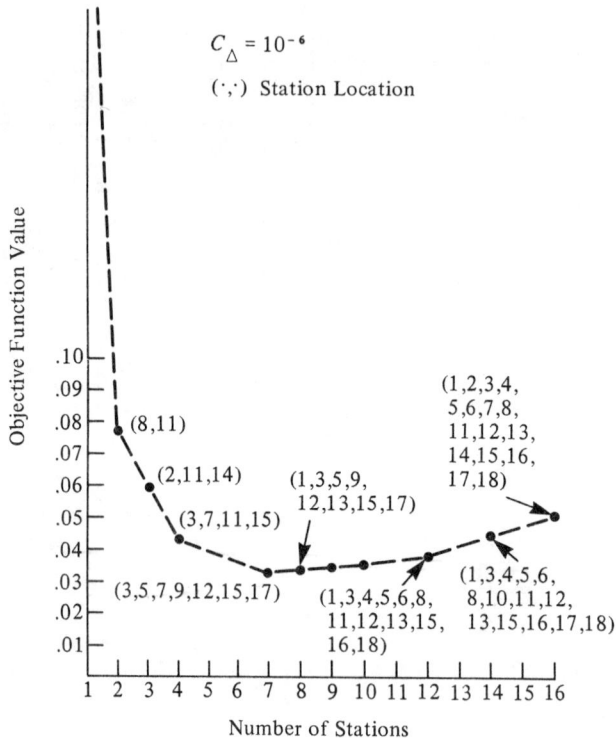

Figure 7.8 Optimal solutions for $C_\Delta = 10^{-6}$, using full measure of accuracy for an 18-point grid pattern (from Bras and Rodríguez-Iturbe, 1976b).

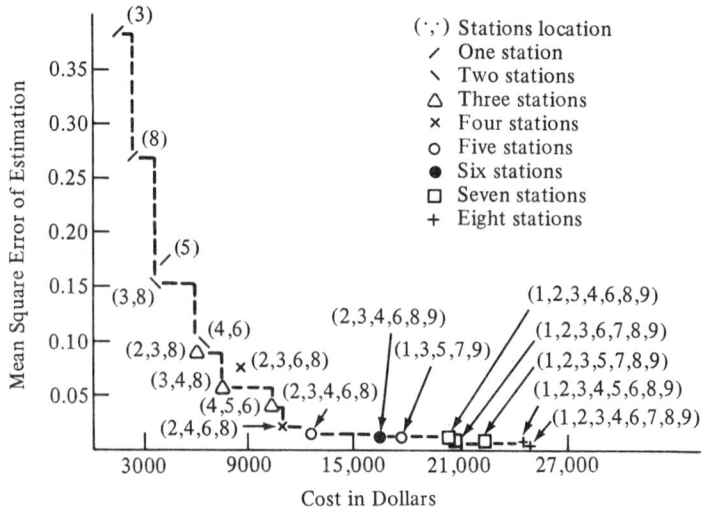

Figure 7.9 Transformation curve of the example problem (from Bras and Rodríguez-Iturbe, 1976b).

Bras and Rodríguez-Iturbe (1975) studied the sensitivity of the design method by varying the following: (1) the value of the point variance, (2) the covariance function parameter, and (3) the form of covariance function.

It was concluded that a reasonable amount of uncertainty in these factors can exist without seriously hampering network-design results.

The above technique compared favorably with other recently suggested network-design procedures. Again, the work of Bras and Rodríguez-Iturbe (1975) details a comparison with methods suggested by Rodríguez-Iturbe and Mejía (1974) and Lenton and Rodríguez-Iturbe (1974).

By combining Eqs. (7.26) and (7.34) (with $\mathbf{Z}_0 = \mathbf{0}$), the suggested mean area precipitation estimator is obtained:

$$\tilde{P} = \gamma^T \left[\Sigma \mathbf{H}^T \mathbf{R}^{-1} \mathbf{Z} + \Sigma \psi^{-1} \mathbf{M} \right], \tag{7.46}$$

which can also be expressed as

$$\tilde{P} = \gamma^T \left[\mathbf{M} + \psi \mathbf{H}^T (\mathbf{H} \psi \mathbf{H}^T + \mathbf{R})^{-1} (\mathbf{Z} - \mathbf{H} \mathbf{M}) \right], \tag{7.47}$$

where $\mathbf{M} = E[\mathbf{f}]$.

The reader may point out that the prior mean vector $\mathbf{M}$ is very difficult to obtain. Fortunately, for small $\mathbf{R}$ it can be shown that Eq. (7.47) reduces to

$$\tilde{P} = \gamma^T \left[\psi \mathbf{H}^T (\mathbf{H} \psi \mathbf{H}^T + \mathbf{R})^{-1} \mathbf{Z} \right] = \gamma^T \Sigma \mathbf{H}^T \mathbf{R}^{-1} \mathbf{Z}, \tag{7.48}$$

which is an estimator of the desired mean area precipitation. Equation (7.48) defines the weights to be assigned to each observation or each element of vector $\mathbf{Z}$.

7.7 OPTIMAL ESTIMATION USING KRIGING

Since late 1960 and early 1970, mining engineers, geologists, hydrologists, and geophysicists in general have frequently used a method of estimation of random fields called Kriging. As in Sections 7.1 through 7.5, the object here is to estimate values of a field (or linear functions of the field) at a point (or points) from a limited set of observed values. In fact, Chirlin and Wood (1982) show that basic Kriging is but a subset (no observation error) of the theory discussed in Sections 7.1 through 7.5. With judicious definition of the state vector $\mathbf{X}$, Kriging can be shown to be equivalent to Eq. (7.4).

Nevertheless, the following sections will develop Kriging along the lines originally set by one of its main expositors, G. Matheron (1971, 1973, 1976a, 1976b). This approach will emphasize the major extensions to the analysis of nonhomogeneous random fields.

Kriging has been extensively used in mining engineering and in hydrology. An extensive list of references is found at the end of the book. The presentation will follow the work of Chua and Bras (1980, 1982). As will be seen, Kriging can be used to estimate linear functions of random fields or point values. However, the derivations to follow will concentrate on obtaining the mean of a random variable over two-dimensional space. This is the area average of precipitation seen in Section 7.6.

7.7.1 Basic Assumptions and Definitions

The mean of the random field $Z(\mathbf{u})$ is defined as

$$m(\mathbf{u}) = E[Z(\mathbf{u})], \tag{7.49}$$

where $E[\;\cdot\;]$ denotes the expectation operator. Like $Z(\mathbf{u})$, $m(\mathbf{u})$ is defined over all points $\mathbf{u}$ in the area of interest A. The mean $m(\mathbf{u})$ is sometimes called the drift, although as will be seen, the notion of the drift is somewhat different when the concept of generalized covariances is considered in Section 7.10. The random field with the mean removed is termed the residual, defined as

$$Y(\mathbf{u}) = Z(\mathbf{u}) - m(\mathbf{u}). \tag{7.50}$$

The spatial dependence of the random process $Z(\mathbf{u})$ at any two points $\mathbf{u}_1$ and $\mathbf{u}_2$ can be modeled by either the covariance or the semivariogram. The covariance, $\mathrm{cov}(\mathbf{u}_1, \mathbf{u}_2)$ is defined as

$$\begin{aligned}
\mathrm{cov}(\mathbf{u}_1, \mathbf{u}_2) &= E\big[\{Z(\mathbf{u}_1) - m(\mathbf{u}_1)\}\{Z(\mathbf{u}_2) - m(\mathbf{u}_2)\}\big] \\
&= E[Y(\mathbf{u}_1)Y(\mathbf{u}_2)] \\
&= E[Z(\mathbf{u}_1)Z(\mathbf{u}_2)] - m(\mathbf{u}_1)m(\mathbf{u}_2).
\end{aligned} \tag{7.51}$$

The semivariogram, denoted $\gamma(\mathbf{u}_1, \mathbf{u}_2)$ is defined as

$$\begin{aligned}
\gamma(\mathbf{u}_1, \mathbf{u}_2) &= \frac{1}{2}\mathrm{var}[Z(\mathbf{u}_1) - Z(\mathbf{u}_2)] \\
&= \frac{1}{2}E\big[(Z(\mathbf{u}_1) - Z(\mathbf{u}_2))^2\big] - \frac{1}{2}\big[(m(\mathbf{u}_1) - m(\mathbf{u}_2))^2\big] \\
&= \frac{1}{2}E\big[(Y(\mathbf{u}_1) - Y(\mathbf{u}_2))^2\big].
\end{aligned} \tag{7.52}$$

As is evident from Eqs. (7.51) and (7.52), the covariance of $Z(\mathbf{u}_1)$ and $Z(\mathbf{u}_2)$ is the expected value of the product of the residuals, whereas the semivariogram is the expected value of the square of the difference of the residuals, halved.

The objective of Kriging is to find the best linear unbiased estimate of a linear function of field $Z(\mathbf{u})$. We will emphasize the estimate $\hat{P}$ of the true mean area precipitation P. The qualifiers of the estimate can be defined as

1. *Linearity*: the estimator $\hat{P}$ is formed from a linear combination of the observed values $Z(\mathbf{u}_i)$

$$\hat{P} = \sum_{i=1}^{n} \lambda_i Z(\mathbf{u}_i), \qquad (7.53)$$

where λ_i, $i = 1, \ldots, n$ are a set of weights to be optimized according to the criterion for "best" estimator.

2. *Unbiasedness*: This condition requires that the expected value of the estimator $\hat{P}$ be equal to the expected value of the true mean area precipitation P, i.e.,

$$E[\hat{P}] = E[P]. \qquad (7.54)$$

3. *Best criterion*: The estimator will be considered "best" if it gives the smallest estimation variance. The estimation variance, also called the mean square error is defined as:

$$\sigma_p^2 = E\left[(P - \hat{P})^2\right] = \text{var}[P - \hat{P}]. \qquad (7.55)$$

7.8 KRIGING OF STATIONARY RANDOM FIELDS

7.8.1 Second-Order Stationarity and the Intrinsic Hypothesis

As stated in Chapter 1, a random field is said to be second-order stationary if it satisfies the following conditions in its mean, variance, and covariance:

$$\textit{mean:} \quad E[Z(\mathbf{u})] = m(\mathbf{u}) = m, \text{ independent of } \mathbf{u} \qquad (7.56)$$

$$\textit{variance:} \quad \text{var}[Z(\mathbf{u})] = \sigma^2(\mathbf{u}) = \sigma^2, \text{ independent of } \mathbf{u} \qquad (7.57)$$

$$\textit{covariance:} \quad \text{cov}(\mathbf{u}_1, \mathbf{u}_2) = \text{cov}(\mathbf{u}_1 - \mathbf{u}_2)$$
$$= \text{cov}(\mathbf{v}), \qquad (7.58)$$

where $\mathbf{v} = \mathbf{u}_1 - \mathbf{u}_2$, i.e., the covariance between the process at points $\mathbf{u}_1$ and $\mathbf{u}_2$ is independent of the individual locations $\mathbf{u}_1$ and $\mathbf{u}_2$ and dependent only on their difference.

The random process $Z(\mathbf{u})$ is said to satisfy the intrinsic hypothesis if its first-order differences $Z(\mathbf{u}_1) - Z(\mathbf{u}_2)$ are stationary in the mean and variance:

$$\textit{mean:} \quad E[Z(\mathbf{u}_1) - Z(\mathbf{u}_2)] = m(\mathbf{v}) \qquad (7.59)$$

$$\textit{variance:} \quad \text{var}[Z(\mathbf{u}_1) - Z(\mathbf{u}_2)] = 2\gamma(\mathbf{v}). \qquad (7.60)$$

The mean and variance of the first-order difference $Z(\mathbf{u}_1) - Z(\mathbf{u}_2)$ are independent of the actual locations of $\mathbf{u}_1$ and $\mathbf{u}_2$ and dependent only on their vector difference $\mathbf{v}$. It is clear that the assumption made in Eq. (7.60) also defines the semivariogram, or more specifically, a stationary semivariogram. If the drift is constant, the semivariogram is defined directly by the difference of $Z(\mathbf{u}_1)$ and $Z(\mathbf{u}_2)$ (see Eq. 7.52).

Second-order stationarity automatically implies intrinsic properties, but not vice versa. Formally, the relationship can be written as

$$\text{var}\big[Z(\mathbf{u}_1) - Z(\mathbf{u}_2) \big] = \text{var}\big[Z(\mathbf{u}_2) \big] + \text{var}\big[Z(\mathbf{u}_1) \big] - 2\,\text{cov}(\mathbf{u}_1, \mathbf{u}_2). \tag{7.61}$$

If second-order stationarity holds,

$$\gamma(\mathbf{v}) = \frac{1}{2}\text{var}\big[Z(\mathbf{u}_1) - Z(\mathbf{u}_2) \big] = \text{cov}(0) - \text{cov}(\mathbf{v}). \tag{7.62}$$

Since $|\text{cov}(\mathbf{v})| \le \text{cov}(0)$ and $\text{cov}(0)$ is bounded, constant, and independent of $\mathbf{u}_1$ and $\mathbf{u}_2$, it is clear that $\gamma(\mathbf{v})$ is finite and dependent only on $\mathbf{v}$ when $Z(\mathbf{u})$ is second-order stationary. The intrinsic hypothesis on the other hand assumes that only first-order differences of $Z(\mathbf{u})$ have finite variances, whereas the variance of $Z(\mathbf{u})$ itself may well be infinite, in which case the process would not be stationary to the second order. In view of this, Eq. (7.61) is clearly not valid unless the variance of $Z(\mathbf{u})$ is finite. The intrinsic hypothesis is evidently a less restrictive assumption and was in fact developed to deal with infinite-variance processes. In addition, following Matheron (1971), several properties of linear combinations of the random function $Z(\mathbf{u})$, of the form

$$\omega = \sum_{\alpha=1}^{n} \lambda_\alpha Z(\mathbf{u}_\alpha) \tag{7.63}$$

are important depending on whether second-order stationarity or an intrinsic hypothesis is assumed. Linear combinations such as the above are those used in forming linear estimates of mean area precipitation.

In the case of second-order stationarity (the covariance function exists by definition) and from elementary statistics, the following relation is evident:

$$\text{var}[\omega] = \sum_{\alpha=1}^{n} \sum_{\beta=1}^{n} \lambda_\alpha \lambda_\beta \text{cov}(\mathbf{u}_\alpha - \mathbf{u}_\beta). \tag{7.64}$$

This, of course, also implies that the variance of the linear combinations is finite.

In the intrinsic case, the hypothesis is that only differences such as $Z(\mathbf{u}_\alpha) - Z(\mathbf{u}_\beta)$ of the random variable have a finite variance equal to the semivariogram. $\text{Var}[Z(\mathbf{u})]$ and $\text{cov}(\mathbf{u}_\alpha, \mathbf{u}_\beta)$ do not necessarily exist. However, if differences are considered, say with $Z(\mathbf{u}_0)$ as some arbitrary reference value, the following important relationship can be developed:

$$2\gamma(\mathbf{u}_\alpha - \mathbf{u}_\beta) = \mathrm{var}\big[Z(\mathbf{u}_\alpha) - Z(\mathbf{u}_0) - Z(\mathbf{u}_\beta) + Z(\mathbf{u}_0)\big]$$
$$= \mathrm{var}\big[Z(\mathbf{u}_\alpha) - Z(\mathbf{u}_0)\big] + \mathrm{var}\big[Z(\mathbf{u}_\beta) - Z(\mathbf{u}_0)\big]$$
$$- 2\,\mathrm{cov}\big[\{Z(\mathbf{u}_\alpha) - Z(\mathbf{u}_0)\}\{Z(\mathbf{u}_\beta) - Z(\mathbf{u}_0)\}\big]$$

$$2\gamma(\mathbf{u}_\alpha - \mathbf{u}_\beta) = 2\gamma(\mathbf{u}_\alpha - \mathbf{u}_0) + 2\gamma(\mathbf{u}_\beta - \mathbf{u}_0) - 2\,\mathrm{cov}(\mathbf{u}_\alpha - \mathbf{u}_0, \mathbf{u}_\beta - \mathbf{u}_0)$$

or

$$\mathrm{cov}(\mathbf{u}_\alpha - \mathbf{u}_0, \mathbf{u}_\beta - \mathbf{u}_0) = \gamma(\mathbf{u}_\alpha - \mathbf{u}_0) + \gamma(\mathbf{u}_\beta - \mathbf{u}_0) - \gamma(\mathbf{u}_\alpha - \mathbf{u}_\beta). \quad (7.65)$$

Since all the terms on the right-hand side of the above expression are finite (being variances of first-order differences), the covariance $\mathrm{cov}(\mathbf{u}_\alpha - \mathbf{u}_0, \mathbf{u}_\beta - \mathbf{u}_0)$ is finite.

Consider again the variance of linear combinations:

$$\mathrm{var}[\omega] = \mathrm{var}\left[\sum_{\alpha=1}^{n} \lambda_\alpha Z(\mathbf{u}_\alpha)\right].$$

If the condition $\sum_{\alpha=1}^{n}\lambda_\alpha = 0$ holds, the above can be rewritten:

$$\mathrm{var}[\omega] = \mathrm{var}\left[\sum_{\alpha=1}^{n} \lambda_\alpha \{Z(\mathbf{u}_\alpha) - Z(\mathbf{u}_0)\}\right],$$

and it follows that

$$\mathrm{var}[\omega] = \sum_{\alpha=1}^{n} \sum_{\beta=1}^{n} \lambda_\alpha \lambda_\beta \mathrm{cov}(\mathbf{u}_\alpha - \mathbf{u}_0, \mathbf{u}_\beta - \mathbf{u}_0). \quad (7.66)$$

As has been proven, the covariance, $\mathrm{cov}(\mathbf{u}_\alpha - \mathbf{u}_0, \mathbf{u}_\beta - \mathbf{u}_0)$ is finite. The above shows that in the case where the weights λ_α add up to zero, the variance of linear combinations in the form of Eq. (7.63) is finite. Following Matheron (1971), such linear combinations where $\sum_{\alpha=1}^{n}\lambda_\alpha = 0$ are termed authorized linear combinations and, under the intrinsic hypothesis assumption, only authorized linear combinations possess a finite variance. As will be seen, an extension of the idea of authorized linear combination leads to the concept of intrinsic random function of order k (IRF-k), which will be discussed in Section 7.10.

Another important property of authorized linear combinations must be considered. Using the relation in Eq. (7.65), Eq. (7.66) can be rewritten in terms of semivariograms as follows:

$$\text{var}[\omega] = \sum_{\alpha=1}^{n} \sum_{\beta=1}^{n} \lambda_\alpha \lambda_\beta \gamma(\mathbf{u}_\alpha - \mathbf{u}_0) + \sum_{\alpha=1}^{n} \sum_{\beta=1}^{n} \lambda_\alpha \lambda_\beta \gamma(\mathbf{u}_\beta - \mathbf{u}_0)$$
$$- \sum_{\alpha=1}^{n} \sum_{\beta=1}^{n} \lambda_\alpha \lambda_\beta \gamma(\mathbf{u}_\alpha - \mathbf{u}_\beta) \tag{7.67}$$

or

$$\text{var}[\omega] = \sum_{\alpha=1}^{n} \lambda_\alpha \gamma(\mathbf{u}_\alpha - \mathbf{u}_0) \sum_{\beta=1}^{n} \lambda_\beta + \sum_{\beta=1}^{n} \lambda_\beta \gamma(\mathbf{u}_\beta - \mathbf{u}_0) \sum_{\alpha=1}^{n} \lambda_\alpha$$
$$- \sum_{\alpha=1}^{n} \sum_{\beta=1}^{n} \lambda_\alpha \lambda_\beta \gamma(\mathbf{u}_\alpha - \mathbf{u}_\beta).$$

Since the λ_αs sum to zero, the only nonzero term is the third, leaving

$$\text{var}[\omega] = - \sum_{\alpha=1}^{n} \sum_{\beta=1}^{n} \lambda_\alpha \lambda_\beta \gamma(\mathbf{u}_\alpha - \mathbf{u}_\beta). \tag{7.68}$$

Comparing this with the corresponding expression in the second-order stationary case, it is clear that the covariance in Eq. (7.64) can simply be replaced with the negative of the semivariogram in the intrinsic case, provided that the condition $\sum_{\alpha=1}^{n} \lambda_\alpha = 0$ is satisfied in the latter case. This property will be important in the development of the Kriging equations.

An extension to continuous linear combinations of the form

$$\omega = \int_D \lambda(d\mathbf{u}) Z(\mathbf{u}), \tag{7.69}$$

where D is an arbitrary region of integration, gives the following analogous properties.

Second-order stationary case:

$$\text{var}[\omega] = \int_D \int_D \lambda(d\mathbf{u}_\alpha) \text{cov}(\mathbf{u}_\alpha - \mathbf{u}_\beta) \gamma(d\mathbf{u}_\beta) \tag{7.70}$$

Intrinsic case:

$$\text{var}[\omega] = - \int_D \int_D \lambda(d\mathbf{u}_\alpha) \gamma(\mathbf{u}_\alpha - \mathbf{u}_\beta) \gamma(d\mathbf{u}_\beta) \tag{7.71}$$

provided

$$\int_D \lambda(d\mathbf{u}) = 0. \tag{7.72}$$

7.8.2 Kriging with Covariances

This section develops the Kriging equations for the optimal weights, λ_i ($i = 1, 2, \ldots, n$) and the minimum mean square error, given that the covariance, cov($\mathbf{v}$), and a constant mean are known. This is the "no-drift" case.

The true, unknown, mean area precipitation is

$$P = \frac{1}{A} \int_A Z(\mathbf{u}) \, d\mathbf{u} \qquad (7.73)$$

and a linear estimate is formed; thus

$$\hat{P} = \sum_{i=1}^{n} \lambda_i Z(\mathbf{u}_i). \qquad (7.74)$$

The unbiasedness condition gives

$$E\left[\sum_{i=1}^{n} \lambda_i Z(\mathbf{u}_i) \right] = E\left[\frac{1}{A} \int_A Z(\mathbf{u}) \, d\mathbf{u} \right]. \qquad (7.75)$$

Linearity on both sides of Eq. (7.75) allows the following simplification,

$$\sum_{i=1}^{n} \lambda_i E[Z(\mathbf{u}_i)] = \frac{1}{A} \int_A E[Z(\mathbf{u})] \, d\mathbf{u}. \qquad (7.76)$$

Since the mean is a constant, the above can be rewritten as

$$m \sum_{i=1}^{n} \lambda_i = m \frac{1}{A} \int_A d\mathbf{u},$$

which reduces to the condition

$$\sum_{i=1}^{n} \lambda_i = 1. \qquad (7.77)$$

The mean square error of estimation is developed as follows:

$$\sigma_P^2 = E\left[(P - \hat{P})^2 \right]$$

$$= E[P^2] - 2E[P\hat{P}] + E[\hat{P}^2]$$

$$= E\left[\frac{1}{A^2} \int_A \int_A Z(\mathbf{u}_1) Z(\mathbf{u}_2) \, d\mathbf{u}_1 \, d\mathbf{u}_2 \right] - 2E\left[\frac{1}{A} \int_A \sum_{i=1}^{n} \lambda_i Z(\mathbf{u}) Z(\mathbf{u}_i) \, d\mathbf{u} \right]$$

$$+ E\left[\sum_{i=1}^{n} \sum_{j=1}^{n} \lambda_i \lambda_j Z(\mathbf{u}_i) Z(\mathbf{u}_j) \right].$$

Again, linearity leads to the following,

$$\sigma_P^2 = \frac{1}{A^2} \int_A \int_A E[Z(\mathbf{u}_1)Z(\mathbf{u}_2)] \, d\mathbf{u}_1 \, d\mathbf{u}_2$$

$$- \frac{2}{A} \int_A \sum_{i=1}^{n} \lambda_i E[Z(\mathbf{u})Z(\mathbf{u}_i)] \, d\mathbf{u}$$

$$+ \sum_{i=1}^{n} \sum_{j=1}^{n} \lambda_i \lambda_j E[Z(\mathbf{u}_i)Z(\mathbf{u}_j)]. \tag{7.78}$$

For a second-order stationary process,

$$\text{cov}(\mathbf{v}) = E[Z(\mathbf{u}_1)Z(\mathbf{u}_2)] - m^2 = \text{cov}(\mathbf{u}_1 - \mathbf{u}_2)$$

Hence, Eq. (7.78) can be rewritten as

$$\sigma_P^2 = \frac{1}{A^2} \int_A \int_A \text{cov}(\mathbf{u}_1 - \mathbf{u}_2) \, d\mathbf{u}_1 \, d\mathbf{u}_2$$

$$- \frac{2}{A} \int_A \sum_{i=1}^{n} \lambda_i \text{cov}(\mathbf{u} - \mathbf{u}_i) \, d\mathbf{u} + \sum_{i=1}^{n} \sum_{j=1}^{n} \lambda_i \lambda_j \text{cov}(\mathbf{u}_i - \mathbf{u}_j)$$

$$+ m^2 \left[\frac{1}{A^2} \int_A \int_A d\mathbf{u}_1 \, d\mathbf{u}_2 - 2 \sum_{i=1}^{n} \lambda_i \frac{1}{A} \int_A d\mathbf{u} + \sum_{i=1}^{n} \sum_{j=1}^{n} \lambda_i \lambda_j \right].$$

From Eq. (7.77), it is clear that in the above equation, the expression multiplying m^2 reduces to zero. Hence, the estimation variance equation is

$$\sigma_P^2 = \frac{1}{A^2} \int_A \int_A \text{cov}(\mathbf{u}_1 - \mathbf{u}_2) \, d\mathbf{u}_1 \, d\mathbf{u}_2 - \frac{2}{A} \int_A \sum_{i=1}^{n} \lambda_i \text{cov}(\mathbf{u} - \mathbf{u}_i) \, d\mathbf{u}$$

$$+ \sum_{i=1}^{n} \sum_{j=1}^{n} \lambda_i \lambda_j \text{cov}(\mathbf{u}_i - \mathbf{u}_j). \tag{7.79}$$

To find the set of weights that will give the minimum mean square error (MSE), the expression on the right of Eq. (7.79) must be minimized subject to the constraint that all the weights add to 1, Eq. (7.77). This is accomplished through the Lagrangian method of multipliers. Incorporating the constraint into an auxiliary function,

$$F = \sigma_P^2 + 2\mu \left(\sum_{i=1}^{n} \lambda_i - 1 \right)$$

$$= \frac{1}{A^2} \int_A \int_A \text{cov}(\mathbf{u}_1 - \mathbf{u}_2) \, d\mathbf{u}_1 \, d\mathbf{u}_2 - \frac{2}{A} \int_A \sum_{i=1}^{n} \lambda_i \text{cov}(\mathbf{u} - \mathbf{u}_i) \, d\mathbf{u}$$

$$+ \sum_{i=1}^{n} \sum_{j=1}^{n} \lambda_i \lambda_j \text{cov}(\mathbf{u}_i - \mathbf{u}_j) + 2\mu \left(\sum_{i=1}^{n} \lambda_i - 1 \right). \tag{7.80}$$

The above is then minimized with respect to λ_is and the multiplier μ, which implies solving the following set of equations,

$$\frac{\partial F}{\partial \lambda_i} = -\frac{2}{A} \int_A \text{cov}(\mathbf{u} - \mathbf{u}_i) \, d\mathbf{u} + 2 \sum_{j=1}^{n} \lambda_j \text{cov}(\mathbf{u}_i - \mathbf{u}_j) + 2\mu = 0 \qquad i = 1, \dots, n$$

$$\frac{\partial F}{\partial \mu} = \sum_{i=1}^{n} \lambda_i - 1 = 0$$

or, rewriting,

$$\sum_{j=1}^{n} \lambda_j \text{cov}(\mathbf{u}_i - \mathbf{u}_j) + \mu = \frac{1}{A} \int_A \text{cov}(\mathbf{u} - \mathbf{u}_i) \, d\mathbf{u} \qquad i = 1, 2, \dots, n.$$

$$\sum_{i=1}^{n} \lambda_i = 1 \tag{7.81}$$

This is a set of $n+1$ linear equations in the n weights λ_i, and the Lagrange multiplier μ. The solution gives the optimal weights, which will be denoted λ_i^*.

The minimum MSE, σ_P^2* can be derived as follows. If the first n equations in the set, Eq. (7.81), are each multiplied by λ_i^* and then summed, the following equation is obtained:

$$\sum_{i=1}^{n} \sum_{j=1}^{n} \lambda_i^* \lambda_j^* \text{cov}(\mathbf{u}_i - \mathbf{u}_j) + \mu \sum_{i=1}^{n} \lambda_i^* = \frac{1}{A} \sum_{i=1}^{n} \lambda_i^* \int_A \text{cov}(\mathbf{u} - \mathbf{u}_i) \, d\mathbf{u}. \tag{7.82}$$

The asterisks on the λ_is indicate that Eq. (7.82) applies only to the optimal weights. Substituting for the last term in the MSE equation (Eq. 7.79), using the expression above, gives

$$\sigma_P^2* = \frac{1}{A^2} \int_A \int_A \text{cov}(\mathbf{u}_1 - \mathbf{u}_2) \, d\mathbf{u}_1 \, d\mathbf{u}_2$$

$$- \frac{2}{A} \int_A \sum_{i=1}^{n} \lambda_i^* \text{cov}(\mathbf{u} - \mathbf{u}_i) \, d\mathbf{u}$$

$$+ \frac{1}{A} \int_A \sum_{i=1}^{n} \lambda_i^* \text{cov}(\mathbf{u} - \mathbf{u}_i) \, d\mathbf{u} - \mu \sum_{i=1}^{n} \lambda_i^*,$$

which reduces to the following expression for the minimum MSE,

$$\sigma_P^{2*} = \frac{1}{A^2} \int_A \int_A \text{cov}(\mathbf{u}_1 - \mathbf{u}_2) \, d\mathbf{u}_1 \, d\mathbf{u}_2$$
$$- \frac{1}{A} \int_A \sum_{i=1}^{n} \lambda_i^* \text{cov}(\mathbf{u} - \mathbf{u}_i) \, d\mathbf{u} - \mu. \tag{7.83}$$

Another general approach to developing the MSE equation (Eq. 7.79) is to consider the continuous weighting function,

$$\int_D \lambda(d\mathbf{u}) \equiv \sum_{i=1}^{n} \lambda_i \delta_{\mathbf{u}_i} - \int_A \lambda(d\mathbf{u}), \tag{7.84}$$

where

$$\delta_{\mathbf{u}_i} = \begin{cases} 1 & \mathbf{u} = \mathbf{u}_i \\ 0 & \text{otherwise}, \end{cases} \tag{7.85}$$

which can be thought of as comprising a continuous weighted area integral applied over all differential elements $d\mathbf{u}$ and a "moving window" of unknown weights λ_i applied over the data points $\mathbf{u}_i$. The estimation-error term can be written

$$\hat{P} - P = \sum_{i=1}^{n} \lambda_i Z(\mathbf{u}_i) - \frac{1}{A} \int_A Z(\mathbf{u}) \, d\mathbf{u}$$
$$= \int_A Z(\mathbf{u}) \lambda(d\mathbf{u}), \tag{7.86}$$

where in this case

$$\int_A \lambda(d\mathbf{u}) = \sum_{i=1}^{n} \lambda_i \delta_{\mathbf{u}_i} - \frac{1}{A} \int_A d\mathbf{u}. \tag{7.87}$$

The variance of Eq. (7.86) gives the desired MSE,

$$\text{var}[\hat{P} - P] = \sigma_P^2 = \text{var}\left[\int_A Z(\mathbf{u}) \lambda(d\mathbf{u})\right],$$

which by analogy to the discrete case can be shown to be

$$\text{var}[\hat{P} - P] = \int_A \int_A \lambda(d\mathbf{u}_1) \text{cov}(\mathbf{u}_1 - \mathbf{u}_2) \lambda(d\mathbf{u}_2)$$
$$= \left[\sum_{i=1}^{n} \lambda_i \delta_{\mathbf{u}_i} - \frac{1}{A} \int_A d\mathbf{u}_1\right] \text{cov}(\mathbf{u}_1 - \mathbf{u}_2) \left[\sum_{j=1}^{n} \lambda_j \delta_{\mathbf{u}_j} - \frac{1}{A} \int_A d\mathbf{u}_2\right]$$
$$= \frac{1}{A^2} \int_A \int_A \text{cov}(\mathbf{u}_1 - \mathbf{u}_2) \, d\mathbf{u}_1 \, d\mathbf{u}_2 - \frac{2}{A} \int_A \sum_{i=1}^{n} \lambda_i \text{cov}(\mathbf{u} - \mathbf{u}_i) \, d\mathbf{u}$$
$$+ \sum_{i=1}^{n} \sum_{j=1}^{n} \lambda_i \lambda_j \text{cov}(\mathbf{u}_i - \mathbf{u}_j). \tag{7.88}$$

Equations (7.88) and (7.79) are the same.

7.8.3 Kriging with Semivariograms

Again it is assumed that the semivariogram is known and the mean is stationary (no drift). The unbiasedness condition on the mean gives

$$E[P] = E[\hat{P}]$$

$$E\left[\frac{1}{A} \int_A Z(\mathbf{u}) \, d\mathbf{u}\right] = E\left[\sum_{i=1}^{n} \lambda_i Z(\mathbf{u}_i)\right].$$

Rewriting,

$$E\left[\sum_{i=1}^{n} \lambda_i Z(\mathbf{u}_i) - \frac{1}{A} \int_A Z(\mathbf{u}) \, d\mathbf{u}\right] = 0.$$

Using the continuous weighting function notation of Eq. (7.84), the above can be written,

$$E\left[\int_A Z(\mathbf{u}) \lambda(d\mathbf{u})\right] = 0$$

and

$$\int_A E[Z(\mathbf{u})] \lambda(d\mathbf{u}) = 0.$$

Since the mean is constant

$$m \int_A \lambda(d\mathbf{u}) = 0$$

and

$$\int_A \lambda(d\mathbf{u}) = 0. \tag{7.89}$$

Eq. (7.89) immediately implies that the "continuous linear combination"

$$\int_A Z(\mathbf{u}) \lambda(d\mathbf{u}) = \sum_{i=1}^{n} \lambda_i Z(\mathbf{u}_i) - \frac{1}{A} \int_A Z(\mathbf{u}) \, d\mathbf{u}$$

is an authorized linear combination. This authorized linear combination is the estimation error. The theory in Section 7.8.1 suggests that the variance of this authorized linear combination may be obtained by replacing the covariance functions in the MSE expression (Eq. 7.88 in the second-order stationary case) by the negative of the semivariogram. This leads to

$$
\sigma_P^2 = \text{var}\left[\int_A Z(\mathbf{u})\lambda(d\mathbf{u})\right]
$$

$$
= -\int_A\int_A \lambda(d\mathbf{u}_1)\gamma(\mathbf{u}_1-\mathbf{u}_2)\lambda(d\mathbf{u}_2)
$$

$$
= -\frac{1}{A^2}\int_A\int_A \gamma(\mathbf{u}_1-\mathbf{u}_2)\,d\mathbf{u}_1\,d\mathbf{u}_2
$$

$$
+\frac{2}{A}\int_A\sum_{i=1}^n \lambda_i\gamma(\mathbf{u}_i-\mathbf{u})\,d\mathbf{u}-\sum_{i=1}^n\sum_{j=1}^n \lambda_i\lambda_j\gamma(\mathbf{u}_i-\mathbf{u}_j).
$$

Again by analogy, the covariance functions in Eq. (7.81) can be replaced to obtain the Kriging equations in terms of semivariograms,

$$
\sum_{j=1}^n \lambda_j\gamma(\mathbf{u}_i-\mathbf{u}_j)+\mu' = \frac{1}{A}\int_A \gamma(\mathbf{u}-\mathbf{u}_i)\,d\mathbf{u} \qquad i=1,2,\ldots,n.
$$

$$
\sum_{i=1}^n \lambda_i = 1
$$

(7.90)

This is again a set of $n+1$ linear equations with solutions for the λs and μ'.

Finally, the minimum MSE is given by:

$$
\sigma_P^{2*} = -\frac{1}{A^2}\int_A\int_A \gamma(\mathbf{u}_1-\mathbf{u}_2)\,d\mathbf{u}_1\,d\mathbf{u}_2
$$

$$
+\frac{1}{A}\int_A\sum_{i=1}^n \lambda_i^*\gamma(\mathbf{u}-\mathbf{u}_i)\,d\mathbf{u}+\mu'.
$$

(7.91)

7.9 KRIGING OF NONSTATIONARY FIELDS: UNIVERSAL KRIGING

Nonstationarity is here limited to those cases where the mean cannot be assumed constant, and it becomes necessary to describe and account for the mean function $m(\mathbf{u})$ in some manner. An unknown nonstationary drift also implies that the covariogram, whether semivariogram or covariance, cannot be estimated, since the residuals are unknown. Hence, estimating the mean and covariogram are related processes. Universal Kriging proposes a particular functional form for modeling the mean and treats the related processes of mean and covariogram estimation:

1. as an iterative procedure estimating first one statistic, then the other, and
2. by invoking the theory of intrinsic random function of order k.

The first case will be treated in this section and the theory of IRF-k in Section 7.10.

Developed by Matheron and others at the Fontainebleau School, universal Kriging proposes that the mean (or drift) can be modeled as a linear combination of ν basic functions $f^\ell(\mathbf{u})$ as follows:

$$m(\mathbf{u}) = \sum_{\ell=0}^{\nu} a_\ell f^\ell(\mathbf{u}). \tag{7.92}$$

The basic functions are known functions, but the coefficients, a_ℓ, $\ell = 0,1,2,\ldots,\nu$ are unknown and have to be estimated. The iterative procedure involves finding the best unbiased linear estimate of the coefficients, a_ℓ, of the drift by first assuming some covariogram structure, then forming the residuals from the estimated drift in an attempt to reconstitute the original assumed covariogram. The solution is obtained when the initial assumed covariogram and the reconstituted covariogram agree. The individual cases of drift coefficient estimation assuming the existence of either the covariance or the semivariogram are now presented.

7.9.1 Drift Coefficient Estimation with Covariance

The estimator A_ℓ of the drift coefficient a_ℓ is taken as a linear combination of the n observations $Z(\mathbf{u}_i)$

$$A_\ell = \sum_{i=1}^{n} P_\ell^i Z(\mathbf{u}_i) \qquad \ell = 0,1,\ldots,\nu, \tag{7.93}$$

where P_ℓ^i, $i=1,2,\ldots,n$, is the set of weights corresponding to the ℓth drift coefficient A_ℓ. Once the A_ℓs are estimated, the estimated drift $M(\mathbf{u})$ is formed as:

$$M(\mathbf{u}) = \sum_{\ell=0}^{\nu} A_\ell f^\ell(\mathbf{u}). \tag{7.94}$$

Consider estimating the sth coefficient a_s. The unbiased condition requires that

$$E[A_s] = a_s.$$

Hence,

$$E\left[\sum_{i=1}^{n} P_s^i Z(\mathbf{u}_i)\right] = a_s$$

$$\sum_{i=1}^{n} P_s^i E[Z(\mathbf{u}_i)] = a_s.$$

By Eq. (7.92) the previous equation can be rewritten

$$\sum_{i=1}^{n} P_s^i \sum_{\ell=0}^{v} a_\ell f^\ell(\mathbf{u}_i) = a_s$$

or

$$\sum_{\ell=0}^{v} a_\ell \sum_{i=1}^{n} P_s^i f^\ell(\mathbf{u}_i) = a_s.$$

The above is a condition that must be true individually for each a_ℓ. Hence,

$$\sum_{i=1}^{n} P_s^i f^\ell(\mathbf{u}_i) = \begin{cases} 0 & \ell \neq s \\ 1 & \ell = s \end{cases}$$

or denoting

$$\delta(\ell, s) = \begin{cases} 0 & \ell \neq s \\ 1 & \ell = s \end{cases}$$

$$\sum_{i=1}^{n} P_s^i f^\ell(\mathbf{u}_i) = \delta(\ell, s) \qquad \ell = 0, 1, 2, \ldots, v. \tag{7.95}$$

Note that this is a set of $v + 1$ conditions.

It should be noted that the unbiased condition on the drift coefficients implies that the estimated drift is also unbiased since it is a linear combination of the estimated coefficients.

The estimation variance can be derived as

$$\begin{aligned}
\mathrm{var}[A_s] &= E\left[(A_s - a_s)^2\right] \\
&= E[A_s^2] - a_s^2 \\
&= E\left[\sum_{i=1}^{n} P_s^i Z(\mathbf{u}_i) \sum_{j=1}^{n} P_s^j Z(\mathbf{u}_j)\right] - a_s^2 \\
&= \sum_{i=1}^{n} \sum_{j=1}^{n} P_s^i P_s^j E\left[Z(\mathbf{u}_i) Z(\mathbf{u}_j)\right] - a_s^2 \\
&= \sum_{i=1}^{n} \sum_{j=1}^{n} P_s^i P_s^j \mathrm{cov}(\mathbf{u}_i - \mathbf{u}_j) + \sum_{i=1}^{n} \sum_{j=1}^{n} P_s^i P_s^j m(\mathbf{u}_i) m(\mathbf{u}_j) - a_s^2.
\end{aligned}$$

But $a_s = \sum_{i=1}^{n} P_s^i m(\mathbf{u}_i) = E[A_s]$, which is the unbiased condition, hence the variance equation reduces to

$$\mathrm{var}[A_s] = \sum_{i=1}^{n} \sum_{j=1}^{n} P_s^i P_s^j \mathrm{cov}(\mathbf{u}_i - \mathbf{u}_j). \tag{7.96}$$

Again, using the Lagrangian method of multipliers, the objective function,

$$\omega = \sum_{i=1}^{n} \sum_{j=1}^{n} P_s^i P_s^j \text{cov}(\mathbf{u}_i - \mathbf{u}_j) - 2 \sum_{\ell=0}^{\nu} \mu_s^\ell \sum_{i=1}^{n} \left\{ P_s^i f^\ell(\mathbf{u}_i) - \delta(\ell, s) \right\},$$

is minimized with respect to the n weights, P_s^i, and the Lagrangian multipliers, μ_s^ℓ, resulting in the following equations:

$$\sum_{j=1}^{n} P_s^j \text{cov}(\mathbf{u}_i - \mathbf{u}_j) - \sum_{\ell=0}^{\nu} \mu_s^\ell f^\ell(\mathbf{u}_i) = 0 \qquad i = 1, 2, \ldots, n$$

$$\sum_{i=1}^{n} P_s^i f^\ell(\mathbf{u}_i) = \delta(\ell, s) \qquad \ell = 0, 1, \ldots, \nu. \tag{7.97}$$

This is a set of $n + \nu + 1$ equations, which are solved to give the optimal weights P_s^{i*}. A unique solution exists as long as the functions $f^\ell(\mathbf{u}_i)$ are linearly independent.

To find the minimum MSE equation, the first n equations of the set in Eq. (7.97) are multiplied by P_s^i and added, giving

$$\sum_{j=1}^{n} \sum_{i=1}^{n} P_s^j P_s^i \text{cov}(\mathbf{u}_i - \mathbf{u}_j) = \sum_{\ell=0}^{\nu} \mu_s^\ell \sum_{i=1}^{n} P_s^i f^\ell(\mathbf{u}_i).$$

Also, the last $\nu + 1$ equations in Eq. (7.97) require that $\sum_{i=1}^{n} P_s^i f^\ell(\mathbf{u}_i)$ be nonzero only when $\ell = s$. Hence, the above expression reduces to

$$\sum_{j=1}^{n} \sum_{i=1}^{n} P_s^j P_s^i \text{cov}(\mathbf{u}_i - \mathbf{u}_j) = \mu_s^s. \tag{7.98}$$

From Eqs. (7.96) and (7.98) results

$$\text{var}[A_s] = \mu_s^s. \tag{7.99}$$

7.9.2 Drift Coefficient Estimation with Semivariogram

When just the semivariogram is assumed to exist, the drift can only be estimated up to a constant as will be evident later. Rewriting the definition given by Eq. (7.92) with $f^0(\mathbf{u})$ assumed to be equal to 1, the mean is now defined as

$$m(\mathbf{u}) = a_0 + \sum_{\ell=1}^{\nu} a_\ell f^\ell(\mathbf{u}). \tag{7.100}$$

Defining $m'(\mathbf{u}) = m(\mathbf{u}) - a_0$, then

$$m'(\mathbf{u}) = \sum_{\ell=1}^{\nu} a_\ell f^\ell(\mathbf{u}), \tag{7.101}$$

where $m'(\mathbf{u})$ is the "drift" less the constant a_0 that cannot be obtained when only the semivariogram exists.

The unbiased condition gives the same set of equations as in the case of the covariance, except that the equation with $\ell = 0$ is left out. Unbiasedness requires that

$$\sum_{i=1}^{n} P_s^i f^\ell(\mathbf{u}_i) = \delta(\ell, s) \qquad \ell = 1, 2, \ldots, \nu. \tag{7.102}$$

Consider the variance of the estimator A_s:

$$\mathrm{var}[A_s] = \mathrm{var}\left[\sum_{i=1}^{n} P_s^i Z(\mathbf{u}_i) \right]. \tag{7.103}$$

As is evident, this is the variance of a finite linear combination of the random variables $Z(\mathbf{u}_i)$. As the discussion in Section 7.8.1 showed, under the intrinsic hypothesis such a linear combination has a finite variance only when the sum of weights is zero. Hence, the condition

$$\sum_{i=1}^{n} P_s^i = 0 \tag{7.104}$$

must be invoked in this case.

In addition, since the variance of A_s is also the estimation variance, the MSE can now be simply written in terms of the semivariogram

$$\mathrm{var}[A_s] = - \sum_{i=1}^{n} \sum_{j=1}^{n} P_s^i P_s^j \gamma(\mathbf{u}_i - \mathbf{u}_j). \tag{7.105}$$

It should now be clear why a_0 cannot be estimated. The unbiased condition, setting $\ell = 0$ in Eq. (7.102) (and recalling that $f^0(\mathbf{u}) = 1$) would require that

$$\sum_{i=1}^{n} P_s^i = 1,$$

which contradicts the condition required for an authorized linear combination.

To find the optimal weights, the auxiliary function ω is formed as before,

$$\omega = -\sum_{i=1}^{n}\sum_{j=1}^{n} P_s^i P_s^j \gamma(\mathbf{u}_i - \mathbf{u}_j) - 2\mu_s^0 \sum_{i=1}^{n} P_s^i$$

$$-2\sum_{\ell=1}^{\nu}\mu_s^{\ell}\sum_{i=1}^{n}\left[P_s^i f^{\ell}(\mathbf{u}_i) - \delta(\ell,s)\right], \qquad (7.106)$$

and minimized with respect to the n values of P_s^i, μ_s^0, and the ν values of μ_s^{ℓ}s. The result is the following set of $n + \nu + 1$ equations:

$$\sum_{j=1}^{n} P_s^j \gamma(\mathbf{u}_i - \mathbf{u}_j) + \mu_s^0 + \sum_{\ell=1}^{\nu} \mu_s^{\ell} f^{\ell}(\mathbf{u}_i) = 0 \qquad i = 1, 2, \ldots, n$$

$$\sum_{i=1}^{n} P_s^i = 0$$

$$\sum_{j=1}^{n} P_s^j f^{\ell}(\mathbf{u}_j) = \delta(\ell, s) \qquad \ell = 1, 2, \ldots, \nu.$$

These equations are easily solved for the optimal weights P_s^{i*}. It can also be shown that the minimum estimation variance is given by

$$\text{var}[A_s] = \mu_s^s. \qquad (7.107)$$

After having found the ν sets of optimal weights P_s^{j*}, $s = 1, 2, \ldots, \nu$, and hence the ν coefficients A_s the estimated "drift" can be formed as

$$M'(\mathbf{u}) = \sum_{\ell=1}^{\nu} A_{\ell} f^{\ell}(\mathbf{u}).$$

This is of course the "drift" except for a constant. However, since here the purpose of the estimation of the drift coefficients is to estimate the semivariogram, the unknown constant a_0 disappears once first-order differences are taken. In other words, the true residuals cannot be found because of unknown constant a_0, but their difference can be estimated.

In the iterative universal Kriging procedure, an assumed semivariogram is used to estimate the drift coefficients from which a final variogram is computed. If "satisfactory" correspondence is obtained between the assumed and final variograms, the resultant variogram can then be used in the Kriging equations to estimate the mean area precipitation, or other linear functions of the field. These equations are derived below for the semivariogram case.

7.9.3 Kriging to Estimate Mean Area Precipitation When Semivariogram and Order of Polynomial Drift Are Known

The drift is modeled as

$$m(\mathbf{u}) = \sum_{\ell=0}^{\nu} a_\ell f^\ell(\mathbf{u})$$

$$f^0(\mathbf{u}) = 1.$$

The number ν of basic functions (assumed to be monomials) are known, but the coefficients a_ℓ need not be known in the following derivation.

Unbiasedness requires that

$$E[\hat{P} - P] = 0$$

$$E\left[\sum_i^n \lambda_i Z(\mathbf{u}_i) - \frac{1}{A}\int_A Z(\mathbf{u})\,d\mathbf{u}\right] = 0$$

$$\sum_i^n \lambda_i E[Z(\mathbf{u}_i)] - \frac{1}{A}\int_A E[Z(\mathbf{u})]\,d\mathbf{u} = 0$$

Substituting the definition of the mean,

$$\sum_{i=1}^n \lambda_i \sum_{\ell=0}^\nu a_\ell f^\ell(\mathbf{u}_i) - \frac{1}{A}\int_A \sum_{\ell=0}^\nu a_\ell f^\ell(\mathbf{u})\,d\mathbf{u} = 0$$

or, rewriting,

$$\sum_{\ell=0}^\nu a_\ell \left[\sum_{i=1}^n \lambda_i f^\ell(\mathbf{u}_i) - \frac{1}{A}\int_A f^\ell(\mathbf{u})\,d\mathbf{u}\right] = 0.$$

This unbiasedness condition must be satisfied for any value of the unknown coefficients a_ℓ. Thus, the unbiasedness condition is fully expressed in the following $\nu + 1$ equations:

$$\sum_{i=1}^n \lambda_i f^\ell(\mathbf{u}_i) - \frac{1}{A}\int_A f^\ell(\mathbf{u})\,d\mathbf{u} = 0 \qquad \ell = 0, 1, \ldots, \nu. \qquad (7.108)$$

In the particular case where $\ell = 0$,

$$\sum_{i=1}^n \lambda_i = 1,$$

since

$$f^0(\mathbf{u}) = 1.$$

Alternatively, rewriting in terms of the continuous weighting function,

$$\int_A \lambda(d\mathbf{u}) = \sum_{i=1}^{n} \lambda_i \delta_{\mathbf{u}_i} - \frac{1}{A}\int_A d\mathbf{u}$$

reduces the set of equations in Eq. (7.108) to

$$\sum_{i=1}^{n} \lambda_i f^\ell(\mathbf{u}_i) - \frac{1}{A}\int_A f^\ell(\mathbf{u})\, d\mathbf{u} = \int_A f^\ell(\mathbf{u})\lambda(d\mathbf{u}) = 0. \qquad (7.109)$$

In particular, $\ell = 0$ in Eq. (7.109) gives

$$\int_A \lambda(d\mathbf{u}) = 0.$$

The above condition again implies that

$$\int_A Z(\mathbf{u})\lambda(d\mathbf{u}) = \sum_{i=1}^{n} \lambda_i Z(\mathbf{u}_i) - \frac{1}{A}\int_A Z(\mathbf{u})\, d\mathbf{u}$$

is an authorized linear combination, which allows the variance to be written in terms of the semivariogram $\gamma(\mathbf{u}_1 - \mathbf{u}_2)$ as follows:

$$\begin{aligned}
\sigma_p^2 &= \text{var}\left[\sum_{i=1}^{n} \lambda_i Z(\mathbf{u}_i) - \frac{1}{A}\int_A Z(\mathbf{u})\, d\mathbf{u}\right] \\
&= \text{var}\left[\int_A Z(\mathbf{u})\lambda(d\mathbf{u})\right] \\
&= -\int_A \int_A \lambda(d\mathbf{u}_1)\gamma(\mathbf{u}_1 - \mathbf{u}_2)\lambda(d\mathbf{u}_2).
\end{aligned}$$

As in the stationary case, this can be expanded to give

$$\sigma_p^2 = -\frac{1}{A^2}\int_A\int_A \gamma(\mathbf{u}_1 - \mathbf{u}_2)\, d\mathbf{u}_1\, d\mathbf{u}_2 + \frac{2}{A}\int_A \sum_{i=1}^{n} \lambda_i \gamma(\mathbf{u}_i - \mathbf{u})\, d\mathbf{u}$$

$$- \sum_{i=1}^{n}\sum_{j=1}^{n} \lambda_i \lambda_j \gamma(\mathbf{u}_i - \mathbf{u}_j). \qquad (7.110)$$

The auxiliary function to be minimized is now

$$\omega = \sigma_P^2 - 2\sum_{\ell=0}^{\nu} \mu_\ell\left[\sum_{i=1}^{n} \lambda_i f^\ell(\mathbf{u}_i) - \frac{1}{A}\int_A f^\ell(\mathbf{u})\, d\mathbf{u}\right].$$

The resulting Kriging equations are

$$\sum_{j=1}^{n} \lambda_j \gamma(\mathbf{u}_i - \mathbf{u}_j) + \mu_0 + \sum_{\ell=1}^{\nu} \mu_\ell f^\ell(\mathbf{u}_i)$$

$$= \frac{1}{A} \int_A \gamma(\mathbf{u}_i - \mathbf{u}) \, d\mathbf{u} \qquad i = 1, 2, \ldots, n$$

$$\sum_{i=1}^{n} \lambda_i = 1 \qquad\qquad\qquad\qquad (7.111)$$

$$\sum_{i=1}^{n} \lambda_i f^\ell(\mathbf{u}_i) = \frac{1}{A} \int_A f^\ell(\mathbf{u}) \, d\mathbf{u} \qquad \ell = 1, 2, \ldots, \nu.$$

The solution of these $n + \nu + 1$ equations gives the n optimal weights λ_i^* and the $\nu + 1$ Lagrangian multipliers μ_ℓ.

The minimum MSE σ_P^{2*} comes from multiplying the first n equations in Eq. (7.111) by λ_i^* and then adding

$$\sum_{i=1}^{n} \sum_{j=1}^{n} \lambda_i^* \lambda_j^* \gamma(\mathbf{u}_i - \mathbf{u}_j) + \mu_0 + \sum_{\ell=1}^{\nu} \mu_\ell \sum_{i=1}^{n} \lambda_i^* f^\ell(\mathbf{u}_i) = \frac{1}{A} \int_A \sum_{i=1}^{n} \lambda_i^* \gamma(\mathbf{u}_i - \mathbf{u}) \, d\mathbf{u}.$$

Using the above expression to substitute for the term

$$\sum_{i=1}^{n} \sum_{j=1}^{n} \lambda_i \lambda_j \gamma(\mathbf{u}_i - \mathbf{u}_j)$$

in Eq. (7.110), gives the following result for the minimum MSE:

$$\sigma_P^{2*} = -\frac{1}{A^2} \int_A \int_A \gamma(\mathbf{u}_1 - \mathbf{u}_2) \, d\mathbf{u}_1 \, d\mathbf{u}_2 + \frac{1}{A} \int_A \sum_{i=1}^{n} \lambda_i^* \gamma(\mathbf{u}_i - \mathbf{u}) \, d\mathbf{u}$$

$$+ \frac{1}{A} \sum_{\ell=0}^{\nu} \mu_\ell \int_A f^\ell(\mathbf{u}) \, d\mathbf{u}. \qquad\qquad (7.112)$$

The case where the covariance is used is completely analogous. The semivariogram in Eqs. (7.111) and (7.112) is replaced by the negative of $\mathrm{cov}(\mathbf{u}_1 - \mathbf{u}_2)$.

7.9.4 Problems with the Iterative Algorithm

Many problems remain unresolved in the practical use of the algorithm to iteratively estimate the drift and the semivariogram. Questions arise as to what basic functions are to be used. A polynomial drift is often suggested, but the order k of the drift remains a question. As has been pointed out (Olea, 1975; David, 1977), there is no unique solution. Several combinations of drift and semivariograms may meet the requirements of the procedure. Furthermore, the criterion of a "satisfactory" agreement between initial and final variograms have generally been only vaguely alluded to by those who propose the iterative algorithm (David, 1977; Olea, 1975). There is no indication that the iteration converges to a solution. Applications of the algorithm in the literature are predictably rare. Olea (1975) gives a simple example in the case of point kriging.

Objections such as the above have led some investigators (Gambolati and Volpi, 1979a) to propose a priori known "drifts" based on known physical characteristics of the process instead of on exclusive reliance on statistical inference.

The development of the theory of IRF-k and the use of generalized covariances, which dispenses with the need to estimate the unknown drift coefficients have generally superseded the use of the iterative process described in this section.

7.10   INTRINSIC RANDOM FUNCTIONS OF ORDER k (IRF-k): KRIGING WITH GENERALIZED COVARIANCES

The theory of intrinsic random functions of order k (IRF-k) is an extension of the intrinsic hypothesis discussed in Section 7.8.1. Reasonably lucid presentations of the basic theory can be found in Delfiner (1976) and Davis and David (1978).

Matheron (1973) presents a rigorous mathematical treatment of the theory in terms of vector spaces. In this theory the "drift" will again be assumed to be of the form

$$m(\mathbf{u}) = \sum_{\ell=0}^{\nu} a_\ell f^\ell(\mathbf{u}), \tag{7.113}$$

where the $f^\ell(\mathbf{u})$ are generally assumed to be monomials, in which case $m(\mathbf{u})$ is modeled as a polynomial drift. The order of the drift will be denoted by k, while ν is the number of monomials. For example, in a plane defined by Cartesian coordinates x and y,

$$
\begin{array}{lll}
k = 0 & m(\mathbf{u}) = a_0 & \text{(constant drift, } \nu = 1) \\
k = 1 & m(\mathbf{u}) = a_0 + a_1 x + a_2 y & \text{(linear drift, } \nu = 3) \\
k = 2 & m(\mathbf{u}) = a_0 + a_1 x + a_2 y & \\
& \quad + a_3 x^2 + a_4 y^2 + a_5 xy & \text{(quadratic drift, } \nu = 6).
\end{array}
$$

Corresponding to the idea of authorized linear combinations discussed earlier, a linear combination

$$Z(\lambda) = \sum_{\alpha=1}^{n} \lambda_\alpha Z(\mathbf{u}_\alpha) \tag{7.114}$$

is said to be a generalized increment of order k, if the following conditions for the weights hold:

$$\sum_{\alpha=1}^{n} \lambda_\alpha f^\ell(\mathbf{u}_\alpha) = 0 \tag{7.115}$$

for all monomials of order $\ell \leq k$.

For example:

$$k = 0 \qquad \sum_{\alpha=1}^{n} \lambda_\alpha = 0$$

$$k = 1 \qquad \sum_{\alpha=1}^{n} \lambda_\alpha = 0$$

$$\sum_{\alpha=1}^{n} \lambda_\alpha x_\alpha = 0; \qquad \sum_{\alpha=1}^{n} \lambda_\alpha y_\alpha = 0$$

$$k = 2 \qquad \sum_{\alpha=1}^{n} \lambda_\alpha = 0; \qquad \sum_{\alpha=1}^{n} \lambda_\alpha x_\alpha = 0; \qquad \sum_{\alpha=1}^{n} \lambda_\alpha y_\alpha = 0$$

$$\sum_{\alpha=1}^{n} \lambda_\alpha x_\alpha^2 = 0; \qquad \sum_{\alpha=1}^{n} \lambda_\alpha y_\alpha^2 = 0; \qquad \sum_{\alpha=1}^{n} \lambda_\alpha x_\alpha y_\alpha = 0.$$

If the condition of Eq. (7.115) holds for some value of k, then the random function $Z(\mathbf{u})$ is said to be an intrinsic random function of order k (IRF-k). An important property of an IRF-k follows from this definition. Recall that in the case where only the semivariogram is assumed to exist (which, in IRF-k theory is the $k = 0$ case), the first coefficient of the drift, a_0, is undefined. Analogously and as pointed out by Delfiner (1976), working with generalized increments of order k implies dealing with a whole class of equivalent functions equal to $Z(\mathbf{u})$ up to a polynomial of a degree $\leq k$. It is this class of equivalent functions that is referred to as IRF-k.

Again in analogy to the semivariogram case, where the authorized linear combinations are said to have a finite variance when the condition $\sum_{\alpha=1}^{n} \lambda_\alpha = 0$ is satisfied, in the IRF-k case the variance of the kth-order generalized increments are assumed finite and equal to

$$\mathrm{var}\left[\sum_{\alpha=1}^{n} \lambda_\alpha Z(\mathbf{u}_\alpha) \right] = \sum_{\alpha=1}^{n} \sum_{\beta=1}^{n} \lambda_\alpha \lambda_\beta K(\mathbf{u}_\alpha - \mathbf{u}_\beta), \tag{7.116}$$

where $K(\mathbf{u}_\alpha - \mathbf{u}_\beta)$ is the generalized covariance of order k. Stationarity under translation is also assumed for the generalized covariance,

$$K(\mathbf{u}_\alpha - \mathbf{u}_\beta) \equiv K(\mathbf{v}),$$

where as before, $\mathbf{v} = \mathbf{u}_\alpha - \mathbf{u}_\beta$.

An extension to the continuous case can be made. If the condition

$$\int_D f^\ell(\mathbf{u})\lambda(d\mathbf{u}) = 0 \qquad (7.117)$$

holds for all $f^\ell(\mathbf{u})$ of order $\ell \leq k$, then

$$\int_D Z(\mathbf{u})\lambda(d\mathbf{u}) \qquad (7.118)$$

is a generalized increment of order k and $Z(\mathbf{u})$ is an IRF-k. In addition, the variance of the generalized increment can be written

$$\mathrm{var}\int_D Z(\mathbf{u})\lambda(d\mathbf{u}) = \int_D\int_D \lambda(d\mathbf{u}_\alpha)K(\mathbf{u}_\alpha - \mathbf{u}_\beta)\lambda(d\mathbf{u}_\beta), \qquad (7.119)$$

where $K(\mathbf{u}_\alpha - \mathbf{u}_\beta)$ is a stationary generalized covariance of order k as defined previously.

The techniques for estimating the mean area precipitation directly without estimating the drift coefficients will now be discussed. Assume for the moment that the generalized covariance is known. Referring to the case of Kriging with nonstationary polynomial drift in Section 7.9.3, it is clear that the unbiased condition, Eqs. (7.109) written as

$$\sum_{i=1}^n \lambda_i f^\ell(\mathbf{u}_i) - \frac{1}{A}\int_A f^\ell(\mathbf{u})\, d\mathbf{u} = 0 \qquad (7.120)$$

implies that the continuous "increment"

$$\int_A Z(\mathbf{u})\lambda(d\mathbf{u})$$

is a generalized increment of order k, where, as before,

$$\int_A \lambda(d\mathbf{u}) \equiv \sum_{i=1}^n \lambda_i \delta_{\mathbf{u}_i} - \frac{1}{A}\int_A d\mathbf{u}.$$

Equation (7.120) can be rewritten as

$$\int_A f^\ell(\mathbf{u})\lambda(d\mathbf{u}) = 0.$$

According to IRF-k theory, then, the variance of the continuous generalized increments can be written in terms of the generalized covariance of order k.

$$\text{var}\left[\int_A Z(\mathbf{u})\lambda(d\mathbf{u})\right] = \int_A\int_A \lambda(d\mathbf{u}_1)K(\mathbf{u}_1-\mathbf{u}_2)\lambda(d\mathbf{u}_2)$$

or

$$\text{var}\left[\sum_{i=1}^n \lambda_i Z(\mathbf{u}_i) - \frac{1}{A}\int_A Z(\mathbf{u})\,d\mathbf{u}\right] = \int_A\int_A \lambda(d\mathbf{u}_1)K(\mathbf{u}_1-\mathbf{u}_2)\lambda(d\mathbf{u}_2).$$

$$(7.121)$$

By writing out the weighting function $\int_A\lambda(d\mathbf{u})$ in full, Eq. (7.121) becomes

$$\sigma_p^2 = \frac{1}{A^2}\int_A\int_A K(\mathbf{u}_1-\mathbf{u}_2)\,d\mathbf{u}_1\,d\mathbf{u}_2 - \frac{2}{A}\int_A\sum_{i=1}^n \lambda_i K(\mathbf{u}_i-\mathbf{u})\,d\mathbf{u}$$

$$+ \sum_{i=1}^n\sum_{j=1}^n \lambda_i\lambda_j K(\mathbf{u}_i-\mathbf{u}_j). \qquad (7.122)$$

This equation is completely analogous to the stationary mean case with covariance, where the covariance is now replaced by the generalized covariance (hence the name). Comparing the stationary mean case with the semivariogram, it is clear that in the $k=0$ case, the generalized covariance reduces to the negative of the semivariogram. For $k=0$,

$$K(\mathbf{u}_1-\mathbf{u}_2) \equiv -\gamma(\mathbf{u}_1-\mathbf{u}_2).$$

As before, minimizing the mean-square-error expression subject to the unbiased constraint, Eq. (7.120) gives the Kriging equations with generalized covariance.

$$\sum_{j=1}^n \lambda_j K(\mathbf{u}_i-\mathbf{u}_j) - \mu_0 - \sum_{\ell=1}^\nu \mu_\ell f^\ell(\mathbf{u}_i) = \frac{1}{A}\int_A K(\mathbf{u}_i-\mathbf{u})\,d\mathbf{u}$$

$$i=1,2,\dots,n$$

$$\sum_{i=1}^n \lambda_i = 1$$

$$(7.123)$$

$$\sum_{i=1}^n \lambda_i f^\ell(\mathbf{u}_i) = \frac{1}{A}\int_A f^\ell(\mathbf{u})\,d\mathbf{u} \qquad \ell=1,2,\dots,\nu$$

with minimum variance given by

$$\sigma_p^{2*} = \frac{1}{A^2} \int_A \int_A K(\mathbf{u}_1 - \mathbf{u}_2) \, d\mathbf{u}_1 \, d\mathbf{u}_2 - \frac{1}{A} \sum_{i=1}^{n} \lambda_i^* \int_A K(\mathbf{u}_i - \mathbf{u}) \, d\mathbf{u}$$

$$+ \frac{1}{A} \sum_{\ell=0}^{v} \mu_\ell \int_A f^\ell(\mathbf{u}) \, d\mathbf{u}. \tag{7.124}$$

The above equations are analogous to Eqs. (7.111) and (7.112). In that case only the $k = 0$ condition, i.e., $\int_A \lambda(d\mathbf{u}) = 0$, was invoked to allow the use of the semivariogram, whereas in the development of IRF-k theory the full unbiased condition

$$\int_A f^\ell(\mathbf{u}) \lambda(d\mathbf{u}) = 0$$

for order $\ell \leq k$ was used to hypothesize the existence of the generalized covariance, $K(\mathbf{u}_\alpha - \mathbf{u}_\beta)$.

In the iterative case, the coefficients of the drift had to be estimated in order to compute the semivariogram of residuals. IRF-k theory allows the generalized covariance to be estimated directly. How this is accomplished will be briefly presented in the following section.

7.10.1   Structure Identification under IRF-k Assumptions

Matheron (1973) and Delfiner (1976) have suggested the following class of polynomials as useful, practical models of isotropic generalized covariance in $\mathcal{R}^n$

$$K(v) = \sum_{p=0}^{k} (-1)^{p+1} \frac{\alpha_p}{(2p+1)!} \frac{\Gamma\left(\frac{n}{2}\right) p!}{\sqrt{\pi}\, \Gamma\left(\frac{2p+n+1}{2}\right)} (v)^{2p+1} \tag{7.125}$$

provided the coefficients α_p satisfy the following conditions:

$$\sum_{p=0}^{k} \alpha_p x^{k-p} \geq 0 \qquad \forall x \geq 0. \tag{7.126}$$

Table 7.5, taken from Delfiner (1976) summarizes valid polynomial generalized covariances in the cases where $k \leq 2$, and the corresponding constraints on the coefficients, α_i. The possible presence of a nugget effect, $C\delta$, is taken into account in these models.

Having decided on the form of a model of the generalized covariance, the problem remains to determine the order k of the drift and the coefficients of

Table 7.5
Models of polynomial generalized covariances in $\mathcal{R}^2$

Drift	k	f^ℓ in $\mathcal{R}^2$	Models of generalized covariances
Constant	0	0	$K(v) = C\delta + \alpha_0|v|$
Linear	1	$1, x, y$	$K(v) = C\delta + \alpha_0|v| + \alpha_1|v|^3$
Quadratic	2	$1, x, y, x^2, y^2, xy$	$K(v) = C\delta + \alpha_0|v| + \alpha_1|v|^3 + \alpha_2|v|^5$
Constraints on the coefficients		$\alpha_0 \le 0 \qquad \alpha_2 \le 0$ $\alpha_1 \ge -\dfrac{10}{3}\sqrt{\alpha_0\alpha_2}$	$\delta = \begin{cases} 0 & v \ne 0 \\ 1 & v = 0 \end{cases}$

Source: Delfiner (1976). (Copyright © 1976 by D. Reidel Publishing Company, Dordrecht-Holland.)

the generalized covariance model. This process is termed "structure identification."

The point to note in these structure identification procedures under IRF-k assumptions is that only the order k of the drift needs to be determined, but not the drift coefficients themselves. Structure identification is generally performed in two distinct stages: (1) determination of the order k, and (2) estimation of the coefficients of the polynomial generalized covariance.

In stage one, a number of data values are eliminated and estimated from neighboring points, keeping everything the same except for the number of unbiasedness conditions imposed on the estimators (on the weights), which correspond to the assumed order k of the drift. The estimation procedure used at this stage is an ordinary least-squares procedure. Points are divided into an inner and outer ring, and one set is used to estimate the other. This is to compensate for possible symmetry in the configuration of points, which tends to filter polynomials and lead to underestimation of k (Delfiner, 1976). A ranking procedure based on the errors of estimation at each point is then used to pick the order k for which the estimator performs best.

For the second stage consider the notation,

$$Z(\lambda^i) = \sum_{\alpha=1}^n \lambda_\alpha^i Z_\alpha \tag{7.127}$$

$$\sum_{\alpha=0}^n \lambda_\alpha^i f^\ell(\mathbf{u}_\alpha) = 0 \qquad \ell = 0, 1, \ldots, \nu, \tag{7.128}$$

where $Z(\lambda^i)$ is a generalized increment of order k formed by a particular set of λ_α^is, $\alpha = 1, 2, \ldots, n$, satisfying the conditions in Eq. (7.128). Knowing the order k, it is possible to pick several sets of λ_α^is, $\alpha = 1, 2, \ldots, n$ to satisfy Eq. (7.128) and thus form the corresponding sets of $Z(\lambda^i)$.

For each $Z(\lambda^i)$

$$E\left[Z(\lambda^i)^2\right] = \sum_\alpha \sum_\beta \lambda^i_\alpha \lambda^i_\beta K(\mathbf{u}_\alpha - \mathbf{u}_\beta).$$

Given a polynomial generalized covariance of order k

$$K(v) = C\delta + \sum_{p=0}^{k} \alpha_p |v|^{2p+1} \qquad (7.129)$$

it is possible to write

$$E\left[Z(\lambda^i)^2\right] = C\sum_{\alpha=1}^{n} \left(\lambda^i_\alpha\right)^2 + \sum_{p=0}^{k} \alpha_p \sum_{\alpha=1}^{n} \sum_{\beta=1}^{n} \lambda^i_\alpha \lambda^i_\beta |\mathbf{u}_\alpha - \mathbf{u}_\beta|^{2p+1}.$$

If

$$T_i^0 = \sum_{\alpha=1}^{n} \left(\lambda^i_\alpha\right)^2$$

$$T_i^{2p+1} = \sum_{\alpha=1}^{n} \sum_{\beta=1}^{n} \lambda^i_\alpha \lambda^i_\beta |\mathbf{u}_\alpha - \mathbf{u}_\beta|^{2p+1} \qquad \text{for } p = 0,1,\ldots,k,$$

then

$$E\left[Z(\lambda^i)^2\right] = CT_i^0 + \sum_{p=0}^{k} \alpha_p T_i^{2p+1}. \qquad (7.130)$$

Equation (7.130) is a regression equation of $Z(\lambda^i)^2$ on the $k+2$ variables $T_i^0, T_i^1, \ldots, T_i^{2k+1}$. A weighted least-squares regression can now be performed where the weighted sum of squares to be minimized is expressed as

$$Q(C,\boldsymbol{\alpha}) = \sum_{i=1}^{r} \omega_i^2 \left[Z(\lambda^i)^2 - CT_i^0 - \sum_{p=0}^{k} \alpha_p T_i^{2p+1}\right]^2,$$

where the weights suggested are of the form

$$\omega_i^2 = \left(T_i^1\right)^{-2} \quad \text{or} \quad \omega_i^2 = \left(T_i^3\right)^{-2} \quad \text{(when } k > 0)$$

(see Delfiner, 1976, for justification).

The number of possible models of the generalized covariance in Eq. (7.129) that are fed into the regression equation (Eq. 7.130) is restricted by the order k of the drift already determined.

After the regression is performed those models whose coefficients meet the constraints in Table 7.5 are retained. To pick the best model, the criterion used is based on a modification of the residual sum of squares from the regression equations. (Again, refer to Delfiner, 1976, for details.)

A third step is recommended (Puente and Bras, 1982). Known points are now kriged with the generalized covariance determined in stage two, their errors compared with the theoretical standard deviations and the parameters of the generalized covariance readjusted if necessary.

Kitanidis (1983) has evaluated the above and other identification procedures, including a more versatile Maximum Likelihood Approach.

7.11 KRIGING TO ESTIMATE THE MEAN AREA PRECIPITATION IN MOUNTAINOUS REGIONS

Storms occurring over mountainous regions where orographic influences are known to contribute to complex spatial precipitation patterns are thought to be inadequately modeled with the "no-drift" assumption. This fact motivated Chua and Bras (1980, 1982) to apply nonstationary-drift techniques in the estimation of the mean area precipitation of storms occurring over terrains exhibiting large vertical elevation differences.

7.11.1 Case-Study Area

The data used was originally collected under the Colorado River Basin Pilot Project organized and sponsored by the U.S. Bureau of Reclamation and conducted in the San Juan Mountains in northwestern Colorado for five winter seasons from 1970–71 to 1974–75. Data from the months of October, November, and December in 1970 and 1973 are used here.

The area bounded by latitudes 37°21′N and 38°6′N and longitudes 108°W and 107°24′W was chosen for the study because it contained a suitably dense and well-distributed network of stations although it did not conform to any natural boundaries. Defining the point 37°21′N and 108°W as the origin of a cartesian coordinate system, the area and station locations can be illustrated as in Fig. 7.10. The total area was 1703 mi^2. Twenty-one stations were used for the 1970 data; they are listed in Table 7.6. Twenty-nine recording stations were in operation for the 1973 period and are listed in Table 7.7.

The original data consisted of hourly precipitation measurements at each gage. The total amounts accumulated hourly for the 24-hour period prior to 11:00 a.m. Mountain Standard Time (MST) of a particular day was recorded as the daily precipitation for the particular gage. Using the classification of the collecting agency, whereby each day was denoted as either a "no storm" day, a southwest (SW) storm day, general (G) storm day or northwest (NW) storm

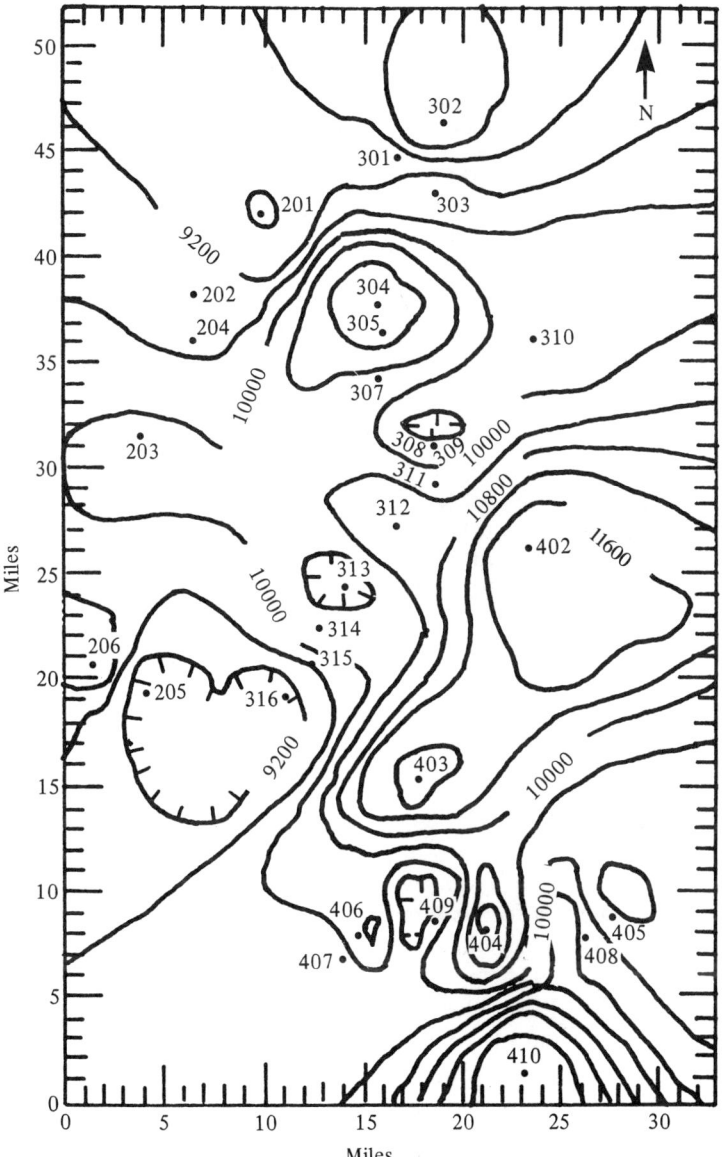

Figure 7.10 San Juan Mountains observation network during 1970 and 1973 together with elevation contours.

Table 7.6
Recording stations for San Juan Mountain storms: 1970 storms

Station number	Latitude (degrees)	Longitude (degrees)	X coordinate	Y coordinate	Elevation (ft)
201	37.95	107.82	9.86	41.50	8,756
202	37.90	107.88	6.57	38.04	9,455
204	37.88	107.88	6.58	36.66	8,842
206	37.65	107.97	1.65	20.75	10,300
301	38.00	107.70	16.43	44.96	9,000
302	38.02	107.67	18.07	46.35	7,740
303	37.97	107.67	18.07	42.89	9,590
304	37.90	107.72	15.34	38.04	11,020
307	37.85	107.72	15.35	34.59	10,070
309	37.80	107.67	18.09	31.13	9,322
310	37.88	107.57	23.56	36.66	9,880
312	37.75	107.70	16.45	27.67	10,880
316	37.63	107.80	10.98	19.37	8,780
402	37.73	107.58	23.04	26.28	11,900
403	37.57	107.68	17.57	15.22	11,700
405	37.48	107.50	27.47	8.99	10,760
406	37.47	107.73	14.84	8.30	10,660
407	37.45	107.75	13.74	6.92	9,510
408	37.47	107.53	25.82	8.30	9,060
409	37.48	107.67	18.13	8.99	9,000
410	37.37	107.58	23.09	1.38	7,650

Source: Chua and Bras (1980).

day, the daily amounts for consecutive days designated with the same storm type were added up to obtain total storm depth. (See Western Scientific Services, Inc., 1971, 1974, for details of classification.) In this manner, four storms were identified from the 1970 data and five storms from the 1973 data. These are listed in Table 7.8 together with their duration and the storm type.

Tables 7.6 and 7.7 also contain the elevations of the recording stations. It can be seen that the area under study lies within an elevation range from about 7,700 ft to 12,000 ft. Contours of elevation are also shown in Fig. 7.10.

7.11.2 Structure Identification: Estimating the Generalized Covariance of Order k

Despite the fact that the theory of IRF-k has been comprehensively developed, examples of structure identification procedures using IRF-k theory and Kriging using generalized covariances have been few. More generally, applications assuming nonstationary, unknown drifts have been rare.

Table 7.7
Recording stations for San Juan mountain storms: 1973 storms

Station number	Latitude (degrees)	Longitude (degrees)	X coordinate	Y coordinate	Elevation (ft)
201	37.95	107.82	9.86	41.50	8,756
202	37.90	107.88	6.57	38.04	9,455
203	37.80	107.93	3.84	31.13	10,040
205	37.63	107.93	3.84	19.37	8,800
206	37.65	107.97	1.65	20.75	10,300
301	38.00	107.70	16.43	44.96	9,000
302	38.02	107.67	18.07	46.35	7,740
303	37.97	107.67	18.07	42.89	9,590
304	37.90	107.72	15.34	38.04	11,020
305	37.88	107.72	15.34	36.66	11,150
307	37.85	107.72	15.35	34.59	10,070
308	37.80	107.67	18.09	31.13	9,280
309	37.80	107.67	18.09	31.13	9,322
310	37.88	107.57	23.56	36.66	9,880
311	37.77	107.67	18.09	29.05	10,440
312	37.75	107.70	16.45	27.67	10,880
313	37.70	107.75	13.71	24.21	9,800
314	37.67	107.77	12.62	22.13	10,600
315	37.65	107.78	12.07	20.75	9,680
316	37.63	107.80	10.98	19.37	8,780
402	37.73	107.58	23.04	26.28	11,900
403	37.57	107.68	17.57	15.22	11,700
404	37.47	107.62	20.88	8.30	11,500
405	37.48	107.50	27.47	8.99	10,760
406	37.47	107.73	14.82	8.30	10,660
407	37.45	107.75	13.74	6.92	9,510
408	37.47	107.53	25.82	8.30	9,060
409	37.48	107.67	18.13	8.99	9,000
410	37.37	107.58	23.04	1.38	7,650

Source: Chua and Bras (1980).

To recapitulate, the process of structure identification involves first the recognition of the order k of the drift and the estimation of the "best" polynomial generalized covariance corresponding to that order. Contrary to expectation, the order of the drift estimated for all the storms was 0 except for the storm denoted November 21. In other words, only the November 21 storm had a nonstationary drift of order one (linear drift), all other storms were stationary in the mean. Such a result was unexpected since it is generally thought that precipitation in mountainous regions and under the influence of various orographic effects would exhibit considerable spatial complexity, which would be reflected in a nonstationary drift. However, under the criteria and

Table 7.8

San Juan- mountain storms: Dates, duration, and storm type

Assigned name	Date 1st day of storm	Duration (no. of days)	Storm type*
6	Oct. 21, 1970	3	SW
9	Oct. 25, 1970	3	G
10	Nov. 1, 1970	1	NW
11	Nov. 7, 1970	2	SW
Nov. 14	Nov. 14, 1973	1	G
Nov. 19	Nov. 19, 1973	2	G
Nov. 21	Nov. 21, 1973	6	SW
Dec. 2	Dec. 2, 1973	1	SW
Dec. 3	Dec. 3, 1973	1	G

*SW denotes southwest, G general, and NW northwest.

Source: Chua and Bras (1982).

assumptions used in structure identification procedures, the point precipitation data of eight out of the nine San Juan storms do not show a nonstationary polynomial drift.

Of the corresponding generalized covariances identified in Table 7.9, all but one, that of order 1, show a nugget effect. Three storms yielded a pure nugget effect. That is, the variance of precipitation values between any two points is constant, irrespective of the particular two points chosen. The results

Table 7.9

Structure identification results: parameters of polynomial generalized covariance in $\mathscr{R}^2$

Storm no.	Order of drift (k)	$K(v) = C\delta + a_1 v + a_3 v^3 + a_5 v^5$				No. of data points
		C	a_1	a_3	a_5	
6	0	0.0351	-3.47×10^{-3}	NA*	NA	21
9	0	0.2210	-0.25×10^{-3}	NA	NA	21
10	0	0.2759	0.	NA	NA	21
11	0	0.0399	0.	NA	NA	21
Nov. 14	0	0.0023	-0.15×10^{-3}	NA	NA	29
Nov. 19	0	0.0430	-0.54×10^{-3}	NA	NA	29
Nov. 21	1	0.0	-68.16×10^{-3}	0.0	NA	29
Dec. 2	0	0.0023	0.065×10^{-3}	NA	NA	29
Dec. 3	0	0.0126	0.	NA	NA	29

*NA denotes not applicable.

Source: Chua and Bras (1982).

from using these generalized covariances to estimate mean area precipitation will be presented in a later section where they are compared with mean area precipitation estimates from the "detrending" procedures.

7.11.3 Detrending: Estimating the Drift from Elevation

The idea behind "detrending" is that data other than the observed values of the phenomenon under study can be related to and used to estimate the statistics of the spatial process. Switzer (1979) has suggested that elevation data should be used as a covariate together with the two-dimensional spatial coordinates to estimate both the drift and the semivariogram. The form of the semivariogram and drift he suggests are

$$\gamma(\mathbf{u}_1, \mathbf{u}_2) = \left[c_1 + c_2 |e(\mathbf{u}_1) - e(\mathbf{u}_2)| \right] \cdot |\mathbf{u}_1 - \mathbf{u}_2| \tag{7.131}$$

$$m(\mathbf{u}) = b_0 + b_1 e(\mathbf{u}), \tag{7.132}$$

where $e(\mathbf{u})$ is the elevation of point $\mathbf{u}$, c_1, c_2 the parameters of the semivariogram, and b_0, b_1 the parameters of drift. The linear relationship for the drift was postulated because it was the simplest, and the available studies on precipitation–elevation relationships did not seem to warrant anything more complex.

Precipitation can now be thought of as

$$Z(\mathbf{u}) = b_0 + b_1 e(\mathbf{u}) + Y(\mathbf{u}).$$

Table 7.10
"Detrending" results: Slope of trend and parameters of variogram (spherical) of residuals

		Parameters: Variogram of residuals		
Storm no.	**Slope of trend (in. per ft of elevation)**	*Nugget* c_0	*Sill* $-C_o$ c	*Range* a
6	13.5×10^{-5}	0.0	0.0696	20.25
9	13.8×10^{-5}	0.0	0.1922	13.50
10	14.1×10^{-5}	0.0	0.2294	13.50
11	5.3×10^{-5}	0.0	0.0323	9.00
Nov. 14	4.2×10^{-5}	0.0	0.0027	9.00
Nov. 19	7.9×10^{-5}	0.0020	0.0412	9.00
Nov. 21	18.6×10^{-5}	0.0	0.2189	15.00
Dec. 2	1.1×10^{-5}	0.0006	0.0032	16.50
Dec. 3	2.05×10^{-5}	0.0030	0.0158	13.50

Source: Chua and Bras (1982).

Ordinary least-squares regression of $Z(\mathbf{u})$ on the elevation $e(\mathbf{u})$ gives the parameters b_0 and b_1, and the residuals from the regression are the variables $Y(\mathbf{u})$. The set of residuals can then be used to estimate their underlying variogram. The value of the mean and the variogram thus estimated can now be introduced directly into the Kriging optimization equations.

The variograms for the resulting residuals for the nine storms available are shown in Table 7.10. The slope of the trend b_1 shows a precipitation increase of between 0.01 and 0.19 in. per 1000-ft rise in elevation. Variograms of residuals were fitted with a spherical model, defined as

$$\gamma(v) = \begin{cases} c_0 + c\left(1.5v/a - 0.5(v/a)^3\right) & v \le a \\ c_0 + c & v > a, \end{cases}$$

where c_0, c, and a are parameters. The nugget effect is c_0, $c_0 + c$ is called the sill, and a is the range of the variogram. Figures 7.11 and 7.12 show the results of the least-squares analysis and variogram fit for Storm 10, and Figs. 7.13 and 7.14 show the corresponding exercises for the storm on November 21.

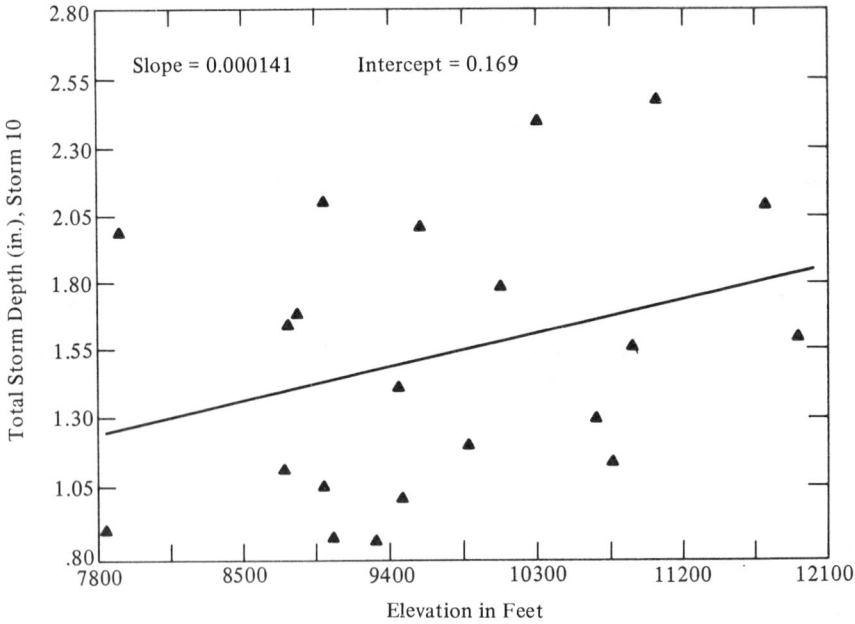

Figure 7.11 Ordinary least-squares fit of drift–elevation relationship—Storm 10 (from Chua and Bras, 1982).

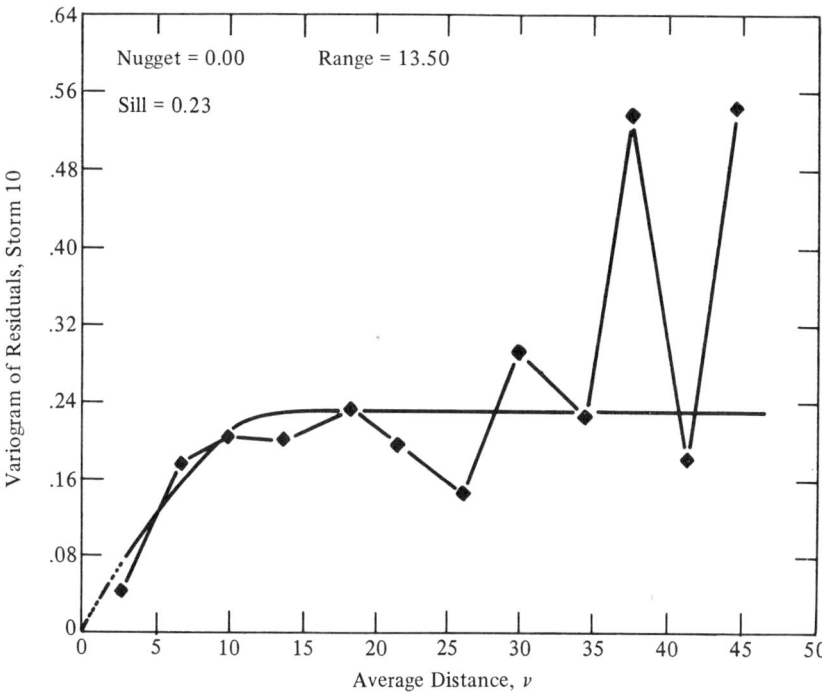

Figure 7.12   Variogram of residuals from "detrending"—Storm 10 (from Chua and Bras, 1982).

7.11.4   Mean Area Precipitation Estimates from Kriging under IRF-*k* Assumptions and under "Detrending" Assumptions

The results from "structure identification" under IRF-*k* assumptions and from "detrending" are now introduced into the appropriate Kriging optimization equation to estimate the mean area precipitation (MAP) of each of the nine storms.

Table 7.11 shows the MAP estimates and the standard deviation of the estimation error from the two Kriging procedures. The MAP values using the Thiessen polygon method have also been included for comparison.

It is clear from the table that the "detrending" method gives consistently higher MAP estimates than the estimates from Kriging with generalized covariances. The generalized covariance MAP estimates and the Thiessen polygon estimates, on the other hand, closely agree. These results are expected if you recall that in all but one case, the storm on November 21, structure identification using the generalized-covariance procedures implied stationarity in the mean. The "detrending" procedure, assuming the existence of a drift, would be expected to give a higher MAP estimate than the generalized covariance procedures, which identified no drift in all but one storm.

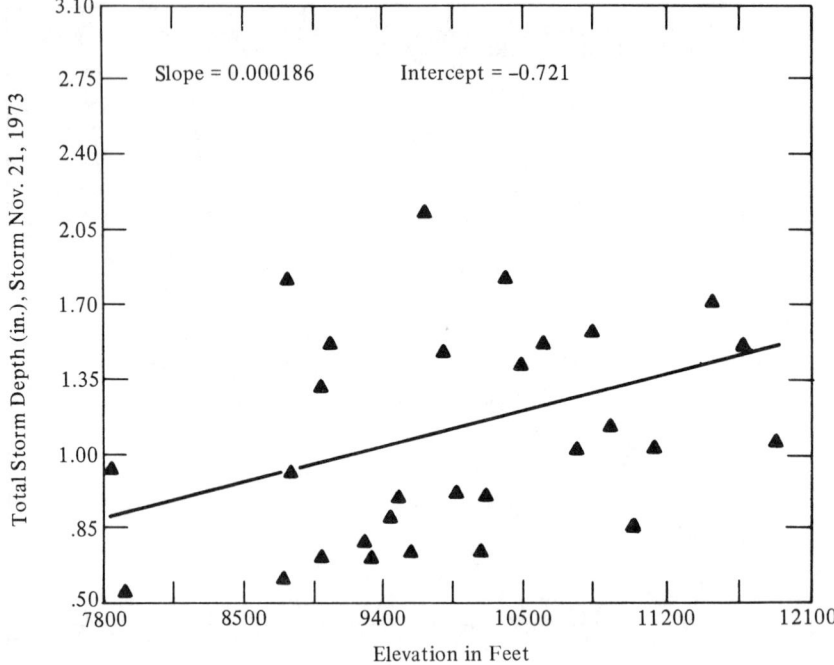

Figure 7.13 Ordinary least-squares fit of drift–elevation relationship—storm on November 21 (from Chua and Bras, 1982).

In terms of estimation errors, the "detrending" procedure again gives a larger value for the standard deviation than the generalized covariance results. In the case of the storm on November 21, where a linear drift was identified and a generalized covariance of order 1 was estimated, the "detrending" standard deviation value was smaller than the corresponding generalized covariance value.

None of the above comparisons, however, offer sufficient criteria to assess the validity and relative merits of the two estimation procedures. The question remains as to how well the MAP estimates from either procedure approximate the true unknown MAP of the process. In addition it remains to be determined whether the values of the estimation variance obtained adequately reflect the true uncertainty associated with each of the estimation procedures. To address these issues, the following experiment was performed.

The experiment consisted of Kriging each observation point using neighboring points and then using various measures of the difference between kriged and known values as criteria for assessing the validity of the assumed underlying drift and covariogram models. In this case, each observation from the set $Z(\mathbf{u}_i)$, $i = 1, 2, \ldots, n$ for each storm is removed and sequentially estimated

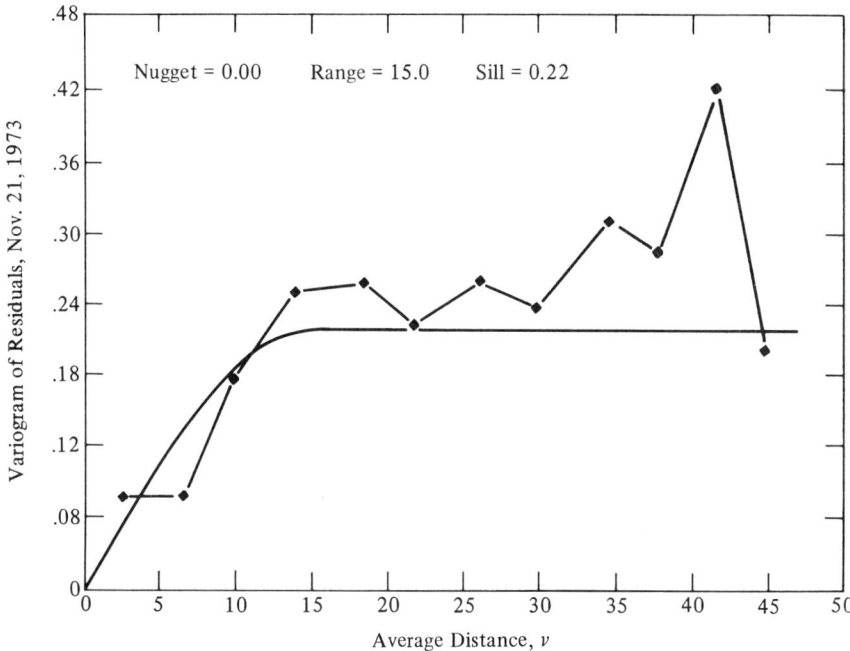

Figure 7.14 Variogram of residuals from "detrending"—storm on November 21 (from Chua and Bras, 1982).

Table 7.11
Mean area precipitation (MAP) estimate and standard deviation of estimation error from detrending and generalized covariance (G.C.)

Storm no.	P (MAP estimate)			σ_P	
	"Detrending"	*G.C.*	*Thiessen polygon*	*"Detrending"*	*G.C.*
6	1.17	1.04	1.03	0.088	0.058
9	1.65	1.48	1.47	0.182	0.103
10	1.69	1.53	1.52	0.198	0.115
11	0.74	0.70	0.68	0.088	0.044
Nov 14	0.11	0.08	0.07	0.011	0.013
Nov 19	0.55	0.51	0.46	0.045	0.042
Nov 21	1.17	0.97	0.96	0.093	0.160
Dec 2	0.076	0.063	0.063	0.013	0.011
Dec 3	0.37	0.35	0.34	0.029	0.021

Source: Chua and Bras (1982).

(using Kriging) from the $n-1$ observations using:

1. the generalized covariance of order k estimated beforehand for each storm, and
2. the estimated trend and variogram of residuals computed from the "detrending" procedure.

In both cases above the difference between the kriged and known point value ε_i and the corresponding theoretical Kriging estimation variance σ^2 are computed for each point. Unbiasedness of the estimator would suggest

$$\frac{1}{n} \sum_{i=1}^{n} \varepsilon_i \simeq 0.$$

The mean square error obviously is a measure of the accuracy of the estimation procedure. If the value

$$\frac{1}{n} \sum_{i=1}^{n} \varepsilon_i^2 \simeq 0$$

is small in the case of estimating point values, it might be expected that Kriging with the same assumptions of drift and variogram will also be accurate for MAP (and other linear functions of the random process). Finally, the standardized mean square error is a measure of the consistency of the standard deviation σ_i and the corresponding error ε_i. If σ_i truly reflects the estimation error from Kriging with the particular assumed drift and variogram models, then it should be expected that

$$\frac{1}{n} \sum_{i=1}^{n} \frac{\varepsilon_i^2}{\sigma_i^2} \simeq 1.$$

These three measures have been used by Gambolati and Volpi (1979b) as criteria to adjust the parameters of the variogram of residuals to obtain the "best" parameters and as an a posteriori validation test for the variogram models chosen.

The values computed in Table 7.12 for a mean error, mean square error, and standardized mean square error (MSE) can now be used to compare the relative merits of the two estimation procedures. The values of the mean error show that neither the "detrending" nor generalized covariance procedures appear to systematically overestimate or underestimate the point values. The largest absolute mean error value occurs for the storm on November 21 with the "detrending" method giving a value of 0.0773. All other values are in the range -0.05 to 0.05.

Table 7.12
Point Kriging test: "Detrending" vs. generalized covariance (G.C.)

Storm no.	n	Mean error $\frac{1}{n}\sum_{i}^{n}\varepsilon_i$		MSE $\frac{1}{n}\sum_{i}^{n}\varepsilon_i^2$		Standardized MSE $\frac{1}{n}\sum_{i}^{n}\frac{\varepsilon_i^2}{\sigma_i^2}$		Order k of G.C.
		Detrending	*G.C.*	*Detrending*	*G.C.*	*Detrending*	*G.C.*	
6	20	0.0241	−0.0410	0.0414	0.0608	1.972	1.112	0
9	21	−0.0071	0.0104	0.1131	0.2160	1.008	0.981	0
10	21	−0.0048	0.0138	0.1304	0.2470	1.049	0.911	0
11	21	−0.0005	0.0187	0.0259	0.0389	0.928	0.992	0
Nov. 14	28	0.0053	0.0013	0.0023	0.0029	1.579	0.934	0
Nov. 19	29	0.0164	−0.0099	0.0247	0.0493	0.898	1.066	0
Nov. 21	28	0.0773	0.0009	0.1360	0.1166	1.702	0.605	1
Dec. 2	29	0.0004	−0.0039	0.0021	0.0287	1.046	1.033	0
Dec. 3	29	0.0045	0.0131	0.0080	0.0139	0.683	1.118	0

Source: Chua and Bras (1982).

The mean-square-error results show that for all storms, except that of November 21, Kriging with the drift and variogram of residuals from detrending gave more accurate point estimates on the average than Kriging with generalized covariances. It may then be inferred that the MAP estimates from "detrending" should be more accurate than the estimates from the generalized-covariance procedures. The exception in this case is the storm on November 21, the only linear (and nonzero)-drift storm identified by the structure-identification procedures. This indicates that for this storm, modeling the mean as a linear drift gives more accurate results than modeling the trend as a linear function of elevation.

The final comparison of the standardized mean square error shows that in general the use of generalized covariance gives values closer to 1 than the results from "detrending." In three cases, Storms 6, November 14, and November 21, the "detrending" values were greater than 1.5 indicating an apparent divergence between estimated and actual error statistics. In the particular case of the storm on November 21 (the case with the generalized covariance of order 1) the standardized-mean-square-error values of 1.702 for "detrending" and 0.605 for generalized covariance both reflect poor agreement between Kriging-estimation variance and the actual kriged point errors. The above results indicate that although the "detrending" procedure may give more accurate point (and MAP) estimates for all the storms, the estimation variance that results from the Kriging equations in some cases, may in fact not reflect the true errors in the estimation procedure. The values of standardized mean square error for the "detrending" procedure are in fact greater than 1 for most of the storms, indicating an underestimation of the actual estimation-error variance. This is not unexpected considering that the procedure is ignoring errors in the assumed trend.

The generalized covariance of order k, on the other hand, seems more consistent in its ability to represent the Kriging-estimation variance. It is clear that the MAP estimate from Kriging (whether with generalized covariances or with some a priori postulated mean function such as "detrending") will be in error due to

1. incomplete observation of the process at only a finite number of points, and
2. inadequate modeling of the drift and spatial correlation of the process.

It is not clear that the value of the estimation variance that is computed by the Kriging system adequately reflects the combination of the above two types of errors. The Kriging-estimation–variance value is directly a function of whatever model of drift and variogram is adopted, which itself may be in error. This issue is of utmost importance in the interpretation of Kriging results, but has received scant attention in applications of Kriging thus far. Chua and Bras (1980) have presented some preliminary investigations into the effect of "model" errors on the Kriging-estimation variance.

To summarize the results of this comparison, it can be concluded from the point Kriging tests that both the procedures of Kriging with "detrending" and Kriging with polynomial generalized covariance have the capability of providing reasonable estimates of point (and MAP) values with no systematic bias. The "detrending" method could well be more accurate in terms of the ability to estimate the known data points and inferring from this it should perform better in estimating MAP than the generalized covariance procedure. In terms of providing an accurate value of the true variance of estimation error, the generalized covariance procedure performs consistently better, except in the case of an order-1 (linear) polynomial drift. The "detrending" procedure tends to underestimate the Kriging-estimation-error variance despite performing better in terms of estimating the kriged value.

CHAPTER 8

Estimation of Dynamic Hydrologic Systems

8.1 THE STATE–SPACE REPRESENTATION OF A STOCHASTIC LINEAR DYNAMIC SYSTEM

Many physical and geophysical systems can be represented by linear differential equations of the form

$$\frac{d^n X(t)}{dt^n} + a_{n-1}(t)\frac{d^{n-1} X(t)}{dt^{n-1}} + \cdots + a_1(t)\frac{dX(t)}{dt} + a_0(t)X(t) = L(t)U(t).$$

$$(8.1)$$

In hydrology, Eq. (8.1) represents a generalized linear, lumped rainfall–runoff model, where X is the basin storage and $U(t)$ is a time-varying input. Commonly, the hydrologic representation is further simplified by making it time invariant; the coefficients $a_0 \ldots a_n$ are assumed constant. In such cases, the solution to Eq. (8.1) is of the form discussed in Chapter 4—a time-invariant response function or instantaneous unit hydrograph. Such time-invariant formulations are easily studied in the frequency domain. This chapter will not be limited to time-invariant systems or to deterministic systems; it will illustrate methods to analyze the stochastic analogy to Eq. (8.1). The solutions presented will effectively solve, for particular cases, the filtering forecasting cycle discussed in Chapter 1. Examples will illustrate applications in rainfall network design and flood forecasting.

The basin storage (X in Eq. 8.1) embodies all the information required to define the development of the basin when the system is unforced, $U(t) = 0$. A variable with such characteristics is called a state variable. Storage is the state of the lumped basin, because its value and that of its derivatives defines the condition of the system at one time.

From now on all system equations will be converted to the form of a multivariate first-order differential equation. This canonical form can be achieved for any linear system by using the concept of state augmentation. Define a state vector

$$\mathbf{X}(t) = \begin{bmatrix} X_1(t) \\ X_2(t) \\ \vdots \\ X_n(t) \end{bmatrix}, \tag{8.2}$$

where

$$X_1(t) = X(t); \quad X_2(t) = \dot{X}_1(t); \quad X_3(t) = \dot{X}_2(t);$$
$$X_4(t) = \dot{X}_3(t); \quad X_5(t) = \dot{X}_4(t); \dots; \quad X_n(t) = \dot{X}_{n-1}(t).$$

(A dot over the variable implies a time derivative.)

Equation (8.1) can be then written in matrix form as

$$\begin{bmatrix} \dot{X}_1(t) \\ \dot{X}_2(t) \\ \vdots \\ \dot{X}_{n-1}(t) \\ \dot{X}_n(t) \end{bmatrix} = \begin{bmatrix} 0 & 1 & 0 & \cdots & 0 & 0 \\ 0 & 0 & 1 & \cdots & 0 & 0 \\ & \cdots & \cdots & \cdots & & \\ & \cdots & \cdots & \cdots & & \\ 0 & 0 & 0 & \cdots & 0 & 1 \\ -a_0(t) & & \cdots & & -a_{n-1}(t) \end{bmatrix} \begin{bmatrix} X_1(t) \\ X_2(t) \\ \vdots \\ X_{n-1}(t) \\ X_n(t) \end{bmatrix} + \begin{bmatrix} 0 \\ 0 \\ \vdots \\ 0 \\ L(t) \end{bmatrix} U(t) \tag{8.3}$$

or

$$\dot{\mathbf{X}}(t) = \mathbf{F}(t)\mathbf{X}(t) + \mathbf{L}(t)U(t). \tag{8.4}$$

Note that $\mathbf{F}(t)$ is an $n \times n$ matrix, $\mathbf{X}(t)$ is an $n \times 1$ vector, but $U(t)$ is a scalar. Correct input dimensions are achieved by multiplying by the $n \times 1$ vector $\mathbf{L}(t)$.

Equation (8.4) is a deterministic expression of a physical system. As it stands, its solution is easily obtained, given some initial conditions. The hydrologic reality is that Eq. (8.4) is at best an imperfect representation of the river basin. Reality is known to be nonlinear and distributed in space. Consequently, Eq. (8.4) is in error. Such is the situation in most mathematical representations of physical systems. In the canonical formulation this error is assumed to be additive white noise, leading to the generalized state–space equation

$$\dot{\mathbf{X}}(t) = \mathbf{F}(t)\mathbf{X}(t) + \mathbf{L}(t)\mathbf{U}(t) + \mathbf{G}(t)\mathbf{W}(t), \tag{8.5}$$

where now the vector $\mathbf{U}(t)$ has arbitrary dimensions, as long as $\mathbf{L}(t)\mathbf{U}(t)$ is an $n \times 1$ vector. The added random term must have the following mean and covariance properties:

$$E[\mathbf{G}(t)\mathbf{W}(t)] = 0$$

$$E\left[(\mathbf{G}(t)\mathbf{W}(t))(\mathbf{G}(\tau)\mathbf{W}(\tau))^T\right] = \mathbf{G}(t)\mathbf{Q}(t)\mathbf{G}^T(t)\delta(t - \tau).$$

The dimension of $\mathbf{G}(t)\mathbf{W}(t)$ must be $n \times 1$. The matrix $\mathbf{Q}(t)$ is called a spectral-density matrix, which becomes a covariance matrix when multiplied by the Dirac delta function (which has units of inverse time). The matrix $\mathbf{Q}(t)\delta(t - \tau)$ has a value of infinity, but its integral over a finite time is finite, hence the density descriptor.

The canonical form can always be achieved, given an additive error form. Assume, for example, that the river basin is represented by

$$\dot{\mathbf{X}}(t) = \mathbf{F}(t)\mathbf{X}(t) + \mathbf{L}(t)\mathbf{U}(t) + \mathbf{G}(t)\mathbf{Y}(t), \tag{8.6}$$

where $\mathbf{Y}(t)$ is a correlated, zero-mean, continuous random process. The noise $\mathbf{Y}(t)$ can then be represented by an equation similar to Eq. (8.5),

$$\dot{\mathbf{Y}}(t) = \mathbf{A}(t)\mathbf{Y}(t) + \mathbf{B}(t)\mathbf{W}(t).$$

By state augmentation, the above equation is combined with Eq. (8.6):

$$\begin{bmatrix} \dot{\mathbf{X}}(t) \\ \dot{\mathbf{Y}}(t) \end{bmatrix} = \begin{bmatrix} \mathbf{F}(t) & \vdots & \mathbf{G}(t) \\ \cdots & \cdots & \cdots \\ \mathbf{0} & \vdots & \mathbf{A}(t) \end{bmatrix} \begin{bmatrix} \mathbf{X}(t) \\ \mathbf{Y}(t) \end{bmatrix} + \begin{bmatrix} \mathbf{L}(t) \\ \cdots \\ \mathbf{0} \end{bmatrix} \mathbf{U}(t) + \begin{bmatrix} \mathbf{0} \\ \cdots \\ \mathbf{B}(t) \end{bmatrix} \mathbf{W}(t),$$

which is in the desired canonical form.

In hydrology, the water storage of the basin (or river stretch) is never directly observed. Observed quantities are discharge or river stage. Assume, for the sake of argument, that discharge is the measured quantity. A common assumption is that discharge is linearly related to storage. An accepted fact is that recording instruments are frequently in error (hopefully of unbiased random nature). It is then possible to define an observation equation

$$\mathbf{Z}(t) = \mathbf{H}(t)\mathbf{X}(t) + \mathbf{V}(t), \tag{8.7}$$

where $\mathbf{Z}(t)$ is an $m \times 1$ vector of observations of the $(n \times 1)$ state vector $\mathbf{X}(t)$, and the dimensions of $\mathbf{Z}(t)$, m, may be less than those of $\mathbf{X}(t)$, n. Matrix $\mathbf{H}(t)(m \times n)$ defines the observation experiment, and thus the basic relationships between the state and the observed value. Finally, the $(m \times 1)$ vector $\mathbf{V}(t)$ is white noise with zero mean and covariance, $E[\mathbf{V}(t)\mathbf{V}^T(\tau)] = \mathbf{R}(t)\delta(t - \tau)$.

The system of Eqs. (8.5) and (8.7) are a stochastic dynamic formulation of the type studied in this chapter. The solution of the stochastic system alludes to the filtering–forecasting cycle discussed in Chapter 1. Given that the model is imperfect and that observations are in error, the goal is to estimate the true state of the system. Estimation is based on the relative confidence on the model and on the observations. Once a "best" state estimate is obtained, the stochastic model can be integrated to obtain forecasts of the state.

8.2 SYSTEM CHARACTERISTICS

8.2.1 The Discrete State–Space Model

Equation (8.5) represents a continuous state–space equation. In hydrology, as well as in many other geophysical sciences, availability of data on natural processes requires a discrete state–space formulation. Since in fact most processes are continuous, it is best to arrive at the discrete formulation by integration of Eq. (8.5). As pointed out by Gelb (1974), this approach also avoids some theoretical pitfalls.

Begin by studying the unforced system:

$$\dot{\mathbf{X}}(t) = \mathbf{F}(t)\mathbf{X}(t). \tag{8.8}$$

Given an initial condition $\mathbf{X}(t_0)$, the state value at time t, $\mathbf{X}(t)$ results from the solution to the above homogeneous differential equation. The solution is in terms of a transition matrix $\mathbf{\Phi}(t, t_0)$:

$$\mathbf{X}(t) = \mathbf{\Phi}(t, t_0)\mathbf{X}(t_0). \tag{8.9}$$

The transition matrix is in turn the solution to the system equation, assuming that each element of the state initially and independently takes a value of unity. In other words,

$$\frac{d\mathbf{\Phi}(t, t_0)}{dt} = \mathbf{F}(t)\mathbf{\Phi}(t, t_0); \qquad \mathbf{\Phi}(t_0, t_0) = \mathbf{I}, \tag{8.10}$$

where $\mathbf{I}$ is the $n \times n$ identity matrix (if n is the state-vector dimensions).

Equation (8.9) is valid because in a linear system the response is proportional to the initial condition. Some valuable properties of $\Phi(t, t_0)$ are

$$\Phi(t_2, t_0) = \Phi(t_2, t_1)\Phi(t_1, t_0) \tag{8.11}$$

$$\Phi(t_0, t) = \Phi^{-1}(t, t_0). \tag{8.12}$$

The last condition implies that $\Phi(t, t_0)$ is nonsingular.

Of particular interest is the time-invariant model, where $\mathbf{F}$ is not a function of time. In that case (which implies stationarity for the stochastic formulation), $\Phi(t, t_0)$ is not a function of absolute time but only of the relative time,

$$\Phi(t, t_0) = \Phi(t - t_0). \tag{8.13}$$

In such cases the solution is easily achieved. Expanding $\mathbf{X}(t)$ in a Taylor series around the initial condition $\mathbf{X}(t_0)$ yields

$$\mathbf{X}(t) = \mathbf{X}(t_0) + \dot{\mathbf{X}}(t_0)(t - t_0) + \frac{\ddot{\mathbf{X}}(t_0)}{2!}(t - t_0)^2 + \cdots, \tag{8.14}$$

but using Eq. (8.8),

$$\dot{\mathbf{X}}(t_0) = \mathbf{FX}(t_0)$$
$$\ddot{\mathbf{X}}(t_0) = \mathbf{F} \cdot \mathbf{FX}(t_0) = \mathbf{F}^2\mathbf{X}(t_0)$$
$$\dddot{\mathbf{X}}(t_0) = \mathbf{F} \cdot \mathbf{F} \cdot \mathbf{FX}(t_0) = \mathbf{F}^3\mathbf{X}(t_0).$$

In the above, "powers" of a matrix imply the sequential multiplication of the matrix (the number of multiplications specified by the power). Substituting the above in Eq. (8.14) results in:

$$\mathbf{X}(t) = \left[\mathbf{I} + \mathbf{F}(t - t_0) + \frac{\mathbf{F}^2}{2!}(t - t_0)^2 + \cdots\right]\mathbf{X}(t_0). \tag{8.15}$$

The matrix-series expansion in brackets is defined as the exponential form of a matrix

$$e^{\mathbf{A}} = \mathbf{I} + \mathbf{A} + \frac{\mathbf{A}^2}{2!} + \frac{\mathbf{A}^3}{3!} + \cdots. \tag{8.16}$$

Therefore

$$\Phi(t - t_0) = e^{\mathbf{F} \cdot (t - t_0)}. \tag{8.17}$$

The solution in Eq. (8.17) is analogous and equivalent to that of scalar differential equations. A simple hydrologic example is the linear reservoir,

$$\frac{dS(t)}{dt} = -kS(t),$$

where S is storage. Assuming an initial storage of $S(t_0)$ the storage at arbitrary time t is

$$S(t) = S(t_0)e^{-k(t-t_0)}$$

or

$$\Phi(t - t_0) = e^{-k(t-t_0)}. \tag{8.18}$$

Now that the transition matrix is defined, it is easier to find the solution to the forced system:

$$\dot{\mathbf{X}}(t) = \mathbf{F}(t)\mathbf{X}(t) + \mathbf{L}(t)\mathbf{U}(t). \tag{8.19}$$

Again, the approach is like that corresponding to scalar first-order differential equations. The total response is the sum of the homogeneous and particular solutions. The homogeneous response to the initial conditions was already seen. The particular solution corresponds to the effect of the input history, $\mathbf{U}(t)$; $t \in (t_0, t)$, on the state vector at time t. As seen in Chapter 4, linear-systems response to inputs is determined via the convolution equation. In hydrology, the convolution is between the rainfall and the instantaneous unit hydrograph, which is nothing more than the transition function $\Phi(t, t_0)$. Therefore, the solution to Eq. (8.19) is

$$\mathbf{X}(t) = \Phi(t, t_0)\mathbf{X}(t_0) + \int_{t_0}^{t} \Phi(t, \tau)\mathbf{L}(\tau)\mathbf{U}(\tau)\, d\tau. \tag{8.20}$$

Following the linear-reservoir example, given a constant, infinite duration, input I, the storage obeys the equation

$$\frac{dS(t)}{dt} = -kS(t) + I.$$

The value of $S(t)$, given an initial storage $S(t_0)$, is then

$$S(t) = e^{-k(t-t_0)}S(t_0) + I\int_{t_0}^{t} e^{-k(t-\tau)}\, d\tau$$

$$= S(t_0)e^{-k(t-t_0)} + \frac{I}{k}e^{-kt}[e^{kt} - e^{kt_0}]$$

$$= S(t_0)e^{-k(t-t_0)} + \frac{I}{k}[1 - e^{-k(t-t_0)}].$$

Finally, it is necessary to address the effect of a stochastic forcing function, $\mathbf{G}(t)\mathbf{W}(t)$, in

$$\dot{\mathbf{X}}(t) = \mathbf{F}(t)\mathbf{X}(t) + \mathbf{L}(t)\mathbf{U}(t) + \mathbf{G}(t)\mathbf{W}(t). \tag{8.21}$$

The effect is conceptually another convolution leading to

$$\mathbf{X}(t) = \mathbf{\Phi}(t, t_0)\mathbf{X}(t_0) + \int_{t_0}^{t} \mathbf{\Phi}(t, \tau)\mathbf{L}(\tau)\mathbf{U}(\tau) \, d\tau$$

$$+ \int_{t_0}^{t} \mathbf{\Phi}(t, \tau)\mathbf{G}(\tau)\mathbf{W}(\tau) \, d\tau. \tag{8.22}$$

The integral of a stochastic process, as in the last term of Eq. (8.22), is not defined within the classic Riemann sense. In fact, it requires limit and convergence definitions of a stochastic nature. Itô calculus is the necessary tool (Jazwinski, 1970). We will simply express Eq. (8.22) as

$$\mathbf{X}(t) = \mathbf{\Phi}(t, t_0)\mathbf{X}(t_0) + \mathbf{\Lambda}(t_0)\mathbf{U}(t_0) + \mathbf{\Gamma}(t_0)\mathbf{W}(t_0), \tag{8.23}$$

where the last two terms on the right-hand side of the equation correspond to the integrals in Eq. (8.22). This implies that the factors $\mathbf{\Lambda}(t_0)$, $\mathbf{U}(t_0)$, $\mathbf{\Gamma}(t_0)$, and $\mathbf{W}(t_0)$ are not separately defined. This peculiarity is a result of deriving the discrete formulation from the original state–space continuous equation (Eq. 8.6). In practice, though, the discrete formulation is usually obtained directly, without first requiring the continuous equivalent. Generally, then, all factors in Eq. (8.23) are well defined. From now on, we will assume that this is the case.

If the discrete equation is written over an interval Δt, Eq. (8.23) becomes

$$\mathbf{X}(t + \Delta t) = \mathbf{\Phi}(t + \Delta t, t)\mathbf{X}(t) + \mathbf{\Lambda}(t)\mathbf{U}(t) + \mathbf{\Gamma}(t)\mathbf{W}(t), \tag{8.24}$$

which is the more familiar difference equation of a discrete state–space model. For the sake of brevity, Eq. (8.24) is commonly and interchangeably presented as

$$\mathbf{X}(k + 1) = \mathbf{\Phi}(k)\mathbf{X}(k) + \mathbf{\Lambda}(k)\mathbf{U}(k) + \mathbf{\Gamma}(k)\mathbf{W}(k), \tag{8.25}$$

where $t = k \cdot \Delta t$ and $\mathbf{\Phi}(k)$ is taken to be always valid over a one-time period transition.

8.2.2 Observability and Controllability

The properties of observability and controllability play a major role in modern estimation and control theory. The stability of the estimator, or filter, of the state implies that the variance of estimation will remain finite and convergent independently of the initial assumptions made regarding the variance of the state. As will become clear in Section 8.3, this is a particularly desirable property given our difficulties in defining state variance. Complete observability ensures stable filters. Observable time-invariant models will in fact achieve steady-state filtering algorithms. Complete controllability is required to drive a dynamic system to any solution.

It is impossible to cover all of the nuances of observability and controllability within the scope of this work. This section will, nevertheless, give the necessary conditions of deterministic linear systems to satisfy both properties. Only on rare occasions should a user violate these properties.

It is easier to derive observability conditions for time-invariant linear deterministic systems. Following Gelb (1974), let us assume an undriven dynamic model

$$\mathbf{X}(k+1) = \mathbf{\Phi}\mathbf{X}(k) \qquad k = 0, 1, \ldots, n-1 \qquad (8.26)$$

observed without error

$$\mathbf{Z}(k) = \mathbf{H}\mathbf{X}(k) \qquad k = 0, 1, 2, \ldots, n-1. \qquad (8.27)$$

Observability is the capacity to determine the values of the state from observations, in particular, the initial state $\mathbf{X}(0)$. Using Eqs. (8.26) and (8.27) it is clear that

$$\mathbf{Z}(0) = \mathbf{H}\mathbf{X}(0)$$
$$\mathbf{Z}(1) = \mathbf{H}\mathbf{X}(1) = \mathbf{H}\mathbf{\Phi}\mathbf{X}(0)$$
$$\mathbf{Z}(2) = \mathbf{H}\mathbf{X}(2) = \mathbf{H}\mathbf{\Phi}\mathbf{\Phi}\mathbf{X}(0) = \mathbf{H}\mathbf{\Phi}^2\mathbf{X}(0) \qquad (8.28)$$
$$\vdots$$
$$\mathbf{Z}(n-1) = \mathbf{H}\mathbf{\Phi}^{n-1}\mathbf{X}(0).$$

A solution for $\mathbf{X}(0)$ is obtained from the above system of equations as long as the matrix*

$$\left[\mathbf{H}^T \ \vdots \ \mathbf{\Phi}^T\mathbf{H}^T \ \vdots \ldots \vdots \ (\mathbf{\Phi}^T)^{n-1}\mathbf{H}^T \right] \qquad (8.29)$$

is of rank n. The previous condition guarantees observability for time-invariant linear systems.

Sage and Melsa (1971a) state that in time-variant discrete linear systems complete observability implies that

$$\sum_{k=k_0}^{k=k_f} \mathbf{\Phi}^T(k, k_0)\mathbf{H}^T(k)\mathbf{H}(k)\mathbf{\Phi}(k, k_0) \qquad (8.30)$$

is a positive definite matrix in the interval $[k_0, k_f]$.

Continuous systems have corresponding conditions. A time-variant deterministic system is observable in the time interval $[t_0, t_f]$ where $t_f > t_0$, if the

---

*Matrix $\mathbf{C} = [\mathbf{A} \ \vdots \ \mathbf{B}]$ represents a compound matrix resulting from matrices $\mathbf{A}$ and $\mathbf{B}$. Therefore, $\mathbf{C}$ has a number of rows equal to the rows of $\mathbf{A}$ and a number of columns equal to the sum of those of $\mathbf{A}$ and $\mathbf{B}$.

matrix

$$\int_{t_0}^{t_f} \Phi^T(t, t_0) H^T(t) H(t) \Phi(t, t_0)\, dt \tag{8.31}$$

is positive definite. The condition on constant coefficient systems is that the matrix

$$\left[H^T \; \vdots \; F^T H^T \; \vdots \; (F^T)^2 H^T \; \vdots \dots \vdots \; (F^T)^{n-1} H^T \right] \tag{8.32}$$

is of rank n.

Controllability is the ability to manipulate inputs into a dynamic system in order to drive it to a desired state. The time-invariant deterministic model

$$X(k+1) = \Phi X(k) + \Lambda U(k) \qquad k = 0, \dots, n-1 \tag{8.33}$$

is controllable if it can be driven to a particular state $X(n)$ from an initial state $X(0)$ with judicious selection of $U(0), U(1), \dots, U(n-1)$. Using Eq. (8.33)

$$\begin{aligned}
X(1) &= \Phi X(0) + \Lambda U(0) \\
X(2) &= \Phi X(1) + \Lambda U(1) \\
&= \Phi^2 X(0) + \Phi \Lambda U(0) + \Lambda U(1) \\
&\vdots
\end{aligned} \tag{8.34}$$

$$X(n) = \Phi^n X(0) + \Phi^{n-1} \Lambda U(0) + \Phi^{n-2} \Lambda U(1) + \cdots + \Lambda U(n-1).$$

The first term of the last equation in Eq. (8.34) is known. Therefore, to obtain a desired terminal state $X(n)$ the following equation must have a solution:

$$X(n) - \Phi^n X(0) = \Phi^{n-1} \Lambda U(0) + \Phi^{n-2} \Lambda U(1) + \cdots + \Lambda U(n-1), \tag{8.35}$$

which requires that the matrix

$$\left[\Phi^{n-1} \Lambda \; \vdots \; \Phi^{n-2} \Lambda \; \vdots \dots \vdots \; \Phi \Lambda \; \vdots \; \Lambda \right] \tag{8.36}$$

be of rank n.

Complete controllability of a time-varying discrete linear system, as given by Sage and Melsa (1971a), requires the following matrix to be positive definite in the interval $[k_0, k_f]$, where $k_f > k_0$

$$\sum_{k=k_0}^{k=k_f} \Phi(k_0, k) \Lambda(k) \Lambda^T(k) \Phi^T(k_0, k). \tag{8.37}$$

Controllable continuous linear systems have equivalent criteria. Time-invariant systems must have the matrix

$$\left[L \; \vdots \; FL \; \vdots \; F^2 L \; \vdots \dots \vdots \; F^{n-1} L \right] \tag{8.38}$$

of rank n.

For dynamic time-varying continuous systems, controllability in the interval $[t_0, t_f]$ requires

$$\int_{t_0}^{t_f} \Phi(t_0, t) \Lambda(t) \Lambda^T(t) \Phi^T(t_0, t) \, dt \tag{8.39}$$

to be positive definite.

8.2.3 Propagation of Variance: Forecasting

Chapter 1 characterized the estimation and forecasting of hydrologic systems as a cycle. This section will derive the forecasting cycle corresponding to the continuous and discrete dynamic equations given in Eq. (8.5) and Eq. (8.25), respectively. For the sake of brevity, let us deal with the zero-mean equivalents of Eqs. (8.5) and (8.25), after known deterministic inputs $L(t)U(t)$ and $\Lambda(k)U(k)$ are subtracted from both sides of the equations.

The Discrete Model

The zero-mean dynamic equation is

$$X(k+1) = \Phi(k)X(k) + \Gamma(k)W(k). \tag{8.40}$$

The second moment properties of the noise term are

$$E\left[\Gamma(k)W(k)W^T(k)\Gamma^T(k)\right] = \Gamma(k)Q(k)\Gamma^T(k), \tag{8.41}$$

where $Q(k)$ is the covariance matrix of $W(k)$. It was shown in Chapter 7 that the state estimate minimizing variance or any other quadratic loss is in fact the conditional mean of the process. Assume that the conditional mean of $X(k)$ given the latest observation $Z(k)$ is known and represented by $\hat{X}(k|k)$. The mean square error of estimation is correspondingly assumed to be known and represented by $\Sigma(k|k)$. The expectation, conditional on $Z(k)$, of the linear system in Eq. (8.40) is then

$$\hat{X}(k+1|k) = \Phi(k)\hat{X}(k|k), \tag{8.42}$$

a direct consequence of whiteness and the zero-mean property of $W(k)$. The notation $\hat{X}(k+1|k)$ stands for the minimum-variance estimate of $X(k+1)$ given $Z(k)$ up to time interval k. The above is the conditional forecasted mean value of the state.

The forecasting error incurred by Eq. (8.42) is clearly

$$X(k+1) - \hat{X}(k+1|k) = \Phi(k)[X(k) - \hat{X}(k|k)] + \Gamma(k)W(k)$$

or

$$\delta \mathbf{X}(k+1|k) = \mathbf{\Phi}(k)\delta \mathbf{X}(k|k) + \mathbf{\Gamma}(k)\mathbf{W}(k).$$

The mean square error of forecasting is then

$$E\left[\delta \mathbf{X}(k+1|k)\delta \mathbf{X}^T(k+1|k)\right]$$
$$= \mathbf{\Phi}(k)E\left[\delta \mathbf{X}(k|k)\delta \mathbf{X}^T(k|k)\right]\mathbf{\Phi}^T(k) + \mathbf{\Gamma}(k)\mathbf{Q}(k)\mathbf{\Gamma}^T(k) \quad (8.43)$$
$$\mathbf{\Sigma}(k+1|k) = \mathbf{\Phi}(k)\mathbf{\Sigma}(k|k)\mathbf{\Phi}^T(k) + \mathbf{\Gamma}(k)\mathbf{Q}(k)\mathbf{\Gamma}^T(k).$$

If the model error term $\mathbf{\Gamma}(k)\mathbf{W}(k)$ is Gaussian, Eqs. (8.42) and (8.43) are the complete solution to the one time-step forecasting problem. The forecast is the conditional multivariate Gaussian distribution with mean $\hat{\mathbf{X}}(k+1|k)$ and covariance $\mathbf{\Sigma}(k+1|k)$.

The Continuous Model

The zero-mean continuous dynamic equation is

$$\dot{\mathbf{X}}(t) = \mathbf{F}(t)\mathbf{X}(t) + \mathbf{G}(t)\mathbf{W}(t). \quad (8.44)$$

Given $\hat{\mathbf{X}}(t|t)$, the conditional mean at any time t in the future is the solution to

$$\dot{\hat{\mathbf{X}}}(t|t) = \mathbf{F}(t)\hat{\mathbf{X}}(t|t). \quad (8.45)$$

The mean-square-error propagation equation will be obtained following the heuristic procedure presented by Gelb (1974). The idea is to arrive at the continuous formulation as a limit of the corresponding discrete model.

First, the second-moment properties of the noise term of the continuous system are

$$E\left[(\mathbf{G}(t)\mathbf{W}(t))(\mathbf{G}(\tau)\mathbf{W}(\tau))^T\right] = \mathbf{G}(t)\mathbf{Q}(t)\mathbf{G}^T(t)\delta(t-\tau), \quad (8.46)$$

where $\delta(t-\tau)$ is the Dirac delta function and $\mathbf{Q}(t)$ is the spectral density of $\mathbf{W}(t)$. Equation (8.22) defined the discrete-model noise component as a convolution of the continuous one,

$$\mathbf{\Gamma}(k)\mathbf{W}(k) = \int_t^{t+\Delta t}\mathbf{\Phi}(t+\Delta t, \tau)\mathbf{G}(\tau)\mathbf{W}(\tau)\, d\tau, \quad (8.47)$$

from which second-moment properties can be expressed as

$$\mathbf{\Gamma}(k)\mathbf{Q}(k)\mathbf{\Gamma}^T(k) = \int_t^{t+\Delta t}\int_t^{t+\Delta t}\mathbf{\Phi}(t+\Delta t, \tau)\mathbf{G}(\tau)$$
$$\cdot E\left[\mathbf{W}(\tau)\mathbf{W}^T(\sigma)\right]\mathbf{G}^T(\sigma)\mathbf{\Phi}^T(t+\Delta t, \sigma)\, d\tau\, d\sigma.$$

Using Eq. (8.46),

$$
\begin{aligned}
\Gamma(k)\mathbf{Q}(k)\Gamma^T(k) &= \int_t^{t+\Delta t}\int_t^{t+\Delta t}\mathbf{\Phi}(t+\Delta t,\tau)\mathbf{G}(\tau)\mathbf{Q}(\tau)\delta(\tau-\sigma)\\
&\quad\cdot\mathbf{G}^T(\sigma)\mathbf{\Phi}^T(t+\Delta t,\sigma)\,d\tau\,d\sigma\\
&= \int_t^{t+\Delta t}\mathbf{\Phi}(t+\Delta t,\tau)\mathbf{G}(\tau)\mathbf{Q}(\tau)\mathbf{G}^T(\tau)\mathbf{\Phi}^T(t+\Delta t,\tau)\,d\tau.
\end{aligned}
$$

$$(8.48)$$

As Δt goes to zero (small Δt),

$$
\mathbf{\Phi}(t+\Delta t,t)\to\mathbf{I}
$$
$$
\Delta t\to 0.
$$

For small Δt Eq. (8.48) collapses to

$$
\Gamma(k)\mathbf{Q}(k)\Gamma^T(k)\to\mathbf{G}(t)\mathbf{Q}(t)\mathbf{G}^T(t)\cdot\Delta t. \tag{8.49}
$$

The behavior of the transition matrix can also be studied in more detail. Section 8.2.1 established

$$
\frac{d\mathbf{\Phi}(t+\Delta t,t)}{d(t+\Delta t)}=\mathbf{F}(t+\Delta t)\mathbf{\Phi}(t+\Delta t,t). \tag{8.50}
$$

The derivative in the above equation can be approximated by the following difference as $\Delta t'$ goes to zero:

$$
\frac{d\mathbf{\Phi}(t+\Delta t,t)}{d(t+\Delta t)}=\frac{\mathbf{\Phi}(t+\Delta t+\Delta t',t)-\mathbf{\Phi}(t+\Delta t,t)}{\Delta t'}.
$$

Substituting in Eq. (8.50)

$$
\frac{\mathbf{\Phi}(t+\Delta t+\Delta t',t)-\mathbf{\Phi}(t+\Delta t,t)}{\Delta t'}=\mathbf{F}(t+\Delta t)\mathbf{\Phi}(t+\Delta t,t),
$$

which, on letting $\Delta t\to 0$, becomes

$$
\begin{aligned}
\mathbf{\Phi}(t+\Delta t',t) &= \Delta t'\mathbf{F}(t)\mathbf{\Phi}(t,t)+\mathbf{\Phi}(t,t)\\
&= \mathbf{I}+\mathbf{F}(t)\Delta t'.
\end{aligned} \tag{8.51}
$$

Substituting Eq. (8.51) for the transition matrix in the expression for the mean square error of forecast, Eq. (8.43), and using Eq. (8.49) and letting $t=k\,\Delta t$ results in

$$
\mathbf{\Sigma}(t+\Delta t|t)=\bigl(\mathbf{I}+\mathbf{F}(t)\,\Delta t'\bigr)\mathbf{\Sigma}(t|t)\bigl(\mathbf{I}+\mathbf{F}^T(t)\,\Delta t'\bigr)+\mathbf{G}(t)\mathbf{Q}(t)\mathbf{G}^T(t)\,\Delta t.
$$

Rearranging

$$\Sigma(t + \Delta t|t) = \Sigma(t|t) + \left[\mathbf{F}(t)\Sigma(t|t) + \Sigma(t|t)\mathbf{F}^T(t) \right] \Delta t'$$
$$+ \left[\mathbf{G}(t)\mathbf{Q}(t)\mathbf{G}^T(t) \right] \Delta t + \mathbf{F}(t)\Sigma(t|t)\mathbf{F}^T(t)(\Delta t')^2. \quad (8.52)$$

The first term on the right-hand side of Eq. (8.52) can be moved to the left-hand side and all terms divided by Δt:

$$\frac{\Sigma(t + \Delta t|t) - \Sigma(t|t)}{\Delta t} = \left[\mathbf{F}(t)\Sigma(t|t) + \Sigma(t|t)\mathbf{F}^T(t) \right] \frac{\Delta t'}{\Delta t}$$
$$+ \left[\mathbf{G}(t)\mathbf{Q}(t)\mathbf{G}^T(t) \right] + \mathbf{F}(t)\Sigma(t|t)\mathbf{F}^T(t) \frac{(\Delta t')^2}{\Delta t}.$$

Taking limits as both Δt and $\Delta t'$ go to zero, and assuming they do so at the same rate, leads to the desired nonlinear differential equation on the mean square error of forecasting in the absence of measurements:

$$\frac{d\Sigma(t|t)}{dt} = \mathbf{F}(t)\Sigma(t|t) + \Sigma(t|t)\mathbf{F}^T(t) + \mathbf{G}(t)\mathbf{Q}(t)\mathbf{G}^T(t). \quad (8.53)$$

Equations (8.45) and (8.53) then form the basis for forecasting given a continuous time-varying linear dynamic system.

8.3 THE KALMAN FILTER

8.3.1 A Bayesian Approach for the Discrete Filter

Schweppe (1973) derives the filtering algorithm for the dynamic discrete linear system of Eq. (8.25) with discrete observations $\mathbf{Z}(k) = \mathbf{H}(k)\mathbf{X}(k) + \mathbf{V}(k)$ using the static-filter results of Chapter 7. The idea is to combine repetitive observations of a state vector $\mathbf{X}(k)$.

Assume there are two observations of a vector $\mathbf{X}$

$$\mathbf{Z}_1 = \mathbf{H}_1\mathbf{X} + \mathbf{V}_1 \tag{8.54}$$

$$\mathbf{Z}_2 = \mathbf{H}_2\mathbf{X} + \mathbf{V}_2, \tag{8.55}$$

where

$$E[\mathbf{X}] = \mathbf{m}_x \quad \text{and} \quad E\left[(\mathbf{X} - \mathbf{m}_x)(\mathbf{X} - \mathbf{m}_x)^T \right] = \mathbf{\Psi}$$
$$E[\mathbf{V}_1\mathbf{V}_1^T] = \mathbf{R}_1; \quad E[\mathbf{V}_2\mathbf{V}_2^T] = \mathbf{R}_2; \quad E[\mathbf{V}_1\mathbf{V}_2^T] = \mathbf{0}$$
$$E[\mathbf{V}_1] = E[\mathbf{V}_2] = \mathbf{0}.$$

Augmenting the observation vector yields,

$$\mathbf{Z} = \mathbf{HX} + \mathbf{V}, \qquad (8.56)$$

where

$$\mathbf{Z} = \begin{bmatrix} \mathbf{Z}_1 \\ \cdots \\ \mathbf{Z}_2 \end{bmatrix} \qquad \mathbf{H} = \begin{bmatrix} \mathbf{H}_1 \\ \cdots \\ \mathbf{H}_2 \end{bmatrix} \qquad \mathbf{V} = \begin{bmatrix} \mathbf{V}_1 \\ \cdots \\ \mathbf{V}_2 \end{bmatrix}.$$

Using the Bayesian estimation equation of Chapter 7 with Eq. (8.56) yields,

$$\hat{\mathbf{X}} = \Sigma \left[\mathbf{H}_1^T \mathbf{R}_1^{-1} \mathbf{Z}_1 + \mathbf{H}_2^T \mathbf{R}_2^{-1} \mathbf{Z}_2 + \boldsymbol{\Psi}^{-1} \mathbf{m}_x \right] \qquad (8.57)$$

$$\Sigma = \left[\mathbf{H}_1^T \mathbf{R}_1^{-1} \mathbf{H}_1 + \mathbf{H}_2^T \mathbf{R}_2^{-1} \mathbf{H}_2 + \boldsymbol{\Psi}^{-1} \right]^{-1}. \qquad (8.58)$$

The above can also be viewed as the combination of two estimates of $\mathbf{X}$, one Bayesian and one Fisher, which are in turn interpreted as observations on the state $\mathbf{X}$. Essentially, the Bayesian estimate of $\mathbf{X}$ using Eq. (8.54) is

$$\hat{\mathbf{X}}_1 = \Sigma_1 \left[\mathbf{H}_1^T \mathbf{R}_1^{-1} \mathbf{Z}_1 + \boldsymbol{\Psi}^{-1} \mathbf{m}_x \right] \qquad (8.59)$$

$$\Sigma_1 = \left[\mathbf{H}_1^T \mathbf{R}_1^{-1} \mathbf{H}_1 + \boldsymbol{\Psi}^{-1} \right]^{-1}. \qquad (8.60)$$

The Fisher estimate of $\mathbf{X}$ using the second observation is

$$\hat{\mathbf{X}}_2 = \Sigma_2 \left[\mathbf{H}_2^T \mathbf{R}_2^{-1} \mathbf{Z}_2 \right] \qquad (8.61)$$

$$\Sigma_2 = \left[\mathbf{H}_2^T \mathbf{R}_2^{-1} \mathbf{H}_2 \right]^{-1}. \qquad (8.62)$$

The estimates $\hat{\mathbf{X}}_1$ and $\hat{\mathbf{X}}_2$ can be interpreted as observations on $\mathbf{X}$,

$$\hat{\mathbf{X}}_1 = \mathbf{X} + \delta \mathbf{X}_1 \qquad (8.63)$$

$$\hat{\mathbf{X}}_2 = \mathbf{X} + \delta \mathbf{X}_2 \qquad (8.64)$$

with error covariances

$$E\left[\delta \mathbf{X}_1 \delta \mathbf{X}_1^T \right] = \Sigma_1 \qquad \text{and} \qquad E\left[\delta \mathbf{X}_2 \delta \mathbf{X}_2^T \right] = \Sigma_2.$$

The Fisher estimate of $\mathbf{X}$ from Eqs. (8.63) and (8.64) results in

$$\hat{\mathbf{X}} = \left[\Sigma_1^{-1} + \Sigma_2^{-1} \right]^{-1} \left[\Sigma_1^{-1} \hat{\mathbf{X}}_1 + \Sigma_2^{-1} \hat{\mathbf{X}}_2 \right], \qquad (8.65)$$

which is equivalent to Eq. (8.57) derived from the direct augmentation of the observation vector.

In summary, prior information is utilized only once in the processing of multiple observations. This concept is exploited to derive the dynamic Kalman filter.

Given the discrete dynamic model

$$\mathbf{X}(k+1) = \mathbf{\Phi}(k)\mathbf{X}(k) + \mathbf{\Gamma}(k)\mathbf{W}(k) \tag{8.66}$$

$$\mathbf{Z}(k) = \mathbf{H}(k)\mathbf{X}(k) + \mathbf{V}(k) \tag{8.67}$$

assume that an estimate $\hat{\mathbf{X}}(k|k)$ is available with error covariance

$$\mathbf{\Sigma}(k|k) = E\left[(\mathbf{X}(k) - \hat{\mathbf{X}}(k|k))(\mathbf{X}(k) - \hat{\mathbf{X}}(k|k))^T\right]. \tag{8.68}$$

The minimum variance one-step-ahead prediction, $\hat{\mathbf{X}}(k+1|k)$, was given by Eq. (8.42) as

$$\hat{\mathbf{X}}(k+1|k) = \mathbf{\Phi}(k)\hat{\mathbf{X}}(k|k). \tag{8.69}$$

The corresponding error covariance (Eq. 8.43) is repeated here,

$$\mathbf{\Sigma}(k+1|k) = \mathbf{\Phi}(k)\mathbf{\Sigma}(k|k)\mathbf{\Phi}^T(k) + \mathbf{\Gamma}(k)\mathbf{Q}(k)\mathbf{\Gamma}^T(k). \tag{8.70}$$

If a new observation,

$$\mathbf{Z}(k+1) = \mathbf{H}(k+1)\mathbf{X}(k+1) + \mathbf{V}(k+1), \tag{8.71}$$

becomes available it can be combined with the latest estimate in the manner discussed at the beginning of this section. Let the first observation be

$$\hat{\mathbf{X}}(k+1|k) = \mathbf{X}(k+1) - \delta\mathbf{X}(k+1|k),$$

where

$$\delta\mathbf{X}(k+1|k) = \mathbf{X}(k+1) - \hat{\mathbf{X}}(k+1|k)$$
$$E\left[\delta\mathbf{X}(k+1|k)\delta\mathbf{X}^T(k+1|k)\right] = \mathbf{\Sigma}(k+1|k).$$

The second observation, $\mathbf{Z}(k+1)$, is given by Eq. (8.71). Augmenting the observation vector,

$$\tilde{\mathbf{Z}}(k+1) = \tilde{\mathbf{H}}(k+1)\mathbf{X}(k+1) + \tilde{\mathbf{V}}(k+1), \tag{8.72}$$

where

$$\tilde{\mathbf{Z}}(k+1) = \begin{bmatrix} \mathbf{Z}(k+1) \\ \cdots\cdots\cdots \\ \hat{\mathbf{X}}(k+1|k) \end{bmatrix} \tag{8.73}$$

$$\tilde{\mathbf{H}}(k+1) = \begin{bmatrix} \mathbf{H}(k+1) \\ \cdots\cdots \\ \mathbf{I} \end{bmatrix} \tag{8.74}$$

$$\tilde{\mathbf{V}}(k+1) = \begin{bmatrix} \mathbf{V}(k+1) \\ \cdots\cdots\cdots\cdots \\ -\delta\mathbf{X}(k+1|k) \end{bmatrix} \tag{8.75}$$

$$\tilde{\mathbf{R}}(k+1) = E\left[\tilde{\mathbf{V}}(k+1)\tilde{\mathbf{V}}^T(k+1)\right]$$
$$= \begin{bmatrix} \mathbf{R}(k+1) & \vdots & \mathbf{0} \\ \cdots\cdots\cdots & \cdots\cdots\cdots \\ \mathbf{0} & \vdots & \Sigma(k+1|k) \end{bmatrix} \tag{8.76}$$

and using a static Fisher estimate to obtain $\hat{\mathbf{X}}(k+1|k+1)$ results in

$$\hat{\mathbf{X}}(k+1|k+1)$$
$$= \left[\tilde{\mathbf{H}}^T(k+1)\tilde{\mathbf{R}}^{-1}(k+1)\tilde{\mathbf{H}}(k+1)\right]^{-1}\tilde{\mathbf{H}}^T(k+1)\tilde{\mathbf{R}}^{-1}(k+1)\tilde{\mathbf{Z}}(k+1), \tag{8.77}$$

which on substituting Eqs. (8.69), (8.70), and (8.72) through (8.76) yields the dynamic Kalman filter (Kalman, 1960, 1963; Kalman and Bucy, 1961)

$$\hat{\mathbf{X}}(k+1|k+1) = \Sigma(k+1|k+1)\left[\mathbf{H}^T(k+1)\mathbf{R}^{-1}(k+1)\mathbf{Z}(k+1)\right.$$
$$\left. + \Sigma^{-1}(k+1|k)\Phi(k)\hat{\mathbf{X}}(k|k)\right] \tag{8.78}$$

$$\hat{\mathbf{X}}(k+1|k) = \Phi(k)\hat{\mathbf{X}}(k|k) \tag{8.79}$$

$$\Sigma(k+1|k) = \Phi(k)\Sigma(k|k)\Phi^T(k) + \Gamma(k)\mathbf{Q}(k)\Gamma^T(k) \tag{8.80}$$

$$\Sigma(k+1|k+1) = \left[\mathbf{H}^T(k+1)\mathbf{R}^{-1}(k+1)\mathbf{H}(k+1)\right.$$
$$\left. + \Sigma^{-1}(k+1|k)\right]^{-1} \tag{8.81}$$

$$\Sigma(0|0) = \Psi \tag{8.82}$$

$$\hat{\mathbf{X}}(0|0) = \mathbf{0}. \tag{8.83}$$

The last two equations give necessary initial conditions to start the iterative algorithm. The zero initial condition is logical given the zero-mean, no deterministic input system being studied. In the case of nonzero deterministic inputs, Eqs. (8.78) and (8.79) are modified to

$$\hat{\mathbf{X}}(k+1|k+1) = \Sigma(k+1|k+1)\left[\mathbf{H}^T(k+1)\mathbf{R}^{-1}(k+1)\mathbf{Z}(k+1)\right.$$
$$\left. + \Sigma^{-1}(k+1|k)(\Phi(k)\hat{\mathbf{X}}(k|k) + \Lambda(k)\mathbf{U}(k))\right] \tag{8.84}$$

$$\hat{\mathbf{X}}(k+1|k) = \Phi(k)\hat{\mathbf{X}}(k|k) + \Lambda(k)\mathbf{U}(k). \tag{8.85}$$

Initial conditions are then generally set to nonzero value

$$\hat{\mathbf{X}}(0|0) = \mathbf{X}_0.$$

Equations (8.78) through (8.83) can be algebraically manipulated to various different forms. Most popular variations follow. The estimated covariance can be expressed as

$$\boldsymbol{\Sigma}(k+1|k+1) = \boldsymbol{\Sigma}(k+1|k) - \boldsymbol{\Sigma}(k+1|k)\mathbf{H}^T(k+1)$$
$$\cdot\left[\mathbf{R}(k+1) + \mathbf{H}(k+1)\boldsymbol{\Sigma}(k+1|k)\mathbf{H}^T(k+1)\right]^{-1}$$
$$\cdot\mathbf{H}(k+1)\boldsymbol{\Sigma}(k+1|k). \tag{8.86}$$

The numerical advantage of Eq. (8.86) over Eq. (8.81) is that it requires the inversion of an $m \times m$ matrix instead of the generally larger $n \times n$ matrix that appears in parentheses in Eq. (8.81). Intuitively it also shows that the updated covariance is the result of the forecasted covariance minus a positive definite matrix (a function of the observations through $\mathbf{R}(k)$). Therefore, observations reduce the forecast uncertainty. This will be further discussed in a forthcoming section.

The most popular and appealing form of the filter is the following:

$$\hat{\mathbf{X}}(k+1|k+1) = \hat{\mathbf{X}}(k+1|k) + \mathbf{K}(k+1)\left[\mathbf{Z}(k+1)\right.$$
$$\left. - \mathbf{H}(k+1)\hat{\mathbf{X}}(k+1|k)\right] \tag{8.87}$$
$$\hat{\mathbf{X}}(k+1|k) = \boldsymbol{\Phi}(k)\hat{\mathbf{X}}(k|k) + \boldsymbol{\Lambda}(k)\mathbf{U}(k) \tag{8.88}$$
$$\mathbf{K}(k+1) = \boldsymbol{\Sigma}(k+1|k)\mathbf{H}^T(k+1)\left[\mathbf{R}(k+1)\right.$$
$$\left. + \mathbf{H}(k+1)\boldsymbol{\Sigma}(k+1|k)\mathbf{H}^T(k+1)\right]^{-1}$$
$$= \boldsymbol{\Sigma}(k+1|k+1)\mathbf{H}^T(k+1)\mathbf{R}^{-1}(k+1) \tag{8.89}$$
$$\boldsymbol{\Sigma}(k+1|k+1) = \left[\mathbf{I} - \mathbf{K}(k+1)\mathbf{H}(k+1)\right]\boldsymbol{\Sigma}(k+1|k), \tag{8.90}$$

where $\boldsymbol{\Sigma}(k+1|k)$, $\mathbf{X}(0)$, and $\boldsymbol{\Sigma}(0|0)$ are as previously defined. The important variation is in the introduction of the gain matrix $\mathbf{K}(t)$. The updated estimate is now expressed as a linear combination of the best estimate before making an observation, i.e. the forecast, and the deviation of the observation $\mathbf{Z}(t)$ from the "expected" observation, $\mathbf{H}(k+1)\hat{\mathbf{X}}(k+1|k)$. The proportionality factor is the Kalman gain matrix $\mathbf{K}(k)$. The difference $\mathbf{Z}(k+1) - \mathbf{H}(k+1)\hat{\mathbf{X}}(k+1|k)$ is called the innovation sequence and will be discussed later in detail.

8.3.2 The Continuous Filter

Following Gelb (1974) the continuous Kalman filter is derived as the limit of the discrete formulation as Δt goes to zero. To do so it is necessary to make use of the following relationships, as developed under "The Continuous Model" Section 8.2.3. As $\Delta t \to 0$

$$\Phi(t + \Delta t, t) \to \mathbf{I} + \mathbf{F}(t)\,\Delta t \tag{8.91}$$

$$\mathbf{\Gamma}(k)\mathbf{Q}(k)\mathbf{\Gamma}(k) \to \mathbf{G}(t)\mathbf{Q}(t)\mathbf{G}(t)\,\Delta t. \tag{8.92}$$

Furthermore, the discrete covariance of the measurement error approaches

$$\mathbf{R}(k+1) = \mathbf{R}(t + \Delta t) \to \frac{\mathbf{R}(t)}{\Delta t} \tag{8.93}$$

as

$$\Delta t \to 0.$$

The implication of Eq. (8.93) is that the finite variance of discrete noise goes to infinity in the continuous representation. In continuous form $\mathbf{R}(t)\delta(t - \tau)$ is a covariance with infinite values at t, and only finite variance over a finite Δt. Schweppe (1973) argues how the inverse time dependence of the discrete error covariance and the continuous equivalent is the only relationship yielding well-behaved, finite variance processes.

Substituting Eqs. (8.91) and (8.92) in the discrete-error–propagation equation was already shown to yield Eq. (8.52),

$$\Sigma(t + \Delta t|t) = \Sigma(t|t) + \left[\mathbf{F}(t)\Sigma(t|t) + \Sigma(t|t)\mathbf{F}^T(t)\right]\Delta t$$
$$+ \left[\mathbf{G}(t)\mathbf{Q}(t)\mathbf{G}^T(t)\right]\Delta t + 0(\Delta t), \tag{8.94}$$

where $0(\Delta t)$ implies zero-order terms in Δt, which are ignored.

Inserting Eq. (8.94) in the updated covariance expression given in Eq. (8.90) and dividing by Δt yields

$$\frac{\Sigma(t + \Delta t|t + \Delta t) - \Sigma(t|t)}{\Delta t} = \mathbf{F}(t)\Sigma(t|t) + \Sigma(t|t)\mathbf{F}^T(t) + \mathbf{G}(t)\mathbf{Q}(t)\mathbf{G}^T(t)$$

$$- \frac{1}{\Delta t}\mathbf{K}(t + \Delta t)\mathbf{H}(t + \Delta t)\Sigma(t|t)$$

$$- \mathbf{K}(t + \Delta t)\mathbf{H}(t + \Delta t)\mathbf{F}(t)\Sigma(t|t)$$

$$- \mathbf{K}(t + \Delta t)\mathbf{H}(t + \Delta t)\Sigma(t|t)\mathbf{F}^T(t)$$

$$- \mathbf{K}(t + \Delta t)\mathbf{H}(t + \Delta t)\mathbf{G}(t)\mathbf{Q}(t)\mathbf{G}^T(t) + 0(\Delta t). \tag{8.95}$$

The term $\mathbf{K}(t + \Delta t)/\Delta t$ is expanded using Eq. (8.89)

$$\frac{1}{\Delta t}\mathbf{K}(t + \Delta t) = \frac{1}{\Delta t}\Sigma(t + \Delta t|t)\mathbf{H}^T(t + \Delta t)$$
$$\cdot\left[\mathbf{R}(t + \Delta t) + \mathbf{H}(t + \Delta t)\Sigma(t + \Delta t|t)\mathbf{H}^T(t + \Delta t)\right]^{-1}$$
$$= \Sigma(t + \Delta t|t)\mathbf{H}^T(t + \Delta t)$$
$$\cdot\left[\mathbf{R}(t + \Delta t)\Delta t + \mathbf{H}(t + \Delta t)\Sigma(t + \Delta t|t)\mathbf{H}^T(t + \Delta t)\Delta t\right]^{-1},$$

which as Δt approaches zero, using Eq. (8.93), results in

$$\lim_{\Delta t \to 0} \frac{\mathbf{K}(t + \Delta t)}{\Delta t} \to \Sigma(t|t)\mathbf{H}^T(t)\mathbf{R}^{-1}(t), \tag{8.96}$$

where $\mathbf{R}(t)$ is now the spectral distribution of the continuous measurement noise. Using Eq. (8.96) in (8.95) when Δt approaches zero yields

$$\dot{\Sigma}(t|t) = \mathbf{F}(t)\Sigma(t|t) + \Sigma(t|t)\mathbf{F}^T(t) + \mathbf{G}(t)\mathbf{Q}(t)\mathbf{G}^T(t)$$
$$- \Sigma(t|t)\mathbf{H}^T(t)\mathbf{R}^{-1}(t)\mathbf{H}(t)\Sigma(t|t), \tag{8.97}$$

which is the continuous-error–propagation equation when a continuous observation with spectral distribution $\mathbf{R}(t)$ is taken.

Using Eq. (8.91) in the state update equation, Eq. (8.87), gives

$$\frac{\hat{\mathbf{X}}(t + \Delta t|t + \Delta t) - \hat{\mathbf{X}}(t|t)}{\Delta t} = \mathbf{F}(t)\hat{\mathbf{X}}(t|t) + \frac{\mathbf{K}(t + \Delta t)}{\Delta t}$$
$$\cdot\left[\mathbf{Z}(t + \Delta t) - \mathbf{H}(t + \Delta t)\hat{\mathbf{X}}(t|t)\right.$$
$$\left. - \mathbf{H}(t + \Delta t)\mathbf{F}(t)\hat{\mathbf{X}}(t|t)\Delta t\right],$$

which as $\Delta t \to 0$ (using Eq. 8.96) leads to the continuous update equation for zero-mean processes,

$$\dot{\hat{\mathbf{X}}}(t|t) = \mathbf{F}(t)\hat{\mathbf{X}}(t|t) + \Sigma(t|t)\mathbf{H}^T(t)\mathbf{R}^{-1}(t)\left[\mathbf{Z}(t) - \mathbf{H}(t)\hat{\mathbf{X}}(t|t)\right]$$
$$= \mathbf{F}(t)\hat{\mathbf{X}}(t|t) + \mathbf{K}(t)\left[\mathbf{Z}(t) - \mathbf{H}(t)\hat{\mathbf{X}}(t|t)\right] \tag{8.98}$$
$$\mathbf{K}(t) = \Sigma(t|t)\mathbf{H}^T(t)\mathbf{R}^{-1}(t). \tag{8.99}$$

Equations (8.97) through (8.99) are the Kalman filter for continuous systems and continuous observations. The state update equation is modified as follows for the nonzero-mean, deterministic input case,

$$\dot{\hat{\mathbf{X}}}(t|t) = \mathbf{F}(t)\hat{\mathbf{X}}(t|t) + \mathbf{L}(t)\mathbf{U}(t) + \mathbf{K}(t)\left[\mathbf{Z}(t) - \mathbf{H}(t)\hat{\mathbf{X}}(t|t)\right]. \tag{8.100}$$

8.3.3   Observability and Controllability of Stochastic Systems

Section 8.2.2 discussed controllability of deterministic systems. In stochastic systems controllability depends on the level of added noise to the different model states. Models with errors resulting in unbounded or ill-conditioned state covariances are not controllable. Gelb (1974) defines the controllability condition of nonstationary continuous stochastic systems as the achievement of bounded and positive definite error covariances for some time $t > 0$. For the unobserved system the error covariance is given by the solution of Eq. (8.53), which Gelb gives as

$$\Sigma(t|t) = \int_0^t \Phi(t, \tau) \mathbf{G}(\tau) \mathbf{Q}(\tau) \mathbf{G}^T(\tau) \Phi^T(t, \tau) \, d\tau, \qquad (8.101)$$

where $\Phi(t, \tau)$ is the transition matrix of the system. For a uniformly completely controllable system $\Sigma(t|t)$ must be bounded and positive definite, which results from every state being excited by the model noise.

The condition for discrete systems is that

$$\beta_1 \mathbf{I} \leq \sum_{i=k-N}^{k-1} \Phi(k, i+1) \mathbf{Q}(i) \Phi^T(k, i+1) \leq \beta_2 \mathbf{I} \qquad (8.102)$$

for some $N > 0$ and for positive values of β_1 and β_2.

Uniform and complete observability of linear, nonstationary, stochastic systems ensures that given enough observations it is possible to estimate, with finite variance, every element of the state vector even if the initial state of knowledge was absolute ignorance (infinite variance). This condition is expressed in terms of the information matrix, or the inverse of the variance of estimation, given certain observations. Gelb (1974) gives the solution to Eq. (8.97), given $\Sigma(0|0) \to \infty$, as

$$\Sigma^{-1}(t|t) = \int_0^t \Phi^T(\tau, t) \mathbf{H}^T(\tau) \mathbf{R}^{-1}(\tau) \mathbf{H}(\tau) \Phi(\tau, t) \, d\tau. \qquad (8.103)$$

Finite $\Sigma(t|t)$ requires the existence and boundedness of $\Sigma^{-1}(t|t)$ (positive definiteness) for some $t > 0$. The equivalent observability criteria for discrete systems is

$$\alpha_1 \mathbf{I} \leq \sum_{i=k-N}^{k} \Phi^T(i, k) \mathbf{H}^T(i) \mathbf{R}^{-1}(i) \mathbf{H}(i) \Phi(i, k) \leq \alpha_2 \mathbf{I} \qquad (8.104)$$

for some $N > 0$, $\alpha_1 > 0$, and $\alpha_2 > 0$.

Complete and uniform stochastic observability and controllability, bounded error covariances, and stable dynamics ensure filter stability. This desirable feature implies that even in unobserved systems, filter estimates converge to process mean values and finite variances. In practice, stability may not exist or may be hard to prove but reasonable filter performance may still be observed. Frequent observation is the best insurance against the undesirable divergence of filter results.

8.3.4 Comments on Filter Form and Behavior

Before proceeding with some hydrologic applications of the Kalman filter, it is instructive to study its form and expected behavior. From Eq. (8.78), it is clear that the best linear estimate of the state is a linear combination of the observation $\mathbf{Z}(k+1)$ and the model prediction $\hat{\mathbf{X}}(k+1|k)$. If observations are highly unreliable $\mathbf{R}(k+1)$ will be large, implying that $\mathbf{R}^{-1}(k+1)$ goes to zero, at least relative to the model covariance $\Sigma(k+1|k+1)$. The weight on the observation, $\Sigma(k+1|k+1)\mathbf{H}^T(k+1)\mathbf{R}^{-1}(k+1)$ is then nearly zero and the best estimate is a linear function of the model forecast. If the model error covariance $\mathbf{Q}(k)$ goes to infinity, Eq. (8.80) indicates that $\Sigma(k+1|k)$ also approaches infinity, leading to zero weight on the model prediction during estimation (see Eq. 8.78). In that case the updating step relies exclusively on the observation.

One of the valuable properties of the Kalman filter (and linear estimators in general) is the independence of the state covariance matrices of the observations. Equations (8.80) and (8.81), which propagate and update state covariance, respectively, do not depend on state observations. That property allows a priori knowledge of filter behavior and accuracy, an invaluable characteristic in system design.

The value of information is explicitly given in Eq. (8.86). The updated covariances are given as differences of positive definite matrices. The result must also be positive definite, which then implies that estimation covariance $\Sigma(k+1|k+1)$ must be "smaller" or equal to forecast covariance $\Sigma(k+1|k)$. The covariance-reduction value of each piece of information is embodied in the second term of Eq. (8.86). Note that if $\mathbf{R}(k+1)$ goes to infinity, the second term in Eq. (8.86) will go to the null matrix, implying no information content in the observation. It should not be ignored that $\Gamma(k)$ and $\mathbf{H}(k)$ play major roles in the propagation of covariances. Model error can be unacceptable given large $\Gamma(k)$, even if $\mathbf{Q}(k)$ is small. Similarly, observations may be useless if the measurement matrix $\mathbf{H}(k)$ is so small that observations lack power to help in the estimation of the state.

A final comment on the form of the filter is required. Equation (8.87) emphasizes the linear combination of the two sources of information used in Kalman filtering. One source is the model, the other consists of the observations. Note that the new estimate is the latest (and hopefully best) state estimate plus a correction proportional to the deviation of observations from expected observations. This filter structure can and has been used in innumerable filter formulations, all suboptimal for the problem discussed in this chapter. Suboptimal filtering may nevertheless be desirable for the sake of numerical efficiency or to solve problems not fitting the canonical description of this chapter. The "art" of filter design lies in the selection of appropriate gains with some minimum stability and convergence characteristics. What makes the Kalman filter optimal for the canonical model discussed here is the form of the Kalman gain matrix $\mathbf{K}(k)$. Some of the implications of this optimality will be discussed under "Optimality and Filter Parameters" (following).

Finally, the real value of the Kalman filter is the explicit incorporation of model structure (represented by a stochastic differential or difference equation) in the estimation problem. The estimation and forecasting will be improved with model formulation. Hopefully this formulation includes basic physical knowledge of the process. A priori information must be exploited to a maximum. If the model structure is poor or lacks information beyond that exclusively available from data, the procedure then takes a purely statistical meaning, nothing more than a least-squares procedure.

Optimality and Filter Parameters

Optimality of the Kalman filter depends on complete adherence to the system canonical form and complete knowledge of all parameters. That implies exact information, e.g., for the discrete case, on $\mathbf{\Phi}(k)$, $\mathbf{H}(k)$, $\mathbf{Q}(k)$, $\mathbf{R}(k)$, $\mathbf{\Lambda}(k)$, and $\mathbf{\Gamma}(k)$. Optimality is simply not guaranteed unless the above is true.

The optimal filter will exhibit an uncorrelated innovation sequence. The innovations were previously defined as

$$\mathbf{v}(k) = \mathbf{Z}(k) - \mathbf{H}(k)\hat{\mathbf{X}}(k|k-1). \qquad (8.105)$$

Under optimality, $\mathbf{v}(k)$ is a white-noise sequence with the following properties,

$$E[\mathbf{v}(k)] = \mathbf{0}$$
$$E[\mathbf{v}(k)\mathbf{v}^T(m)] = [\mathbf{H}(k)\mathbf{\Sigma}(k|k-1)\mathbf{H}^T(k) + \mathbf{R}(k)]\delta(k-m). \qquad (8.106)$$

Although the above properties help determine if a given filter formulation is optimally processing data (failure to do so implies that improvement is possible with better parameter selection) it does not give information about the relative value of different models and filters. A suboptimally operating model–filter combination may, for example, be a better forecasting tool than another model–filter set.

If a filter is observed to be suboptimal, it is conceptually possible to adjust model–filter parameters in order to achieve optimality. In other words, observed deviations from the theoretical-innovations property can be used to drive a search procedure for model–filter parameters. Such techniques are called adaptive filtering and will be discussed in Chapter 9.

Incorrect specification of parameters, particularly $\mathbf{Q}(k)$ and $\mathbf{R}(k)$, commonly leads to an inconsistent filter behavior called divergence. In such cases the actual observed errors of prediction and estimation are larger than the theoretically expected ones. For example, too small a $\mathbf{Q}(k)$ matrix will underestimate the actual innovation variance as given by Eq. (8.106). Remember that $\Sigma(k|k-1)$ is directly a function of $\mathbf{Q}(k)$. Solutions to divergence range from trial-and-error procedures, which increase noise covariances consistently, to sophisticated adaptive filtering techniques, which consider model and filter parameters as unknowns to be identified. Again, Chapter 9 will discuss divergence in some detail.

8.4 APPLICATIONS AND EXAMPLES

8.4.1 Parameter Estimation and Forecasting

In 1973 Hino published one of the first applications of Kalman filtering to water resources and in particular to hydrology. The problem was to identify the parameters of a linear, lumped, discharge model from sequentially available data points. Updated parameters are then used to forecast streamflow between measurements. His formulation follows.

Runoff or discharge can be generally represented by an equation of the following form:

$$
\begin{aligned}
Q_{1,k} = {} & \alpha_{11}^1 Q_{1,k-1} + \alpha_{12}^1 Q_{1,k-2} + \cdots + \alpha_{1j}^1 Q_{1,k-j} \\
& + \alpha_{21}^1 Q_{2,k-1} + \alpha_{22}^1 Q_{2,k-2} + \cdots + \alpha_{2j}^1 Q_{2,k-j} \\
& + \cdots + a_{11}^1 I_{1,k-1} + a_{12}^1 I_{1,k-2} + \cdots + a_{1\ell}^1 I_{1,k-\ell} \\
& + a_{21}^1 I_{2,k-1} + a_{22}^1 I_{2,k-2} + \cdots + a_{2\ell}^1 I_{2,k-\ell} \\
& + \cdots + W_{1,k-1}
\end{aligned}
\tag{8.107}
$$

$$
\begin{aligned}
\vdots \\
Q_{m,k} = {} & \alpha_{11}^m Q_{1,k-1} + \alpha_{12}^m Q_{1,k-2} + \cdots + \alpha_{1j}^m Q_{1,k-j} \\
& + \cdots + a_{11}^m I_{1,k-1} + a_{12}^m I_{1,k-2} + \cdots + a_{1\ell}^m I_{1,k-\ell} \\
& + \cdots + W_{m,k-1}.
\end{aligned}
$$

The above is the general representation of discharge Q at $m(i=1$ to $m)$ locations at time k. It is clear that in this general formulation discharge at a point i at time k can be a function of j previous discharges and ℓ previous rainfall inputs I at any of m different locations. In reality, this relation can be considerably simplified since discharge at any location is usually assumed to be a function of previous discharges at that location and rainfall at contributing overland segments.

Equation (8.107) can be expressed in matrix notation as

$$Q(k) = M(k-1)h(k-1) + W(k-1), \tag{8.108}$$

where $Q(k)$ is the m-dimensional vector of discharges

$$Q(k) = [Q_{1,k}, Q_{2,k}, \ldots, Q_{m,k}] \tag{8.109}$$

and $h(k-1)$ is a vector with $(j \times m + \ell \times n) \times m$ elements. The form of this vector is

$$h^T(k-1) = \left[\alpha_{11}^1 \alpha_{12}^1 \ldots \alpha_{1j}^1; \; \alpha_{21}^1 \ldots \alpha_{2j}^1 \ldots a_{11}^1 a_{12}^1 \ldots a_{1\ell}^1; \right.$$
$$a_{21}^1 a_{22}^1 \ldots a_{2\ell}^1 \ldots \alpha_{11}^2 \alpha_{12}^2 \ldots \alpha_{1j}^2; \; \alpha_{21}^2 \ldots$$
$$\left. \alpha_{2j}^2 \ldots a_{11}^2 a_{12}^2 \ldots a_{1\ell}^2; \; a_{21}^2 a_{22}^2 \ldots a_{2\ell}^2; \ldots \right]. \tag{8.110}$$

The elements of the vector are the coefficients of the discharge and input variables that appear in Eq. (8.107).

M is a matrix of m rows and $(j \times m + \ell \times n) \times m$ columns. The matrix is diagonal with terms of the form

$$y = \left[Q_{1,k-1} Q_{1,k-2} \cdots Q_{1,k-j} Q_{2,k-1} \cdots Q_{2,k-j} \cdots \right.$$
$$Q_{m,k-1} \cdots Q_{m,k-j} I_{1,k-1} I_{1,k-2} \cdots I_{1,k-\ell}$$
$$\left. I_{2,k-1} \cdots I_{2,k-\ell} \cdots I_{n,k-1} \cdots I_{n,k-\ell} \right]. \tag{8.111}$$

$W(k-1)$ is a vector of m elements equal to the white-noise component $W_{1,k-1} \ldots W_{m,k-1}$ shown in Eq. (8.107).

Substituting a variable vector $Z(k)$ for $Q(k+1)$ in Eq. (8.108) results in an equation of the form

$$Z(k) = M(k)h(k) + W(k). \tag{8.112}$$

The above can be interpreted as a measurement system where $Z(k)$ is the observation on the state vector $h(k)$.

$Z(k)$ is the vector of observed discharges at time $k+1$. The dynamics of the state vector $h(k)$ are simply of the form

$$h(k+1) = h(k) + V(k), \tag{8.113}$$

where $V(k)$ is a white noise component that adds uncertainty to the otherwise invariant state.

Equations (8.112) and (8.113) define a dynamic system that can be studied with Kalman filter techniques. The formulation becomes

$$\hat{\mathbf{h}}(k+1|k+1) = \hat{\mathbf{h}}(k|k) + \mathbf{K}(k+1)\{\mathbf{Z}(k+1) - \mathbf{M}(k+1)\hat{\mathbf{h}}(k|k)\}$$

$$\mathbf{K}(k+1) = \boldsymbol{\Sigma}(k+1|k+1)\mathbf{M}^T(k+1)\mathbf{R}^{-1}(k+1)$$

$$\boldsymbol{\Sigma}(k+1|k+1) = \boldsymbol{\Sigma}(k+1|k) - \boldsymbol{\Sigma}(k+1|k)\mathbf{M}^T(k+1)\big[\mathbf{R}(k+1) + \mathbf{M}(k+1)$$

$$\cdot \boldsymbol{\Sigma}(k+1|k)\mathbf{M}^T(k+1)\big]^{-1}\mathbf{M}(k+1)\boldsymbol{\Sigma}(k+1|k) \qquad (8.114)$$

$$\boldsymbol{\Sigma}(k+1|k) = \boldsymbol{\Sigma}(k|k) + \mathbf{Q}(k),$$

where, in this case,

$$E\big[\mathbf{W}(k)\mathbf{W}^T(k)\big] = \mathbf{R} \qquad \text{and} \qquad E\big[\mathbf{V}(k)\mathbf{V}^T(k)\big] = \mathbf{Q}.$$

Equation (8.114) gives an estimation of the runoff-model parameters sequentially using available information. These new parameters can then be used to forecast until a new observation becomes available.

The above lumped formulation is particularly attractive for use with the traditional unit hydrograph approach. For such a case, the runoff model simplifies considerably because discharge is only a function of past inputs. Matrix $\mathbf{M}(k)$ then consists of observed rainfall and the elements of vector $\mathbf{h}(k)$ become the unit hydrograph ordinates.

In his original work Hino (1973) presented an example of identifying the ordinates of a unit hydrograph. In that case Eq. (8.107) is simplified to one location and exclusive dependence of discharge on present and past rainfalls. Figure 8.1 presents the various unit-hydrograph ordinates (elements of h) identified at various time steps with different initial conditions on the state. As more information becomes available the identification improves. Figure 8.2 gives predicted hydrographs based on the latest unit-hydrograph estimates at various times using poor initial estimates of $\mathbf{h}(k)$, which were given in Fig. 8.1a. Figure 8.2a assumes no rainfall beyond the latest observed point. Figure 8.2b corresponds to the case when the future rainfall history is perfectly known. Figure 8.3 again provides forecasted information, this time using the $\mathbf{h}(k)$ estimated from a good initial condition. As expected, forecasted values improve with $\mathbf{h}(k)$. In his analysis Hino assumed values of $\mathbf{R}$ and $\mathbf{Q}$.

It is important to conclude this example with a commentary on some of the unattractive features of the system.

1. The lumped, black-box–system approach does not allow an explicit relation between the identified unit hydrograph and the physical parameters of the basin.

2. The interpretation of the observation equation (Eq. 8.112) eliminates the explicit consideration of true observation errors due to instruments or techniques.

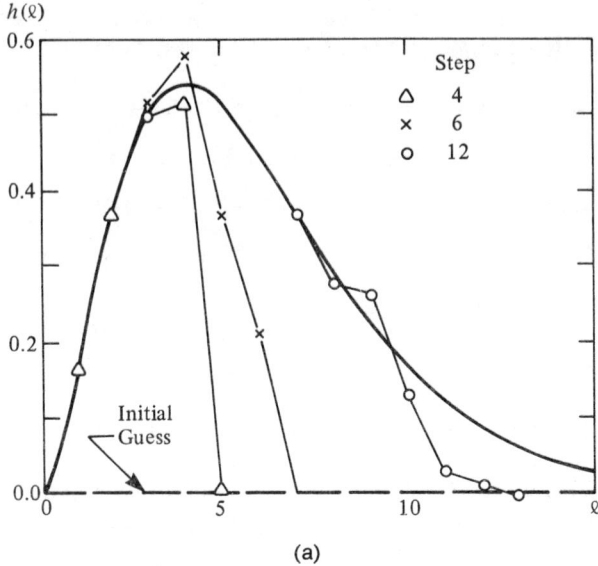

(a)

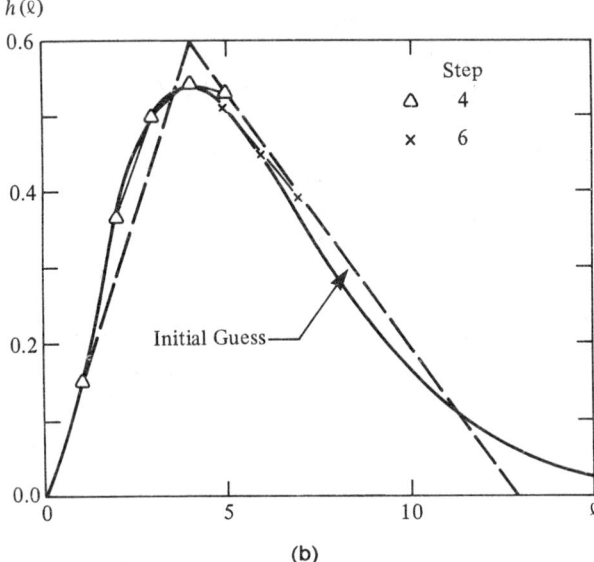

(b)

Figure 8.1   (a) Unit-hydrograph identification from the zero initial guess (from Hino, 1973). (b) Unit-hydrograph identification from a relatively accurate initial guess (from Hino, 1973).

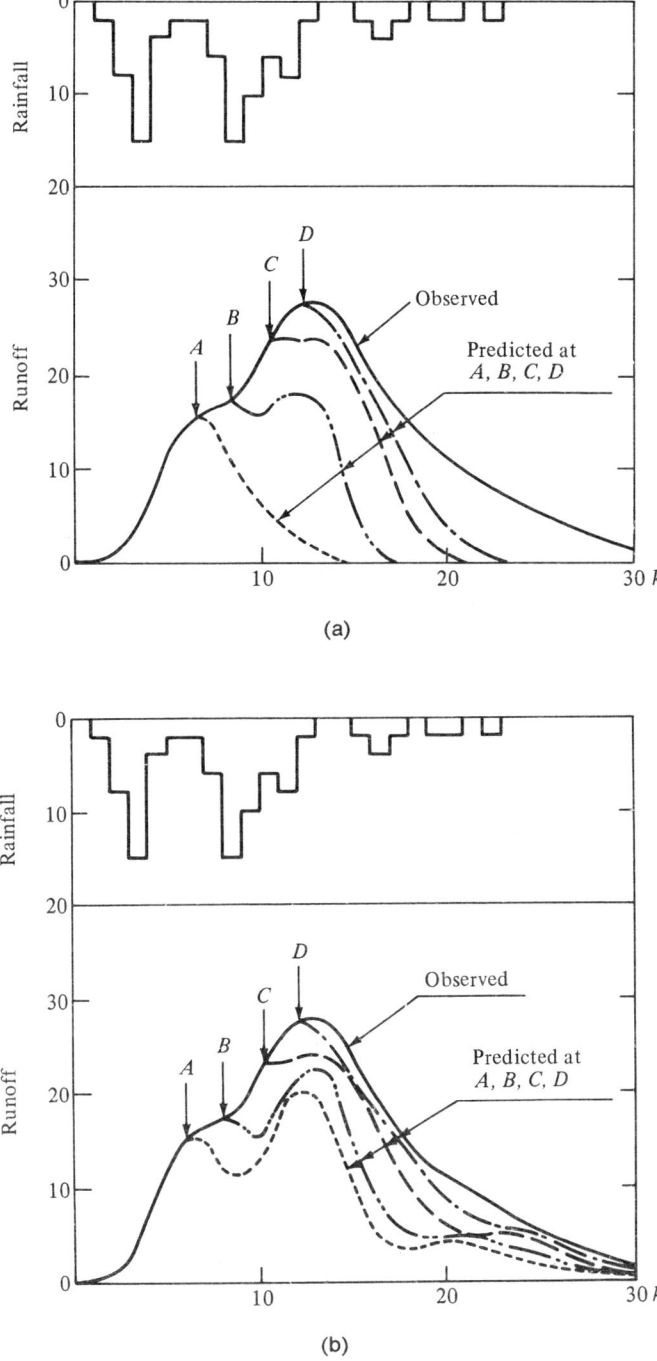

Figure 8.2 (a) On-line runoff prediction from the zero initial guess of $\mathbf{h}(k)$ when future rainfall data are not given (from Hino, 1973). (b) On-line runoff prediction from the zero initial guess of $\mathbf{h}(k)$ when future rainfall data are given (from Hino, 1973).

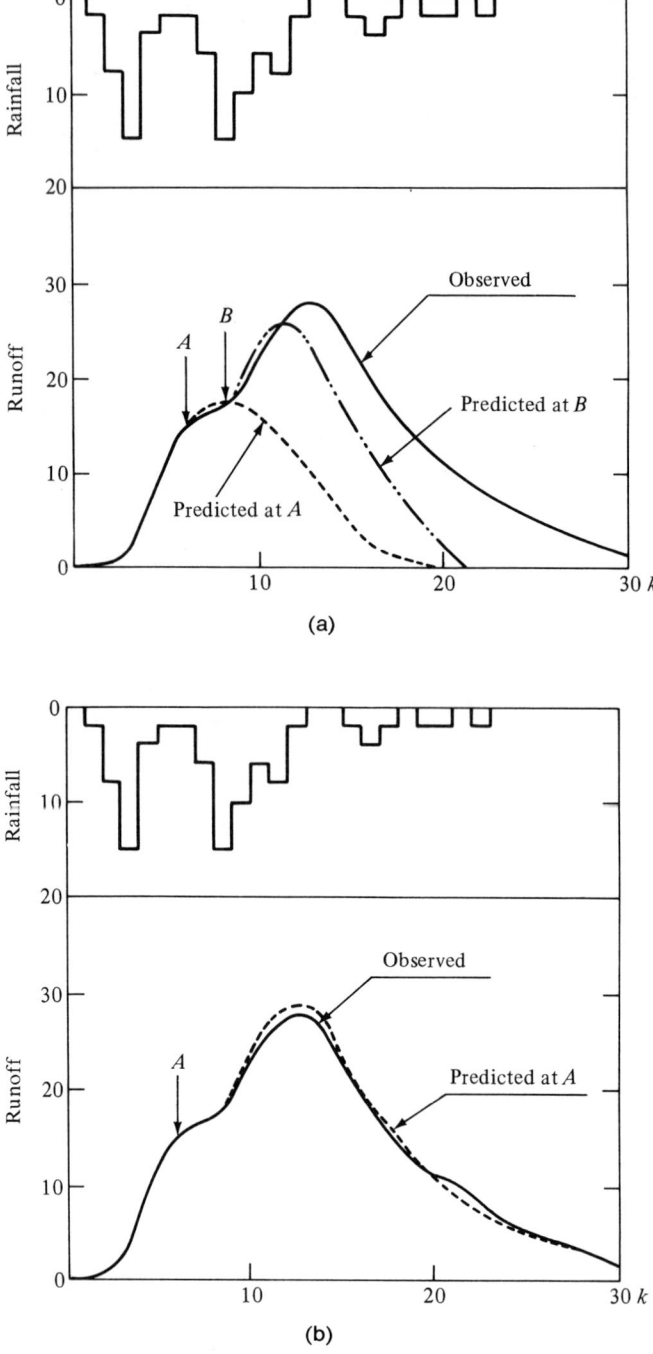

Figure 8.3 (a) On-line forecast from a relatively accurate initial curve of $\mathbf{h}(k)$ (Fig. 8.1b) when future rainfall data are not given (from Hino, 1973). (b) On-line runoff forecast from a relatively accurate initial curve of $\mathbf{h}(k)$ (Fig. 8.1b) when future rainfall data are given (from Hino, 1973).

3. The model used is a time-invariant linear system. For basins exhibiting large nonlinearities we cannot expect to ever achieve a steady-state solution for the catchment response function, $\mathbf{h}(k)$.

4. Error terms $\mathbf{W}(k)$ and $\mathbf{V}(k)$ are hard to describe or to quantify in terms of their covariances. Particularly, the term $\mathbf{V}(k)$ has no obvious physical significance that could be used in its estimation.

5. The dimensions of the involved matrices and vectors can be unmanageable if fine resolution of the unit hydrograph is required, if the "memory" ℓ is large and if we consider discharge at more than one location.

Kitanidis and Bras (1979) studied the stability of parameter estimates based on Kalman filtering techniques. Their results also illustrate some of the common numerical-accuracy problems encountered in filtering techniques. The objective was to fit a hydrologic model of the form

$$Q(k) = \beta_1 Q(k-1) + \beta_2 Q(k-2) + \beta_3 i(k) + \beta_4 i(k-1) + \beta_5 i(k-2),$$
$$(8.115)$$

where $Q(k)$ is the mean daily flow in cubic feet per second (cfs); $i(k)$ is the mean area precipitation (cfs); β_i is the ith coefficient of the discharge relationship. To use the Kalman filter in parameter estimation a procedure similar to that of Hino (1973) is used. A measurement equation is suggested,

$$y(k+1) = \mathbf{X}^T(k+1)\beta(k+1) + \varepsilon(k+1), \qquad (8.116)$$

where $\varepsilon(k+1) \sim N(0, \sigma^2)$, while $E[\varepsilon(k)\cdot\varepsilon(m)] = 0$, $k \neq m$.

In the above, $y(k)$ are observed daily average discharges. The measurement matrix $\mathbf{X}^T$ is of dimensions (1×5) with elements the observed values of $Q(k-1)$, $Q(k-2)$, $i(k)$, $i(k-1)$ and $i(k-2)$. The state vector $\beta(k)$ consists of the unknown coefficients. $\beta(k)$ is a 5×1 vector. Since the suggested model is assumed to be time invariant, the coefficients are assumed to be unknown constants with dynamics

$$\beta(k+1) = \beta(k). \qquad (8.117)$$

In the Kalman filter, which is the minimum posterior variance estimator, it is usually assumed that the measurement variance σ^2 is known. Although there are ways to get estimates of this quantity, Kitanidis and Bras (1979) use an arbitrary estimate of σ^2, since their purpose is to estimate the parameters and not the error covariance matrix of the estimates. It is convenient to divide the measured quantities $\mathbf{X}^T(k)$ and $y(k)$ by σ, in order to obtain a measurement error with variance 1. In this case, the Kalman filter provides the best estimate, after $k+1$ measurements:

$$\hat{\beta}(k+1|k+1) = \hat{\beta}(k|k) + \mathbf{K}(k+1)\left[y(k+1) - \mathbf{X}^T(k+1)\beta(k|k)\right],$$
$$(8.118)$$

where

$$\mathbf{K}(k+1) = \Sigma(k+1|k)\mathbf{X}(k+1)\left[1+\mathbf{X}^T(k+1)\Sigma(k+1|k)\mathbf{X}(k+1)\right]^{-1}$$
$$\Sigma(k+1|k) = \Sigma(k|k), \tag{8.119}$$

where $\hat{\beta}(k|k)$ is the best estimate of β at time k given information to time k.

As initial conditions, $\hat{\beta}(0|0) = \mathbf{0}$ and $\Sigma(0|0) = 10^{20}$ were assumed. This is essentially diffuse prior information about the parameters. The results are given in Table 8.1.

It is noted that even though only five parameters are estimated, it takes almost three years of data until $\hat{\beta}_3$, $\hat{\beta}_4$, and $\hat{\beta}_5$ are stabilized near their final values. $\hat{\beta}_1$ seems to continuously increase even for more than six years of data.

Another interesting problem is the numerical accuracy of the Kalman filter. Because of the fact that Kalman filtering is susceptible to accuracy degradation due to sensitivity to computer roundoff errors, a high accuracy and stability version of the Kalman filter, known as the Potter algorithm, was used (Potter and Stern, 1963). Additionally, all measurements $y(t)$ and the elements of $\mathbf{X}(t)$ were normalized so that all numbers were approximately of the same order of magnitude. The final result was checked through a high-accuracy off-line ordinary-least-squares estimation using singular value decomposition and double-precision arithmetic. The relative difference was found to be less than 10^{-3}. However, without conditioning of the measurements to the same order of magnitude the algorithm failed to converge to the right solution.

Table 8.1
Kalman filter estimates for the parameters of the hydrologic model of Eq. (8.115)

Number of observations	$\hat{\beta}_1$	$\hat{\beta}_2$	$\hat{\beta}_3$	$\hat{\beta}_4$	$\hat{\beta}_5$
200	0.7502	−0.0168	26.24	56.21	−20.24
400	0.7727	−0.0232	19.78	36.69	−12.16
600	0.7976	−0.0256	17.87	29.45	−7.35
800	0.8320	−0.0411	15.26	24.97	−7.20
1000	0.8483	−0.0327	10.84	23.15	−8.08
1200	0.8701	−0.0411	9.86	23.74	−9.75
1400	0.8621	−0.0295	9.34	23.50	−10.85
1600	0.8719	−0.0296	10.47	23.10	−11.11
1800	0.8730	−0.0319	9.93	22.51	−10.45
2000	0.8794	−0.0370	9.40	23.58	−10.36
2200	0.8824	−0.0395	8.66	22.89	−9.95
2400	0.9004	−0.0460	8.69	22.46	−10.52
2600	0.9078	−0.0517	8.41	20.37	−8.88
2800	0.9314	−0.0672	8.21	19.60	−9.32

Source: Kitanidis and Bras (1979).

For 2800 observations parameter estimates from the unconditioned algorithm were $\hat{\beta}_1 = 0.7005$, $\hat{\beta}_2 = 0.0959$, $\hat{\beta}_3 = 9.15$, $\hat{\beta}_4 = 21.65$, and $\hat{\beta}_5 = -5.42$.

This is indicative of the numerical problems that can appear in Kalman filter estimation and the need to take special measures to improve the numerical accuracy of estimation algorithms.

8.4.2 Rainfall-Network Design

Chapter 6 discussed the results of Rodríguez-Iturbe and Mejía (1974) in designing networks to estimate the long-term area average of rainfall,

$$P = \lim_{T \to \infty} \frac{1}{TA} \sum_{k=1}^{T} \int_A f(\mathbf{x}, k) \, d\mathbf{x} \qquad (8.120)$$

where $f(\mathbf{x}, k)$ is the rainfall accumulation at location $\mathbf{x}$ during period k (i.e., weekly, monthly, yearly); and T is the total number of time periods accounted for.

Rodríguez-Iturbe and Mejía (1974) obtained an expression for the mean square error of, approximating Eq. (8.120),

$$E\left[(P - \hat{P})^2\right] \qquad (8.121)$$

when

$$\hat{P} = \frac{1}{T} \frac{1}{N} \sum_{k=1}^{T} \sum_{j=1}^{N} f(\mathbf{x}_j, k).$$

Their work assumed random or stratified random sampling. Covariance functions were known, isotropic, and homogeneous. Furthermore, the covariance was assumed to be of separable form

$$\text{cov}\left(f(\mathbf{x}_1, k_1); f(\mathbf{x}_2, k_2)\right) = \sigma^2 \rho^{|\tau|} r(\tilde{\mathbf{x}}), \qquad (8.122)$$

where $\tau = k_2 - k_1$, $\tilde{\mathbf{x}} = \mathbf{x}_2 - \mathbf{x}_1$ is the distance between points 2 and 1, $r(\cdot)$ is the spatial covariance function, ρ is the lag-one autocorrelation coefficient, and σ^2 is the point variance of the process.

Rodríguez-Iturbe and Mejía proved that for a stationary process in space and time, P (Eq. 8.120) has zero variance, and therefore is a constant. Using that result, Bras and Colón (1978) modeled the rainfall process as

$$f(\mathbf{x}_i, k) = P + \varepsilon(\mathbf{x}_i, k), \qquad (8.123)$$

where P is the mean area time average of the process, as defined in Eq. (8.120), and $\varepsilon(\mathbf{x}_i, k)$ is a deviation from the mean at the point with coordinate vector $\mathbf{x}_i$ and time k. The mean of $\varepsilon(\mathbf{x}_i, k)$ is zero.

Bras and Colón suggested modeling the values of $\varepsilon(\mathbf{x}_i, k)$ as a multivariate autoregressive model (see Chapter 3), an assumption that corresponds exactly to Rodríguez-Iturbe and Mejía's separable covariance structure (Eq. 8.122), but is not limited to it. The rainfall process was then represented by

$$
\begin{aligned}
\mathbf{f}(k) &= \mathbf{1}P + \mathbf{A}\{\mathbf{f}(k-1) - \mathbf{1}P\} + \mathbf{B}\mathbf{W}(k) \\
&= \mathbf{1}P + \mathbf{A}\varepsilon(k-1) + \mathbf{B}\mathbf{W}(k) \\
&= [\mathbf{I} - \mathbf{A}]\mathbf{1}P + \mathbf{A}\mathbf{f}(k-1) + \mathbf{B}\mathbf{W}(k),
\end{aligned} \tag{8.124}
$$

where $\mathbf{f}^T(k) = [f(\mathbf{x}_1, k) \ldots f(\mathbf{x}_N, k)]$, $\varepsilon^T(k) = [\varepsilon(\mathbf{x}_1, k) \ldots \varepsilon(\mathbf{x}_N, k)]$, $\mathbf{I}$ is the identity matrix, $\mathbf{1}$ is an $N \times 1$ vector of 1s, $\mathbf{A}$ is an $N \times N$ matrix, $\mathbf{B}$ is an $N \times N$ matrix, and $\mathbf{W}(t)$ is an $N \times 1$ vector of random variables:

$$
E\left[\mathbf{W}(k_1)\mathbf{W}^T(k_2)\right] = \begin{cases} \mathbf{I} & k_1 = k_2 \\ \mathbf{0} & k_1 \neq k_2. \end{cases}
$$

Matrices $\mathbf{A}$ and $\mathbf{B}$ are easily obtained (see Chapter 3) given the covariance (or an estimate) function of the rainfall process (i.e., Eq. 8.122).

N is the number of rain gages in the basin. Observations are again of the form

$$
\mathbf{Z}(k) = \mathbf{H}\mathbf{f}(k) + \mathbf{V}(k), \tag{8.125}
$$

where $\mathbf{Z}(k)$ is the $N \times 1$ vector of imperfect observations at time k, $\mathbf{f}(k)$ is the $N \times 1$ vector of rainfall at time k and at the N spatial locations as given by Eq. (8.124), $\mathbf{V}(k)$ is the $N \times 1$ vector of measurement error, zero mean, and uncorrelated with $\mathbf{f}(k)$:

$$
E\left[\mathbf{V}(k_1)\mathbf{V}^T(k_2)\right] = \begin{cases} \mathbf{R} & k_1 = k_2 \\ \mathbf{0} & k_1 \neq k_2, \end{cases}
$$

$\mathbf{R}$ is the measurement error covariance matrix, assumed to be of diagonal form, and $\mathbf{H}$ is the $N \times N$ matrix defining the rain-gage network.

The mean P in Eq. (8.124) is an unknown variable. Augmenting the state vector, the following dynamics and observation equations were suggested

$$
\mathbf{f}'(k) = \mathbf{A}'\mathbf{f}'(k-1) + \mathbf{B}'\mathbf{W}(k) \tag{8.126}
$$
$$
\mathbf{Z}(k) = \mathbf{H}'\mathbf{f}'(k) + \mathbf{V}(k), \tag{8.127}
$$

where

$$
\mathbf{f}'(k) = \begin{bmatrix} \varepsilon(k) \\ \cdots \\ P(k) \end{bmatrix} \qquad \mathbf{A}' = \begin{bmatrix} \mathbf{A} & \vdots & 0 \\ \cdots & \cdots & \cdots \\ 0 & \vdots & 1 \end{bmatrix} \qquad \mathbf{B}' = \begin{bmatrix} \mathbf{B} \\ \cdots \\ 0 \end{bmatrix} \qquad \mathbf{H}' = \mathbf{H}\begin{bmatrix} \mathbf{I} & \vdots & \mathbf{1} \end{bmatrix},
$$

where $\mathbf{1}$ is an $N \times 1$ column vector of 1s.

The system of Eqs. (8.126) and (8.127) can be studied under a Kalman filter formulation to obtain linear estimates of the state vector $\mathbf{f}'(k)$ given observations $\mathbf{Z}(k)$. Furthermore, the linear solution yields the mean square error of estimating $\mathbf{f}'(k)$ independently of the observations. The above implies having the ability to obtain the error of estimating P for a given network, in terms of number and locations of observations, and given the number of time periods.

The mean-square-error matrix is given by the following pair of iterative equations,

$$\Sigma(k|k) = \Sigma(k|k-1) - \Sigma(k|k-1)\mathbf{H}'^{T}\{\mathbf{R} + \mathbf{H}'\Sigma(k|k-1)\mathbf{H}'^{T}\}^{-1}\mathbf{H}'\Sigma(k|k-1)$$

$$(8.128)$$

$$\Sigma(k|k-1) = \mathbf{A}'\Sigma(k-1|k-1)\mathbf{A}'^{T} + \mathbf{B}'\mathbf{B}'^{T},$$

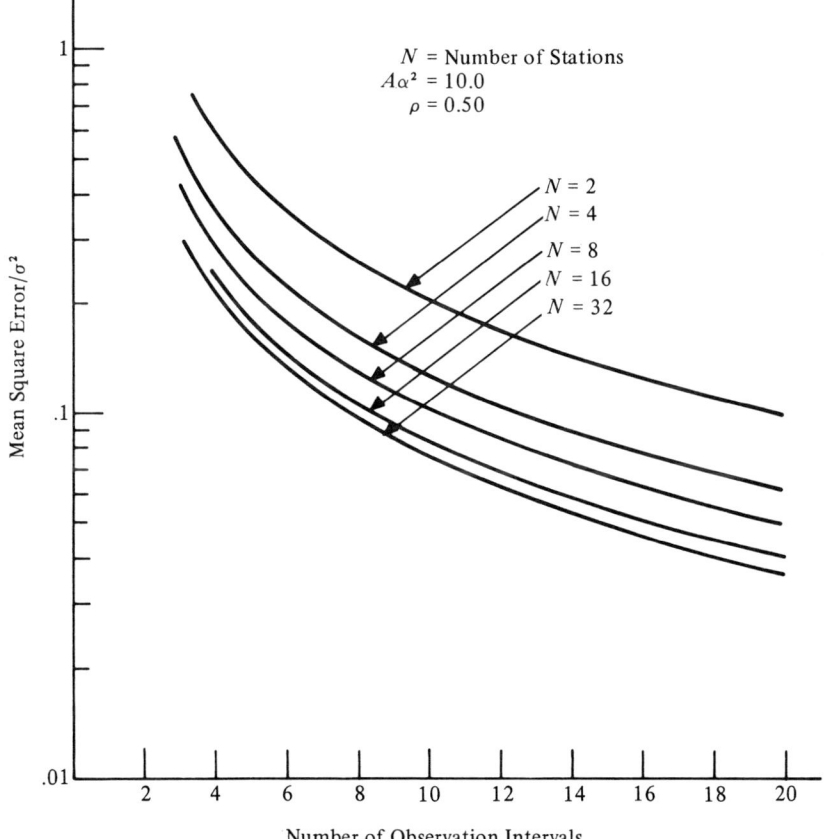

Figure 8.4 Normalized mean square error as a function of observation intervals and number of stations for a uniformly distributed network (from Bras and Colón, 1978).

where

$$\Sigma(k|k) = E\left[\{\mathbf{f}'(k)-\hat{\mathbf{f}}'(k|k)\}\{\mathbf{f}'(k)-\hat{\mathbf{f}}'(k|k)\}^T\right]$$

is the mean square error of estimation matrix at time k, given observations to time k.

Bras and Colón (1978) evaluated the set of equations in Eq. (8.128) for uniform networks of different densities and over different areas. Various autocorrelation coefficients and spatial covariance forms were tested. The solution of Eq. (8.128) requires prior information on the state. Bras and Colón

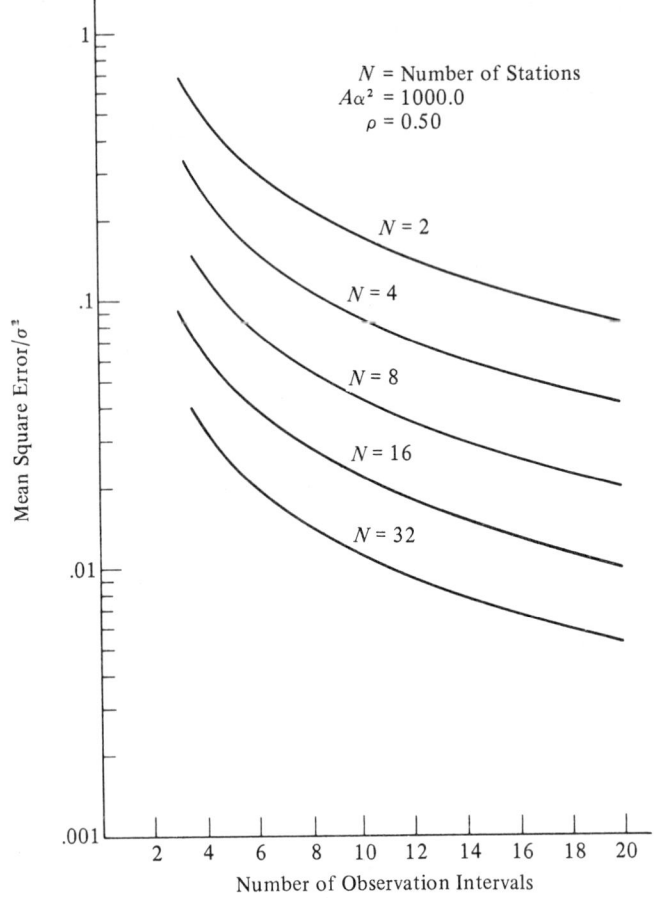

Figure 8.5 Normalized mean square error as a function of observation intervals and number of stations for a uniformly distributed network (from Bras and Colón, 1978).

assumed diffuse prior knowledge as initial conditions. The initial mean square error matrix, $\Sigma(0|0)$, was assumed to be of the form $Q\mathbf{I}$, where $\mathbf{I}$ is the identity matrix and Q a very large number.

Typical results are shown in Figs. 8.4 and 8.5. For an exponential spatial covariance of parameter α and for a lag-one autocorrelation of 0.5, Figs. 8.4 and 8.5 give the mean square error of estimating the long-term area mean for nondimensional areas $A\alpha^2 = 10.0$ and $A\alpha^2 = 1000.0$, respectively. The mean square error is given as a function of N and the number of periods of observation in each figure.

Figure 8.6, again from Bras and Colón (1978), points out the trade-off existing between the number of stations and the number of observation intervals. The figure shows curves of equal mean square error for a given area ($A\alpha^2 = 1000.0$) and autocorrelation ($\rho = 0.0$). This space–time trade-off curve leads to a very important observation. Note the sharp elbow of the curves. At any of the extremes, reduction in the number of stations (time) implies tremendous increases in the number of time intervals (stations) required to achieve a given accuracy. Clearly then, an efficient and logical design for a

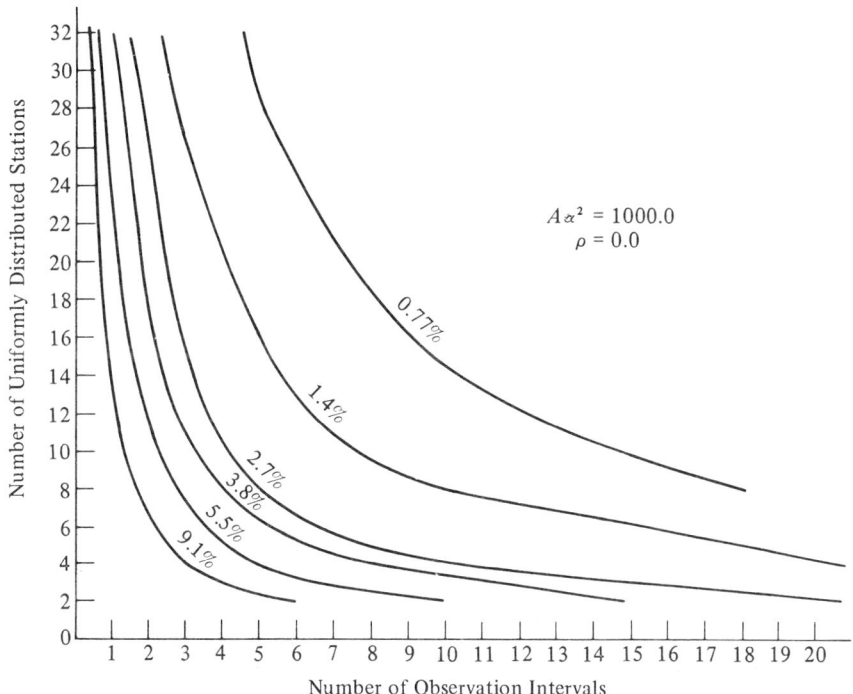

Figure 8.6 Trade-off curves between number of stations and intervals of observations for different mean square error as a percentage of point variance (from Bras and Colón, 1978).

given accuracy level lies in the curve's elbow. Movements to either side of this point imply large changes in design parameters.

The use of the Kalman filter in this network-design problem improves over all previously available techniques. The method considers measurement error and particular placement of observations in a systematic, not random, manner. Any network configuration can be studied, with spatially varying measurement errors. The procedure is general and simple. It is not limited to areas of any shape and it is not, theoretically, limited to isotropic, homogeneous, and separable covariance structures. Finally, the procedure explicitly acknowledges an unknown mean for the process and allows introduction of prior information on the state vector (and the mean).

8.4.3 Estimation, Control, and Forecasting of Biochemical Oxygen Demand (BOD) and Dissolved Oxygen (DO)

This example follows from Szöllösi-Nagy (1976) and Lettenmaier and Burges (1976). Past examples have mainly dealt with system identification and network design. This example will illustrate how a well-known water-quality model can be cast within the framework of the Kalman filter.

The relationship and time development of BOD and DO in a water body have long been represented by the so-called Streeter–Phelps equations:

$$\frac{dB}{dt} = -K_1 B \tag{8.129}$$

$$\frac{dD}{dt} = -K_2 D + K_1 B, \tag{8.130}$$

where D is oxygen deficit and B is BOD remaining at time t. Two constant parameters appear in the equation; K_1 is a BOD decay coefficient and K_2 is the re-aeration coefficient.

Defining states $X_1(t)$ and $X_2(t)$ as B and D, respectively, the state–space representation of the above system is simply

$$\frac{d}{dt}\begin{bmatrix} X_1(t) \\ X_2(t) \end{bmatrix} = \begin{bmatrix} -K_1 & 0 \\ K_1 & -K_2 \end{bmatrix}\begin{bmatrix} X_1(t) \\ X_2(t) \end{bmatrix} \tag{8.131}$$

or

$$\dot{\mathbf{X}}(t) = \mathbf{F}\mathbf{X}(t).$$

The above system might be controlled by BOD additions (or reductions) and by re-aeration. Representing these activities as $U_1(t)$ and $U_2(t)$, Eq. (8.131) is modified to

$$\dot{\mathbf{X}}(t) = \mathbf{F}\mathbf{X}(t) + \mathbf{L}\mathbf{U}(t), \tag{8.132}$$

where

$$\mathbf{L} = \begin{bmatrix} 1 & 0 \\ 0 & -1 \end{bmatrix}.$$

The negative sign is to emphasize that re-areation reduces oxygen deficit.

The transition matrix of the discrete equivalent of Eq. (8.132) can be obtained directly from Eq. (8.17)

$$\mathbf{\Phi}(t - t_0) = e^{\mathbf{F}(t - t_0)},$$

which over one time step Δt, leads to

$$\mathbf{\Phi}(t + \Delta t, t) = \begin{bmatrix} e^{-K_1 \Delta t} & \vdots & 0 \\ \cdots\cdots\cdots\cdots\cdots & \vdots & \cdots \\ \frac{K_1}{K_2 - K_1}\{e^{-K_1 \Delta t} - e^{-K_2 \Delta t}\} & \vdots & e^{-K_2 \Delta t} \end{bmatrix}. \qquad (8.133)$$

From Eq. (8.22) the discrete controls are given by

$$\int_{t-\Delta t}^{t} \mathbf{\Phi}(t, \tau)\mathbf{L}\mathbf{U}(\tau)\, d\tau.$$

For simplicity, let us assume that the controls are fixed and time invariant. The previous equation then yields simply $\mathbf{\Lambda}\mathbf{U}$, where

$$\mathbf{\Lambda} = \int_{t-\Delta t}^{t} \mathbf{\Phi}(t, \tau)\, d\tau\mathbf{L}$$

$$= \begin{bmatrix} \frac{1}{K_1}[1 - e^{-K_1 \Delta t}] & \vdots & 0 \\ \cdots\cdots\cdots\cdots\cdots\cdots\cdots\cdots & \vdots & \cdots\cdots\cdots \\ \frac{K_1}{K_2 - K_1}\left[\frac{1}{K_1}(1 - e^{-K_1 \Delta t}) - \frac{1}{K_2}(1 - e^{-K_2 \Delta t})\right] & \vdots & \frac{-1}{K_2}[1 - e^{-K_2 \Delta t}] \end{bmatrix}.$$

$$(8.134)$$

The discrete model is of the form,

$$\mathbf{X}(n + 1) = \mathbf{\Phi}\mathbf{X}(n) + \mathbf{\Lambda}\mathbf{U}, \qquad (8.135)$$

where the time index n is used instead of the traditional k to avoid confusion with constants K_1 and K_2.

Let us first study the deterministic observability of the states in Eq. (8.134). Define an observation of the form

$$\mathbf{Z} = \mathbf{H}\mathbf{X}. \qquad (8.136)$$

If **H** is the identity matrix, implying that both $X_1(\text{BOD})$ and $X_2(\text{D})$ are observed, then, from Eq. (8.29), observability requires that the matrix

$$
\mathcal{O} = \begin{bmatrix} 1 & 0 & e^{-K_1\Delta t} & \dfrac{K_1}{K_2-K_1}\left(e^{-K_1\Delta t} - e^{-K_2\Delta t}\right) \\ 0 & 1 & 0 & e^{-K_2\Delta t} \end{bmatrix}
$$

must be at least of rank 2. When $K_1 \neq K_2$, observability is clear. When $K_1 = K_2$ the upper term on the right approaches $K_1\Delta t e^{-K_1\Delta t}$ (by L'Hospital's rule). The matrix $\mathcal{O}$ becomes

$$
\mathcal{O} = \begin{bmatrix} 1 & 0 & e^{-K_1\Delta t} & K_1\Delta t e^{-K_1\Delta t} \\ 0 & 1 & 0 & e^{-K_1\Delta t} \end{bmatrix},
$$

which is still of rank 2, so the system is observable. Nevertheless, $K_1 = K_2$ is an unusual, uninteresting case.

If **H** is [1,0], implying observation of BOD only, the observability matrix is

$$
\mathcal{O} = \begin{bmatrix} 1 & e^{-K_1\Delta t} \\ 0 & 0 \end{bmatrix},
$$

which is clearly singular and of rank 1. So observations of BOD only cannot define the system states.

If D (or DO) is observed, **H** is [0,1] leading to

$$
\mathcal{O} = \begin{bmatrix} 0 & \dfrac{K_1}{K_2-K_1}\left(e^{-K_1\Delta t} - e^{-K_2\Delta t}\right) \\ 1 & e^{-K_2\Delta t} \end{bmatrix},
$$

which if $K_1 \neq K_2$ is of rank 2, leading to observability. If $K_1 = K_2$ then

$$
\mathcal{O} = \begin{bmatrix} 0 & K_1\Delta t e^{-K_1\Delta t} \\ 1 & e^{-K_1\Delta t} \end{bmatrix},
$$

which is of rank 2, and the system is observable.

Deterministic controllability depends on the rank of the matrix shown in Eq. (8.36). If U_1 and U_2, leading to the matrix in Eq. (8.134), are nonzero, the system can be shown to be controllable. If $U_1 \neq 0$ and $U_2 = 0$, matrix Λ^T is of the form $[\Lambda_{11}, \Lambda_{21}]$, where Λ_{ij} is the element in the ith row, jth column of Λ. The controllability matrix becomes

$$
\mathcal{C} = \begin{bmatrix} \Lambda_{11} & \phi_{11}\Lambda_{11} \\ \Lambda_{21} & \phi_{21}\Lambda_{11} + \phi_{22}\Lambda_{21} \end{bmatrix},
$$

where ϕ_{ij} is an element of matrix $\boldsymbol{\Phi}$. Matrix $\mathscr{C}$ is of rank 2 even when $K_1 = K_2$.

If $U_2 \neq 0$ and $U_1 = 0$,

$$\mathscr{C} = \begin{bmatrix} 0 & 0 \\ \Lambda_{22} & \phi_{22}\Lambda_{22} \end{bmatrix},$$

which is always singular. The system cannot be controlled exclusively by re-aeration.

Lettenmaier and Burges (1976) converted the dynamic equation, Eq. (8.135), and the observations, Eq. (8.136), into a stochastic discrete system of the form

$$\mathbf{X}(n+1) = \boldsymbol{\Phi}\mathbf{X}(n) \tag{8.137}$$

$$\mathbf{Z}(n) = \mathbf{HX}(n) + \mathbf{V}(n). \tag{8.138}$$

The matrix $\mathbf{H}$ is of the form $[0,1]$ implying that only DO is observed with an error of known variance $\mathbf{R}$. The dynamic model is assumed perfect, that is, no error with $\mathbf{Q} = \mathbf{0}$. The coefficients K_1 and K_2 appearing in $\boldsymbol{\Phi}$ are assumed to be known. Clearly this is a trivial situation where no filtering is necessary if initial conditions $\hat{\mathbf{X}}(0|0)$ were perfectly known ($\boldsymbol{\Sigma}(0|0) = \mathbf{0}$). Lettenmaier and Burges assume that $\hat{\mathbf{X}}(0|0)$ has effectively infinite covariance; $\boldsymbol{\Sigma}(0|0)$ is taken to be very large, discounting the initial value of the state. Therefore the question is just how fast the noisy observations are filtered so as to put the perfect model back on track.

Observations corresponding to Eq. (8.138) were processed by Lettenmaier and Burges (1976) using a Kalman filter operating on Eqs. (8.137) and (8.138). Figure 8.7 shows the actual, predicted, and measured absolute errors during

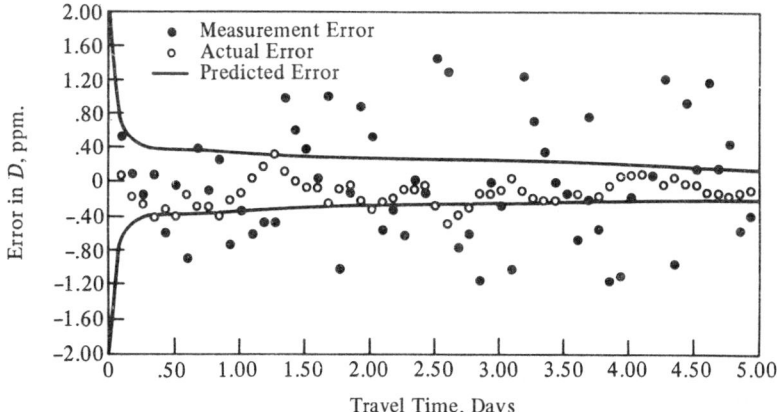

Figure 8.7 Predicted, actual, and measurement errors in oxygen deficit (D) for a linear Kalman filter (from Lettenmaier and Burges, 1976).

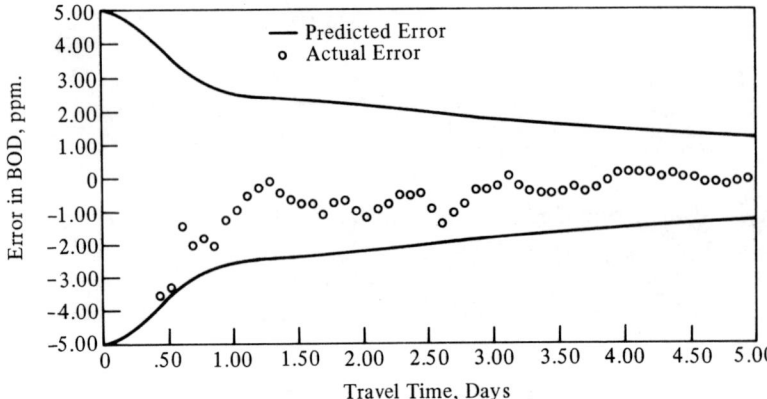

Figure 8.8   Predicted and actual errors in BOD for a linear Kalman filter (from Lettenmaier and Burges, 1976).

oxygen deficit analysis. The actual errors are seen to envelop the predicted errors, which in turn are much smaller than the errors of measurements of oxygen deficit. In the limit the forecast errors will reduce to a point where further observations would be unnecessary. Figure 8.8 shows the same convergence to zero of absolute errors in BOD estimation. Remember that no BOD measurements were taken. Figure 8.9 shows how the standard deviations of errors, square root of diagonals of $\Sigma(n|n)$, get monotonically smaller with time. Given that $\mathbf{Q} = \mathbf{0}$ in the example, such behavior is consistent with Eqs.

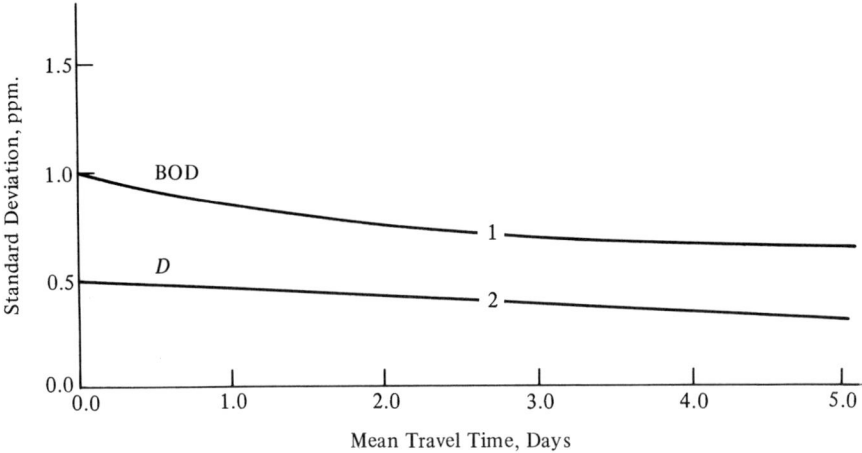

Figure 8.9   Predicted errors in BOD and D, conditioned upon initial measurement for linear Kalman filter models of a BOD–D system (from Lettenmaier and Burges, 1976).

(8.80) and (8.81) for a stable dynamic system, where matrix $\mathbf{\Phi}$ has eigenvalues less than one. Although such behavior is expected given the example, the reality of field data is usually different. In particular $\mathbf{Q}$ is rarely zero, $\mathbf{R}$ is not perfectly known, the dynamic model parameters $\mathbf{\Phi}$ are never exactly known, and the dynamic model structure may not represent the data well. All these issues complicate the use of filtering techniques and will be discussed in Chapter 9.

We should point out that although this is a good illustrative example, it is limited by the assumptions inherent in the very simple dynamics of Eq. (8.132). Results would possibly not be valid in more realistic situations involving issues like stratified water bodies and time varying controls.

CHAPTER 9

Additional Dynamic Filtering Concepts

9.1 INTRODUCTION

Chapter 8 introduced the Kalman filter and presented some reasonably straightforward applications. The basic assumptions of the development were

1. The system and observation equation fit the canonical form.
2. The system and observation equation are linear.
3. All parameters of the model and filter are perfectly known. This includes covariance matrices of dynamic models and measurement errors.

The reformulation of models in canonical form to a large extent is an art. Redefinition of states and state augmentation are the most basic tools of this art.

This chapter will first discuss how to handle nonlinearities within the Kalman-filter framework. Then the estimation of model dynamics and filter parameters will be discussed. Within this framework, the techniques of state augmentation, maximum likelihood, and suboptimal filtering will be presented.

9.2 LINEARIZATION OF NONLINEAR SYSTEMS
(after Georgakakos and Bras, 1982a)

Many geophysical processes are described by nonlinear differential equations not directly amenable to the concepts discussed in Chapter 8. Nonlinearity also commonly arises in parameter estimation of linear systems. This will be illustrated in Section 9.3.

Treatment of nonlinear systems within a Kalman-filter formulation is an approximate approach. The exact solution would require the definition of full probability density functions to perform the Bayes integrals presented in Eqs. (1.25) and (1.26). The approximate approaches are all predicated on piece-wise linearizations of the nonlinear dynamics and observations equations. The differences in methodologies lie in the nature of the linearization. It is nevertheless possible to present the most common procedures within a single general formulation. What follows is the treatment of a nonlinear continuous dynamic system with a nonlinear discrete observation. This is probably the most realistic representation of geophysical situations. All other combinations of discrete and continuous representations are similarly handled.

It is assumed that the random variables are approximately Gaussian in order not to compromise the accuracy of the solution. The system-dynamics equation is written as

$$\frac{d\mathbf{X}(t)}{dt} = \mathbf{F}(\mathbf{X}(t), \mathbf{U}(t), t) + \mathbf{w}(t) \tag{9.1}$$

and the observation equation as

$$\mathbf{Z}(k) = \mathbf{G}(\mathbf{X}(k), \mathbf{U}(k), k) + \mathbf{v}(k); \qquad k = 1, 2, 3, \ldots, \tag{9.2}$$

where $\mathbf{X}(t)$ is an $(n \times 1)$ state vector at time t, $\mathbf{U}(t)$ is a $(p \times 1)$ vector of random inputs to the system that are uncorrelated to the state variable errors, $\mathbf{Z}(k)$ is an $(m \times 1)$ vector of observations of the system output, k is the discrete-time index, $\mathbf{F}(\cdot)$ is a nonlinear $(n \times 1)$ vector function, $\mathbf{G}(\cdot)$ is a nonlinear $(m \times 1)$ vector function, $\mathbf{w}(t)$ is a continuous-time white-noise process of zero mean and covariance parameter (time-varying spectral density) $\mathbf{Q}(t)$, and $\mathbf{v}(k)$ is a discrete-time white-noise sequence with covariance matrix $\mathbf{R}(k)$ and zero-mean.

The second-moment properties of the noise terms are

$$E\{\mathbf{w}(t)\mathbf{w}^T(\tau)\} = \mathbf{Q}(t) \cdot \delta(t - \tau) \tag{9.3}$$

and

$$E\{\mathbf{v}(k)\mathbf{v}^T(j)\} = \mathbf{R}(k) \cdot \delta_{kj}, \tag{9.4}$$

where δ_{kj} is Kronecker's delta, nonzero at $k = j$, where it takes a value of 1;

$\delta(t)$ is the Dirac delta function, being zero everywhere except at $t = 0$, where it becomes infinite; and $E\{\cdot\}$ is the expectation operator.

In general, one needs a prediction estimate (to be used for the hydrologic forecast) $\hat{\mathbf{X}}(t|k\,\Delta t)$ with associated covariance matrix $\Sigma(t|k\,\Delta t)$, for times after the observation at time $k\,\Delta t$ and before the one at $(k+1)\,\Delta t$. The filtered estimate $\hat{\mathbf{X}}(k+1|k+1)$ is obtained at time $(k+1)\,\Delta t$ when the new observation becomes available. The associated covariance matrix is denoted by $\Sigma(k+1|k+1)$. The current mean of $\mathbf{X}(t)$ is obtained from

$$\hat{\mathbf{X}}(t) = E\{\mathbf{X}(t)\}, \tag{9.5}$$

where, depending on the available measurements, the expectation is over the appropriate conditional density. Also,

$$\Sigma(t) = E\left\{(\mathbf{X}(t)-\hat{\mathbf{X}}(t))(\mathbf{X}(t)-\hat{\mathbf{X}}(t))^T\right\} \tag{9.6}$$

with the same convention for the expectation operator.

The mean vector and the covariance parameter matrix of the continuous-input vector are denoted by $\hat{\mathbf{U}}(t)$ and $\mathbf{Q}_u(t)$, respectively.

Let a general linear approximation of $\mathbf{F}(\cdot)$ and $\mathbf{G}(\cdot)$ be

$$\begin{aligned}
\mathbf{F}(\mathbf{X}(t),\mathbf{U}(t),t) \approx &\ \mathbf{F}_0\big(\mathbf{d}_{x(t)},\mathbf{d}_{u(t)},t\big) \\
&+ \mathbf{N}_F\big(\mathbf{d}_{x(t)},\mathbf{d}_{u(t)},t\big)(\mathbf{X}(t)-\hat{\mathbf{X}}(t)) \\
&+ \mathbf{M}_F\big(\mathbf{d}_{x(t)},\mathbf{d}_{u(t)},t\big)(\mathbf{U}(t)-\hat{\mathbf{U}}(t))
\end{aligned} \tag{9.7}$$

and

$$\begin{aligned}
\mathbf{G}(\mathbf{X}(k),\mathbf{U}(k),k) = &\ \mathbf{G}_0\big(\mathbf{d}_{x(k)},\mathbf{d}_{u(k)},k\big) \\
&+ \mathbf{N}_G\big(\mathbf{d}_{x(k)},\mathbf{d}_{u(k)},k\big)(\mathbf{X}(k)-\hat{\mathbf{X}}(k)) \\
&+ \mathbf{M}_G\big(\mathbf{d}_{x(k)},\mathbf{d}_{u(k)},k\big)(\mathbf{U}(k)-\hat{\mathbf{U}}(k)),
\end{aligned} \tag{9.8}$$

where $\mathbf{d}_{x(t)},\mathbf{d}_{u(t)}$ represent vectors of parameters of the probability distributions of the random processes $\mathbf{X}(t)$ and $\mathbf{U}(t)$, respectively, at time t. For instance, for ordinary Taylor's series expansion $\mathbf{d}_{x(t)},\mathbf{d}_{u(t)}$ will be the mean vectors $\hat{\mathbf{X}}(t)$ and $\hat{\mathbf{U}}(t)$. As will be seen, for the method of statistical linearization with Gaussian distributions, they will include the elements of the respective second-moment matrices, provided they exist. For statistical linearization with non-Gaussian distributions, higher-order moments will be added to the vectors $\mathbf{d}_{x(t)}$ and $\mathbf{d}_{u(t)}$. Depending on the kind of linearization procedure, different expressions for the vectors $\mathbf{F}_0(\cdot),\mathbf{G}_0(\cdot)$ and the matrices $\mathbf{N}_F(\cdot)$, $\mathbf{M}_F(\cdot),\mathbf{N}_G(\cdot),\mathbf{M}_G(\cdot)$ will result.

At this point, we should distinguish between the continuous process $\mathbf{U}(t)$ and its discrete counterpart $\mathbf{U}(k)$. Even though they are both uncorrelated in time, they have, in general, different means and covariance matrices. The mean of $\mathbf{U}(t)$ can be estimated at each time t with $k\,\Delta t \le t \le (k+1)\,\Delta t$, $\forall k$, from

$$\hat{\mathbf{U}}(t) = \hat{\mathbf{U}}(k) + \left(\frac{t - k\,\Delta t}{\Delta t} \right)(\hat{\mathbf{U}}(k+1) - \hat{\mathbf{U}}(k)), \tag{9.9}$$

where $\hat{\mathbf{U}}(k), \hat{\mathbf{U}}(k+1)$ are equal to the mean vectors of the discrete-time process $\mathbf{U}(k)$ and Δt is a discrete-time interval.

It is assumed that $(\mathbf{U}(t) - \hat{\mathbf{U}}(t))$ is a continuous white noise with covariance parameter $\mathbf{Q}_u(t)$. If the corresponding discrete-time covariance matrix is $\mathbf{Q}_u(k)$, then, according to Gelb (1974),

$$\mathbf{Q}_u(k) \approx \Delta t \cdot \mathbf{Q}_u(t) \tag{9.10}$$

with Δt assumed constant and small.

Note that by assuming no time correlation for $\mathbf{U}(k) - \hat{\mathbf{U}}(k)$, we are assuming that we obtain the mean vector $\hat{\mathbf{U}}(k)$ by a forecast procedure with uncorrelated time-prediction residuals.

Due to the assumption of no correlation between the input and the system states, the terms

$$\mathbf{W}(t) = \mathbf{M}_F\big(\mathbf{d}_{x(t)}, \mathbf{d}_{u(t)}, t\big)(\mathbf{U}(t) - \hat{\mathbf{U}}(t)) + \mathbf{w}(t) \tag{9.11}$$

and

$$\mathbf{V}(k) = \mathbf{M}_G\big(\mathbf{d}_{x(k)}, \mathbf{d}_{u(k)}, k\big)(\mathbf{U}(k) - \hat{\mathbf{U}}(k)) + \mathbf{v}(k) \tag{9.12}$$

are white noises, with zero means and second moments defined by

$$\begin{aligned} E\{\mathbf{W}(t) \cdot \mathbf{W}^T(\tau)\} &= \Big[\mathbf{M}_F\big(\mathbf{d}_{x(t)}, \mathbf{d}_{u(t)}, t\big) \cdot \mathbf{Q}_u(t) \\ &\quad \cdot \mathbf{M}_F^T\big(\mathbf{d}_{x(t)}, \mathbf{d}_{u(t)}, t\big) + \mathbf{Q}(t) \Big] \cdot \delta(t - \tau) \qquad (9.13) \\ &= \mathbf{Q}'(t) \cdot \delta(t - \tau) \end{aligned}$$

and

$$\begin{aligned} E\{\mathbf{V}(k) \cdot \mathbf{V}^T(j)\} &= \Big[\mathbf{M}_G\big(\mathbf{d}_{x(k)}, \mathbf{d}_{u(k)}, k\big) \cdot \mathbf{Q}_u(k) \\ &\quad \cdot \mathbf{M}_G^T\big(\mathbf{d}_{x(k)}, \mathbf{d}_{u(k)}, k\big) + \mathbf{R}(k) \Big] \cdot \delta_{jk} \qquad (9.14) \\ &= \mathbf{R}'(k) \cdot \delta_{kj}. \end{aligned}$$

In addition, since $\mathbf{U}(t), \mathbf{w}(t), \mathbf{v}(k)$ have been taken as approximately Gaussian for all times, linear combinations of them, such as $\mathbf{W}(t)$ or $\mathbf{V}(k)$, also approach normality.

The filter equations for the interval $[k, k+1]$ then take the forms shown as follows:

State-Estimate Propagation

$$\frac{d\hat{\mathbf{X}}(t|k\,\Delta t)}{dt} = \mathbf{F}_0\big(\mathbf{d}_{x(t)},\mathbf{d}_{u(t)},t\big), \qquad t \in [k\,\Delta t,(k+1)\,\Delta t] \quad (9.15)$$

Error-Covariance Propagation

$$\frac{d\Sigma(t|k\,\Delta t)}{dt} = \mathbf{N}_F\big(\mathbf{d}_{x(t)},\mathbf{d}_{u(t)},t\big)\Sigma(t|k\,\Delta t) + \Sigma(t|k\,\Delta t)$$
$$\cdot\mathbf{N}_F^T\big(\mathbf{d}_{x(t)},\mathbf{d}_{u(t)},t\big) + \mathbf{Q}'(t|k\,\Delta t),$$
$$t \in [k\,\Delta t,(k+1)\,\Delta t] \quad (9.16)$$

State-Estimate Update

$$\hat{\mathbf{X}}(k+1|k+1) = \hat{\mathbf{X}}(k+1|k) + \mathbf{K}(k+1)$$
$$\cdot\big[\mathbf{Z}(k+1) - \mathbf{G}_0\big(\mathbf{d}_{x(k+1)},\mathbf{d}_{u(k+1)},k+1\big)\big] \quad (9.17)$$

Error-Covariance Update

$$\Sigma(k+1|k+1) = \big[\mathbf{I} - \mathbf{K}(k+1)\cdot\mathbf{N}_G\big(\mathbf{d}_{x(k+1)},\mathbf{d}_{u(k+1)},k+1\big)\big]\cdot\Sigma(k+1|k)$$
$$\cdot\big[\mathbf{I} - \mathbf{K}(k+1)\cdot\mathbf{N}_G\big(\mathbf{d}_{x(k+1)},\mathbf{d}_{u(k+1)},k+1\big)\big]^T$$
$$+ \mathbf{K}(k+1)\cdot\mathbf{R}'(k+1)\cdot\mathbf{K}^T(k+1) \quad (9.18)$$

Filter Gain

$$\mathbf{K}(k+1) = \Sigma(k+1|k)\cdot\mathbf{N}_G^T\big(\mathbf{d}_{x(k+1)},\mathbf{d}_{u(k+1)},k+1\big)$$
$$\cdot\big[\mathbf{N}_G\big(\mathbf{d}_{x(k+1)},\mathbf{d}_{u(k+1)},k+1\big)\Sigma(k+1|k)$$
$$\cdot\mathbf{N}_G^T\big(\mathbf{d}_{x(k+1)},\mathbf{d}_{u(k+1)},k+1\big) + \mathbf{R}'(k+1)\big]^{-1}, \quad (9.19)$$

where $\mathbf{I}$ is the $n \times n$ identity matrix.

The following are some comments regarding the filter equations.

1. Due to the dependence of $\mathbf{F}_0(\cdot)$ and $\mathbf{N}_F(\cdot)$ on the best available distributions of the system state, Eqs. (9.15) and (9.16) are coupled differential equations.

2. Care should be exercised to prevent the covariance matrix $\Sigma(t|k\,\Delta t)$ from becoming negative at the propagation step. Positive definiteness should be enforced.

3. To ensure the positive definiteness of $\Sigma(k+1|k+1)$, Eq. (9.18) is recommended, rather than the simpler expression:

$$\Sigma(k+1|k+1) = \big[\mathbf{I} - \mathbf{K}(k+1)\cdot\mathbf{N}_G\big(\mathbf{d}_{x(k+1)},\mathbf{d}_{u(k+1)},k+1\big)\big]\cdot\Sigma(k+1|k).$$
$$(9.20)$$

4. The linearization vectors $\mathbf{F}_0(\cdot), \mathbf{G}_0(\cdot)$ and matrices $\mathbf{N}_F(\cdot), \mathbf{N}_G(\cdot)$ always use the parameters of the current probability distribution for the vector $\mathbf{X}(t)$. This is made explicit through the notation.

5. The innovation sequence

$$\nu(k+1) = \mathbf{Z}(k+1) - \mathbf{G}_0\left(\mathbf{d}_{x(k+1)}, \mathbf{d}_{u(k+1)}, k+1\right) \qquad (9.21)$$

should be uncorrelated, with zero mean and variance $\mathbf{Q}_\nu(k+1)$, given by

$$\mathbf{Q}_\nu(k+1) = \mathbf{N}_G\left(\mathbf{d}_{x(k+1)}, \mathbf{d}_{u(k+1)}, k+1\right) \cdot \Sigma(k+1|k)$$
$$\cdot \mathbf{N}_G^T\left(\mathbf{d}_{x(k+1)}, \mathbf{d}_{u(k+1)}, k+1\right) + \mathbf{R}'(k+1). \qquad (9.22)$$

In the following, expressions for the vectors $\mathbf{F}_0(\cdot), \mathbf{G}_0(\cdot)$ and for the matrices $\mathbf{N}_F(\cdot), \mathbf{M}_F(\cdot), \mathbf{N}_G(\cdot), \mathbf{M}_G(\cdot)$ are presented for linearizations based on Taylor's series expansion and for statistical linearization.

9.2.1 Taylor's Series Expansion, the Extended Kalman Filter

The idea here is to expand the nonlinear functions about the current best estimates of the relevant random variables and then to keep the first two terms in the Taylor expansion:

$$\mathbf{F}(\mathbf{X}(t), \mathbf{U}(t), t) = \mathbf{F}_0(\hat{\mathbf{X}}(t), \hat{\mathbf{U}}(t), t)$$
$$+ \mathbf{N}_F(\hat{\mathbf{X}}(t), \hat{\mathbf{U}}(t), t)(\mathbf{X}(t) - \hat{\mathbf{X}}(t))$$
$$+ \mathbf{M}_F(\hat{\mathbf{X}}(t), \hat{\mathbf{U}}(t), t)(\mathbf{U}(t) - \hat{\mathbf{U}}(t)) \qquad (9.23)$$

with the ijth elements of $\mathbf{N}_F(\cdot)$ and $\mathbf{M}_F(\cdot)$ given by

$$\left\{\mathbf{N}_F(\hat{\mathbf{X}}(t), \hat{\mathbf{U}}(t), t)\right\}_{ij} = \left.\frac{\partial F_i(\mathbf{X}(t), \mathbf{U}(t), t)}{\partial X_j(t)}\right|_{\substack{\mathbf{X}(t) = \hat{\mathbf{X}}(t) \\ \mathbf{U}(t) = \hat{\mathbf{U}}(t)}} \qquad (9.24)$$

$$\left\{\mathbf{M}_F(\hat{\mathbf{X}}(t), \hat{\mathbf{U}}(t), t)\right\}_{ij} = \left.\frac{\partial F_i(\mathbf{X}(t), \mathbf{U}(t), t)}{\partial U_j(t)}\right|_{\substack{\mathbf{X}(t) = \hat{\mathbf{X}}(t) \\ \mathbf{U}(t) = \hat{\mathbf{U}}(t)}} \qquad (9.25)$$

where ∂ / ∂ denotes partial derivative; $F_i(\cdot)$ is the ith element of vector function $\mathbf{F}(\cdot)$; $X_j(t)$ is the jth element of the state vector $\mathbf{X}(t)$; and $U_j(t)$ is the jth element of the input vector $\mathbf{U}(t)$. Following convention, $\hat{\mathbf{X}}(t)$ stands for the latest available conditional mean of the state. Also,

$$\mathbf{F}_0(\hat{\mathbf{X}}(t), \hat{\mathbf{U}}(t), t) = \mathbf{F}(\hat{\mathbf{X}}(t), \hat{\mathbf{U}}(t), t). \qquad (9.26)$$

Similarly,

$$\{\mathbf{N}_G(\hat{\mathbf{X}}(k),\hat{\mathbf{U}}(k),k)\}_{ij} = \left. \frac{\partial G_i(\mathbf{X}(k),\mathbf{U}(k),k)}{\partial X_j(k)} \right|_{\substack{\mathbf{X}(k)=\hat{\mathbf{X}}(k) \\ \mathbf{U}(k)=\hat{\mathbf{U}}(k)}} \quad (9.27)$$

$$\{\mathbf{M}_G(\hat{\mathbf{X}}(k),\hat{\mathbf{U}}(k),k)\}_{ij} = \left. \frac{\partial G_i(\mathbf{X}(k),\mathbf{U}(k),k)}{\partial U_j(k)} \right|_{\substack{\mathbf{X}(k)=\hat{\mathbf{X}}(k) \\ \mathbf{U}(k)=\hat{\mathbf{U}}(k)}} \quad (9.28)$$

$$\mathbf{G}_0(\mathbf{X}(k),\mathbf{U}(k),k) = \mathbf{G}(\hat{\mathbf{X}}(k),\hat{\mathbf{U}}(k),k) \quad (9.29)$$

The important features of the linearization procedure are:

1. Linearization matrices and vectors are only functions of the state and input mean vectors. That is,

$$\mathbf{d}_{x(t)} = \hat{\mathbf{X}}(t) \quad (9.30)$$

$$\mathbf{d}_{u(t)} = \hat{\mathbf{U}}(t) \quad (9.31)$$

2. When used with the filter equations (the so-called extended Kalman filter [Gelb, 1974]), they produce biased estimates because in general

$$E\{\mathbf{F}(\mathbf{X}(t),\mathbf{U}(t),t)\} \neq \mathbf{F}(\hat{\mathbf{X}}(t),\hat{\mathbf{U}}(t),t).$$

3. Taylor expansions are easy if the nonlinear functions of interest are differentiable. If not, this procedure cannot be applied.

As a summary and illustration, the extended Kalman filter in a continuous-discrete system would be

State - Estimate Propagation

$$\dot{\hat{\mathbf{X}}}(t|k\,\Delta t) = \mathbf{F}(\hat{\mathbf{X}}(k\,\Delta t|k\,\Delta t),\hat{\mathbf{U}}(k\,\Delta t),t); t \in \left[k\,\Delta t,(k+1)\,\Delta t\right] \quad (9.32)$$

Error - Covariance Propagation

$$\dot{\boldsymbol{\Sigma}}(t|k\,\Delta t) = \mathbf{N}_F(\hat{\mathbf{X}}(k\,\Delta t|k\,\Delta t),\hat{\mathbf{U}}(k\,\Delta t),t)\boldsymbol{\Sigma}(t|k\,\Delta t)+\boldsymbol{\Sigma}(t|k\,\Delta t)$$
$$\cdot\mathbf{N}_F^T(\hat{\mathbf{X}}(k\,\Delta t|k\,\Delta t),\hat{\mathbf{U}}(k\,\Delta t),t)+\mathbf{M}_F(\hat{\mathbf{X}}(k\,\Delta t|k\,\Delta t),\hat{\mathbf{U}}(k\,\Delta t),t)$$
$$\cdot\mathbf{Q}_u(t)\mathbf{M}_F^T(\hat{\mathbf{X}}(k\,\Delta t|k\,\Delta t),\hat{\mathbf{U}}(k\,\Delta t),t)+\mathbf{Q}(t);$$
$$t \in \left[k\,\Delta t,(k+1)\,\Delta t\right] \quad (9.33)$$

State - Estimate Update

$$\hat{\mathbf{X}}(k+1|k+1) = \hat{\mathbf{X}}(k+1|k)+\mathbf{K}(k+1)$$
$$\cdot\{\mathbf{Z}(k+1)-\mathbf{G}_0(\hat{\mathbf{X}}(k+1|k),\hat{\mathbf{U}}(k),k+1)\} \quad (9.34)$$

Error-Covariance Update

$$
\begin{aligned}
\Sigma(k+1|k+1) = & \left[\mathbf{I}-\mathbf{K}(k+1)\mathbf{N}_G(\hat{\mathbf{X}}(k+1|k),\hat{\mathbf{U}}(k+1),k+1)\right]\cdot\Sigma(k+1|k) \\
& \cdot\left[\mathbf{I}-\mathbf{K}(k+1)\mathbf{N}_G(\hat{\mathbf{X}}(k+1|k),\hat{\mathbf{U}}(k+1),k+1)\right]^T \\
& +\mathbf{K}(k+1)\left[\mathbf{M}_G(\hat{\mathbf{X}}(k+1|k),\hat{\mathbf{U}}(k+1),k+1)\right. \\
& \left.\cdot\mathbf{Q}_u(k+1)\mathbf{M}_G^T(\hat{\mathbf{X}}(k+1|k),\hat{\mathbf{U}}(k+1),k+1)+\mathbf{R}(k+1)\right] \\
& \cdot\mathbf{K}^T(k+1) \quad\quad\quad (9.35)
\end{aligned}
$$

Filter Gain

$$
\begin{aligned}
\mathbf{K}(k+1) = & \Sigma(k+1|k)\mathbf{N}_G^T(\hat{\mathbf{X}}(k+1|k),\hat{\mathbf{U}}(k+1),k+1) \\
& \cdot\left[\mathbf{N}_G(\hat{\mathbf{X}}(k+1|k),\hat{\mathbf{U}}(k+1),k+1)\Sigma(k+1|k)\right. \\
& \cdot\mathbf{N}_G^T(\hat{\mathbf{X}}(k+1|k),\hat{\mathbf{U}}(k+1),k+1)+\mathbf{M}_G(\hat{\mathbf{X}}(k+1|k),\hat{\mathbf{U}}(k+1),k+1) \\
& \left.\cdot\mathbf{Q}_u(k+1)\mathbf{M}_G^T(\hat{\mathbf{X}}(k+1|k),\hat{\mathbf{U}}(k+1),k+1)+\mathbf{R}(k+1)\right]^{-1} \quad (9.36)
\end{aligned}
$$

9.2.2 Statistical Linearization

Define errors of approximation:

$$
\begin{aligned}
\mathbf{e}_F(t) = & \mathbf{F}(\mathbf{X}(t),\mathbf{U}(t),t)-\mathbf{F}_0(\cdot)-\mathbf{N}_F(\cdot)(\mathbf{X}(t)-\hat{\mathbf{X}}(t)) \\
& -\mathbf{M}_F(\cdot)(\mathbf{U}(t)-\hat{\mathbf{U}}(t)) \\
\mathbf{e}_G(k) = & \mathbf{G}(\mathbf{X}(k),\mathbf{U}(k),k)-\mathbf{G}_0(\cdot)-\mathbf{N}_G(\cdot)(\mathbf{X}(k)-\hat{\mathbf{X}}(k)) \\
& -\mathbf{M}_G(\cdot)(\mathbf{U}(k)-\hat{\mathbf{U}}(k)).
\end{aligned}
$$

Minimizing the quadratic forms

$$
E\left[\mathbf{e}_F^T(t)\mathbf{e}_F(t)\right] \quad\text{and}\quad E\left[\mathbf{e}_G^T(k)\mathbf{e}_G(k)\right]
$$

over expressions for $\mathbf{F}_0(\cdot)$, $\mathbf{N}_F(\cdot)$, $\mathbf{M}_F(\cdot)$, $\mathbf{G}_0(\cdot)$, $\mathbf{N}_G(\cdot)$, and $\mathbf{M}_G(\cdot)$ leads to

$$
\mathbf{N}_F(\mathbf{d}_{x(t)},\mathbf{d}_{u(t)},t) = E\left\{\mathbf{F}(\mathbf{X}(t),\mathbf{U}(t),t)\cdot\mathbf{r}_x^T(t)\right\}\cdot\Sigma^{-1}(t) \quad (9.37)
$$

$$
\mathbf{N}_G(\mathbf{d}_{x(k)},\mathbf{d}_{u(k)},k) = E\left\{\mathbf{G}(\mathbf{X}(k),\mathbf{U}(k),k)\cdot\mathbf{r}_x^T(k)\right\}\cdot\Sigma^{-1}(k) \quad (9.38)
$$

$$
\mathbf{M}_G(\mathbf{d}_{x(k)},\mathbf{d}_{u(k)},k) = E\left\{\mathbf{G}(\mathbf{X}(k),\mathbf{U}(k),k)\cdot\mathbf{r}_u^T(k)\right\}\cdot\mathbf{Q}_u^{-1}(k) \quad (9.39)
$$

$$
\mathbf{F}_0(\mathbf{d}_{x(t)},\mathbf{d}_{u(t)},t) = E\left\{\mathbf{F}(\mathbf{X}(t),\mathbf{U}(t),t)\right\} \quad (9.40)
$$

$$
\mathbf{G}_0(\mathbf{d}_{x(k)},\mathbf{d}_{u(k)},k) = E\left\{\mathbf{G}(\mathbf{X}(k),\mathbf{U}(k),k)\right\}, \quad (9.41)
$$

where

$$\mathbf{r}_x(t) = \mathbf{X}(t) - \hat{\mathbf{X}}(t) \qquad \text{and} \qquad \mathbf{r}_u(t) = \mathbf{U}(t) - \hat{\mathbf{U}}(t).$$

Due to the assumed white-noise nature of the continuous process $\mathbf{U}(t)$, its covariance matrix does not exist; therefore $\mathbf{M}_F(\cdot)$ is defined as in Eq. (9.25). For this linearization procedure, it is important to note the following:

1. The linearization matrices and vectors are functions of the state and input first- *and* second-moment properties given the Gaussian assumption. In general they are functions of all the parameters of their probability distributions.

2. Equations (9.15) and (9.17) together with Eqs. (9.40) and (9.41) show that this procedure, used with a linear filter, produces unbiased estimates.

3. An assumption about the underlying probability distribution is necessary to compute the expectations yielding the coefficient matrices. Usually a Gaussian assumption suffices for good performance, while providing the state mean and covariance recursively if this procedure is used with the linear filter.

4. No assumption regarding function differentiability is made for the statistical linearization procedure.

Point 3 above is the biggest drawback of statistical linearization. In effect, to compute the coefficient matrices it is necessary to know the full distribution of the state at every time step. In practice, it is commonly assumed that the distribution is approximately Gaussian. In such cases the coefficient matrices are computed numerically using the latest available state estimate $\hat{\mathbf{X}}(t|k\,\Delta t)$ and the latest available covariance matrix $\Sigma(t|k\,\Delta t)$ as parameters of the distribution. Keep in mind that $\hat{\mathbf{X}}(t|k\,\Delta t)$ and $\Sigma(t|k\,\Delta t)$ are the results of the Kalman-filter algorithm. So, coefficient matrices are computed piece-wise and in a feedback mode from results of the filter.

When the nonlinear functions in question are differentiable, Georgakakos and Bras (1982b) provide analytical expressions for the expectations of nonlinear functions appearing in Eqs. (9.37) through (9.41). If $f(\mathbf{X})$ is a nonlinear function of vector $\mathbf{X}$ with elements $\{x_1, \ldots, x_m\}$, then

$$E\{f(\mathbf{X})\} = \sum_{n=0}^{N-1} \sum_{\substack{n_1, \ldots, n_m \\ n_1 + \cdots + n_m = n}} \frac{1}{n_1! n_2! \cdots n_m!}$$

$$\cdot \left. \frac{\partial^n f(\mathbf{X})}{(\partial x_1)^{n_1} (\partial x_2)^{n_2} \cdots (\partial x_m)^{n_m}} \right|_{\mathbf{X} = \hat{\mathbf{X}}} E\{r_1^{n_1} \cdots r_m^{n_m}\} \qquad (9.42)$$

where the second summation implies summing over all combinations $n_1, \ldots, n_m$ such that $n_1 + \cdots + n_m = n$.

In Eq. (9.42) the nth-order central moment of the random vector $\mathbf{X} = [x_1, x_2, \ldots, x_m]^T$ is involved. If this moment can be expressed as a function of a finite number of parameters, characterizing the joint probability density of $\mathbf{X}$, then the number of terms in the expansion, $N = N^*$, that provides the best approximation to the $E\{f(\mathbf{X})\}$ can be determined by comparison with numerical-integration schemes. Experiments indicate that N^* is generally small; Georgakakos and Bras (1982b) found that an N^* of 4 to 8 seemed sufficient.

Expressions for the nth-order central moment of the common multivariate normal distribution are

$$E\{r_{i_1}r_{i_2} \cdots r_{i_L}\} = 0 \qquad L = 2k+1 \qquad k = 0,1,\ldots$$

$$E\{r_{i_1}r_{i_2} \cdots r_{i_L}\} = \sum_{\substack{j_1, j_2, \ldots, j_L \\ P_e\{i_1, i_2, \ldots, i_L\}}} \sigma^2_{j_1 j_2}\sigma^2_{j_3 j_4} \cdots \sigma^2_{j_{L-1} j_L} \tag{9.43}$$

$$L = 2k, \, k = 0,1,\ldots$$

where $r_i = x_i - E[x_i]$.

The set $j_1, j_2, \ldots, j_L$ is a permutation of the subscripts $i_1, i_2, \ldots, i_L$ such that distinct pairs $j_1 j_2, j_3 j_4, \ldots, j_{L-1} j_L$ result. The number of such permutations is indicated by $P_e\{i_1, i_2, \ldots, i_L\}$, where $i_1, i_2, \ldots, i_L$ are integers selected from the set $\{1, 2, \ldots, n\}$ with repetition allowed. The covariance between the j_i and j_k elements of X is $\sigma^2_{j_i j_k}$. For example,

$$E\{(x_1 - \mu_1)(x_2 - \mu_2)(x_3 - \mu_3)(x_4 - \mu_4)\} = \sigma^2_{12}\sigma^2_{34} + \sigma^2_{13}\sigma^2_{24} + \sigma^2_{14}\sigma^2_{23}.$$

Therefore, assuming normality, Eq. (9.42) is easily evaluated in terms of the elements of the error-covariance matrix given by the Kalman filter itself. Georgakakos and Bras (1982b) call the above analytical approximation to expectation of nonlinear functions the Gauss–Taylor procedure. For nonlinear vector functions, it can be used on a term-by-term basis.

9.2.3 Statistical Linearization in Flood Routing: An Example
(after Bras and Georgakakos, 1980)

Mein, Laurenson, and McMahon (1974) presented a simple but general and flexible routing model based on nonlinear reservoirs. Denote by $S_i(t)$ and $Q_i(t)$ the volume of water in storage in the ith reservoir and its corresponding discharge, respectively. A relationship between $S_i(t)$ and $Q_i(t)$ of the type

$$Q_i(t) = a_i S_i^m(t) \qquad \forall i \tag{9.44}$$

is assumed. The continuity equation is

$$\frac{dS_i(t)}{dt} = I_i(t) - Q_i(t), \tag{9.45}$$

where $I_i(t)$ is the input discharge in the ith reservoir at time t; Eqs. (9.44) and (9.45) are the building blocks of a routing model based on nonlinear reservoirs.

In a differential form, the equations of motion of a system of n nonlinear reservoirs in series of the type described are

$$\frac{dS_1(t)}{dt} = u_1(t) - a_1 \cdot S_1^m(t)$$

$$\frac{dS_2(t)}{dt} = u_2(t) + a_1 \cdot S_1^m(t) - a_2 \cdot S_2^m(t)$$

$$\vdots \qquad\qquad \vdots$$

$$\frac{dS_i(t)}{dt} = u_i(t) + a_{i-1} \cdot S_{i-1}^m(t) - a_i \cdot S_i^m(t) \qquad (9.46)$$

$$\vdots \qquad\qquad \vdots$$

$$\frac{dS_n(t)}{dt} = u_n(t) + a_{n-1} \cdot S_{n-1}^m(t) - a_n \cdot S_n^m(t).$$

In Eq. (9.46), u_i, $\forall i$, is the portion of the total channel inflow (TCI) $(u_i = P_i \cdot (\text{TCI}))$ that serves as input to the ith reservoir and a_i, $\forall i$, and m are the parameters of the model. The output of the system is the discharge:

$$Q_n(t) = a_n \cdot S_n^m(t). \qquad (9.47)$$

It provides the basis of an observation equation relating observed discharge to storage.

Use of the Kalman filter with the above system required linearization of the nonlinear functions involved. Concentrate on the nonlinear function: $f(x_i) = x_i^m$ with m not an integer. The objective is to derive an "equivalent" linear function to the function $f(x_i)$. For the purposes of this development the dependence of x_i and related quantities on time t is not explicitly shown.

Assume that x_i can be considered to be the sum of a bias term (a mean level) μ_i and a Gaussian random process r_i with zero mean and variance equal to σ_i^2:

$$x_i = \mu_i + r_i. \qquad (9.48)$$

Denote by $f_a(x_i)$ the "equivalent" linear function desired. Then,

$$f_a(x_i) = N_{\mu_i} \cdot \mu_i + N_{r_i} \cdot r_i. \qquad (9.49)$$

Use of Eqs. (9.37) and (9.40) yield

$$N_{\mu_i} = \frac{E\{(\mu_i + r_i)^m\}}{\mu_i} \qquad (9.50)$$

$$N_{r_i} = \frac{E\{r_i(\mu_i + r_i)^m\}}{E\{r_i^2\}}, \qquad (9.51)$$

with a resultant mean square error ($\varepsilon_i = f(x_i) - f_a(x_i)$) of approximation:

$$E\{\varepsilon_i^2\} = E\{(\mu_i + r_i)^{2m}\} - E^2\{(\mu_i + r_i)^m\}$$
$$- \frac{E^2\{r_i(\mu_i + r_i)^m\}}{E\{r_i^2\}}. \tag{9.52}$$

For the function x_i^m, where m is not an integer, the Gauss–Taylor procedure (Eq. 9.42) and Eqs. (9.50) and (9.51) result in the following expressions for the linearization coefficients:

$$N_{\mu_i} = \mu_i^{(m-1)} \cdot \left\{ 1 + \sum_{k=1}^{N'} \frac{\prod_{i=0}^{2k-1}(m-i) \prod_{j=1}^{k}(2j-1)}{(2k)!} \cdot V_i^{2k} \right\} \tag{9.53}$$

$$N_{r_i} = \mu_i^{(m-1)} \cdot \sum_{k=1}^{N'} \frac{\prod_{i=0}^{2k-2}(m-i) \prod_{j=1}^{k}(2j-1)}{(2k-1)!} \cdot V_i^{2k-2}, \tag{9.54}$$

where Σ and Π are the summation and product operators respectively, V_i is the coefficient of variation of x_i, $V_i = \mu_i/\sigma_i$. The summation is over $N' = N/2$, since the expected value of all the odd powers of a Gaussian random variable is equal to zero.

The series appearing in Eqs. (9.53) and (9.54) are asymptotic ones; that is, for a certain $N' = N^*$, the residual error is minimum while for N becoming very large the error grows unbounded. N^* decreases rapidly as V_i increases, and it is relatively insensitive to the value of m for $0.6 \leq m \leq 1.5$.

Table 9.1 presents the results of numerical integration for the functions $E\{(\mu_i + r_i)^m\}/\mu_i^m = E_1$, $E\{r_i(\mu_i + r_i)^m\}/\mu_i^{(m+1)} = E_2$ (appearing in Eqs. 9.50 and 9.51), as well as similar results of the Gauss–Taylor approximation, for different values of V_i and m. Two values of N' were used to indicate the asymptotic nature of the series involved in the Gauss–Taylor method: a value of 2 or a value of N^*, where N^* is given in Table 9.2 for the parameters m and V_i used in Table 9.1.

Comparison of the results obtained by numerical integration with those obtained by the Gauss–Taylor approximation indicate that the proposed method gives a good approximation to the functions E_1, E_2. In fact it gives the exact values for small coefficients of variation (e.g., on the order of 0.3) and for integer values of m.

The dependence of N_{μ_i} on V_i is displayed in Fig. 9.1 for different values of m ranging from $m = 0.8$ to $m = 2.0$. Similar results are presented in Fig. 9.2 for N_{r_i}. The following observations are valid. The normalized gain, $N_{\mu_i}/\mu_i^{(m-1)}$, corresponding to the bias term, is relatively insensitive to the value of the

Table 9.1

Numerical integration (NI) and Gauss–Taylor (GT) results for the evaluation of the functions E_1 and E_2

Function		E_1			E_2		
V_i	m	NI	$GT; N_1 = 2$	$GT; N_1 = N_1^*$	NI	$GT; N_2 = 2$	$GT; N_2 = N_2^*$
0.3	1.2	1.011	1.011	1.011	0.107	0.107	0.107
0.8	1.2	1.117	1.078	1.000	0.682	0.730	0.768
0.3	0.8	0.993	0.993	0.993	0.073	0.073	0.073
0.8	0.8	0.980	0.950	1.000	0.475	0.550	0.512

Source: Bras and Georgakakos (1980).

exponent m for small values of $V_i (\leq 0.30)$. This gain is otherwise quadratic in V_i. The normalized gain $N_{r_i}/\mu_i^{(m-1)}$, corresponding to the Gaussian residual term is relatively sensitive to the value of m (is practically equal to m) for small coefficients of variation (≤ 0.30). If V_i is small and m is within the range 0.8–2.0, the results of the statistical linearization become identical to those of the ordinary linearization (using the leading term in the Taylor series expansion). Namely, $N_{\mu_i} = \mu_i^{(m-1)}$ and $N_{r_i} = m \cdot \mu_i^{(m-1)}$.

Following are the Kalman-filter formulas pertinent to the statistically linearized channel-routing model with three nonlinear reservoirs in series. The system dynamics are described by a continuous model while the observations are available every six hours; thus the formulation corresponds to a continuous discrete filter.

State - Estimate Propagation

$$\frac{d}{dt}\mu\left(t|(k-1)\,\Delta t\right) = \mathbf{N}_\mu(t)\cdot\mu\left(t|(k-1)\,\Delta t\right) + \mathbf{P}\cdot\mu_u(t) \tag{9.55}$$

$$\mu\left(t_0|(k-1)\,\Delta t\right) = \mu\left((k-1)\,\Delta t|(k-1)\,\Delta t\right), \tag{9.56}$$

Table 9.2

Values of N_1^*, N_2^* that result in minimum residual errors when used in the Gauss–Taylor approximation of E_1 and E_2, respectively

V_i \ m	0.8	1.2
0.3	$N_1^* = 5; N_2^* = 5$	$N_1^* = 5; N_2^* = 6$
0.8	$N_1^* = 0; N_2^* = 1$	$N_1^* = 0; N_2^* = 1$

Source: Bras and Georgakakos (1980).

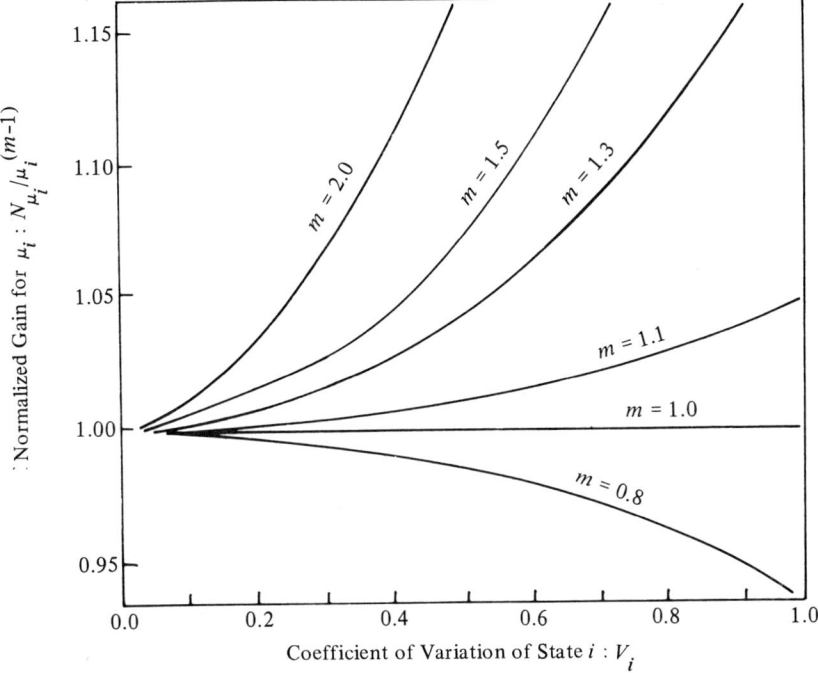

Figure 9.1 Normalized gain for the bias μ_i as a function of the coefficient of variation of the ith state for different values of the exponent m (follows the presentation by Bras and Georgakakos, 1980).

where the mean total channel input, $\mu_u(t)$ is a scalar.

Error-Covariance Propagation

$$\frac{d}{dt}\Sigma(t|(k-1)\Delta t) = \mathbf{N}_r(t)\cdot\Sigma(t|(k-1)\Delta t) + \Sigma(t|(k-1)\Delta t)\cdot\mathbf{N}_r^T(t)$$

$$+\mathbf{P}\cdot\sigma_u^2(t)\cdot\mathbf{P}^T + \mathbf{Q} \tag{9.57}$$

$$\Sigma(t_0|(k-1)\Delta t) = \Sigma((k-1)\Delta t)|(k-1)\Delta t) \tag{9.58}$$

State-Estimate Update

$$\mu(k|k) = \mu(k|k-1) + \mathbf{K}(k)\cdot\nu(k) \tag{9.59}$$

Innovation Sequence

$$\nu(k) = Z(k) - C_3(k)\cdot\mu_3(k|k-1) \tag{9.60}$$

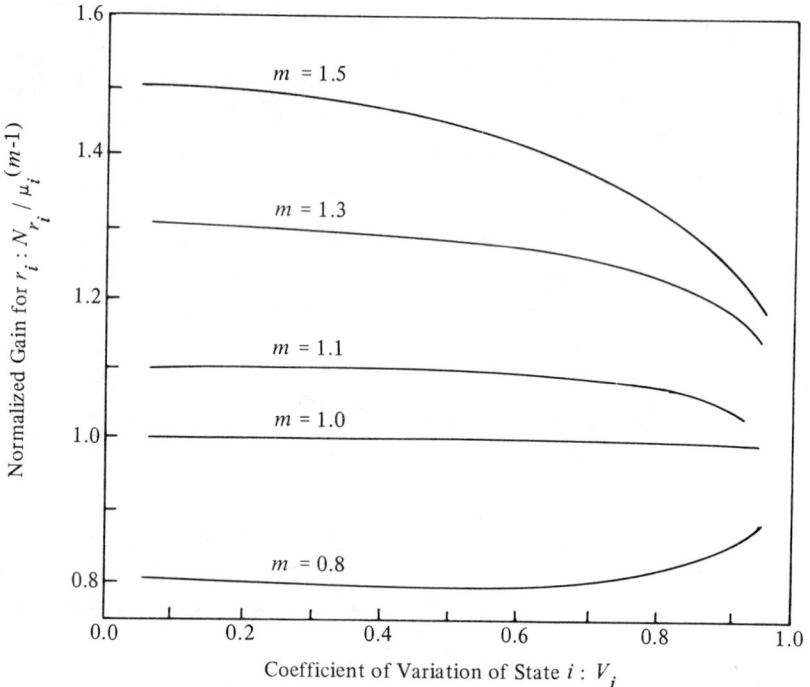

Figure 9.2 Normalized gain for the Gaussian residual r_i as a function of the coefficient of variation of the ith state for different values of the exponent m (follows the presentation by Bras and Georgakakos, 1980).

Error-Covariance Update

$$\Sigma(k|k) = [\mathbf{I} - \mathbf{K}(k) \cdot \mathbf{H}(k)] \cdot \Sigma(k|k-1) \tag{9.61}$$

Gain Vector

$$\mathbf{K}(k) = \Sigma(k|k-1) \cdot \mathbf{H}^T(k) \cdot Q_\nu^{-1}(k) \tag{9.62}$$

Variance of the Innovations Sequence

$$Q_\nu(k) = \mathbf{H}(k) \cdot \Sigma(k|k-1) \cdot \mathbf{H}^T(k) + R(k), \tag{9.63}$$

where

$$\mathbf{\mu}(t) = [\mu_1(t) \quad \mu_2(t) \quad \mu_3(t)]^T \tag{9.64}$$

($\mu_i(t)$ is the mean value of the ith state at time t.)

$$\mathbf{N}_\mu(t) = \begin{bmatrix} -C_1(t) & 0 & 0 \\ C_1(t) & -C_2(t) & 0 \\ 0 & C_2(t) & -C_3(t) \end{bmatrix}$$

$$C_i(t) = a_i \cdot N_{\mu_i}(t); \qquad i = 1, 2, 3 \qquad (9.65)$$

$$\mathbf{P} = [P_1 \quad P_2 \quad P_3]^T$$

(P_i is the portion of the total channel input that serves as an input to the ith reservoir, $\mu_u(t)$ the mean value of the total channel input at time t, t_0 the initial time, and $\Sigma(t|t)$ the state-covariance matrix at time t given information up to time t.)

$$\mathbf{N}_r(t) = \begin{bmatrix} -C_1'(t) & 0 & 0 \\ C_1'(t) & -C_2'(t) & 0 \\ 0 & C_2'(t) & -C_3'(t) \end{bmatrix}$$

$$C_i'(t) = a_i \cdot N_{r_i}(t) \qquad (9.66)$$

$$\sigma_u(t) = V_I \cdot \mu_u(t)$$

(V_I is the coefficient of variation of the input.)

$$\mathbf{Q} = \begin{bmatrix} q_{11} & 0 & 0 \\ 0 & q_{22} & 0 \\ 0 & 0 & q_{33} \end{bmatrix}$$

(q_{ii} is the variance parameter of the model error corresponding to the ith state, assumed constant, and $Z(k)$ the observation of discharge in the outlet of the drainage basin at time k.)

$$\mathbf{I} = \begin{bmatrix} 1 & 0 & 0 \\ 0 & 1 & 0 \\ 0 & 0 & 1 \end{bmatrix}$$

$$\mathbf{H}(t) = \begin{bmatrix} 0 & 0 & C_3'(t) \end{bmatrix} \qquad (9.67)$$

$$R(k) = V_0^2 \cdot Z^2(k)$$

(V_0 is the coefficient of variation of the observation.)

$N_{\mu_i}(t)$ and $N_{r_i}(t)$ are given by Eqs. (9.53) and (9.54), respectively.

Note that the state-estimate as well as the error-covariance propagation equations consist of a nonlinear system of coupled differential equations due to the dependence of $\mathbf{N}_\mu(t), \mathbf{N}_r(t)$ on the elements of the vector $\mu(t|t)$ and the matrix $\Sigma(t|t)$.

In the previous formulation, it was assumed that

1. The model error covariance matrix $\mathbf{Q}$ is diagonal with elements q_{ii}.
2. One discharge observation is taken at the outlet of the third nonlinear reservoir in series.
3. The total channel input is a scalar quantity, uncorrelated in time, with a known constant coefficient of variation V_I. The standard deviation σ_μ is thus obtained by multiplying V_I by the known mean of the input at any one time.
4. The discharge-measurement error variance is taken proportional to the square of the measurement. The constant of proportionality is V_0^2, where V_0 acts as a constant coefficient of variation. This technique is a common attempt to account for nonstationarity in model and measurement errors.

Using the National Weather Service (NWS) soil moisture accounting model (Peck, 1976; Kitanidis and Bras, 1980b) to provide total channel inflow (TCI), the above formulation was used to route streamflow in the Bird Creek Basin (2344 km^2) near Sperry, Oklahoma. The NWS model was used to forecast six-hour discharge values at the gaging station, utilizing six-hour estimates of mean area rainfall and potential evapotranspiration. Water year 1959 was chosen as the testing period since very low as well as high flows were observed (0.5 to 286 m^3/sec) in that year. The parameters of the routing model are 0.59, 1.15, and 1.15 for a_1, a_2, and a_3, respectively. The corresponding channel inflow distribution was $P_1 = 0.95$, $P_2 = 0.05$, and $P_3 = 0.0$. The expo nent m took a value of 0.8. The coefficient of variation of the total channel input was 2.5; that of the observations was 0.1.

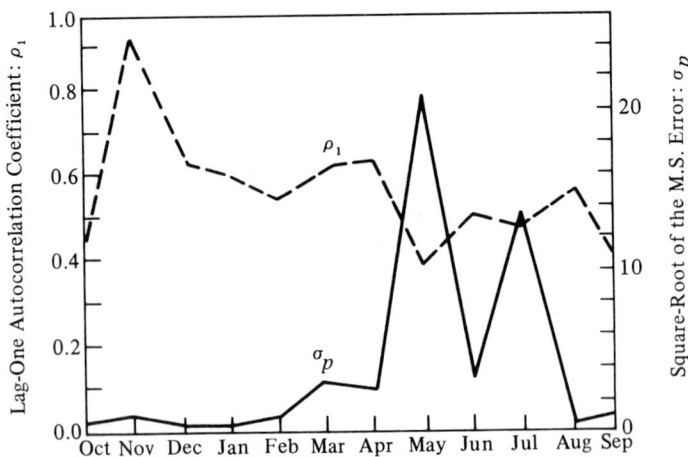

Figure 9.3 Lag-one autocorrelation coefficient and square root of the mean square error of prediction as functions of time (follows the presentation by Bras and Georgakakos, 1980).

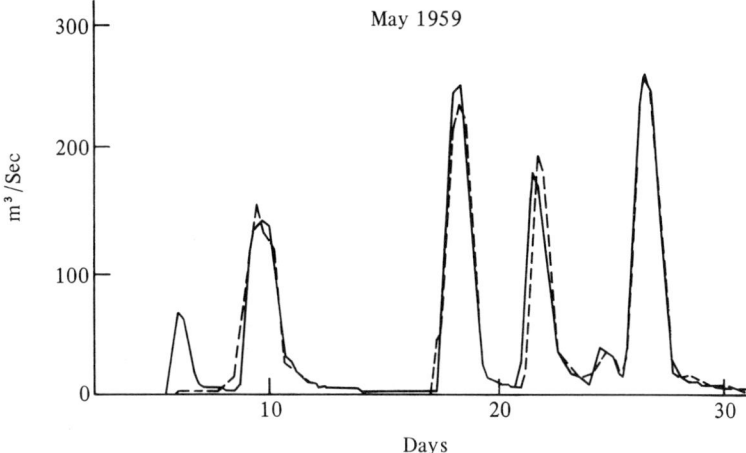

Figure 9.4 Six-hour lead forecasts for May, 1959. Dashed line corresponds to the forecast values, the solid line to the observed values (follows the presentation by Bras and Georgakakos, 1980).

The highly nonstationary nature of the one-step-ahead predicted residuals (innovations) is depicted in Fig. 9.3. At the end of the water year 1959, the correlation coefficient ρ_1 was equal to 0.44. The mean standard error was equal to 4.50 m³/sec. The substantially higher values of σ_p for the months of May and July are due mainly to the high values of the observed discharge in these months. The overall fit can be seen in Figs. 9.4 and 9.5, which display six-hour values of observed versus forecast (six hours ahead) discharge (in m³/sec) for May and July, 1959, respectively.

Of particular importance is the fact that in almost all cases of high-flow hydrographs (> 100 m³/sec), the timing as well as the magnitude of the peak were accurately forecast. Accurate forecasting of the time of rise of the hydrograph was also observed. Also worthy of note is the fact that the estimates of total channel inflow provided by the soil-moisture accounting scheme, running in a deterministic mode, required a very high coefficient of variation. This was due to the fact that although possible, the state of the soil-moisture accounting system was not updated with observations. The relatively high correlation of the residuals implies that the statistics used for the error sequences were not optimal in the least-squares sense. Sensitivity analysis or adaptive schemes can be used to improve the performance of the filter. These will be discussed in following sections.

To end this section on nonlinear filtering, it is necessary to state that the extended Kalman filter and statistical linearization are not the only techniques available to deal with the problem. Next in popularity are the iterative extended Kalman filter and the second-order filter. The former suggests simply that, at every filtering step, you repeatedly update the state and the covariance, always linearizing around the latest estimate. Multiple iterations at each update

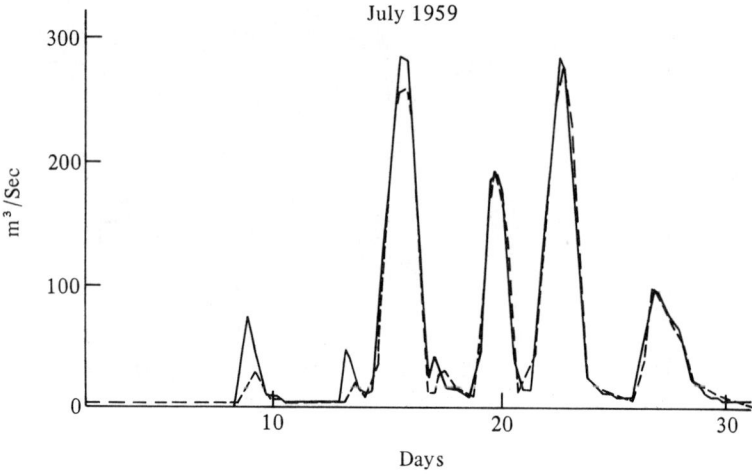

Figure 9.5 Six-hour lead forecasts for July, 1959. Dashed line corresponds to the forecast values, the solid line to the observed values (follows the presentation by Bras and Georgakakos, 1980).

step would improve the linearization and hopefully the results. Second-order filters are predicated on Taylor expansions up to second-order terms. Upon taking expectations this will yield a filter where the state estimate is now a function of the covariance of estimation and the covariance estimate depends on fourth powers of the root mean square errors of estimation. Further details can be found in Gelb (1974) and Schweppe (1973).

9.3 PARAMETER ESTIMATION

Parameter estimation may refer to identification of terms in the model dynamics as well as the identification of model and measurement error-covariance matrices. Real-time parameter estimation within the framework of a filtering algorithm is commonly called adaptive filtering, since the model is constantly adapting to new information. Several techniques will be presented here, all applicable to a variety of parameter-estimation problems. Nevertheless, they will be presented within the context of proven, but limited, examples.

9.3.1 State Augmentation

The canonical dynamic discrete linear model can be expressed as

$$\begin{aligned}
\mathbf{X}(k+1) &= \mathbf{\Phi}(k; \boldsymbol{\alpha})\mathbf{X}(k) + \mathbf{G}(k)\mathbf{W}(k) \\
\mathbf{Z}(k) &= \mathbf{H}(k)\mathbf{X}(k) + \mathbf{V}(k),
\end{aligned} \tag{9.68}$$

where matrix $\boldsymbol{\Phi}(\cdot)$ is explicitly shown to be a function of a parameter vector $\boldsymbol{\alpha}$. It is common for the vector $\boldsymbol{\alpha}$ not to be known with certainty; therefore it must be estimated. One possible approach is to augment the state vector $\mathbf{X}$ by the unknown vector $\boldsymbol{\alpha}$. If the parameters are known to be constants, a parameter "dynamics" equation may be formulated as

$$\boldsymbol{\alpha}(k+1) = \boldsymbol{\alpha}(k). \tag{9.69}$$

A common alternative is to assume that the parameters follow a random walk

$$\boldsymbol{\alpha}(k+1) = \boldsymbol{\alpha}(k) + \mathbf{W}_\alpha(k). \tag{9.70}$$

Defining a state vector

$$\tilde{\mathbf{X}}(t) = \begin{bmatrix} \mathbf{X}(t) \\ \cdots \\ \boldsymbol{\alpha}(t) \end{bmatrix}$$

an augmented system would be

$$\begin{aligned} \tilde{\mathbf{X}}(k+1) &= \tilde{\boldsymbol{\Phi}}(\tilde{\mathbf{X}}(k)) + \tilde{\mathbf{G}}\tilde{\mathbf{W}}(k) \\ \mathbf{Z}(k) &= \tilde{\mathbf{H}}(k)\tilde{\mathbf{X}}(k) + \tilde{\mathbf{V}}(k), \end{aligned} \tag{9.71}$$

where invariably the transition matrix is now a nonlinear function of the new state vector. Other matrices and vectors are defined as

$$\tilde{\mathbf{W}}(k) = \begin{bmatrix} \mathbf{W}(k) \\ \cdots \\ \mathbf{W}_\alpha(k) \end{bmatrix}$$

$$\tilde{\mathbf{G}}(k) = \begin{bmatrix} \mathbf{G}(k) & \vdots & \mathbf{0} \\ \cdots & \vdots & \cdots \\ \mathbf{0} & \vdots & \mathbf{I} \end{bmatrix}$$

$$\tilde{\mathbf{H}}(k) = \begin{bmatrix} \mathbf{H}(k) & \vdots & \mathbf{0} \end{bmatrix}$$

$$\tilde{\mathbf{V}}(k) = \begin{bmatrix} \mathbf{V}(k) \\ \cdots \\ \mathbf{0} \end{bmatrix}.$$

The estimation of the new state vector, including the unknown parameters, can now proceed with one of the methods for handling nonlinear systems discussed in the previous section.

Bras and Restrepo-Posada (1980) illustrated the use of state augmentation in identifying the parameters of a hypothetical conceptual rainfall–runoff model. Their goal was also to show the problems of parameter identification and estimation in general, when the mathematical model does not correspond to the observations. The following paragraphs are adapted from their work.

Conceptual rainfall–runoff models will always have structural errors, due to the nature of the model. In order to learn about the effect of these errors, a simple two-part experiment was performed. In the first part, a model of two linear reservoirs (described below) was used to generate synthetic streamflows. Then, the input and the synthetic output series, contaminated with white Gaussian noise, were used in an attempt to estimate the original model parameters. The initial parameters used to start the estimation algorithm differed widely from those used in the synthetic-generation exercise.

In the second part of the experiment, a model of three linear reservoirs was used to generate the synthetic streamflows, which were contaminated with noise and then used to estimate the parameters of a model of two linear reservoirs. In both cases, state augmentation was the technique chosen for the stochastic state-parameter estimation. The resulting nonlinear filtering problem was solved with an extended Kalman filter.

The two-reservoir catchment model is a simplified conceptual model of a catchment. A schematic diagram of the model is presented in Fig. 9.6. The reservoir represented by the dashed line will become the third linear reservoir in the second part of the experiment.

The upper reservoir approximates the surface runoff and interflow. The lower reservoir represents the base flow. The net input to the model, $u(t)$, is divided between the two reservoirs. This is accomplished by a smooth "S" curve, which is a function of the water content in the lower reservoir. When this water content is high, the lower reservoir becomes saturated and the value for the "S" curve tends asymptotically to 1. The controlling role of the lower reservoir is schematically represented by the dashed line in Fig. 9.6.

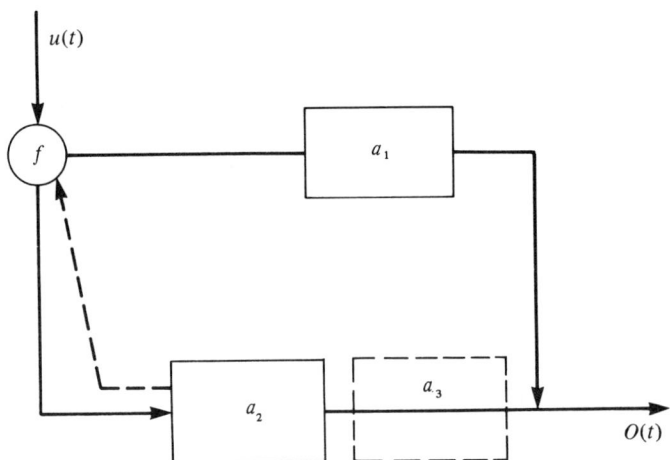

Figure 9.6 Two- and three-reservoir conceptual catchment models (from Bras and Restrepo-Posada, Proceedings of the 3rd International Symposium on Stochastic Hydraulics, 1980, pp. 61–70).

Defining x_i as the outflow from reservoir i, it is possible to express the governing equations as

$$\frac{1}{a_1}\frac{dx_1(t)}{dt} = \left[1 - f\left(\frac{x_2(t)}{a_2}\right)\right]u(t) - x_1(t)$$
$$\frac{1}{a_2}\frac{dx_2(t)}{dt} = f\left(\frac{x_2(t)}{a_2}\right)u(t) - x_2(t),$$
(9.72)

where a_1 is the parameter of the upper (linear) reservoir, a_2 the parameter of the lower (linear) reservoir, and f the smooth "S" curve, defined as

$$f(y) = \frac{1}{2}\left[1 - \tanh\left(a_3(y - a_4)\right)\right].$$

The nonlinear system of equations can be approximated by a forward finite-difference scheme:

$$x_1(k+1) = (1 - a_1\Delta t)\cdot x_1(k) + a_1 u(k)\Delta t\left[1 - f\left(\frac{x_2(k)}{a_2}\right)\right]$$
$$x_2(k+1) = (1 - a_2\Delta t)\cdot x_2(k) + a_2 u(k)\Delta t\left[f\left(\frac{x_2(k)}{a_2}\right)\right].$$
(9.73)

In matrix notation, Eq. (9.73) becomes

$$\mathbf{X}(k+1) = \mathbf{\Phi}[\mathbf{X}(k)].$$
(9.74)

The above deterministic system is converted into a stochastic representation by adding a noise term

$$\mathbf{X}(k+1) = \mathbf{\Phi}[\mathbf{X}(k)] + \mathbf{GW}(k),$$
(9.75)

where $\mathbf{X}^T(k)$ is the state vector, $[x_1(k), x_2(k)]$, $\mathbf{G}$ is the model error matrix, and $\mathbf{W}$ is a white Gaussian noise, $N(\mathbf{0}, \mathbf{Q})$. The measurement equation is

$$Z(k+1) = \mathbf{HX}(k+1) + V(k+1),$$
(9.76)

where $\mathbf{H} = [1, 1]$ and V is a white Gaussian noise, $N(0, R)$.

The state vector was augmented to include the parameters a_1 and a_2. The parameters were assumed constant, implying as dynamics, $\alpha(k+1) = \alpha(k)$. The extended Kalman filter was then used to solve the nonlinear system, including parameters as states.

The result discussed first corresponds to the model free from structural errors. In other words, apart from the fact that the parameters are unknown and the output is contaminated with white noise, the structure of the model (two reservoirs in parallel and an input-distribution function) corresponds exactly to the structure of the "real" system.

The estimates at every time step for the parameters a_1 and a_2 are shown in Fig. 9.7; the real values for the parameter (that is, the values used to generate the series of streamflows) were 0.5 and 0.07, respectively. Convergence of the estimates to the real values was rapid.

The second result corresponds to the estimation of the parameters for a model with structural errors. Several values of the parameter of the third reservoir were used to generate different series of streamflows. The impact of the presence of the third reservoir varies according to the value of the parameter of the third reservoir. In all cases, however, the observed effect was qualitatively similar. Because of this, only the results corresponding to a third-reservoir parameter value of 0.4 are discussed. Figure 9.8 shows the estimates for the parameters a_1 and a_2, respectively, at every time step. In both cases, the estimates for the parameters vary with time. The variability is due to the filter response to the error in the structure. Note that since the third reservoir is directly downstream of the second reservoir, it will affect a_2 more.

The effect of structural errors is that the parameter a_2 diverges (or does not converge) from its true value. Nevertheless, the decaying oscillations in the

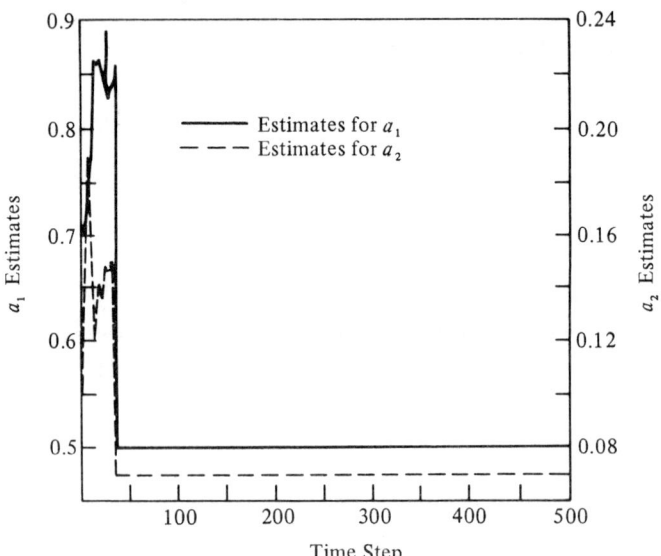

Figure 9.7 Estimates for the parameters of the two-reservoir conceptual catchment model (from Bras and Restrepo-Posada, Proceedings of the 3rd International Symposium on Stochastic Hydraulics, 1980, pp. 61–70).

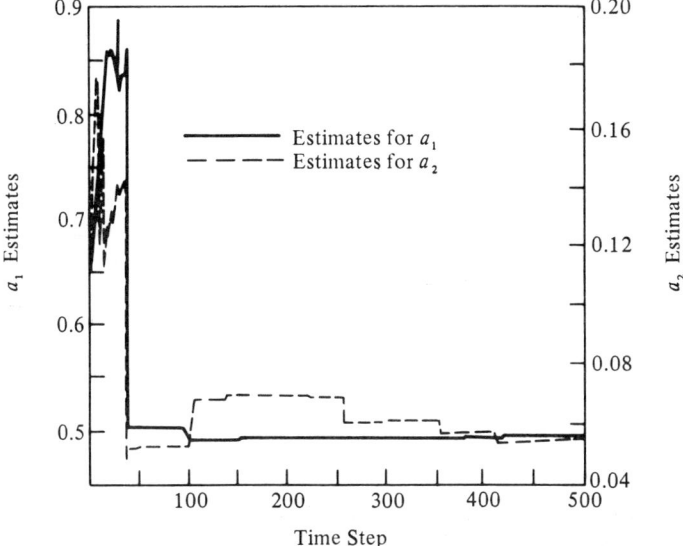

Figure 9.8 Estimates for the parameters of the two-reservoir conceptual catchment model influenced by structural errors (from Bras and Restrepo-Posada, Proceedings of the 3rd International Symposium on Stochastic Hydraulics, 1980, pp. 61–70).

estimate of a_2 indicate that the filter believes it is converging; its theoretical variance of estimation is smaller than the true observed error. This phenomenon is called divergence of filter performance. Divergence is solved by improving the mathematical representation of the system or increasing the model error $\mathbf{Q}(t)$. Divergence is particularly common when solving nonlinear systems through linearization. Although the original model-error representation may be adequate, additional errors are introduced upon linearization. Unless these are accounted for, the filter-error estimates will diverge from true observed values.

In Chapter 8, an estimation example based on Lettenmaier and Burges' (1976) treatment of the Streeter–Phelps equation was presented. It was seen that hypothetically, with known coefficients of BOD decay (K_1) and re-aeration (K_2) and with the perfect model, the predicted errors decayed monotonically (see Fig. 8.9). Lettenmaier and Burges repeated the experiment assuming unknown K_1 and K_2 and river reach velocity (U), which they used to convert travel time to travel distance down the stream. The estimation of states and parameters was then done simultaneously using the extended Kalman filter. Figure 9.9 gives the history of predicted errors. For purposes of comparison, the errors with known parameters are repeated. As can be seen, parameter uncertainty yields errors that are no longer monotonically decreasing, which is a more reasonable result.

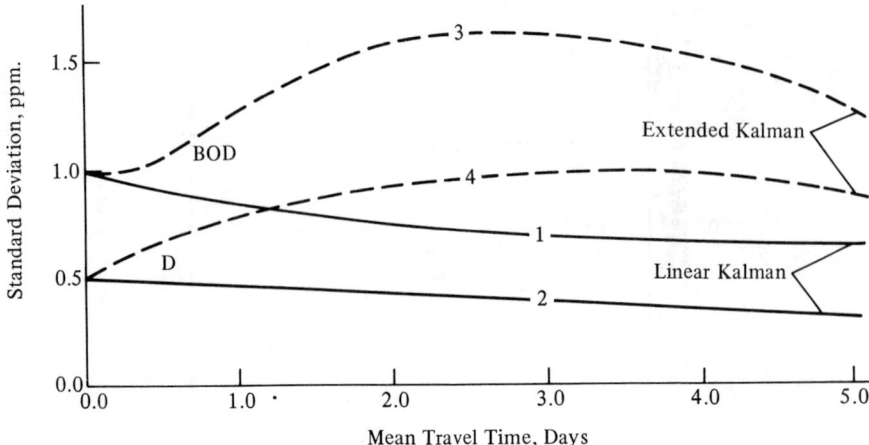

Figure 9.9 Predicted errors in BOD and oxygen deficit, *D*, conditioned upon initial measurements, for linear and extended Kalman-filter models of a BOD–*D* system (from Lettenmaier and Burges, 1976).

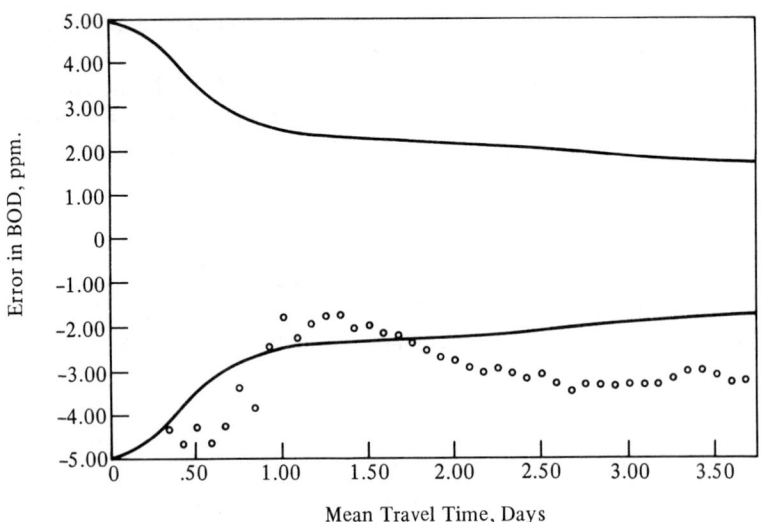

Figure 9.10 Predicted and actual (open circles) errors in BOD for an extended Kalman filter of a BOD–*D* system where sedimentation and non–point-source terms are significant but are neglected in the model dynamics (from Lettenmaier and Burges, 1976).

To study divergence, Lettenmaier and Burges further assumed that the data came from a model with a significant BOD settling rate and a non-point source,

$$\frac{dB}{dt} = -(K_1 + K_3)B + \theta$$

$$\frac{dD}{dt} = K_1 B - K_2 D,$$

where K_3 is a settling rate, D is oxygen deficit, and θ a BOD non-point source. When the data from the above model were processed using the extended Kalman filter with the simpler model of Eqs. (8.129) and (8.130) with unknown coefficients, divergence was observed. Figure 9.10 shows how the actual errors in BOD estimation are widely different from the error bounds (2 standard deviations) that were theoretically expected.

Figure 9.11 shows a normalized mean square error of estimation of D (E_1) and BOD (E_2) as a function of values of the model-error variance for each state, Q_{11} for D and Q_{22} for BOD. Clearly there is a set of model-error terms that yield the minimum overall mean square error.

Georgakakos and Bras (1982b) extended the routing example of Section 9.2.3 and used statistical linearization to estimate simultaneously, through state augmentation, the parameters a_i and the states x_i of the nonlinear routing model of Eq. (9.46). The Taylor–Gauss approximation was again used to

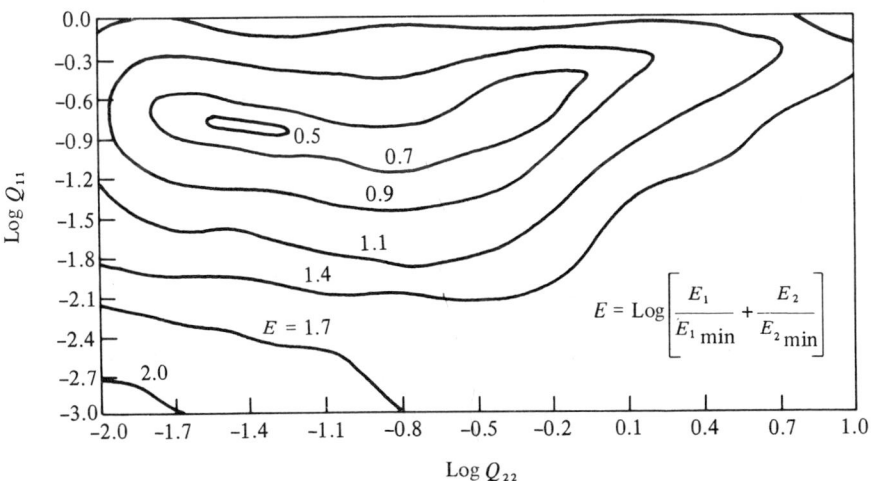

Figure 9.11 Estimated bivariate contours of the rescaled mean square error for a BOD–D system where sedimentation and non–point-source terms are significant but have been neglected in the model dynamics (from Lettenmaier and Burges, 1976).

obtain linearizing matrices. Parameter convergence and overall better behavior in forecasting streamflow was observed.

9.3.2   The Method of Maximum Likelihood
(after Restrepo-Posada and Bras, 1982)

Maximum likelihood is probably the most powerful technique for parameter estimation. The full extent and detail is too much to cover in this book. Nevertheless, it is important to illustrate its application within the Kalman-filtering environment. The following discussion is largely based on a document by Restrepo-Posada and Bras (1982).

Let $\mathbf{Z}$ be a set of identically distributed random variables, and let

$$P_Z(\mathbf{Z}; \boldsymbol{\theta}) \qquad (9.77)$$

be the joint probability density function of that set, in which $\boldsymbol{\theta}$ is the set of unknown parameters of that distribution. Given a set of measurements on $\mathbf{Z}$, say $\mathbf{z}$, the likelihood function of the sample $\mathbf{z}$ is defined by

$$L(\boldsymbol{\theta}|\mathbf{z}) = P_Z(\mathbf{z}; \boldsymbol{\theta}). \qquad (9.78)$$

Having defined the likelihood of the sample by Eq. (9.78), the criterion of maximum likelihood can now be defined:

The maximum-likelihood estimator of $\boldsymbol{\theta}$ is the value $\boldsymbol{\theta}^*$ that maximizes Eq. (9.78).

For some probability density functions, it is more convenient to define the estimation in terms of the logarithm of the likelihood function (log-likelihood function), which will be denoted by ξ. Since the logarithm is a monotonically increasing function, the value $\boldsymbol{\theta}^*$ that maximizes the likelihood function also maximizes the log-likelihood function.

The maximum-likelihood estimators, herein $\boldsymbol{\theta}^*$, have a series of properties that are independent of the underlying distribution. For large N, where N is the number of data values, the estimator $\boldsymbol{\theta}^*$ is approximately normal, with mean asymptotically equal to the true parameter value; that is, $\boldsymbol{\theta}^*$ is asymptotically unbiased. Its mean square error is asymptotically equal to

$$\mathrm{var}\left[\theta_i^*\right] = -\left[N \cdot E\left[\partial^2 \xi / \partial \theta_i^2\right]\right]^{-1},$$

where the subscript i refers to a given element of vector $\boldsymbol{\theta}$.

Also asymptotically, the maximum-likelihood estimator has the minimum expected square error of all possible unbiased estimators: it is efficient. It also is consistent, since with increasingly high probability the estimates will be arbitrarily close to the true values of the parameters. Finally, θ^* makes maximum use of the information contained in the data; that is, it is sufficient.

Within the context of the discrete stationary stochastic dynamic system,

$$\mathbf{X}(k+1) = \boldsymbol{\Phi}\mathbf{X}(k) + \mathbf{B}\mathbf{U}(k) + \mathbf{G}\mathbf{W}(k) \tag{9.79}$$

$$\mathbf{Z}(k) = \mathbf{H}\mathbf{X}(k) + \mathbf{V}(k), \tag{9.80}$$

the likelihood function is given by the distribution of $\mathbf{Z}(k)$, which in turn depends on parameters $\boldsymbol{\Phi}$, $\mathbf{B}$, $\mathbf{G}$, and $\mathbf{H}$, and on the error covariances, $\mathbf{Q} = E[\mathbf{W}(k)\mathbf{W}^T(k)]$ and $\mathbf{R} = E[\mathbf{V}(k)\mathbf{V}^T(k)]$.

If it is desired to estimate unknown elements of matrices $\boldsymbol{\Phi}$, $\mathbf{B}$, $\mathbf{G}$, $\mathbf{H}$, $\mathbf{Q}$, and $\mathbf{R}$, a first step may be to form the likelihood function of the unknown parameters given the available N observations,

$$L[\theta|\mathbf{Z}(1)\dots\mathbf{Z}(N)] = P_Z[\mathbf{Z}_N; \theta],$$

where

$$\mathbf{Z}_N^T = \left[\mathbf{Z}(1) \; \vdots \; \dots \; \vdots \; \mathbf{Z}(N)\right].$$

Using Bayes' rule, the distribution of the observed values, $\mathbf{Z}_N$, becomes

$$P_Z(\mathbf{Z}_N; \theta) = P_Z(\mathbf{Z}_{N-1}; \theta) P_Z(\mathbf{Z}(N)|\mathbf{Z}_{N-1}; \theta), \tag{9.81}$$

which leads to the log-likelihood function,

$$\xi(\theta|\mathbf{Z}_N) = \ln P_Z(\mathbf{Z}_N; \theta).$$

Equation (9.81) becomes

$$\xi(\theta|\mathbf{Z}_N) = \xi(\theta|\mathbf{Z}_{N-1}) + \ln P_Z(\mathbf{Z}(N)|\mathbf{Z}_{N-1}; \theta) \tag{9.82}$$

with $\xi(\theta|0) = 0$.

Within the context of the linear systems of Eqs. (9.79) and (9.80), it is easy to assume a regenerating Gaussian distribution for the random vectors $\mathbf{X}(k)$ and $\mathbf{Z}(k)$. Furthermore,

$$P_Z(\mathbf{Z}(N)|\mathbf{Z}_{N-1}) = P_Z(\mathbf{Z}(N)|\mathbf{Z}(N-1)),$$

which in turn is

$$P_Z(\mathbf{Z}(N)|\mathbf{Z}(N-1)) = \left[(2\pi)^m|\Sigma_Z(N|N-1)|\right]^{-1/2}$$
$$\cdot \exp\left\{-\frac{1}{2}\left[\delta_Z^T(N)\Sigma_Z^{-1}(N|N-1)\delta_Z(N)\right]\right\}, \tag{9.83}$$

where

$$\delta_Z(N) = Z(N) - \hat{Z}(N|N-1)$$
$$\Sigma_Z(N|N-1) = E[\delta_Z(N)\delta_Z^T(N)].$$

The dimension of $Z(k)$ is m and $\hat{Z}(N|N-1)$ stands for the best estimate (conditional mean) of $Z(N)$ given information up to $N-1$.

The use of Eq. (9.83) in Eq. (9.82) then results in

$$2\xi(\theta|Z_N) = 2\xi(\theta|Z_{N-1}) - \ln\{|\Sigma_Z(N|N-1)|\} - m\ln(2\pi)$$
$$- \delta_Z^T(N)\Sigma_Z^{-1}(N|N-1)\delta_Z(N) \qquad (9.84)$$
$$\xi(\theta|0) = 0.$$

Expanding the above equation yields,

$$2\xi(\theta|Z_N) = \xi_b(\theta|Z_N) + \xi_o(\theta|Z_N). \qquad (9.85)$$

The terms ξ_b and ξ_o are known, respectively, as the bias component and the observations component of the log-likelihood function and are defined by

$$\xi_b(\theta|Z_N) = - Nm\ln(2\pi) - \sum_{k=1}^{N} \ln|\Sigma_Z(k|k-1;\theta)| \qquad (9.86)$$

$$\xi_o(\theta|Z_N) = - \sum_{k=1}^{N} \delta_Z^T(k;\theta)\Sigma_Z^{-1}(k|k-1;\theta)\delta_Z(k;\theta). \qquad (9.87)$$

The observations component, ξ_o, is made up of the sum of the squares of the residuals, normalized by their variances. This important term will be used for optimality tests. The values $\hat{Z}(k|k-1)$ and $\Sigma_Z(k|k-1)$ are calculated by means of the Kalman-filter equations. For the sake of completeness, the equations are repeated here. The dependence of the state variables on the unknown set of parameters θ is explicitly shown. However, for shortness in notation, the θ dependence of the matrices Φ, H, B, Q, R is not shown.

$$\hat{X}(k+1|k+1;\theta) = \Phi\hat{X}(k|k;\theta) + BU(k)$$
$$+ \Sigma_X(k+1|k;\theta)H^T\Sigma_Z^{-1}(k+1|k;\theta)\delta_Z(k+1;\theta)$$
$$(9.88)$$

$$\hat{Z}(k+1|k;\theta) = H[\Phi\hat{X}(k|k;\theta) + BU(k)] \qquad (9.89)$$

$$\Sigma_X(k+1|k+1;\theta) = \Sigma_X(k+1|k;\theta)$$
$$- \Sigma_X(k+1|k;\theta)H^T\Sigma_Z^{-1}(k+1|k;\theta)H\Sigma_X(k+1|k;\theta)$$
$$(9.90)$$

$$\Sigma_X(k+1|k;\theta) = \Phi\Sigma_X(k|k;\theta)\Phi^T + GQG^T \qquad (9.91)$$

$$\Sigma_X(k+1|k) = E\left[(X(k+1) - \hat{X}(k+1|k))(X(k+1) - \hat{X}(k+1|k))^T\right]$$

$$\Sigma_Z(k+1|k;\theta) = R + H\Sigma_X(k+1|k;\theta)H^T \qquad (9.92)$$

with initial conditions

$$\Sigma_X(0|0; \theta) = \Sigma_0$$
$$\hat{X}(0|0) = \hat{X}_0.$$

System identification is, per se, an off-line procedure. That is, for some historic data one looks for the best set of parameters that fit a chosen model. In contrast to off-line procedures, some models require the use of "adaptive" parameters, so that short-term inadequacies of the model in representing real life can be corrected on-line, that is, in real time. This is particularly true for the case of linear models that approximate nonlinear processes. This modification of the parameters to account for the most recent errors in the prediction commonly hampers the long-range forecasting capability of the models. It seems natural, then, that models that accurately represent a real process should be calibrated off-line.

Some other considerations, such as the cost of the estimation scheme, play a role in deciding whether to develop an on-line or off-line parameter-estimation procedure. As a rule of thumb, on-line parameter-estimation schemes may be cheaper to use than off-line ones. This is due to the fact that in an on-line scheme the time steps used to calculate the state variables are combined with successive approximations to the parameter's values. In the case of maximum-likelihood estimation of dynamic systems, however, there is no way of calculating exactly the log-likelihood function on-line, at a given time step, without having to redo all the computations from $k = 1$ to $k = N$. Since the cost of doing this will eliminate the cost advantage of the on-line methods, approximations of the log-likelihood function must be used. This approach, however, invalidates the properties of the maximum-likelihood estimators, since the true log-likelihood is not being used. In general, we can say that the better the approximation, the higher the cost. There is, therefore, a trade-off between cost and accuracy. Nevertheless, on-line maximum-likelihood approximations do exist and in fact there are some exact solutions for particular situations. The interested reader is referred to Schweppe (1973) for additional discussions on this topic.

As presented, the maximum-likelihood approach of Eqs. (9.86) and (9.87) is valid only for linear stationary systems. Nonlinearity is handled by using one of the linearization procedures of Section 9.2. The validity of the approach depends on the extent to which the Gaussian assumption is satisfied and on the nondivergent nature of the linearized filter.

Nonstationarity (time-varying parameters) is somewhat harder to treat conceptually. The simplest approach is to assume that the process is stationary over finite periods of time. For example, in streamflow modeling, Eqs. (9.86) and (9.87) may be evaluated by seasonal groupings of data. Similarly, the summations may be limited to finite windows, for example, from $N - M$ to N where M is the "memory" of the process. The size of this memory will depend on the gradualness of the nonstationarity. For rapidly changing conditions, the approach is not possible. It must be realized that approximations of this type do not guarantee the optimality properties of the original procedure.

Maximum-Likelihood Estimation with Prior Information

One of the major problems addressed in the automatic parameter-estimation literature (particularly in hydrology) is the convergence of some parameters to "unrealistic" values. This immediately suggests the idea of the existence of prior knowledge about the range of some of the parameters. If this is the case, this prior information should be incorporated within the parameter-estimation procedure. In a Bayesian framework, this amounts to maximizing the posterior probability density which is proportional to the product of the likelihood function and the prior probability density

$$p''(\theta) = \kappa L(\theta|\mathbf{Z}_N) \cdot p'(\theta), \tag{9.93}$$

in which p'' is the posterior probability density, L the likelihood function, κ the normalizing constant, and p' the prior probability density. Taking logarithms in both sides of Eq. (9.93),

$$\ln p''(\theta) = \ln \kappa + \xi(\theta|\mathbf{Z}_N) + \ln p'(\theta). \tag{9.94}$$

The problem can be formulated as

$$\max\left[\xi(\theta|\mathbf{Z}_N) + \ln(p'(\theta))\right]. \tag{9.95}$$

Note that since $\ln \kappa$ is a constant, it was taken out of Eq. (9.95) without affecting the outcome of the maximization process. The importance of this approach is that not only the initial estimates, but also the quality of these estimates is taken into account. Under the Gaussian assumption, the logarithm of the prior probability density that enters into Eq. (9.95) can be written:

$$2 \cdot \ln p'(\theta) = - n_\theta (2\pi) - \ln|\Sigma_\theta|^{-1} + 2\xi_{\theta'}(\theta|\theta') \tag{9.96}$$

$$\xi_{\theta'}(\theta|\theta') = (1/2) \cdot (\theta - \theta')^T \cdot \Sigma_\theta^{-1} \cdot (\theta - \theta'), \tag{9.97}$$

where n_θ is the number of parameters with prior information θ' and covariance matrix Σ_θ. Of the three terms on the right-hand side of Eq. (9.96), the first two will remain constant under the posterior optimization of Eq. (9.95) and can be dropped without affecting the outcome. The remaining term is the square of the deviations of the prior values from the final ones, weighted by the prior variance of estimation. This term acts as an extended-observation component of the log-likelihood (ξ_o). The extended-observation component will consist of the measurements of the process and n_θ observations on the parameters. We can combine Eqs. (9.95) and (9.97) in the form of Eq. (9.85) to arrive at

$$2\xi_x(\theta|\mathbf{Z}_N, \theta') = \xi_b(\theta|\mathbf{Z}_N) + \xi_o'(\theta|\mathbf{Z}_N, \theta') \tag{9.98}$$

$$\xi_o'(\theta|\mathbf{Z}_N, \theta') = \xi_o(\theta|\mathbf{Z}_N) + \xi_{\theta'}(\theta|\theta'), \tag{9.99}$$

in which $\xi_x(\theta|Z_N, \theta')$ is the extended log-likelihood of the parameters given the discharge measurements and the prior estimates of the parameters, and $\xi'_o(\theta|Z_N, \theta')$ the extended-observation component.

Two final observations can be made regarding $\xi_{\theta'}(\theta|\theta')$. First, if the prior estimates of the parameters form an independent set, Σ_θ will be a diagonal matrix, and $\xi_{\theta'}(\theta|\theta')$ can be written as

$$\xi_{\theta'}(\theta \mid \theta') = \left\{\frac{1}{2} \cdot \left(\theta_i - \theta'_i\right) / \sigma^2_{\theta_i}\right\} \tag{9.100}$$

in which $\sigma^2_{\theta_i}$ is the prior variance of estimation of the parameter θ_i. Second, there are no restrictions on the number of observations of each parameter. Each independent observation can be included, weighted by its variance of estimation.

Optimization Procedures

The optimization of the likelihood function requires the use of nonlinear, unconstrained optimization schemes. These nonlinear optimization algorithms are based on the systematic search for the maximum of the objective function along a sequence of straight-line searches. In each of these searches, a one-dimensional optimization problem is solved. A new direction of search is chosen once an optimum is found. The way in which the direction of these line searches is chosen defines the difference among the nonlinear optimization procedures.

The efficiency of a nonlinear optimization algorithm is measured in terms of the rate of convergence of the solution toward the optimal point. The most efficient algorithms are also the most demanding ones in terms of computational requirements. Therefore, there is a trade-off between cost of computation and efficiency of the algorithm. The ease of implementation of the algorithm is also a factor. The choice of algorithm is also problem-dependent, controlled by the shape of the likelihood surface.

A typical algorithm is the Davidon–Fletcher–Powell (DFP) (see Luenberger, 1973). DFP has been developed for minimization of nonlinear functions. Therefore, it is used to minimize the negative of the log-likelihood function defined in Eq. (9.98). The general steps to follow are

1. Start with any symmetric positive definite estimate of the inverse Hessian (for example, the identity matrix), and any point θ^0. Set $k = 0$.
2. Set $\mathbf{d}^k = -\mathbf{S}^k\mathbf{g}^k$, where $\mathbf{g}^k$ is the gradient of the negative log-likelihood function with respect to the parameters at iteration k, $\mathbf{S}^k$ is the approximation to the inverse Hessian, and $\mathbf{d}^k$ is the direction of search.
3. Minimize $-\xi(\theta^k + \alpha^k\mathbf{d}^k)$ with respect to α^k to obtain a new value of the parameters θ^{k+1}. The corresponding gradient $\mathbf{g}^{k+1}$ is calculated at this

point. The new set of parameters is calculated at the optimal α^k, as

$$\theta^{k+1} = \theta^k + \mathbf{p}^k,$$

where

$$\mathbf{p}^k = \alpha^k \mathbf{d}^k.$$

4. Compute the difference between the gradients at iterations $k+1$ and $k, \mathbf{q}^k$, and use it to approximate the inverse Hessian:

$$\mathbf{q}^k = \mathbf{g}^{k+1} - \mathbf{g}^k.$$

Improve the approximation to the inverse Hessian by means of

$$\mathbf{S}^{k+1} = \mathbf{S}^k + \frac{\mathbf{p}^k \mathbf{p}^{k^T}}{\mathbf{p}^k \mathbf{q}^k} - \frac{\mathbf{S}^k \mathbf{q}^k \mathbf{q}^{k^T} \mathbf{S}^k}{\mathbf{q}^{k^T} \mathbf{S}^k \mathbf{q}^k}.$$

Update k and go to Step 2.

The approximation to the inverse Hessian $\mathbf{S}^k$, computed by DFP is exact for quadratic objective functions. Since the log-likelihood is a nonquadratic function, the approximate inverse Hessian at some point after several iterations may not resemble the real inverse Hessian at that point. The solution is to re-initialize the approximation to the inverse Hessian after a number of linear searches by setting the Hessian equal to the identity matrix.

The linear optimization over α^k in Step 3 needs careful thought to avoid excessive computation. Generally, the optimization is a nonexhaustive search combined with reasonable interpolation algorithms. The reader is referred to Restrepo-Posada and Bras (1982) for one such procedure.

A common occurrence in parameter estimation of hydrologic models is the failure to converge to a feasible parameter set. This may be due simply to problems with the search procedure for the particular likelihood surface, or to structural errors in the model. Due to incompatibility of model and data, the "optimal" parameter set may not be a feasible set. This is illustrated in Fig. 9.12, where the search is seen moving in the direction of nonfeasible negative parameter θ_1. The solution is of necessity a constrained optimization approach. The reader is again referred to Restrepo-Posada and Bras for such a heuristic approach.

In the neighborhood of the optimal point, the log-likelihood function can be approximated by a quadratic surface. It has been shown by Edwards (1972) that the log-likelihood function evaluated at 2 standard deviations of the parameters away from the optimum is exactly two units smaller than the maximum. Therefore, if a linear search fails to improve the log-likelihood function by more than two units, probably the parameters are closer than 2 standard deviations from their maximum-likelihood values. Since any improvement in the log-likelihood function thereafter will be very small, the last point in the linear search can be taken as the maximum-likelihood estimate of the parameters. This criterion has been used successfully in the general purpose

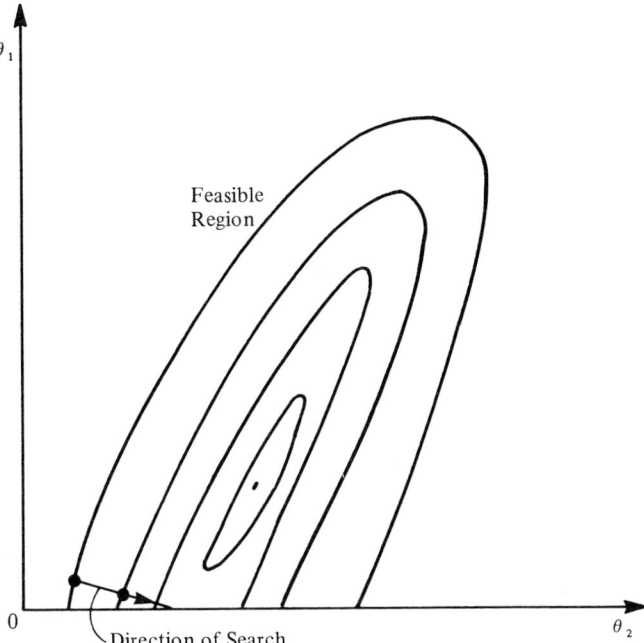

Figure 9.12 Contours of a log-likelihood function showing convergence to nonfeasible regions (from Restrepo-Posada and Bras, 1982).

system identifier and evaluator (GPSIE), developed by Peterson (1976). Nevertheless, the experience with estimating the parameters of the conceptual rainfall–runoff models (Restrepo-Posada and Bras, 1982) shows that sometimes the difference between the log-likelihood functions at two consecutive iterations may be less than two units, yet the maximum of the likelihood surface is not near. A stricter criterion would call for the global search to be suspended after the optimization algorithm fails to improve the log-likelihood function by more than a given small number of units. The goodness of this estimate can be measured by several criteria:

1. Positive definiteness of the information matrix $\mathbf{F}$, which is approximated by the negative Hessian. This is a necessary condition for optimality.

2. The elements in the diagonal of the parameter covariance matrix, approximated by the inverse of $\mathbf{F}$, give us the lower bounds on the variances of the parameter's estimates, a measure of the quality of these estimates.

3. The value of the log-likelihood function at 2 standard deviations of the parameters away from the optimal point should be two units less than the value at the optimum.

4. The expected value of the sum of squares of the normalized residuals (Eq. 9.87) equals the number of degrees of freedom, i.e., the number of

measurements minus the number of parameters to be estimated from these data.

5. The lag-zero correlation coefficient of the normalized residuals should be close to its expected value, 1.0, and the lag-j ($j \neq 0$) correlation coefficient should be close to zero. This is a statement of whiteness of residuals. The standard deviations of these correlation coefficients can also be calculated. The normalized correlation coefficients, defined as the estimated correlation coefficients divided by their respective standard deviations, can thus be obtained. These give an idea of the deviation of the correlation coefficients away from their expected value, in units of standard deviations. For scalar measurements, the correlation coefficients, $\rho(i)$, are estimated by $\hat{\rho}(i)$, which is computed by

$$\hat{\rho}(j) = \sum_{t=1}^{N-j} [\nu(t) \cdot \nu(t-j)]/(N-j) \qquad j = 0, 1, \ldots, \qquad (9.101)$$

where N is the total number of measurements and j is the lag.

It can be shown that the standard deviation of the estimates of the correlation coefficients, σ_ρ, is

$$\sigma_\rho(i) = \frac{1 + E[\rho(i)]}{N} - \frac{i-1}{N^2} \qquad (9.102)$$

(Peterson, 1975) in which

$$E[\rho(i)] = \begin{cases} 1 & \text{for } i = 0 \\ 0 & \text{for } i > 0 \end{cases}.$$

A problem associated with the previous convergence criterion may appear when the log-likelihood is very sensitive to some of the parameters, and relatively insensitive to the others. This problem is illustrated, for a two-parameter case, in Fig. 9.13. In that figure, the superscripts f and b denote, respectively, a forward and backward perturbation of the parameters. These perturbations are used to calculate the gradient of the log-likelihood function with respect to the parameters by a central-difference scheme. In Fig. 9.13a, θ^k is the value of the parameter vector at iteration k.

For that case, the gradients computed by finite differences will be such that

$$|g_2^k| \gg |g_1^k|.$$

This causes the direction of linear search, $\mathbf{d}^k$, to be mostly in the direction of θ_2. This can lead to cases where the difference between the values of the

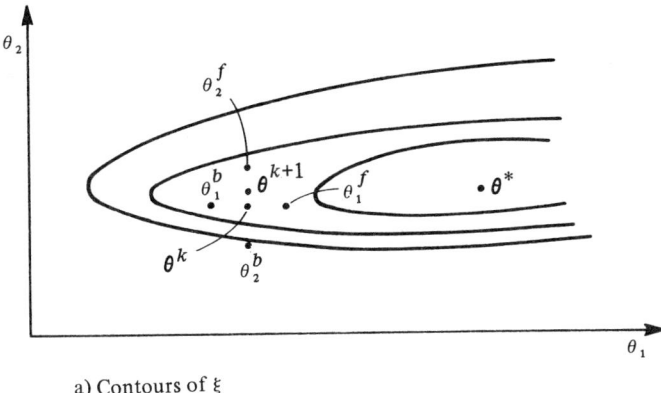

a) Contours of ξ

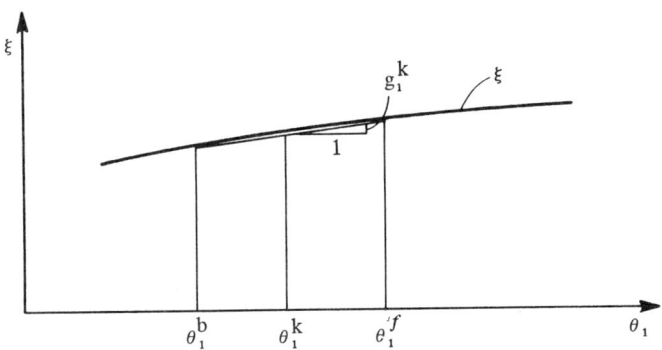

b) ξ in the Direction of θ_1

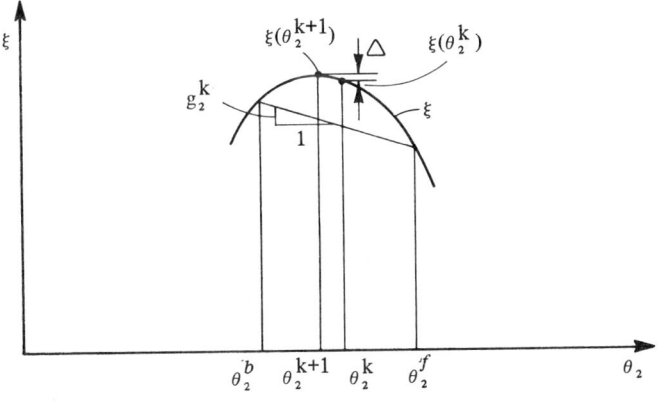

c) ξ in the Direction of θ_2

Figure 9.13 Calculation of the gradient by a central-finite-difference scheme (from Restrepo-Posada and Bras, 1982).

log-likelihood at the end of the linear search, $\xi(\theta^{k+1})$, and that at the beginning, $\xi(\theta^k)$, Δ in Fig. 9.13c, is smaller than the global convergence criterion proposed above. The nonlinear optimization process then stops at points far from the global optimum. A solution may be to include in following linear searches only those parameters for which changes in the forward or backward direction show improvement in the likelihood function. Unfortunately, convergence to the optimal solution cannot be guaranteed for such approaches.

An Example of Parameter Estimation in a Conceptual Base Flow Model

Restrepo-Posada and Bras (1982) illustrate the maximum-likelihood procedure in parameter estimation of a simple conceptual model of dry-period flow in rivers—base flow. Figure 9.14 schematizes the model. It consists of two "reservoirs" with uncorrelated noisy inputs, which yield corresponding discharges that when added constitute river discharge. The discharge is measured with noise leading to observation $\mathbf{Z}(t)$. Such models are frequently used to represent the behavior of lower soil zones within larger-scope moisture-accounting models such as the National Weather Service (NWS) River Forecasting Model (Peck, 1976). Restrepo-Posada and Bras (1982) in fact studied the base-flow model within the NWS model context; that will not be covered here.

The two reservoirs in the system are linear, but generally with different time-delay constants. The goal is to mimic different rates of discharge depend-

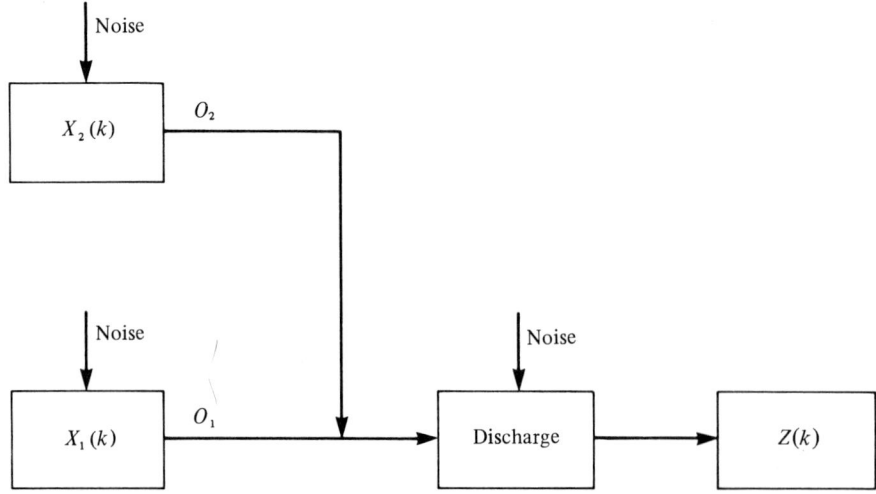

Figure 9.14   Schematic representation of the stochastic model of base flow (from Restrepo-Posada and Bras, 1982).

ing on the amount of water in each reservoir. Such behavior is common in the study of hydrograph-recession curves (Linsley et al., 1982).

The state–space equations of the model can be expressed in continuous or discrete form as well as by using storage or discharge as states. Using discharge as state, with $X_1(k)$ and $X_2(k)$ the discharges of reservoirs 1 and 2, respectively, the discrete system becomes:

$$X_1(k+1) = A_1 X_1(k) + W_1(k) \tag{9.103}$$

$$X_2(k+1) = A_2 X_2(k) + W_2(k) \tag{9.104}$$

$$Z(k) = X_1(k) + X_2(k) + V(k), \tag{9.105}$$

where $A_1 = \exp(-k_1 \Delta t)$, $A_2 = \exp(-k_2 \Delta t)$, and $W_1(k)$, $W_2(k)$, and $V(k)$ are discrete-time independent white Gaussian noises with zero-mean variances

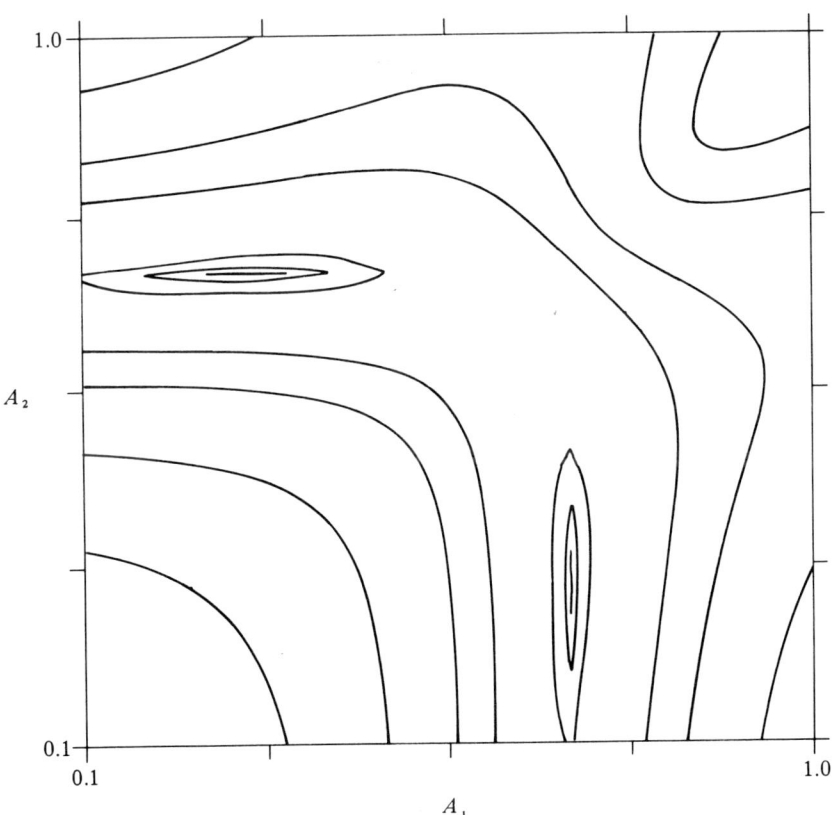

Figure 9.15 Contour lines of the log-likelihood function of the model dynamics parameters (from Restrepo-Posada and Bras, 1982).

Q_{11}, Q_{22}, and $R(k)$, respectively. The model error variances Q_{11} and Q_{22} are assumed to be constants and $R(k) = RZ^2(k)$.

The model can be shown to be generally observable except in the noninteresting situation where $A_1 = A_2$ (Restrepo-Posada and Bras, 1982).

It is possible to visually study the likelihood function of various parameters. By simulation with an assumed set of parameters,

$$\mathbf{A} = \begin{bmatrix} 0.7 & 0 \\ 0 & 0.3 \end{bmatrix}; \qquad \mathbf{X}(0) = \begin{bmatrix} 4 \\ 4 \end{bmatrix}$$

$$Q_{11} = Q_{22} = 10^{-6}; \qquad R = 10^{-4}$$

the likelihood functions for 2 parameters at a time can be evaluated.

Figure 9.15 shows a contour plot of the likelihood of A_1 and A_2 (other parameters constant). The function is symmetrical with respect to the line $A_1 = A_2$. Furthermore, there is a ridge along that line, indicated the previously mentioned lack of observability.

Figure 9.16 is an isometric view of the likelihood function when the model errors Q_{11} and Q_{22} are varying and everything else is constant. The main feature is a large plateau where the function is insensitive to parameter values. This leads to optimization difficulties when starting from points on the plateau, far from the maximum. The maximum itself is ill defined, lying along an

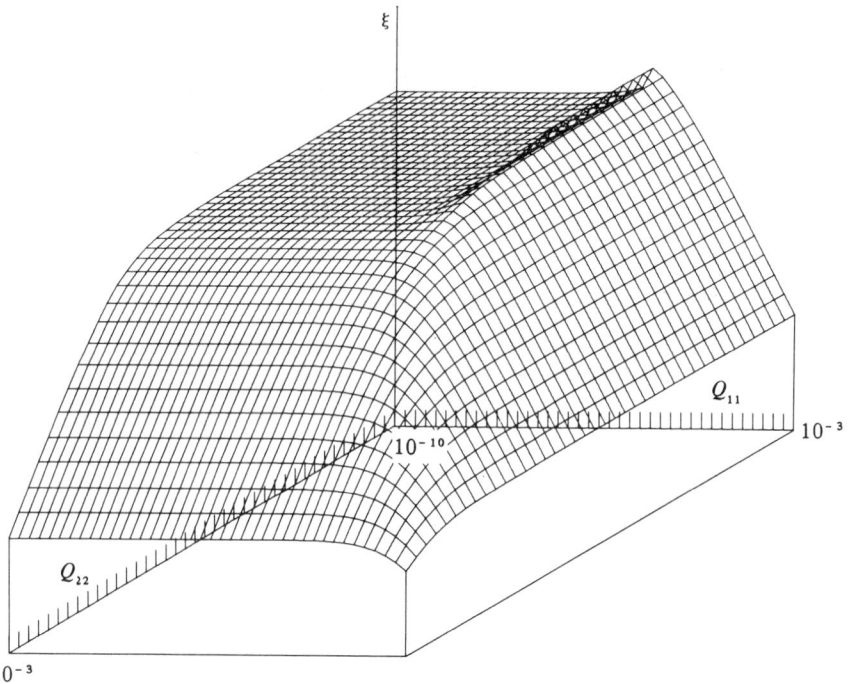

Figure 9.16 Isometric view of the log-likelihood function of the model-error parameters (from Restrepo-Posada and Bras, 1982).

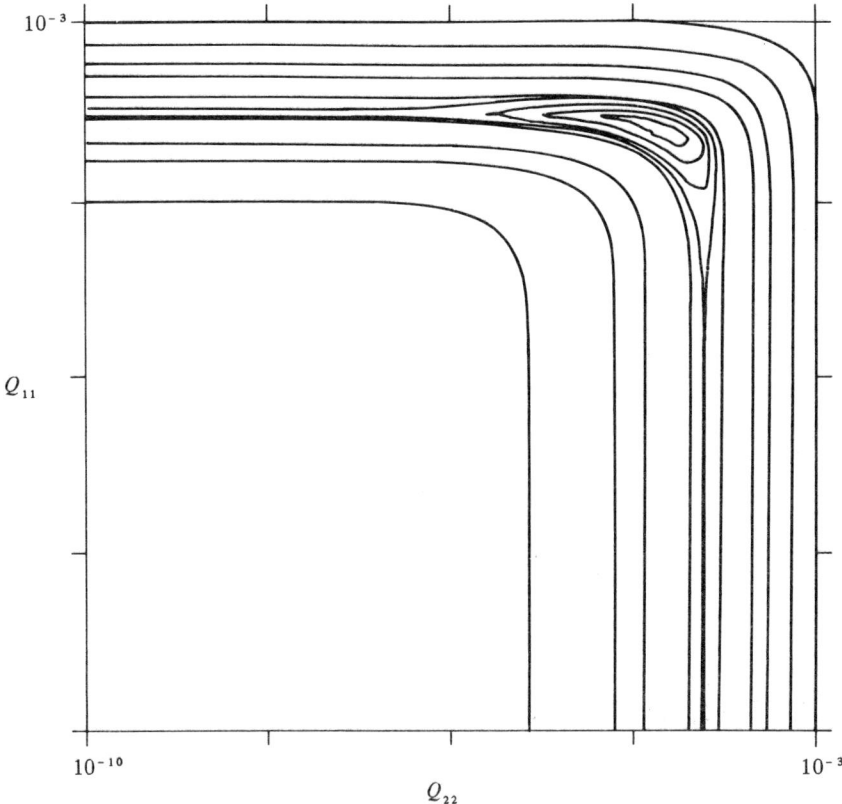

Figure 9.17 Contour lines of the log-likelihood function of the model-error parameters with an edge around the plateau (from Restrepo-Posada and Bras, 1982).

almost horizontal ridge parallel to the Q_{22} axis. This indifference to Q_{22} values along that ridge is a parameter identifiability problem. With a different parameter set ($\mathbf{A}, \mathbf{X}(0)$, and R), the likelihood function took the form contoured in Fig. 9.17. Here a somewhat better-defined maximum is seen, but it is very shallow and at the intersection of two horizontal ridges parallel to the axes.

Parameter calibration was performed using real data from the Bird Creek Basin, Oklahoma, for July and August 1957. The discharge during the period is plotted on semilog paper in Fig. 9.18. In such a graph, the slopes of the lines a and b are rough estimates of the parameters A_1 and A_2, respectively, from which k_1 and k_2 can be obtained. Estimates of A_1, A_2, $X_1(0)$, $X_2(0)$, Q_{11}, and R were made using maximum likelihood and a weighted least-squares procedure that minimized, over parameter set θ,

$$J = R \sum_{k=1}^{N} [(Z(k) - \hat{Z}(k; \theta))/Z(k)]^2.$$

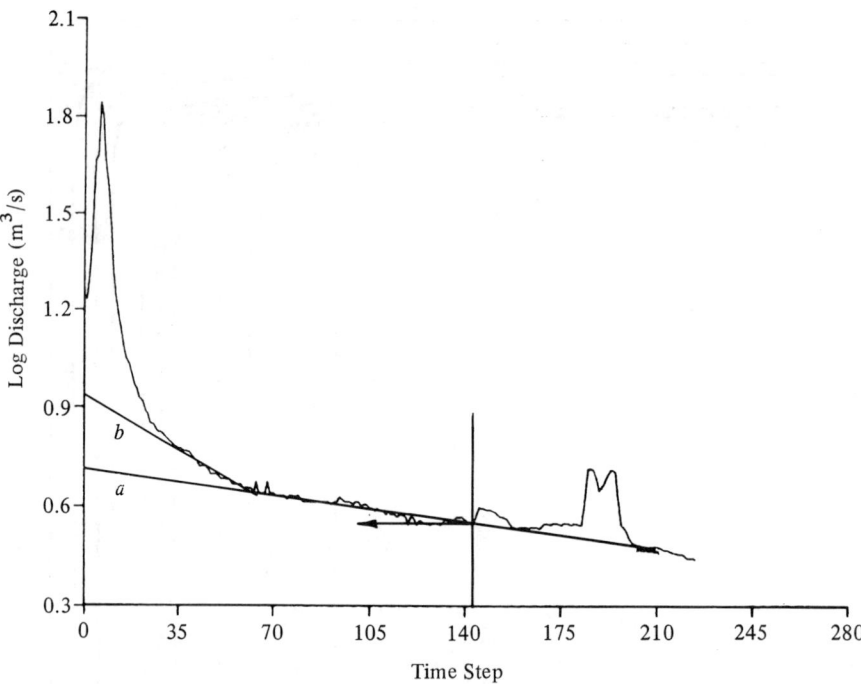

Figure 9.18 Semilogarithmic plot of Bird Creek discharges for July and August, 1957 (from Restrepo-Posada and Bras, 1982).

The parameter estimates and corresponding coefficients of variation obtained with the maximum-likelihood procedure are presented in Table 9.3. The correlation matrix, and the serial correlation of the residuals are included in Table 9.4. Figure 9.19 shows the predicted and the measured discharges, in which the parameters estimated by maximum likelihood have been used.

The lower bounds on the error for the parameter estimates (Table 9.3) are under 15% of the parameter values. The correlations among the parameter estimates are also small. Both results indicate that good estimates of the parameters were obtained. The estimates of k_1 given by the National Weather Service, the estimate from Fig. 9.18, and that obtained by weighted least squares are essentially identical, while differing considerably from the maximum-likelihood estimate. Keep in mind that the method of maximum likelihood includes, in addition to the sum of the squares, the bias term of the likelihood function. A wider range of values was obtained for the estimates of k_2, with the smallest value given by the NWS, and the highest given by maximum likelihood. This wide range of values for that parameter would make us think that that parameter would not be highly observable. But paradoxically, the coefficient of variation in the maximum-likelihood estimate of this parameter is only 6%.

Table 9.3
Parameter estimates by different procedures*

Parameter	NWS	Visual	WLS	ML	CV
k_1	0.013	0.013	0.014	0.020	0.12
k_2	0.126	0.338	0.390	0.550	0.06
$X_1(0)$	—	—	5.35	6.04	0.06
$X_2(0)$	—	—	6.89	7.00	0.05
Q_{11}	—	—	—	1.96×10^{-3}	0.15
R	—	—	—	1.79×10^{-4}	0.13

*NWS denotes National Weather Service, WLS weighted least squares, ML maximum likelihood, and CV coefficient of variation.

Source: Restrepo-Posada and Bras (1982).

Table 9.4
Maximum-likelihood post-optimality analysis

	k_1	k_2	$X_1(0)$	$X_2(0)$	Q_{11}	R
			Correlation matrix			
k_1	1					
k_2	0.41	1				
$X_1(0)$	0.37	0.54	1	Symmetric		
$X_2(0)$	−0.27	−0.10	−0.11	1		
Q_{11}	−0.07	0.54	0.02	0.03	1	
R	0.06	−0.03	−0.01	−0.03	−0.13	1

Serial correlation coefficients of the normalized residuals

Lag	0	1	2	3	4
ρ	1.00	−0.05	0.03	0.00	0.34
r	0.04	−0.54	0.30	0.03	3.72

Sum of squares test

Number of data points......................	117
Number of parameters.....................	6
Expected sum of squares	111
Standard deviation	14.9
Computed sum of squares	116.54
Deviation (units of standard deviation) ...	0.37
Durbin and Watson statistic	1.022

Source: Restrepo-Posada and Bras (1982).

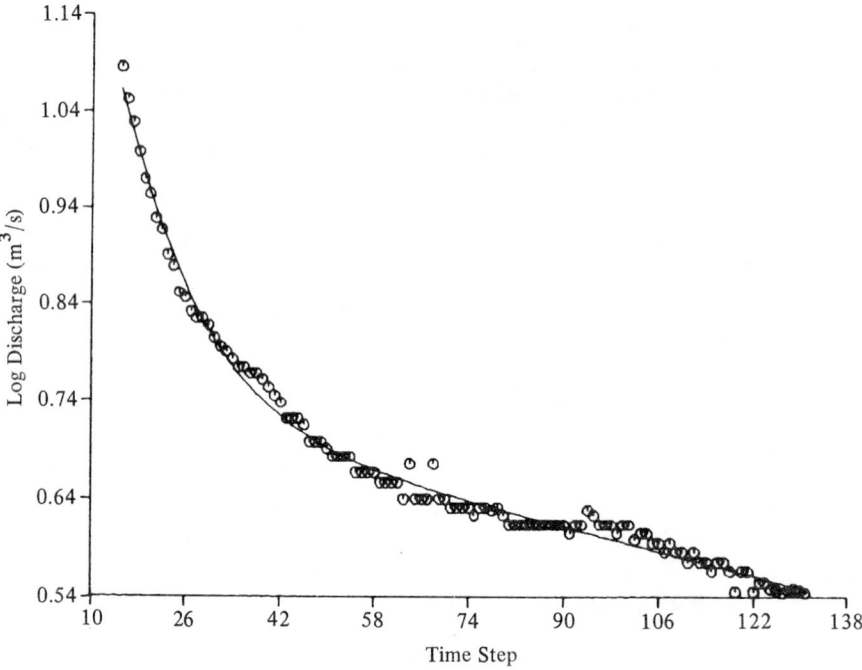

Figure 9.19 Six-hour lead forecast of Bird Creek discharges using a stochastic base-flow model with parameters estimated by maximum likelihood (from Restrepo-Posada and Bras, 1982).

The quality of the convergence point for the maximum-likelihood estimates can be judged by the comparison between the calculated sum of squares of the normalized residuals and its expected value and by the degree of independence of the residuals. The results indicate that the calculated sum of squares of the normalized residual is only 0.37 standard deviation from its expected value. This indicates that the point of convergence is very close to the maximum-likelihood point. The estimates of the serial correlation coefficients of the normalized residuals are excellent, with the exception of $\rho(4)$, which is 3.7 standard deviations off its expected value. The reason for this is that Bird Creek's discharge records are good, except during low flows, (i.e., during the period of base-flow activity) in which the instantaneous discharges are manually estimated from the daily average. This is done by letting the instantaneous discharge be equal to the daily average instantaneous discharge. Since we have four "measurements" per day, the first two measurements, are on the average, underestimated, and the last two measurements are, also on the average, overestimated. This introduces an error with a periodicity of four time steps (24 hours), which is the cause of the very high $\rho(4)$. The Durbin–Watson statistic gives another measure of the independence of the residuals. A value of about 2 for that statistic would indicate a good degree of independence. The value of 1.022 indicates that the residuals are not independent. The problem, again, is blamed on the reconstruction of the instantaneous discharge from the daily average.

The method of maximum likelihood has many other possible uses. Particularly useful is its application within the context of hypothesis testing. Schweppe (1973) covers this topic extensively, particularly its use in discriminating among possible model structures. Kitanidis and Bras (1980a) adapted a hypothesis testing, generalized-likelihood–ratio approach suggested by Willsky and Jones (1976), to identify abrupt transient errors in the input to rainfall–runoff models. This is a common occurrence since the nature of errors during dry flow periods can be very different from those during wet periods when rainfall measurements can be suddenly and catastrophically incorrect.

9.4 SUBOPTIMAL FILTERING: STOCHASTIC APPROXIMATIONS

The optimality of the Kalman filter lies in the choice of the gain matrix, $\mathbf{K}(k)$, given known parameters of the system and perfect correspondence between model dynamics and reality. In hydrology and in most of geophysics, the rule is that if parameters are not known or model structures are not perfect, optimality loses some meaning. At best the discussion should be about reasonable and good results, suboptimal results.

In suboptimal filtering, it is common to exploit the structure of the linear filter,

$$\hat{\mathbf{X}}(k+1|k+1) = \hat{\mathbf{X}}(k+1|k) + \mathbf{P}(k+1)\left[\mathbf{Z}(k+1) - \mathbf{H}(k+1)\hat{\mathbf{X}}(k+1|k)\right],$$

$$(9.106)$$

and concentrate on a reasonable choice of the gain matrix $\mathbf{P}(k+1)$. The level of reasonableness is measured in terms of the stability of the filter and on how well the filter "tracks" the observations.

The use of suboptimal filtering may arise out of convenience. If, for example, the system is stationary, observable, and controllable, it is known that the Kalman gain matrix $\mathbf{K}(k)$ will achieve a steady-state value. This steady-state value may be used even during transients as an approximation in some cases. In other situations a nonstationary system yields very slowly varying gain matrices. Using a representative mean value throughout or several values grouped according to time or latest system state may be a reasonable simplifying assumption. It may be even possible to parameterize $\mathbf{K}(k)$ in terms of simple functions such as power or exponentially decaying forms.

Suboptimal filters may also be used to obtain estimates of states or parameters that otherwise cannot be formulated in a Kalman framework. If the selected gain matrices lead to convergent estimates, within finite time periods, such approaches are called stochastic approximations. One use of such techniques is to estimate model parameters. Section 9.3.1 discussed parameter estimation using state augmentation and the extended Kalman filter or a similar linearization approach to solve the resulting nonlinear system. An alternative is to hypothesize that the parameters can be obtained in a recursive manner, analogous to a Kalman filter:

$$\hat{\alpha}(k+1|k+1) = \hat{\alpha}(k+1|k) + \mathbf{P}(k+1)[\mathbf{Z}(k+1) - \mathbf{H}(k+1)\hat{\mathbf{X}}(k+1|k)]$$
(9.107)

$$\hat{\alpha}(k+1|k) = \hat{\alpha}(k|k).$$
(9.108)

Bras and Restrepo-Posada (1980) illustrated the above approach estimating a limited set of the parameters of the U.S. National Weather Service soil-moisture accounting model. The following is abstracted from that reference.

The core of the NWS model is the soil-moisture accounting routine, which is fully described by Peck (1976). In essence, the soil-moisture accounting model conceptually divides the soil into two zones, upper and lower, interconnected by a percolation function, which is the heart of the model. Both zones contain a tension-water element that is depleted only by evaporation. There is one free-water element in the upper zone that controls surface runoff and interflow. The lower zone contains a tension-water element in addition to two free-water elements, primary and supplementary. These two lower zone free-water elements control the base flow. Each of the two components of the base flow is a linear function of the water content in the respective free-water element. Similarly, the interflow is a linear function of the water content of the upper zone free-water element. Three other terms contribute to the channel input: direct runoff, surface runoff, and additional surface runoff. The first is the part of precipitation that falls on water surfaces or on impervious areas. The second appears when the upper, free-water, zone becomes saturated. The last fulfills the need for a faster model response to the input, under increasing soil-moisture conditions. The example focuses on the on-line estimation of the parameters of three linear terms. These three parameters will be referred to herein as: UZK, upper-zone free-water constant; LZPK, lower-zone free-water (primary) constant; LZSK, lower zone free-water (supplementary) constant.

The efficiency of the algorithm in Eqs. (9.107) and (9.108) depends on how much information about the system's structure is built into the gain term $\mathbf{P}(k)$. As a choice, the following heuristic criterion was adopted. The correction for the ith parameter should be proportional to the output from the ith linear reservoir. Moreover, the correction terms should decrease as the time step increases. The gain terms $P_i(k)$ are thus calculated according to the following expression:

$$P_i(k) = \frac{\beta_i \omega_i(k)}{\sqrt{S_i(t)}},$$
(9.109)

where

$$\omega_i(k) = \frac{0_i(k)}{\sum_{j=1}^{3} 0_j(k)} \qquad (9.110)$$

$$S_i(t) = \sum_{r=1}^{t} \omega_i(r), \qquad (9.111)$$

and $0_i(k)$ is the output from reservoir i at time k and β_i is a scaling factor. When using simulated streamflows, the structure of the model corresponds exactly to the structure of the "real" system and the statistics of the error terms added to the input and output series are perfectly known. In this experiment, the precipitation series corresponding to October, 1958, to September, 1959, at Bird Creek near Sperry, Oklahoma, was used. This series was used as an input to the NWS model as modified by Kitanidis and Bras (1978) and an output series was thus generated. The value of the linear reservoir parameters used corresponded to the parameters fitted for that basin by the NWS. The values were UZK, 0.3 day^{-1}; LZPK, 0.013 day^{-1}; and LZSK, 0.126 day^{-1}. Figures 9.20 and 9.21 show the values for the parameters as estimated by the stochastic-approximations algorithm with the heuristic gain matrix as mentioned above. The final estimate is very close to the real values, although the jagged patterns shown by the estimate of LZPK in Fig. 9.21 indicates that further improvement is still possible.

Estimation for the same three parameters was also attempted using real streamflow data.

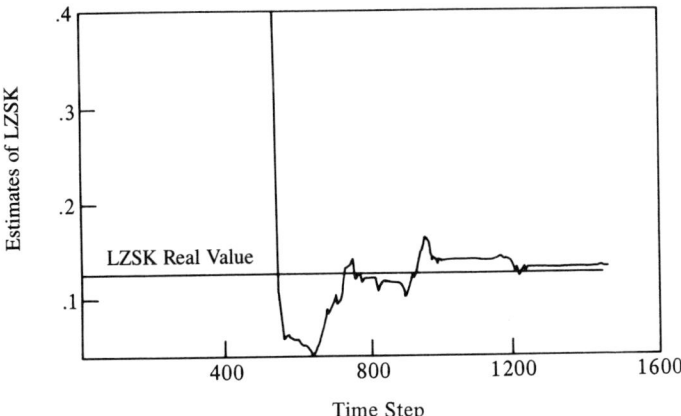

Figure 9.20 Estimates for LZSK at every time step, by the stochastic-approximations algorithm (synthetic data case) (from Bras and Restrepo-Posada, Proceedings of the 3rd International Symposium on Stochastic Hydraulics, 1980, pp. 61–70).

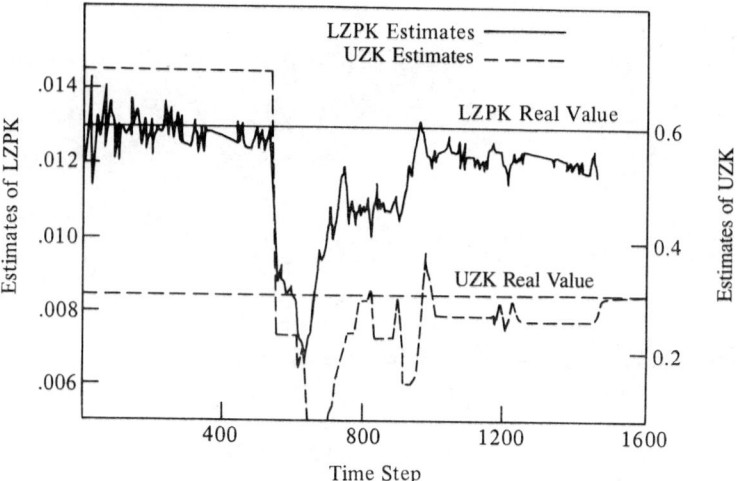

Figure 9.21 Estimates for LZPK and UZK at every time step, by the stochastic-approximations algorithm (synthetic data case) (from Bras and Restrepo-Posada, Proceedings of the 3rd International Symposium on Stochastic Hydraulics, 1980, pp. 61–70).

The estimates for the three parameters at every time step are shown in Figs. 9.22 and 9.23. Also shown are the results of a heuristic calibration to the observations. Although the final value for UZK and LZPK are reasonably close to the "hand-fitted" values, it is not clear whether they would remain for a longer data record. More important, however, is the consistently divergent pattern displayed by the estimate of LZSK. It should be stressed that in a discrete-time formulation of a linear reservoir, a value of the parameter greater than unity is unfeasible. The reason for the problem was traced down to severe structural errors in the model formulation in the routing portion of the model.

There are other more mathematical bases to determine the gain matrices $\mathbf{P}(k)$. An alternative, as presented by Kitanidis and Bras (1979), follows. Assume a general scalar observation model,

$$Z(k) = f(\boldsymbol{\theta}, \mathbf{X}(k)) + \varepsilon(k), \qquad (9.112)$$

where $\boldsymbol{\theta}$ is an unknown vector of parameters, $\mathbf{X}(k)$ is a known or estimated vector, and $\varepsilon(k)$ is a random variable. A two-term Taylor expansion of Eq. (9.112) leads to

$$Z(k) \approx f(\hat{\boldsymbol{\theta}}(k-1), \mathbf{X}(k)) + \mathbf{g}[\boldsymbol{\theta} - \hat{\boldsymbol{\theta}}(k-1)] + \varepsilon(k), \qquad (9.113)$$

where $\mathbf{g}$ is a row vector with elements

$$g_i = \left. \frac{\partial f}{\partial \theta_i} \right|_{\boldsymbol{\theta} = \hat{\boldsymbol{\theta}}(k-1)}.$$

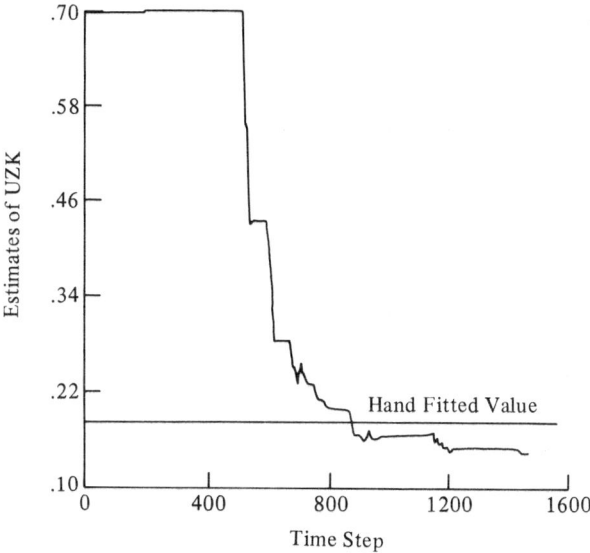

Figure 9.22 Estimates for UZK at every time step, by the stochastic approximation algorithm (real data case) (from Bras and Restrepo-Posada, Proceedings of the 3rd International Symposium on Stochastic Hydraulics, 1980, pp. 61–70).

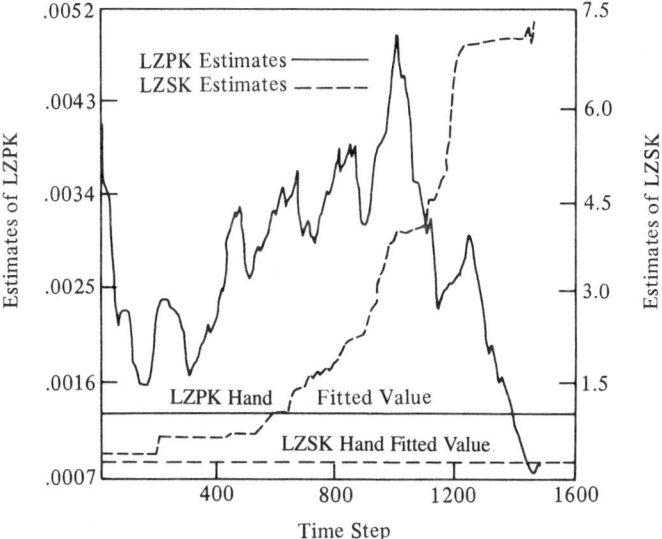

Figure 9.23 Estimates for LZPK and LZSK at every time step, by the stochastic approximations algorithm (real data case) (from Bras and Restrepo-Posada, Proceedings of the 3rd International Symposium on Stochastic Hydraulics, 1980, pp. 61–70).

If the variance of $\varepsilon(k)$ is known to be $\sigma^2(k)$ and $\varepsilon(k)$ is independent of $\hat{\theta}(k-1)$, then a possible recursive estimator of $\theta(k)$ is

$$\hat{\theta}(k) = \hat{\theta}(k-1) + \frac{S(k-1)g^T}{gS(k-1)g^T + \sigma^2(k)} [Z(k) - f(\hat{\theta}(k-1), X(k))],$$

(9.114)

where $S(k)$ is the second moment of estimation and is recursively obtained as

$$S(k) = \left[I - \frac{S(k-1)g^T g}{gS(k-1)g^T + \sigma^2(k)} \right] S(k-1).$$ (9.115)

The above gain is suggested by Schweppe (1973). Other possible choices are

$$P(k) = A/k,$$ (9.116)

where A is a constant matrix (Sakrison, 1966);

$$P(k) = (g(k)g^T(k))^{-1} g^T(k)/k$$ (9.117)

$$P(k) = \frac{g^T(k)}{\sum\limits_{i=1}^{k} \|g(i)\|},$$ (9.118)

where the symbol $\|\cdot\|$ implies the norm of a vector (Albert and Gardner, 1967).

The reader interested in further study on stochastic approximations is referred to Robins and Monro (1951), Albert and Gardner (1967), Sakrison (1966), Ho and Lee (1965), Saridis and Stein (1968), and Nevelšon and Hasmïnskii (1972).

Kitanidis and Bras (1979) used the algorithm of Eq. (9.115) to estimate the error structure $Q = E[\Gamma W(k)W^T(k)\Gamma^T] = \Gamma C \Gamma^T$ of a canonical state–space dynamic system,

$$X(k+1) = \Phi(k)X(k) + \Gamma W(k)$$
$$Z(k) = H(k)X(k) + V(k),$$

with all the necessary properties for use of the Kalman filter. The only unknowns in these examples were elements of matrix Γ. The goal was to estimate Γ with a suboptimal filter that would develop in parallel to a Kalman filter of the original system. The parameter filter would feed estimates of Γ at every time step to the Kalman filter. On the other hand, the Kalman filter would provide the parameter filter with the innovation sequence $v(k)$ and its covariance structure. The method is based on the fact that the innovation sequence should be an uncorrelated random process. Kitanidis and Bras (1979) then hypothesized that they had "observations" of the correlation structure of the innovations. The observations are (for a scalar observation)

$$r_i(k) = v(k)v(k-i) = C_i(k) + \varepsilon_i(k),$$ (9.119)

where $C_i(k) = E[v(k)v(k-i)]$ or the expected lag-i correlation of the residuals, $v(k) = Z(k) - \mathbf{H}\hat{\mathbf{X}}(k|k-1)$. For arbitrary-gain matrices and parameters the correlation function is given by

$$C_0(k) = \mathbf{H}(k)\Sigma(k|k-1)\mathbf{H}^T(k) + R(k)$$
$$C_i(k) = \mathbf{H}(k)\Phi(k-1)[\mathbf{I} - \mathbf{K}(k-1)\mathbf{H}(k-1)]\Phi(k-2)$$
$$[\mathbf{I} - \mathbf{K}(k-2)\mathbf{H}(k-2)] \cdots \Phi(k-i)[\Sigma(k-i|k-i-1)\mathbf{H}^T(k-i)$$
$$- \mathbf{K}(k-i)C_0(k-i)], \tag{9.120}$$

under the assumption of optimal Kalman-filter behavior, i.e., known parameters, $C_i(k) = 0$ for all $i \neq 0$. The statistics of $\varepsilon_i(k)$ can be obtained if the optimality of the Kalman filter, leading to uncorrelated innovations, is assumed. Then,

$$E\big[\varepsilon_i(k)\varepsilon_j(\ell)\big] = \begin{cases} 2C_0^2(k) & k = \ell,\, i = j = 0 \\ C_0(k) \cdot C_0(k-i) & k = \ell,\, i = j \neq 0 \\ 0 & \text{otherwise.} \end{cases} \tag{9.121}$$

Equation (9.119) is then (with supporting Eqs. 9.120 and 9.121) in the form of the general problem represented in Eq. (9.112). A suboptimal filter can then be formulated to obtain estimates of the unknown elements of matrix Γ. Calling these parameters θ, the sequential estimator becomes

$$\hat{\theta}(k,0) = \hat{\theta}(k-1, j) + \mathbf{P}_0(k)[v^2(k) - \mathbf{H}(k)\Sigma(k|k-1)\mathbf{H}^T(k) - R(k)]$$
$$\hat{\theta}(k,1) = \hat{\theta}(k,0) + \mathbf{P}_1(k)[v(k) \cdot v(k-1)]$$
$$\vdots \qquad \vdots \tag{9.122}$$
$$\hat{\theta}(k, j) = \hat{\theta}(k, j-1) + \mathbf{P}_j(k)[v(k) \cdot v(k-j)],$$

where $\hat{\theta}(k, j)$ is the parameter estimate at time k using "observations" of the lag-j correlation of the innovation sequence. To obtain the system of Eqs. (9.122), the linearization discussed in Eq. (9.113) was made around the expected value of the observations given if the optimal parameters were on hand. In such a case

$$E\big[v^2(k)\big] = \mathbf{H}(k)\Sigma(k|k-1)\mathbf{H}^T(k) + R(k)$$

and

$$E\big[v(k)v(k-j)\big] = 0.$$

The gain matrices $\mathbf{P}_j(k)$ are formed as in Eq. (9.114). To obtain the required gradients, a numerical procedure was adopted; it is discussed in Kitanidis and Bras (1979).

The use of the above method is illustrated with the following example from Kitanidis and Bras (1979). Define a time-varying linear system with the following parameters:

$$\Phi(k) = \begin{bmatrix} \Phi_{11}(k) & 0.816 \\ -0.6 & 0.4 \end{bmatrix}.$$

The model-error covariance matrix was

$$Q = \begin{bmatrix} 1 & 0 \\ 0 & 1 \end{bmatrix}.$$

The measurement error variance was $R = 0.4$, and the measurement matrix $H = [1 \quad 0.1]$.

In order to make the transition matrix time-varying but known at each time step,

$$\Phi_{11}(k) = 0.5 + 0.5r,$$

where $r \sim N(0, 1)$.

Note that this model is not stable a given percentage of times.

Assume a model-error covariance,

$$Q = \Gamma C \Gamma^T,$$

where $C = I$.

$\Gamma = \text{diag}[\theta_1, \theta_2]$, θ_1 and θ_2 are to be estimated. True parameters are

$$\theta_1 = 1, \qquad \theta_2 = 1.$$

Initial estimates of θ_1 and θ_2 are taken as

$$\theta(0) = \begin{bmatrix} 8 \\ 8 \end{bmatrix}$$

with a second moment of estimation error

$$S(0) = \begin{bmatrix} 64 & 0 \\ 0 & 64 \end{bmatrix}.$$

The trace of θ_i versus number of time steps is shown in Fig. 9.24, when the estimation is made based only on the lag-zero, C_0, and lag-one, C_1, correlations of residuals. Also shown is the line $\theta_i + 2\sqrt{S_{ii}}$ as a function of iterations. The standard error $\sqrt{S_{ii}}$ reduces exponentially with time. The actual estimate remains within the 2 standard error range most of the time. The use of additional lags in the correlation structure was found to be of marginal benefit in this example.

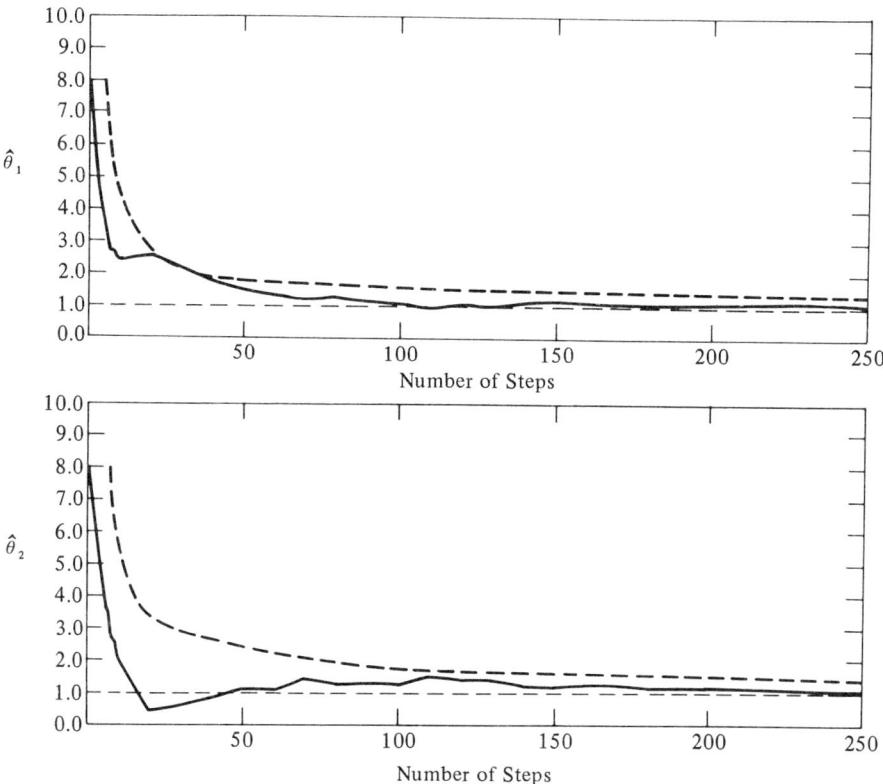

Figure 9.24 Recursive estimation of θ_1 and θ_2 when the variance C_0 and the covariance C_1 of the residuals are fitted. Dashed lines represent the error range (from Kitanidis and Bras, 1979).

In using the above and similar algorithms, it must be realized that it is not possible to estimate all elements of $\mathbf{Q}$ if the dimensions of the observations $\mathbf{Z}(k)$ are less than the dimensions of the state $\mathbf{X}(k)$.

The method of estimation with parallel interactive filters of state and parameters is appealing because of computational economy, linearity of formulation, and conceptual attractiveness. The above stochastic-approximation algorithm is not the only available scheme. In reality infinite alternatives to weights can be suggested. Todini (1979) suggested a similar approach which he called mutually interactive state-parameter (MISP) estimation. Essentially, he interpreted the product $\mathbf{H}(k)\hat{\mathbf{X}}(k|k)$ as an observation on unknown parameters in the system dynamics. For cases where all states are discrete and measurable and for some small limitations on the form of $\mathbf{H}$, he then derived parallel filters for state and dynamics parameters. He also suggested compatible estimates for model and measurement-error statistics. An important contribution of his approach is that the state estimator explicitly acknowledges that model error increases because of uncertainty in the latest parameter estimate.

The reader is referred to Todini (1979) and O'Connell (1980) for details. As a summary, though, define a discrete observable system of the type,

$$\mathbf{X}(k) = \mathbf{\Phi}(k-1)\mathbf{X}(k-1) + \mathbf{\Gamma}(k-1)\mathbf{W}(k-1)$$

$$\mathbf{Z}(k) = \mathbf{H}(k)\mathbf{X}(k) + \mathbf{V}(k),$$

where

$$E[\mathbf{W}(k)] = \overline{\mathbf{W}}$$
$$E[\mathbf{V}(k)] = \overline{\mathbf{V}}$$
$$E\left[(\mathbf{W}(k)-\overline{\mathbf{W}})(\mathbf{W}(j)-\overline{\mathbf{W}})^T\right] = \mathbf{Q}\delta_{kj}$$
$$E\left[(\mathbf{V}(k)-\overline{\mathbf{V}})(\mathbf{V}(j)-\overline{\mathbf{V}})^T\right] = \mathbf{R}\delta_{kj}$$
$$E\left[(\mathbf{W}(k)-\overline{\mathbf{W}})(\mathbf{V}(k)-\overline{\mathbf{V}})^T\right] = \mathbf{0} \qquad \forall t, k$$

and $\mathbf{\Gamma}(k-1) = \mathbf{H}^T(k)$ are $n \times m$ matrices, where m denotes the dimensions of the observation vector. In the above, elements of $\mathbf{\Phi}(k)$, $\mathbf{R}$, $\overline{\mathbf{W}}$, $\overline{\mathbf{V}}$, and $\mathbf{Q}$ can be assumed unknown. Furthermore, the matrix $\mathbf{\Phi}(k)$ can be parameterized, through row decomposition, as

$$\mathbf{\Phi}(k) = \begin{bmatrix} \mathbf{a}_1^T(t) + \mathbf{\theta}^T(k+1)\mathbf{B}_1^T(k) \\ \vdots \\ \mathbf{a}_n^T(k) + \mathbf{\theta}^T(k+1)\mathbf{B}_n^T(k) \end{bmatrix},$$

where $\mathbf{a}_j(k)$ is an $n \times 1$ vector, $\mathbf{B}_j(k)$ is an $n \times p$ matrix, and $\mathbf{\theta}(k)$ is a constant, unknown ($p \times 1$) vector of parameters. Also define the $n \times n$ matrix

$$\mathbf{A}(k) = \begin{bmatrix} \mathbf{a}_1^T(k) \\ \vdots \\ \mathbf{a}_n^T(k) \end{bmatrix}$$

and the $n \times p$ matrix

$$\mathbf{F}(k+1) = \begin{bmatrix} \hat{\mathbf{X}}^T(k|k)\mathbf{B}_1(k) \\ \vdots \\ \hat{\mathbf{X}}^T(k|k)\mathbf{B}_n(k) \end{bmatrix}.$$

The complete MISP algorithm for estimating vector $\mathbf{\theta}(k)$ and the state $\mathbf{X}(k)$

with parallel filters is then

$$\bar{\mathbf{V}}(k) = \frac{k-1}{k}\bar{\mathbf{V}}(k-1) + \frac{1}{k}[\mathbf{Z}(k) - \mathbf{H}(k)\hat{\mathbf{X}}(k|k-1)]$$

$$\boldsymbol{v}(k) = \mathbf{Z}(k) - \mathbf{H}(k)\hat{\mathbf{X}}(k|k-1) - \bar{\mathbf{V}}(k)$$

$$\hat{\mathbf{R}}(k) = \frac{k-1}{k}\hat{\mathbf{R}}(k-1) + \frac{1}{k}\left[\boldsymbol{v}(k)\boldsymbol{v}(k)^T - \mathbf{H}(k)\boldsymbol{\Sigma}(k|k-1)\mathbf{H}(k)^T\right]$$

$$\mathbf{K}(k) = \boldsymbol{\Sigma}(k|k-1)\mathbf{H}(k)^T\left[\mathbf{H}(k)\boldsymbol{\Sigma}(k|k-1)\mathbf{H}(k)^T + \hat{\mathbf{R}}(k)\right]^{-1}$$

$$\hat{\mathbf{X}}(k|k) = \hat{\mathbf{X}}(k|k-1) + \mathbf{K}(k)\boldsymbol{v}(k)$$

$$\boldsymbol{\Sigma}(k|k) = [\mathbf{I} - \mathbf{K}(k)\mathbf{H}(k)]\boldsymbol{\Sigma}(k|k-1)$$

$$\boldsymbol{v}^*(k) = \mathbf{H}(k)\mathbf{K}(k)\boldsymbol{v}(k)$$

$$\mathbf{R}^* = \mathbf{H}(k)\mathbf{K}(k)\mathbf{C}_0(k)\mathbf{K}(k)^T\mathbf{H}(k)^T$$

$$\mathbf{C}_0(k) = \mathbf{H}(k)\boldsymbol{\Sigma}(k|k-1)\mathbf{H}^T(k) + \hat{\mathbf{R}}(k)$$

$$\mathbf{K}^*(k) = \boldsymbol{\Sigma}^*(k|k-1)\mathbf{H}^*(k)^T\left[\mathbf{H}^*(k)\boldsymbol{\Sigma}^*(k|k-1)\mathbf{H}^*(k)^T + \mathbf{R}^*\right]^{-1}$$

$$\hat{\boldsymbol{\theta}}(k|k) = \hat{\boldsymbol{\theta}}(k|k-1) + \mathbf{K}^*(k)\boldsymbol{v}^*(k)$$

$$\boldsymbol{\Sigma}^*(k|k) = [\mathbf{I} - \mathbf{K}^*(k)\mathbf{H}^*(k)]\boldsymbol{\Sigma}^*(k|k-1)$$

$$\hat{\boldsymbol{\theta}}(k+1|k) = \hat{\boldsymbol{\theta}}(k|k)$$

$$\boldsymbol{\Sigma}^*(k+1|k) = \boldsymbol{\Sigma}^*(k|k)$$

$$\hat{\bar{\mathbf{w}}}(k+1|k) = \hat{\bar{\mathbf{w}}}(k|k-1) + \frac{1}{k}\left[\boldsymbol{\Gamma}(k)^T\boldsymbol{\Gamma}(k)\right]^{-1}\boldsymbol{\Gamma}(k)^T\mathbf{K}(k)\boldsymbol{v}(k)$$

$$\hat{\mathbf{Q}}'(k+1|k) = \frac{k-1}{k}\hat{\mathbf{Q}}'(k|k-1) + \frac{1}{k}\left[\boldsymbol{\Gamma}(k)^T\boldsymbol{\Gamma}(k)\right]^{-1}\boldsymbol{\Gamma}(k)^T$$

$$\cdot\left[\mathbf{K}(k)\boldsymbol{v}(k)\boldsymbol{v}(k)^T\mathbf{K}(k)^T + \boldsymbol{\Sigma}(k|k)\right.$$

$$\left. - \hat{\boldsymbol{\Phi}}(k-1)\boldsymbol{\Sigma}(k-1|k-1)\hat{\boldsymbol{\Phi}}(k-1)^T\right]\boldsymbol{\Gamma}(k)\left[\boldsymbol{\Gamma}(k)^T\boldsymbol{\Gamma}(k)\right]^{-1}$$

$$\hat{\mathbf{Q}}(k+1|k) = \hat{\mathbf{Q}}'(k+1|k)$$

$$- \left(\left[\boldsymbol{\Gamma}(k)^T\boldsymbol{\Gamma}(k)\right]^{-1}\boldsymbol{\Gamma}(k)^T\mathbf{E}(k)\boldsymbol{\Gamma}(k)\left[\boldsymbol{\Gamma}(k)^T\boldsymbol{\Gamma}(k)\right]^{-1}\right)$$

$$\hat{\mathbf{X}}(k+1|k) = \boldsymbol{\Phi}(k)\hat{\mathbf{X}}(k|k) + \boldsymbol{\Gamma}(k+1)\hat{\bar{\mathbf{w}}}(k+1|k)$$

$$\boldsymbol{\Sigma}(k+1|k) = \hat{\boldsymbol{\Phi}}(k)\boldsymbol{\Sigma}(k|k)\hat{\boldsymbol{\Phi}}(k)^T$$

$$+ \left\{\boldsymbol{\Gamma}(k+1)\hat{\mathbf{Q}}(k+1|k)\boldsymbol{\Gamma}(k+1)^T + \mathbf{E}(k+1)\right\}.$$

In the above set of equations, terms superscripted with asterisks refer to the model-parameters filter. Other terms correspond to the parallel-state filter. The transfer of information between the two filters is shown in Fig. 9.25, where

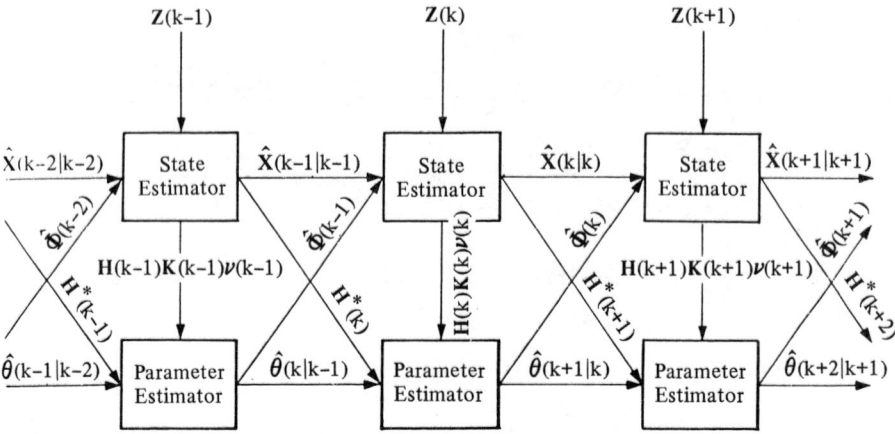

Figure 9.25 Timing diagram for MISP algorithm illustrating alternative use of two Kalman filters for state and parameter estimation (from Todini, 1979).

$\mathbf{H}^*(k) = \mathbf{H}(k)\mathbf{F}(k)$. The term $\mathbf{E}(k+1)$ in the error-covariance estimates $\hat{\mathbf{Q}}(k+1|k)$ and $\boldsymbol{\Sigma}(k+1|k)$ accounts for added uncertainty due to errors in the estimation of $\boldsymbol{\theta}(k)$. The $(n \times n)$ matrix $\mathbf{E}(k+1)$ has elements i, j of the form

$$\{\mathbf{E}(k+1)\}_{ij} = \mathrm{tr}\left\{\mathbf{B}_i^T(k)\left(\boldsymbol{\Sigma}(k|k) + \hat{\mathbf{X}}(k|k)\hat{\mathbf{X}}^T(k|k)\right)\mathbf{B}_j(k)\boldsymbol{\Sigma}^*(k+1|k)\right\},$$

where $\mathrm{tr}\{\cdot\}$ stands for the trace of the matrix in parentheses. O'Connell (1980), in giving the above expression, discusses the practical and simplifying aspects of its numerical evaluation.

References

CHAPTER 1

Benjamin, J. R., and C. A. Cornell [1970]. *Probability, Statistics, and Decision for Civil Engineers*. New York: McGraw-Hill.

Bras, R. L. [1978]. "Sampling Network Design in Hydrology and Water Quality Sampling: A Review of Linear Estimation Theory Applications." In: *Applications of Kalman Filter to Hydrology, Hydraulics and Water Resources*. C. L. Chiu, ed. Pittsburgh: University of Pittsburgh, Department of Civil Engineering, p. 155–200.

Idem [1979]. "Sampling of Interrelated Random Fields: The Rainfall-Runoff Case." *Water Resources Res*. 15(6):1767–80.

Hufschmidt, M. M., and M. B. Fiering [1966]. *Simulation Techniques for Design of Water Resource Systems*. Cambridge, Mass.: Harvard University Press.

Karlin, S., and H. Taylor [1975]. *A First Course in Stochastic Processes*, 2nd ed. New York: Academic Press.

Idem [1981]. *A Second Course in Stochastic Processes*. New York: Academic Press.

Kitanidis, P. K., and R. L. Bras [1980]. "Real-Time Forecasting with a Conceptual Hydrologic Model, 1: Analysis of Uncertainty." *Water Resources Res.* 16(6):1025–34.

Idem [1980]. "Real-Time Forecasting with a Conceptual Hydrologic Model, 2: Applications and Results." *Water Resources Res*, 16(6):1034–45.

Lee, Y. W. [1960]. *Statistical Theory of Communication.* New York: J. Wiley.

Melsa, J. L., and A. P. Sage [1973]. *An Introduction to Probability and Stochastic Processes.* Englewood Cliffs, N. J.: Prentice-Hall.

Moss, M. E., and G. D. Tasker [1979]. "Progress in the Design of Hydrologic Data Networks." *Rev. Geophys. Space Phys.* 17(6):1298–306.

Rodríguez-Iturbe, I., and J. M. Mejía [1974]. "The Design of Rainfall Networks in Time and Space." *Water Resources Res.* 10(4):713–29.

Vicens, G., and J. C. Schaake, Jr. [1972]. "Simulation Criteria for Selecting Water Resource System Alternatives." Cambridge, Mass.: Ralph M. Parsons Laboratory for Water Resources and Hydrodynamics, M. I. T. Technical Report 154.

Yevjevich, V. [1972]. *Stochastic Processes in Hydrology.* Fort Collins, Colo.: Water Resources Publications.

CHAPTER 2

Bartlett, M. S. [1946]. "On the Theoretical Specification of Sampling Properties of Autocorrelated Time Series." *J. Royal Stat. Soc.* B8:27.

Idem [1966]. *Stochastic Processes.* New York: Cambridge University Press.

Bobée, B., and R. Robitaille [1975]. "Correction of Bias in the Estimation of the Coefficient of Skewness." *Water Resources Res.* 11(6):851–4.

Box, G. E., and G. M. Jenkins [1976]. *Time Series Analysis: Forecasting and Control.* San Francisco: Holden Day.

Box, G. E., G. M. Jenkins, and D. W. Bacon [1968]. "Models for Forecasting Seasonal and Non-Seasonal Time Series." In: *Proceedings of the Advanced Seminar on Spectral Analysis of Time Series.* Bernard Harris, ed. New York: Wiley, p. 271–311.

Burges, S. J., D. P. Lettenmaier, and C. L. Bates [1975]. "Properties of the Three-Parameter Log Normal Probability Distribution." *Water Resources Res.* 11(2):229–35.

Carlson, R. E., A. J. A. MacCormick, and D. G. Watts [1970]. "Application of Linear Random Models to Four Annual Streamflow Series." *Water Resources Res.* 6(4):1070–8.

Curry, K., and R. L. Bras [1980]. "Multivariate Seasonal Time Series Forecast with Application to Adaptive Control." Cambridge, Mass.: Ralph M. Parsons Laboratory for Water Resources and Hydrodynamics, M.I.T. Technical Report 253.

Delleur, J. W. [1978]. "Applied Stochastic Modeling of Hydrologic Time Series, Chapter III, Autoregressive-Moving Average Models—ARMA (p, q)." Lecture Notes for the Computer Workshop in Statistical Hydrology. July 17–21, 1978. Fort Collins, Colo.: Colorado State University.

Delleur, J. W., and M. L. Kavvas [1978]. "Stochastic Models for Monthly Rainfall Forecasting and Synthetic Generation." *J. Appl. Meteorology.* 17(10):1528–36.

Draper, N., and H. Smith [1966]. *Applied Regression Analysis.* New York: Wiley.

Durbin, J. [1960]. The Fitting of Time Series Models. *Rev. Inst. Stat.* 28:233.

Fiering, M., and B. Jackson [1971]. "Synthetic Streamflows." *Water Resources Monograph* 1. Washington, D.C.: American Geophysical Union.

Hazen, A. [1914]. "Storage to be Provided in Impounding Reservoirs for Municipal Water Supply." *Trans. A.S.C.E.* 77(1308):1539–640.

Hipel, K. W. [1975]. "Contemporary Box–Jenkins Modelling in Hydrology." Waterloo, Canada: Ph.D. thesis presented to the University of Waterloo.

Hipel, K. W., A. I. McLeod, W. C. Lennox [1977]. "Advances in Box–Jenkins Modelling, 1: Model Construction." *Water Resources Res.* 13(3):567–77.

Hufschmidt, M. M., and M. B. Fiering [1966]. *Simulation Techniques for Design of Water Resource Systems.* Cambridge, Mass.: Harvard University Press.

Hurst, H. E. [1951]. "Long-Term Storage Capacities of Reservoirs." *Trans. A.S.C.E.* 116:776–808.

Jamieson, D. G. [1972]. "River Dee Research Program, I: Operating Multipurpose Reservoir Systems for Water Supply and Flood Alleviation." *Water Resources Res.* 8(4):899–920.

Jamieson, D. G., and J. C. Wilkinson [1972]. "River Dee Research Program, 3: A Short-Term Control Strategy for Multipurpose Reservoir Systems." *Water Resources Res.* 8(4):911–20.

Jamieson, D. G., D. K. Smith, and J. C. Wilkinson [1976]. "Evaluation of Short Term Operational Policies for a Multipurpose Reservoir System." *J. Hydrology.* 28(2–4):191–213.

Kashyap, R. L. and A. R. Rao [1976]. *Dynamic Stochastic Models from Empirical Data.* New York: Academic Press.

Kirby, W. [1974]. "Algebraic Boundedness of Sample Statistics." *Water Resources Res.* 10(2):220–2.

Maass, A., M. M. Hufschmidt, R. Dorfman, H. A. Thomas, Jr., S. A. Marglin, and G. M. Fair [1962]. *Design of Water-Resource Systems.* Cambridge, Mass.: Harvard University Press.

Matalas, N. C. [1967]. "Mathematical Assessment of Synthetic Hydrology." *Water Resources Res.* 3(4):937–45.

McLeod, A. I., and K. W. Hipel [1978]. "Simulation Procedures for Box–Jenkins Models." *Water Resources Res.* 14(5):969–80.

McLeod, A. I., K. W. Hipel, and W. C. Lenox [1977]. "Advances in Box–Jenkins Modeling, 2: Applications." *Water Resources Res.* 13(3):577–86.

Mejía, J. M. [1971]. "On the Generation of Multivariate Sequences Exhibiting the Hurst Phenomenon and Some Flood Frequency Analysis." Fort Collins, Colo.: Ph.D. dissertation presented to the Civil Engineering Department, Colorado State University.

Quimpo, R. G. [1968]. "Stochastic Analysis of Daily Flows." *A.S.C.E. J. Hydraulics.* 94HY(1):43–57.

Rao, R. A., and R. L. Kashyap [1973]. "Analysis, Construction, and Validation of Stochastic Models for Monthly River Flows." West Lafayette, Ind.: School of Civil Engineering, Purdue University.

Idem [1974]. "Stochastic Modelling of River Flows." *I.E.E.E. Trans. Automatic Control.* AC-19(6):874–81.

Rodríguez-Iturbe, I. [1968]. "A Modern Statistical Study of Monthly Levels in the Orinoco River." *Bull. I.A.S.H.* 4:25–41.

Rodríguez-Iturbe, I., and C. F. Nordin [1968]. "Time Series Analysis of Water and Sediment Discharges." *Bull. I.A.S.H.* 2:69–84.

Roesner, L. A. and V. Yevjevich [1966]. "Mathematical Models for Time Series of Monthly Precipitation and Monthly Runoff." Hydrology Papers No. 15. Fort Collins, Colo.: Colorado State University.

Salas, J. D., D. C. Boes, and R. A. Smith [1982]. "Estimation of ARMA Models with Seasonal Parameters." *Water Resources Res.* 18(4):1006–10.

Salas, J. D., D. C. Boes, V. Yevjevich, and G. G. S. Pegram [1979]. "Hurst Phenomenon as a Pre-asymptotic Behavior." *J. Hydrology.* 44:1–15.

Salas, J. D., J. W. Delleur, V. Yevjevich, and W. L. Lane [1980]. *Applied Modeling of Hydrologic Time Series.* Littleton, Colo.: Water Resources Publications.

Salas, J. D., G. G. S. Pegram [1977]. "A Seasonal Multivariate Multilag Autoregressive Model in Hydrology." In: *Modeling Hydrologic Processes.* H. J. Morel-Seytoux, J. D. Salas, T. G. Sanders, and R. E. Smith, eds., Littleton, Colo.: Water Resources Publications, pp. 125–45.

Salas, J. D. and V. Yevjevich [1972]. "Stochastic Structure of Water Use Time Series." Hydrology Papers No. 52. Fort Collins, Colo.: Colorado State University.

Sancholuz, A., S. Carrasquel, and D. Vargas [1981]. "Modelos Matemáticos para la Predicción de Niveles del Río Orinoco." Caracas, Venezuela: Laboratorio Nacional de Hidráulica Ernesto Leon D.

Stedinger, J. R. [1980]. "Fitting Log Normal Distributions to Hydrologic Data." *Water Resources Res.* 16(3):481–90.

Sudler, C. E. [1927]. "Storage Required for the Regulation of Streamflow." *Trans. A.S.C.E.* 91(1651):622–60.

Tao, P. C., and J. W. Delleur [1976]. "Seasonal and Non-Seasonal ARMA Models in Hydrology." *A.S.C.E. J. Hydraulics.* 102(HY10):1541–60.

Valdes, J. B., I. Rodríguez-Iturbe, and G. J. Vicens [1977]. "Bayesian Generation of Synthetic Streamflows, 2: The Multivariate Case." *Water Resources Res.* 13(2):291–5.

Vicens, G. J. , I. Rodríguez-Iturbe, and J. C. Schaake, Jr. [1975]. "Bayesian Generation of Synthetic Streamflows." *Water Resources Res.* 11(6):827–38.

Wilson, E. B., and M. M. Hillerty, [1931]. "Distribution of Chi Square." *Proc. Natl. Acad. Sci.* 17:684–8.

Wilson, G. T. [1969]. "Factorization of the Generating Function of a Pure Moving Average Process." *J. Numerical Analysis.* 6(1):1–7.

Wood, E. F., ed. [1980]. Workshop on *Real Time Forecasting/Control of Water Resource Systems.* New York: Pergamon Press.

Yevjevich, V. M. [1963]. "Fluctuations of Wet and Dry Years, Part I: Research Data

Assembly and Mathematical Models." In: Hydrology Papers No. 1. Fort Collins, Colo.: Colorado State University.

Yevjevich, V. M. [1966]. "Stochastic Problems in the Design of Reservoirs," In: *Water Research*. A. V. Kneese and S. C. Smith, eds. Baltimore: Johns Hopkins Press for Resources for the Future, p. 375–411.

Yevjevich, V. M. [1972]. "Structural Analysis of Hydrologic Time Series." Hydrology Paper No. 46. Fort Collins, Colo.: Colorado State University.

CHAPTER 3

Anderson, R. L. [1942]. "Distribution of the Serial Correlation Coefficient." *Ann. Math. Stat.* 13(1):1–13.

Bras, R. L., R. Buchanan, and K. C. Curry [1983]. "Real Time Adaptive Closed-Loop Control of Reservoirs with the High Aswan Dam as a Case Study." *Water Resources Res.* 19(1):33–52.

Clarke, R. T. [1973]. "Mathematical Models in Hydrology." Rome, Italy: Food and Agriculture Organization Irrigation and Drainage Paper 19.

Crosby, O. S. and T. Maddock [1970]. "Estimating Coefficients of a Flow Generator for Monotone Samples of Data." *Water Resources Res.* 6(4):1079–86.

Curry, K. D., and R. L. Bras [1978]. "Theory and Applications of the Multivariate Broken Line, Disaggregation and Monthly Autoregressive Generators to the Nile River." Cambridge, Mass.: Technology Adaptation Program, M.I.T. Report 78–5.

Idem [1980]. "Multivariate Seasonal Time Series Forecast with Application to Adaptive Control." Cambridge, Mass.: Ralph M. Parsons Laboratory for Water Resources and Hydrodynamics, M.I.T. Technical Report 253.

Deeb, A., and C. Puente. [1979]. "Experimentos con Modelos Multivariados para la Generación Sintética de Datos Hidrológicos." Bogotá, Colombia: Universidad de los Andes.

Draper, N., and H. Smith [1966]. *Applied Regression Analysis*. New York: Wiley.

Durbin, J. [1970]. "Testing for Serial Correlation in Least-Squares Regression When Some of the Regressors are Lagged Dependent Variables." *Econometrica*. 38(3):410–21.

Durbin, J., and G. S. Watson [1950]. "Testing for Serial Correlation in Least Squares Regression. I." *Biometrika*. 37:409–28.

Idem [1951]. "Testing for Serial Correlation in Least Squares Regression. II." *Biometrika*. 38:159–78.

Goldberger, A. S. [1964]. *Econometric Theory*. New York: Wiley.

Granger, C. W. J., and P. Newbold [1977]. *Forecasting Economic Time Series*. New York: Academic Press.

Hoshi, K. S., J. Burges, and I. Yamaoka [1978]. "Reservoir Design Capacities for Various Seasonal Operational Hydrology Models." *Proc. Japanese Soc. Civil Eng.* 273:121–34.

Johnston, J. [1972]. *Econometric Methods*. New York: McGraw-Hill.

Kashyap, R. L. [1971]. "Parameter Estimation in Partially Specified Stochastic Systems." *Proceedings, Ninth Annual Allerton Conference on Circuit and System Theory.* Urbana, Ill.: University of Illinois.

Kashyap, R. L., and A. R. Rao [1976]. *Dynamic Stochastic Models from Empirical Data.* New York: Academic Press.

Lane, W. L. [1983]. "Multivariate Modeling of the Upper Colorado River Basin." Presented at the fall meeting of the American Geophysical Union, December 5–9, 1983, San Francisco, Calif.

Loucks, D. P., J. R. Stedinger, and D. A. Haith [1981]. *Water Resource Systems Planning and Analysis.* Englewood Cliffs, N.J.: Prentice-Hall.

Matalas, N. [1967]. "Mathematical Assessment of Synthetic Hydrology." *Water Resources Res.* 3(4):937–46.

Mejía, J. M., and J. Millán [1974]. "Una Metodología para Tratar el Problema de Matrices Inconsistentes en la Generación Multivariada de Series Hidrológicas." VI Congreso Latino Americano de Hidraúlica. Bogotá, Colombia.

Mejía, J. M., I. Rodríguez-Iturbe, and J. R. Cordova [1974]. "Multivariate Generation of Mixtures of Normal and Log-normal Variables." *Water Resources Res.* 10(4):691–4.

Mejía, J. M., J. Rouselle [1976]. "Disaggregation Models in Hydrology Revisited." *Water Resoures Res.* 12(2):185–6.

O'Connell, P. E. [1974]. "Stochastic Modelling of Long-Term Persistence in Streamflow Sequences." London: Ph.D. thesis presented to the Civil Engineering Department, Imperial College.

Panuska, V. [1969]. "An Adaptive Recursive Least Squares Identification Algorithm." *Proceedings 1969 I.E.E.E. Symposium on Adaptive Processes (8th).* University Park, Penn.: Pennsylvania State University, Paper 6-e.

Puente, C. [1978]. "Algunos Modelos Multivariados de Generación Sintética de Datos." Bogotá, Colombia: Proyecto de Grado, Facultad de Ingeniería, Universidad de los Andes.

Rao, R. A., and R. L. Kashyap [1973]. "Analysis Construction and Validation of Stochastic Models for Monthly River Flows." West Lafayette, Ind.: School of Civil Engineering, Purdue University.

Idem [1974]. "Stochastic Modeling of River Flows." *I.E.E.E. Trans. Automatic Control.* AC-19(6):874–81.

Salas, J. D., and G. G. S. Pegram [1977]. "A Seasonal Multivariate Multilag Autoregressive Model in Hydrology." In: *Modeling Hydrologic Processes.* H. J. Morel-Seytoux, J. D. Salas, T. G. Sanders, and R. E. Smith, eds. Littleton, Colo.: Water Resources Publications, pp. 125–45.

Salas, J. D., J. W. Delleur, V. Yevjevich, and W. L. Lane [1980]. *Applied Modeling of Hydrologic Time Series.* Littleton, Colo.: Water Resources Publications.

Santos, E. G., and J. D. Salas [1983]. "A Parsimonious Step Disaggregation Model for Operational Hydrology." Abstract in *EOS Trans. A.G.U.,* 64(45):706 Presented at the fall meeting of the American Geophysical Union, December 5–9, 1983, San Francisco, Calif.

Stedinger, J. R. [1983]. "Advances in Disaggregation Modeling." Presented at the fall meeting of the American Geophysical Union, December 5–9, 1983, San Francisco, Calif.

Valencia, D., and J. C. Schaake, Jr. [1972]. "A Disaggregation Model for Time Series Analysis and Synthesis." Cambridge, Mass.: Ralph M. Parsons Laboratory for Water Resources and Hydrodynamics, M.I.T. Technical Report 149.

Idem [1973]. "Disaggregation Processes in Stochastic Hydrology." *Water Resources Res.* 9(3): 580–5.

Valencia, D., J. García, and M. T. Berdugo [1983]. "Disaggregation Models in Hydrology—An Evaluation." Presented at the fall meeting of the American Geophysical Union, December 5–9, 1983, San Francisco, Calif.

Young, G. K., and Pisano, W. C. [1968]. "Operational Hydrology Using Residuals." *A.S.C.E. J. Hydraulics.* 94(HY4):909–24.

CHAPTER 4

Bendat, J. S., and A. G. Piersol [1966]. *Measurement and Analysis of Random Data.* New York: Wiley.

Boyer, D. N. [1966]. "The Design of Hydrometeorological Networks." Proceedings of the 7th Regional E.C.A.F.E. United Nations Conference on Water Resources Development. Canberra, Australia.

Bras, R. L. [1979]. "Sampling of Interrelated Random Fields: The Rainfall-Runoff Case." *Water Resources Res.* 15(6):1767–80.

Chatfield, C. [1975]. *The Analysis of Time Series: Theory and Practice.* London: Halsted Press.

Dooge, J. C. I. [1977]. "Problems and Methods of Rainfall-Runoff Modelling." In: *Mathematical Models for Surface Water Hydrology*, T. A. Ciriani, U. Maione, and J. R. Wallis, eds. New York: Wiley.

Eagleson, P. S. [1967a]. "Optimum Density of Rainfall Networks." *Water Resources Res.* 3(4):1021–34.

Idem [1967b]. "A Distributed Linear Model for Peak Catchment Discharge." In: *Proceedings of International Hydrology Symposium.* Fort Collins, Colo.: Colorado State University. Vol. 1, pp. 1–8.

Eagleson, P. S., and M. J. Goodspeed [1973]. "Linear Systems Techniques Applied to Hydrologic Data Analysis and Instrument Evaluation." Melbourne, Australia. Commonwealth Science and Indian Resource Organization, Division of Land Use Research, Technical Report 34.

Eagleson, P. S., and W. J. Schack [1966]. "Some Criteria for the Measurement of Rainfall and Runoff." *Water Resources Res.* 12(3):427–36.

Enochson, L. D. [1964]. Frequency Response Functions and Coherence Functions for Multiple Input Linear Systems. *NASA.* CR-32.

Evans, D., B. M. Harley, and R. L. Bras [1972]. "Application of Linear Routing Systems to Regional Groundwater Problems." Cambridge, Mass.: Ralph M. Parsons Laboratory for Water Resources and Hydrodynamics, M.I.T. Technical Report 155.

Goodman, N. R. [1957]. "On the Joint Estimation of the Spectra, Cospectrum and Quadrature Spectrum of a Two-Dimensional Stationary Gaussian Process." New York: New York University. Engineering Statistics Laboratory, Scientific Paper No. 10.

Idem [1965]. "Measurement of Matrix Frequency Response Functions and Multiple Coherence Function." Wright-Patterson Air Force Base, Ohio: AFFDL, TR65-56. Research and Technology Division, Air Force Systems Command.

Goodman, N. R., S. Katz, B. H. Kramer, and M. T. Kuo [1961]. "Frequency Response from Stationary Noise: Two Case Histories." *Technometrics*. 3(2):245–68.

Granger, C. W. J. in association with M. Hatanaka [1964]. *Spectral Analysis of Economic Time Series*. Princeton, N.J.: Princeton University Press.

Kraijenhoff Van DeLeur, D. A. [1966]. *Recent Trends in Hydrograph Synthesis*. Proceedings of Technical Meeting 21. The Hague, Netherlands: Central Organization for Applied Scientific Research in the Netherlands N.N.O.

Jenkins, G. M., and D. G. Watts [1968]. *Spectral Analysis and Its Applications*. San Francisco: Holden Day.

Laurenson, E. M., and T. O'Donnell [1969]. "Data Error Effects in Unit Hydrograph Derivation." *A.S.C.E. J. Hydraulics*. 95(HY-6):1899–918.

Lee, Y. W. [1960]. *Statistical Theory of Communications*. New York: Wiley.

Nash, J. E. [1960]. "A Unit Hydrograph Study, With Particular Reference to British Catchments." *Proc. Inst. Civil Eng*. 17:249–82.

Nash, J. E. [1960]. "A Note on an Investigation into Two Aspects of the Relation Between Rainfall and Storm Runoff." *Int. Assoc. Sci. Hydrology Publ*. 51:567–78.

O'Donnell, T. [1960]. "Instantaneous Unit Hydrograph Derivation by Harmonic Analysis." *I.A.S.H. Publ*. 51:546–57.

Panofsky, H. A., and G. W. Brier [1958]. "Some Applications of Statistics to Meteorology, Mineral Industries." University Park, Penn.: Pennsylvania State University, Extension Services.

Pierson, W. J. [1960]. "On the Use of Time Series Concepts and Spectral and Cross-Spectral Analyses in the Study of Long Range Forecasting Problems." *J. Marine Res*. 18(2):112–32.

Rodríguez-Iturbe, I. [1967]. "The Application of Cross-Spectral Analysis to Hydrologic Time Series." Fort Collins, Colo.: Colorado State University, Hydrology Paper No. 24.

Rodríguez-Iturbe, I., and C. F. Nordin [1968]. "Time Series Analyses of Water and Sediment Discharges." *Bull. I.A.S.H.* XII(2):69–84.

Idem [1974]. "Frequency Domain Analysis of Hydrological Systems, with Monthly Data." *Proceedings, Warsaw Symposium*. Paris: International Association of Hydrological Sciences, Publication No. 100, pp. 343–54.

Rodríguez-Iturbe, I., and M. M. Siddiqui [1969]. "Coherence Analysis of Stationary Processes with Applications to Hydrology." *Bull. I.A.S.H.* XIV(4):77–94.

Singh, K. P. [1976]. "Unit Hydrographs—A Comparative Study." *Water Resources Bull*. 12(2):381–92.

Wiener, N. [1949]. *The Extrapolation, Interpolation, and Smoothing of Stationary Time Series with Engineering Applications.* New York: Wiley.

Woolhiser, D. A., and H. C. Schwalen [1959]. "Area-Depth-Frequency Relations for Thunderstorm Rainfall in Southern Arizona." Tuscon, Ariz.: University of Arizona, Agricultural Engineering Dept., Arizona Agricultural Experimental Station, Technical Paper 527.

Yevjevich, V. [1963]. "Fluctuations of Wet and Dry Years, Part I: Research Data Assembly and Mathematical Models." Hydrology Paper No. 1. Fort Collins, Colo.: Colorado State University.

CHAPTER 5

Anis, A. A. [1955]. "The Variance of the Maximum of Partial Sums of a Finite Number of Independent Normal Variates." *Biometrika.* 42:96–101.

Idem [1956]. "On the Moments of the Maximum of Partial Sums of a Finite Number of Independent Normal Variates." *Biometrika*, 43:79–84.

Anis, A. A., and E. H. Lloyd [1953]. "On the Range of Partial Sums of a Finite Number of Independent Random Variables." *Biometrika*, 40:35–42.

Barnard, G. A. [1956]. "Discussion of Hurst [1956]." *Proc. Inst. Civil Eng.* 5(5):552–3.

Bhattacharya, R. N., V. K. Gupta, and E. Waymire [1983]. "The Hurst Effect Under Trends." *J. Appl. Probability.* 20(3):649–62.

Bobée, B., and R. Robitaille [1975]. "Correction of Bias in the Estimation of the Coefficient of Skewness." *Water Resources Res.* 11(6):851–4.

Boes, D. C., and J. D. Salas [1978]. "Nonstationarity of the Mean and the Hurst Phenomenon." *Water Resources Res.* 14(1):135–43.

Borgman, L. E., and J. Amorocho [1970]. "Some Statistical Problems in Hydrology." *Int. Stat. Inst. Rev.* 38(1):82–96.

Bras, R. L., K. Curry, and R. Buchanan [1981]. "The Multivariate Broken Line Model Revisited: A Discussion on Capabilities and Limitations." *J. Hydrology.* 53:31–51.

Brophy, J. J. [1970]. "Statistical Randomness and 1/f Noise." *Nav. Res. Rev.* 23(10):8–21.

Burges, S. J., and D. P. Lettenmaier [1975]. "Operational Comparison of Streamflow Generation Procedures." Seattle, Wash.: Harris Hydraulic Laboratory, University of Washington, Department of Civil Engineering, Technical Report 45.

Cramer, H., and M. R. Leadbetter [1967]. *Stationary and Related Stochastic Processes.* New York: Wiley.

Curry, K., and R. L. Bras [1978]. "Theory and Applications of the Multivariate Broken Line, Disaggregation and Monthly Autoregressive Streamflow Generators to the Nile River." Cambridge, Mass.: Technology Adaptation Program, M.I.T. Report No. 78–5.

Ditlevsen, O. D. [1971]. "Extremes and First Passage Times." Lyngby, Denmark: Doctoral Disertation presented to the Technical University of Denmark.

Feller, W. [1951]. "The Asymptotic Distribution of the Range of Sums of Independent Random Variables." *Ann. Math. Stat.* 22:427–32.

Fiering, M. B. [1967]. *Steamflow Synthesis*, Cambridge, Mass.: Harvard University Press.

Gardner, M. [1978]. "Mathematical Games: White and Brown Music, Fractal Curves and One-Over-f Fluctuations." *Scientific American.* 238(4):16–32.

Hurst, H. E. [1951]. "Long Term Storage Capacities of Reservoirs." *Trans. A.S.C.E.* 116:776–808.

Idem [1956]. "Methods of Using Long Term Storage in Reservoirs." *Proc. Inst. Civil Eng.* 5(5):519–43.

Hurst, H. E., R. P. Black, and V. M. Simaika [1965]. *Long Term Storage.* London: Constable.

Klemeš, V. [1974]. "The Hurst Phenomenon: a Puzzle?" *Water Resources Res.* 10(4):675–88.

Klemeš, V., R. Drikanthan, and T. A. McMahon [1981]. "Long Memory Flow Models in Reservoir Analysis: What Is Their Practical Value?" *Water Resources Res.* 17(3):737–51.

Lettenmaier, D. P., and S. J. Burges [1977]. "Operational Assessment of Hydrologic Models of Long-Term Persistence." *Water Resources Res.* 13(1):113–24.

Longuet-Higgins, M. S. [1962]. "The Distribution of Intervals between Zeros of a Stationary Random Function." *Trans. Roy. Soc. London.* A(254):557–600.

Mandelbrot, B. B. [1965]. "Une Class de Processus Stochastiques Homothetiques a Soi: Application a la loi Climatologique de H. E. Hurst." *Compte Rendus Academie Science.* 260:3274–7.

Idem [1971]. "A Fast Fractional Gaussian Noise Generator." *Water Resources Res.* 7(3):543–53.

Idem [1977]. *Fractals: Form, Chance and Dimension.* San Francisco, Calif.: Freeman.

Mandelbrot, B. B., and J. W. Van Ness [1968]. "Fractional Brownian Motions, Fractional Gaussian Noises and Applications." *Siam Rev. Appl. Math,* 10(4):422–37.

Mandelbrot, B. B., and J. R. Wallis [1968]. "Noah, Joseph, and Operational Hydrology." *Water Resources Res.* 4(5):909–18.

Idem [1969a]. "Computer Experiments with Fractional Gaussian Noises. Part 1: Averages and Variances." *Water Resources Res.* 5(1):228–41.

Idem [1969b]. "Computer Experiments with Fractional Gaussian Noises. Part 2: Rescaled Ranges and Spectra." *Water Resources Res.* 5(1):260–7.

Idem [1969c]. "Computer Experiments with Fractional Gaussian Noises. Part 3: Mathematical Appendix." *Water Resources Res.* 5(1):260–7.

Idem [1969d]. "Some Long-Run Properties of Geophysical Records." *Water Resources Res.* 5(2):321–40.

Idem [1969e]. "Robustness of the Rescaled Range R/S in the Measurement of Non-Cyclic Long-Run Statistical Dependence." *Water Resources Res.* 5(5):967–88.

Mejía, J. M. [1971]. "On the Generation of Multivariate Sequences Exhibiting the Hurst Phenomena and Some Flood Frequency Analysis." Fort Collins, Colo.: Ph.D.

dissertation presented to the Civil Engineering Department, Colorado State University.

Mejía, J. M., D. R. Dawdy, and C. F. Nordin [1974]. "Streamflow Simulation, 3. The Broken Line Process and Operational Hydrology." *Water Resources Res.* 10(2):242–5.

Mejía, J. M., I. Rodríguez-Iturbe, and D. R. Dawdy [1972]. "Streamflow Simulation, 2: The Broken Line Process as a Potential Model for Hydrologic Simulation." *Water Resources Res.* 8(4):931–41.

Moran, P. A. R. [1959]. *The Theory of Storage*. London: Methuen.

Nordin, C. F., and D. M. Rosbjerg [1970]. "Applications of Crossing Theory in Hydrology." *Bull. I.A.S.H.* 15(1):27–43.

O'Connell, P. E. [1974]. "Stochastic Modelling of Long-Term Persistence in Streamflow Sequences." London: Ph.D. thesis presented to the Civil Engineering Department, Imperial College.

Potter, K. W. [1976]. "Evidence for Nonstationarity as a Physical Explanation of the Hurst Phenomenon." *Water Resources Res.* 12(5):1047–1052.

Rice, S. O. [1954]. "Mathematical Analysis of Random Noise." In: *Selected Papers on Noise and Stochastic Processes*, N. Wax, ed. New York: Dover, pp. 133–294.

Rodríguez-Iturbe, I. [1968]. "A Modern Statistical Study of Monthly Levels of the Orinoco River." *Bull. I.A.S.H.* 13(4):25–40.

Idem [1969]. "Applications of the Theory of Runs to Hydrology." *Water Resources Res.* 5(6):1422–6.

Rodríguez-Iturbe, I., D. R. Dawdy, and L. E. García [1971]. "Adequacy of Markovian Models with Cyclic Components for Stochastic Streamflow Simulation." *Water Resources Res.* 7(5):1127–43.

Salas, J. D. [1974]. "Range of Cumulative Sums, I. Exact and Approximate Expected Values." *J. Hydrology*. 23(3):39–66.

Salas, J. D., D. C. Boes, V. Yevjevich, and G. G. S. Pegram [1979]. "Hurst Phenomenon as a Pre-asymtotic Behavior." *J. Hydrology*. 44:1–15.

Thomas, H. A., and M. B. Fiering [1962]. "Mathematical Synthesis of Streamflow Sequences for the Analysis of River Basins by Simulation." In: *Design of Water Resources Systems*. A. Maass, et al., eds. Cambridge, Mass.: Harvard University Press, p. 459–93.

Tick, L. J., and P. Shaman [1966]. "Sampling Rate and Appearance of Stationary Gaussian Process." *Technometrics*. 8(1):91–106.

Yevjevich, V. M. [1965]. "The Application of Surplus, Deficits, and Range in Hydrology." Fort Collins, Colo.: Colorado State University, Hydrology Paper No. 10.

Idem [1967]. "Mean Range of Linearly Dependent Normal Variables with Application to Storage Problems." *Water Resources Res.* 3(3):663–71.

Appendix to Chapter 5

Curry, K., and R. L. Bras [1978]. "Theory and Applications of the Multivariate Broken Line, Disaggregation and Monthly Autoregressive Streamflow Generators to the Nile River." Cambridge, Mass.: M.I.T., Technology Adaptation Program, Report No. 78–5.

Mandelbrot, B. B., and J. W. Van Ness [1968]. "Fractional Brownian Motions, Fractional Gaussian Noises and Applications." *Siam Rev. Appl. Math.* 19(4):422–37.

Mejía, J. M. [1971]. "On the Generation of Multivariate Sequences Exhibiting the Hurst Phenomena and Some Flood Frequency Analysis." Fort Collins, Colo.: Ph.D. dissertation presented to the Civil Engineering Department, Colorado State University.

CHAPTER 6

Abramowitz, M., and I. A. Stegun [1965]. *Handbook of Mathematical Functions.* New York: Dover.

Bear, J. [1972]. *Dynamics of Fluids in Porous Media.* New York: Elsevier.

Benson, M. A. [1964]. "Factors Affecting the Occurrence of Floods in the Southwest." Washington, D.C.: U.S. Geological Survey, Paper 1580–D.

Benson, M. A., and N. C. Matalas [1967]. "Synthetic Hydrology Based on Regional Statistical Parameters." *Water Resources Res.* 3(4):931–5.

Boyer, M. C. [1957]. "A Correlation of the Characteristic of Great Storms." *EOS Transactions A.G.U.* 38(2):233–6.

Bras, R. L., and I. Rodríguez-Iturbe [1976]. "Rainfall Generation: A Nonstationary Time-Varying Multidimensional Model." *Water Resources Res.* 12(1):450–6.

Davis, D. R., and W. M. Dvoranchik [1971]. "Evaluation of the Worth of Additional Data." *Water Resources Bull.* 7(4):700–7.

Dawdy, D. R., H. E. Kubik, and E. R. Close [1970]. "Value of Streamflow Data for Project Design—A Pilot Study." *Water Resources Res.* 6(4):1045–50.

Delhomme, J. P. [1979]. "Spatial Variability and Uncertainty in Groundwater Flow Parameters: A Geostatistical Approach." *Water Resources Res.* 15(2):269–80.

Duckstein, L., and C. C. Kisiel [1971]. "Efficiency of Hydrologic Data Collection Systems, Role of Type 1 and 2 Errors." *Water Resources Bull.* 7(3):592–604.

Dunne, T. [1978]. "Field Studies of Hillslope Flow Processes." In: *Hillslope Hydrology.* M. J. Kirkby, ed. New York: Wiley-Interscience, pp. 227–93.

Eagleson, P. S. [1967]. "Optimum Density of Rainfall Network." *Water Resources Res.* 3(4):1021–33.

Idem [1970]. *Dynamic Hydrology.* New York: McGraw-Hill.

Idem [1971]. "The Stochastic Kinematic Wave." In: *Systems Approach to Hydrology.* V. Yevjevich, ed. Fort Collins, Colo.: Colorado State University, Water Resources Publication, pp. 210–25.

Idem [1972]. "Dynamics of Flood Frequency." *Water Resources Res.* 8(4):878–98.

Fogel, M. M., and L. Duckstein [1969]. "Point Rainfall Frequencies in Convective Storms." *Water Resources Res.* 5(6):1229–37.

Freeze, R. A. [1980]. "A Stochastic–Conceptual Analysis of Rainfall-Runoff Processes on a Hillslope." *Water Resources Res.* 16(2):391–408.

Gandin, L. S. [1965]. *Objective Analysis of Meteorological Fields*. Jerusalem: Israel Program for Scientific Translations.

Ghosh, B. [1951]. "Random Distances within a Rectangle and Between Two Rectangles." *Bull. Calcutta Math. Soc.* 43:17–24.

Gradshteyn, I. S., and I. M. Ryzhik [1965]. *Table of Integrals, Series and Products*. New York: Academic Press.

Grayman, W. M., and P. S. Eagleson [1971]. "Evaluation of Radar and Raingage Systems for Flood Forecasting." Cambridge, Mass.: Ralph M. Parsons Laboratory for Water Resources and Hydrodynamics, M.I.T. Technical Report 138.

Harris, R. I. [1971]. "The Nature of Wind, The Modern Design of Wind-Sensitive Structures Paper 3." London: *Construction Industry Research and Information Association*.

Hendrick, R. L., and G. H. Comer [1970]. "Space Variations of Precipitation and Implications for Rain Gage Network Design." *J. Hydrology*. 10:151–63.

Hinze, I. O. [1959]. *Turbulence — An Introduction to its Mechanism and Theory*. New York: McGraw-Hill.

Horton, R. E. [1933]. "The Role of Infiltration in the Hydrologic Cycle." *EOS Transactions A.G.U.* 14:446–60.

Houze, R. A. [1969]. "Characteristics of Mesoscale Precipitation Areas." Cambridge, Mass.: M.S. Thesis presented to the Massachusetts Institute of Technology.

Huff, F. A. [1970]. "Spatial Distribution of Rainfall Rates." *Water Resources Res.* 6(1):254–60.

Journel, A. G. [1974]. "Geostatistics for Conditional Simulation of Ore Bodies." *Economic Geology*. 69:673–87.

Journel, A. G., and Ch.J. Huijbregts [1978]. *Mining Geostatistics*. New York: Academic Press.

Leclerc, G., and J. C. Schaake, Jr. [1972]. "Derivation of Hydrologic Frequency Curves." Cambridge, Mass.: Ralph M. Parsons Laboratory for Water Resources and Hydrodynamics, M.I.T. Technical Report 142.

Idem [1973]. "Methodology for Assessing the Potential Impact of Urban Development on Urban Runoff and the Relative Efficiency of Runoff Control Alternatives." Cambridge, Mass.: Ralph M. Parsons Laboratory for Water Resources and Hydrodynamics, M.I.T. Technical Report 167.

Lenton, R. L., and I. Rodríguez-Iturbe [1974]. "On the Collection, the Analysis, and the Synthesis of Spatial Rainfall Data." Cambridge, Mass.: Ralph M. Parsons Laboratory for Water Resources and Hydrodynamics, M.I.T. Technical Report 194.

Idem [1977]. "A Multidimensional Model for the Synthesis of Areal Rainfall Averages." *Water Resources Res.* 13(3):605–12.

Mantoglou, A., and J. L. Wilson [1981]. "Simulation of Random Fields with the Turning Bands Method." Cambridge, Mass.: Ralph M. Parsons Laboratory for Water Resources and Hydrodynamics, M.I.T. Technical Report 264.

Idem [1982]. "The Turning Bands Method for Simulation of Random Fields Using Line Generation by a Spectral Method." *Water Resources Res.* 18(5):1379–94.

Matalas, N. C. [1969]. "Optimum Gaging Station Location." In: *The Progress of*

Hydrology, Proceedings of the First International Seminar for Hydrology Professors. Urbana, Ill.: Department of Civil Engineering, University of Illinois, pp. 473–89.

Matalas, N. C., and E. J. Gilroy [1968]. "Some Comments on Regionalization in Hydrologic Studies." *Water Resources Res.* 4(6):1361–9.

Matern, B. [1960]. "Spatial Variation." *Comm. Swed. Forestry Res. Inst.* 49:1–144.

Matheron, G. [1973]. "The Intrinsic Random Functions and Their Applications." *Adv. Appl. Prob.* 5:439–68.

Mejía, J. A., and I. Rodríguez-Iturbe [1974]. "On the Synthesis of Random Field Sampling from the Spectrum: An Application to the Generation of Hydrologic Spatial Processes." *Water Resources Res.* 10(4):705–11.

Moss, M. M. [1970]. "Optimum Operating Procedure for a River Gaging Station Established to Provide Data for Design of a Water Supply Project." *Water Resources Res.* 6(4):1051–61.

Panchev, S. [1971]. *Random Functions and Turbulence.* New York: Pergamon Press.

Pilgrim, D. H., I. Cordery, and R. French [1969]. "Temporal Patterns of Design Rainfall for Sydney." Australia: Sydney. *Civil Engineering Transactions, Institution of Engineers.*

Rhenals-Figueredo, A. E., I. Rodríguez-Iturbe, and J. C. Schaake, Jr. [1974]. "Bidimensional Spectral Analysis of Rainfall Events." Cambridge, Mass.: Ralph M. Parsons Laboratory for Water Resources and Hydrodynamics, M.I.T. Technical Report 193.

Roche, M. [1963]. *Hydrologie de Surface.* Paris: Gauthier-Villars.

Rodda, J. C. [1969]. "Hydrologic Network Design—Needs Problems, and Approaches." Geneva: World Meteorological Organization, WMO/IHO Report No. 12.

Rodríguez-Iturbe, I. [1975]. "Hydrologic and Flood Frequency Study of the Fajardo River, Puerto Rico." San Juan, Puerto Rico: Rodríguez Amezquita and Assoc., Architects, and Planners, consulting report.

Rodríguez-Iturbe, I., and J. M. Mejía [1974a]. "The Design of Rainfall Networks in Time and Space." *Water Resources Res.* 10(4):713–28.

Idem [1974b]. "On the Transformation of Point Rainfall to Areal Rainfall." *Water Resources Res.* 10(4):729–36.

Smith, R. E. [1972]. "The Infiltration Envelope: Results from a Theoretical Infiltrometer." *J. Hydrology.* 17:1–21.

Stol, P. T. [1972]. "The Relative Efficiency of the Density of Rain-gage Networks." *J. Hydrology.* 15:193–208.

Taylor, G. I. [1935]. "Statistical Theory of Turbulence." *Proc. Roy. Society Series, A.* 151:421–78.

Tschannerl, G. [1970]. "A Decision Framework for the Efficient Use of Data Sources." Cambridge, Mass.: Center for Population Studies, Harvard University, Report No. 4.

U.S. Soil Conservation Service [1971]. "SCS National Engineering Handbook." Section 4. *Hydrology.* Washington, D.C.: U.S. Government Printing Office.

U.S. Weather Bureau [1957]. "Rainfall Intensity—Frequency Regime, 1: The Ohio Valley." Washington, D.C.: U.S. Department of Commerce, Technical Paper No. 29.

U.S. Weather Bureau [1958]. "Rainfall Intensity—Frequency Regime, 2: Southeastern United States." Washington, D.C.: U.S. Department of Commerce, Technical Paper No. 29.

Wilson, C. B. [1976]. "Effects of Spatially Distributed Rainfall on Storm Runoff from a Small Catchment." Cambridge, Mass.: M.S. thesis presented to the Department of Civil Engineering, Massachusetts Institute of Technology.

Wilson, C. B., J. B. Valdes, and I. Rodríguez-Iturbe [1979]. "On the Influence of the Spatial Distribution of Rainfall on Storm Runoff." *Water Resources Res.* 15(2):321–8.

Woolhiser, D. A., and H. C. Schwalen [1959]. "Area–Depth–Frequency Relations for Thunderstorm Rainfall in Southern Arizona." Tucson, Ariz.: Experimental Station, University of Arizona, Technical Paper No. 527.

Yaglom, A. M. [1962]. *An Introduction to the Theory of Stationary Random Functions.* Englewood Cliffs, N.J.: Prentice-Hall.

Idem [1957]. "Some Classes of Random Fields in *n*-Dimensional Space, Related to Stationary Random Processes." *Theory Prob. Appl.* 2(3):273–320.

Zawadzki, I. I. [1973a]. "Errors and Fluctuations of Raingage Estimates of Areal Rainfall." *J. Hydrology.* 18:243–55.

Idem [1973b]. "Statistical Properties of Precipitation Patterns." *J. Appl. Meteorology.* 12:459–72.

CHAPTER 7

Armstrong, M., and R. Jabin [1981]. "Variogram Models Must be Positive-Definite." *J. Math. Geology.* 13(5):455–9.

Bras, R. L., and I. Rodríguez-Iturbe [1975]. "Rainfall–Runoff as Spatial Stochastic Processes: Data Collection and Synthesis." Cambridge, Mass.: Ralph M. Parsons Laboratory for Water Resources and Hydrodynamics, M.I.T. Technical Report 196.

Idem [1976a]. "Evaluation of Mean Square Error Involved in Approximating the Areal Average of a Rainfall Event by a Discrete Summation." *Water Resources Res.* 12(2):181–4.

Idem [1976b]. "Network Design for the Estimation of Area Mean of Rainfall Events." *Water Resources Res.* 12(6):1185–95.

Chirlin, G. R., and E. F. Wood [1982]. "On the Relationship between Kriging and State Estimation." *Water Resources Res.* 18(2):432–8.

Chua, S. H., and R. L. Bras [1980]. "Estimation of Stationary and Non-Stationary Fields: Kriging in the Analysis of Orographic Precipitation." Cambridge, Mass.: Ralph M. Parsons Laboratory for Water Resources and Hydrodynamics, M.I.T. Technical Report 255.

Idem [1982]. "Optimal Estimators of Mean Area Precipitation in Regions of Orographic Influence." *J. Hydrology.* 57(112):23–48.

Cressie, N., and D. M. Hawkins [1980]. "Robust Estimation of the Variogram: I." *J. Math. Geology.* 12(2):115–25.

David, M. [1976]. "The Practice of Kriging." *Advanced Geostatistics in the Mining*

Industry. M. Guarascio et al., eds. Dordrecht, Holland: D. Reidel, pp. 31–48.

Idem [1977]. *Geostatistical Ore Reserve Estimation*, New York: Elsevier.

Davis, M., and M. David [1978]. "Automatic Kriging and Contouring in the Presence of Trends (Universal Kriging Made Simple)." *J. Canadian Petroleum Tech.* 17(1):90–8.

Delfiner, P. [1976]. "Linear Estimation of Non-Stationary Spatial Phenomena." *Advanced Geostatistics in the Mining Industry.* M. Guarascio et al., eds. Dordrecht, Holland: D. Reidel, pp. 49–68.

Delfiner, P., and J. P. Chiles [1978]. "Conditional Simulations: A New Monte Carlo Approach to Probabilistic Evaluation of Hydrocarbon in Place." Fontainebleau, France: Ecole des Mines de Paris.

Delfiner, P., and J. P. Delhomme [1973]. "Optimum Interpolation by Kriging." In: *Display and Analysis of Spatial Data.* J. C. Davis and M. J. McCullagh, eds. New York: Wiley, pp. 96–115.

Delhomme, J. P. [1976]. "Applications de la Theorie des Variables Regionalisees dans les Sciences de l'eau." Fontainebleau, France: Doctoral Thesis presented to Ecole des Mines de Paris, France.

Delhomme, J. P. [1978]. "Kriging in the Hydrosciences." *Adv. Water Resources.* 1(5):251–6.

Idem [1979]. "Spatial Variability and Uncertainty in Groundwater Flow Parameters: A Geostatistical Approach." *Water Resources Res.* 15(2):269–80.

Delhomme, J. P., and P. Delfiner [1973]. "Application du Krigeage a l'optimisation d'une Campagne Pluviometrique en Zone Aride." In: *Proceedings of the Symposium of the Design of Water Resource Projects with Inadequate Data*, (2):191–210. Madrid: Unesco.

Deutsch, R. [1965]. *Estimation Theory.* Englewood Cliffs, N.J.: Prentice-Hall.

Doctor, P. G. [1979]. "An Evaluation of Kriging Techniques for High Level Radioactive Waste Repository Site Characterization." Richland, Wash.: Pacific Northwest Laboratory. PNL-2903. UC-70. 67 pp.

Doctor, P. G., and R. W. Nelson [1981]. "Geostatistical Estimation of Parameters for Hydrological Transport Modeling." *J. Math. Geology.* 13(5):415–28.

Elliott, R. D. [1977]. "Final Report on Methods for Estimating Area Precipitation in Mountainous Areas." Silver Springs, Md.: National Weather Service, Office of Hydrology, Report No. 77–13.

Federov, V. V. [1972]. *Theory of Optimal Experiments.* New York: Academic Press.

Fogel, M. M., and L. Duckstein. [1969]. "Point Rainfall Frequencies in Convective Storms." *Water Resources Res.* 5(6):1229–37.

Gambolati, G., and G. Volpi [1979a]. "Groundwater Contour Mapping in Venice by Stochastic Interpolators. 1. Theory." *Water Resources Res.* 15(2):281–90.

Idem [1979b]. "A Conceptual Deterministic Analysis of the Kriging Technique in Hydrology." *Water Resources Res.* 15(3):625–9.

Gandin, L. S. [1970]. "The Planning of Meteorological Station Networks." In: *Distribution of Precipitation in Mountainous Areas, Vol. I.* Proceedings of Geilo Symposium,

Norway. WMO Technical Note 111. 25 pp. Geneva: World Meteorological Organization.

Grayman, W. M., and P. S. Eagleson [1971]. "Evaluation of Radar and Raingage Systems for Flood Forecasting." Cambridge, Mass.: Ralph M. Parsons Laboratory for Water Resources and Hydrodynamics, M.I.T. Technical Report 138.

Hughes, J. P., and D. P. Lettenmaier [1980]. "Aquatic Monitoring: Data Analysis and Network Design Using Regionalized Variable Theory." Seattle: Charles W. Harris Hydraulics Laboratory, University of Washington, Technical Report No. 65.

Hughes, J. P., and D. P. Lettenmaier [1981]. "Data Requirements for Kriging Estimation and Network Design." *Water Resources Res.* 17(6):1641–50.

Huijbregts, Ch.J. [1973]. "Regionalized Variables and Application to Quantitative Analysis of Spatial Data." In: *Display and Analysis of Spatial Data*. J. C. Davis and M. J. McCullagh, eds. New York: Wiley, pp. 38–53.

Huijbregts, Ch.J., and G. Matheron [1970]. "Universal Kriging: An Optimum Approach to Trend Surface Analysis." In: *Decision-Making in the Mining Industry, Vol. 12*. CIM International Symposium, pp. 159–60.

Journel, A. G. [1974]. "Geostatistics for Conditional Simulation of Ore Bodies." *Economic Geology.* 69:673–87.

Idem [1977]. "Kriging in Terms of Projections." *J. Math. Geology.* 9(6):563–86.

Idem [1980]. "The Lognormal Approach to Predicting Local Distributions of Selective Mining Unit Grades." *J. Math. Geology.* 12(4):285–303.

Journel, A. G., and Ch.J. Huijbregts [1978]. *Mining Geostatistics.* London: Academic Press.

Kafritsas, J., and R. L. Bras [1981]. "The Practice of Kriging." Cambridge, Mass.: Ralph M. Parsons Laboratory for Water Resources and Hydrodynamics, M.I.T. Technical Report 263.

Karlinger, M., and J. Shrivan [1978]. "An Application of Kriging to a Groundwater Quality Network." Paper presented at American Geophysical Union Chapman Conference on Network Design. Tucson, Ariz.

Kitanidis, P. K. [1983]. "Statistical Estimation of Polynomial Generalized Covariance Functions and Hydrologic Applications." *Water Resources Res.* 19(4):909–21.

Larson, L. W., and E. L. Peck [1974]. "Accuracy and Precipitation Measurements for Hydrologic Modeling." *Water Resources Res.* 10(4):857–64.

Lenton, R. L., and I. Rodríguez-Iturbe [1974]. "On the Collection, the Analysis, and the Synthesis of Spatial Rainfall Data." Cambridge, Mass.: Ralph M. Parsons Laboratory for Water Resources and Hydrodynamics, M.I.T. Technical Report 194.

Marechal, A. [1976a]. "Selecting Mineable Blocks: Experimental Results Observed on a Simulated Orebody." In: *Advanced Geostatistics in the Mining Industry*. M. Guarascio et al., eds. Dordrecht, Holland: D. Reidel, pp. 137–61.

Idem [1976b]. "The Practice of Transfer Functions: Numerical Methods and Their Application." In: *Advanced Geostatistics in the Mining Industry*. M. Guarascio et al., eds. Dordrecht, Holland: D. Reidel, pp. 253–76.

Matern, B. [1960]. "Spatial Variation." *Comm. Swed. Forestry Res. Inst.* 49:1–144.

Matheron, G. [1971]. "The Theory of Regionalized Variables and Its Applications." In:

Les Cahiers du Centre de Morphologie Mathematique, No. 5. Fontainebleau, France: Centre de Geostatistique.

Idem [1973]. "The Intrinsic Random Functions and Their Applications." *Adv. Appl. Prob.* 5:439–68.

Idem [1976a]. "A Simple Substitute for Conditional Expectation: The Disjunctive Kriging." In: *Advanced Geostatistics in the Mining Industry*. M. Guarascio et al., eds. Dordrecht, Holland: D. Reidel, pp. 221–36.

Idem [1976b]. "Forecasting Block Grade Distributions: The Transfer Functions." In: *Advanced Geostatistics in the Mining Industry*. M. Guarascio et al., eds. Dordrecht, Holland: D. Reidel, pp. 237–51.

McLaughlin, D. B. [1979]. "Hanford Groundwater Modeling-Statistical Methods for Evaluating Uncertainty and Assessing Sampling Effectiveness." Rockwell Hanford Operations. RHO-C-18: 75 pp.

Morgan, D. L., and F. J. Lourence. [1969]. "Comparison between Raingage and Lysimeter Measurements." *Water Resources Res.* 5(3):724–28.

Olea, R. A. [1975]. "Optimum Mapping Techniques Using Regionalized Variable Theory." In: *Series on Spatial Analysis*. No. 2. Lawrence, Kansas: Kansas Geological Survey.

Idem [1977]. "Measuring Spatial Dependence with Semivariograms." *Series on Spatial Analysis*. No.3. Lawrence, Kansas: Kansas Geological Survey.

Puente, C. E., and R. L. Bras [1982]. "A Comparison of Linear and Non-Linear Random Fields Estimators." Cambridge, Mass.: Ralph M. Parsons Laboratory for Water Resources and Hydrodynamics, M.I.T. Technical Report 287.

Rendu, J. M. [1980]. "Disjunctive Kriging: Comparison of Theory with Actual Results." *J. Math. Geology.* 12(4):305–20.

Rodda, S. C., et al. [1969]. "Hydrologic Network Design—Needs, Problems, and Approaches." Geneva: World Meteorological Organization, WMO/IHO Report No. 12.

Rodríguez-Iturbe, I., and J. M. Mejía [1974]. "The Design of Rainfall Networks in Time and Space." *Water Resources Res.* 10(4):713–28.

Sage, A. P., and J. L. Melsa [1971]. *Estimation Theory with Applications to Communications and Control*. New York: McGraw-Hill.

Schweppe, F. [1973]. *Uncertain Dynamic Systems*. Englewood Cliffs, N.J.: Prentice-Hall.

Sevruk, B. F. [1971]. "Comparison of Mean Rain Catch at Various Gauge Networks." *Nordic Hydrology.* 5:50–4.

Shaw, E. M., and P. E. O'Connell [1976]. "Design of Networks and Data Transfer for Precipitation." Geneva: World Meteorological Organization Hydrological Network Design and Information Transfer, WMO Operational Hydrology Report No. 8.

Starks, T. H., and J. H. Fang [1982]. "On the Estimation of the Generalized Covariance Functions." *J. Math. Geology.* 14(1):57–64.

Switzer, P. [1979]. "Statistical Considerations in Network Design." *Water Resources Res.* 15(6):1712–6.

Villeneuve, J. P., G. Morin, B. Bobée, and D. Leblanc [1979]. "Kriging in the Design of Streamflow Sampling Networks." *Water Resources Res.* 15(6):1833–40.

Volpi, G., and G. Gambolati [1978]. "On the Use of a Main Trend for the Kriging Technique in Hydrology." *Adv. Water Resources*. 1(6):345–9.

Volpi, G., G. Gambolati, L. Cargognin, P. Gatto, and G. Mozzi [1979]. "Groundwater Contour Mapping in Venice by Stochastic Interpolators. 2. Results." *Water Resources Res*. 15(2):291–7.

Western Scientific Services, Inc. [July 1-June 30, 1970-1974]. Colorado River Basin Pilot-Project. Fort Collins, Colo.: Comprehensive Meteorological Data Reports.

Zawadski, I. I. [1973]. "Errors and Fluctuations of Raingage Estimates and Areal Rainfall." *J. Hydrology*. 18:243–55.

Zubrzycki, S. [1958]. "Remarks on Random, Stratified and Systematic Sampling in a Plane." *Colloquium Mathematicum*. 6:252–64.

CHAPTER 8

Åströn, K. J. [1970]. *Introduction to Stochastic Control Theory*. New York: Academic Press.

Åströn, K. J., and Eykhoff, P. [1970]. "System Identification, A Survey." *Automatica*. 7:123–62.

Athans, M. [1974]. "The Importance of Kalman Filtering Methods for Economic Systems." *Ann. Economic Social Measurement*. 3(1):49–64.

Bras, R. L., and R. Colón [1978]. "Time Averaged Areal Mean of Precipitation: Estimation and Network Design." *Water Resources Res*. 14(5):878–88.

Bras, R. L., and I. Rodríguez-Iturbe [1976]. "Rainfall Network Design for Runoff Prediction." *Water Resources Res*. 12(6):1197–208.

Brewer, J. W., and S. F. Moore [1974]. "Monitoring: An Environmental State Estimation Problem." *A.S.M.E. Journal of Dynamic Systems, Measurement, and Control*, pp. 363–5.

Chiu, G. L., ed. [1978]. *Applications of Kalman Filter to Hydrology, Hydraulics, and Water Resources*. Proceedings of American Geophysical Union Chapman Conference, May 22–24, 1978, Department of Civil Engineering, University of Pittsburgh, Pittsburgh, Pa.

Dandy, G. D. [1976]. "Design of Water Quality Sampling Systems for River Networks." Cambridge, Mass.: Ph.D. Thesis presented to the Department of Civil Engineering, Massachusetts Institute of Technology.

Deutsch, R. [1965]. *Estimation Theory*. Englewood Cliffs, N.J.: Prentice-Hall.

Duong, M., C. Winn, and G. Johnson [1975]. "Modern Control Concepts in Hydrology." *IEEE Transactions of Systems, Man, and Cybernetics*. SMC-5(1):46–53.

Federov, V. V. [1972]. *Theory of Optimal Experiments*. New York: Academic Press.

Gelb, A., ed. [1974]. *Applied Optimal Estimation*. Cambridge, Mass.: M.I.T. Press.

Hino, M. [1970]. "Runoff Forecasts by Linear Predictive Filter." *A.S.C.E. J. Hydraulics*. 96(HY3):681–707.

Idem [1973]. "On-Line Prediction of Hydrologic System." In: Proceedings of the 15th Congress of the International Association for Hydraulic Research, Istanbul, Vol. 4. Ankara, Turkey: State Hydraulics Printing House, pp. 121–9.

Jazwinski, A. H. [1970]. *Stochastic Processes and Filtering Theory*. New York: Academic Press.

Kailath, T. [1974]. "A View of Three Decades of Linear Filtering Theory." *I.E.E.E. Trans. Inform. Theory*. IT-20(2):145–81.

Kalman, R. E. [1960]. "A New Approach to Linear Filtering and Prediction Problems." *A.S.M.E. J. Basic Eng*. 82D:35–45.

Idem [1963]. "New Methods in Wiener Filtering Theory." In: *Proceedings of the 1st Symposium of Engineering Applications of Random Function Theory and Probability*. New York: Wiley, pp. 270–388.

Kalman, R. E., and R. S. Bucy [1961]. "New Results in Linear Filtering and Prediction Theory." *A.S.M.E. J. Basic Eng*. 83D:95–108.

Kitanidis, P. K. [1974]. "A Unified Approach to the Parameter Estimation of Groundwater Models." Cambridge, Mass.: M.S. Thesis presented to the Department of Civil Engineering, Massachusetts Institute of Technology.

Kitanidis, P. K., and R. L. Bras [1979]. "Collinearity and Stability in the Estimation of Rainfall–Runoff Model Parameters. *J. Hydrology*. 42(12):91–108.

Koivo, A. J., and G. Phillips [1976]. "Optimal Estimation of DO, BOD, and Stream Parameters Using a Dynamic Discrete Time Model." *Water Resources Res*. 12(4):705–11.

Lee, Y. W. [1960]. *Statistical Theory of Communications*. New York: Wiley.

Lettenmaier, D. P., and S. J. Burges [1976]. "Use of State Estimation Techniques in Water Resource System Modeling." *Water Resources Bull*. 12(1):83–99.

McLaughlin, D. D. [1980]. "Application of Kalman Filtering to Groundwater Basin Modeling and Prediction." In: *Proceedings of the I.I.A.S.A. Workshop on Recent Developments in Real Time Forecasting/Control of Water Resources Systems*. (Oct. 18–20, 1976.) E. Wood, ed. New York: Pergamon Press.

Melsa, J. L., and A. P. Sage [1973]. *An Introduction to Probability and Stochastic Processes*. Englewood Cliffs, N.J.: Prentice-Hall.

Moore, S. F. [1971]. "The Application of Linear Filter Theory to the Design and Improvement of Measurement Systems for Aquatic Environments." Davis, Calif. Ph.D. Dissertation presented to the University of California.

Idem [1973]. "Estimation Theory Applications to Design of Water Quality Monitoring Systems." *A.S.C.E. J. Hydraulics*. 99(HY5):815–31.

Muzik, I. [1974]. "State Variable Model of Overland Flow." *J. Hydrology*. 22:347–64.

Natale, L., and E. Todini [1976a]. "A Stable Estimator for Linear Models 1." *Water Resources Res*. 12(4):667–71.

Idem [1976b]. "A Stable Estimator for Linear Models 2." *Water Resources Res*. 12(4):672–6.

Potter, J. E., and R. G. Stern [1963]. "Statistical Filtering of Space Navigation Measurements." In: *Proceedings of the American Institute of Aeronautics and Astronautics Guidance and Control Conference*. August 12–14, 1963, Cambridge, Mass. New York: AIAA.

Rodríguez-Iturbe, I., and J. M. Mejía [1974]. "On the Design of Rainfall Networks in Time and Space." *Water Resources Res*. 10(4):713–28.

Sage, A. P., and J. L. Melsa [1971a]. *Estimation Theory with Applications to Communications and Control*. New York: McGraw-Hill.

Idem [1971b]. *System Identification*. New York: Academic Press.

Schweppe, F. [1973]. *Uncertain Dynamic Systems*. Englewood Cliffs, N.J.: Prentice-Hall.

Szöllösi-Nagy, A. [1976a]. "An Adaptive Identification and Prediction Algorithm for the Real Time Forecasting of Hydrologic Time Series." *Hydrological Science Bull.* 21(3):163–76.

Idem [1976b]. "Introductory Remarks on the State Space Modeling of Water Resources Systems." Research Memorandum, RM-76-73. Laxenburg, Austria: International Institute for Applied Systems Analysis.

Szöllösi-Nagy, A., E. Todini, and E. Wood [1977]. "A State–Space Model for Real Time Forecasting of Hydrological Time Series." *J. Hydrological Sciences.* 4(1):61–76.

Todini, E., and D. Bouillot [1975]. "A Rainfall–Runoff Kalman Filter Model." In: *System Simulation in Water Resources*. G. C. Vansteenkiste, ed. Amsterdam: North Holland.

Young, P. C. [1974]. "Recursive Approaches to Time Series Analysis." *Bull. Inst. Math Appl.* 10:209–24.

Young, P., and P. Whitehead [1975]. "A Recursive Approach to Time Series Analysis for Multivariate Systems." In: *International Federation for Information Processing Working Conference on Computer Simulation of Water Resource Systems*. July 30–Aug. 2, 1972, Ghent, Belgium. G. C. Vansteenkiste, ed. Amsterdam: North Holland, pp. 39–52.

CHAPTER 9

Albert, A. E., and L. A. Gardner [1967]. *Stochastic Approximations and Nonlinear Regression*. Research Monograph No. 42. Cambridge, Mass: M.I.T. Press.

Boozer, D. D., and W. L. McDaniel [1972]. "On Innovation Sequence Testing of the Kalman Filter." *I.E.E.E. Trans. Automatic Control.* AC-17(1):158–60.

Bras, R. L., and K. P. Georgakakos [1980]. "Real Time Non-Linear Filtering Techniques in Streamflow Forecasting: A Statistical Linearization Approach." In: Proceedings of the Third International Symposium on Stochastic Hydraulics, August 5–7, 1980, Tokyo, pp. 95–105.

Bras, R. L., and P. Restrepo-Posada [1980]. "Real Time, Automatic Parameter Calibration in Conceptual Runoff Forecasting Models." In: Proceedings of the Third International Symposium on Stochastic Hydraulics, August 5–7, 1980, Tokyo, pp. 61–70.

Dawdy, D. R., and T. O'Donnell [1965]. "Mathematical Model of Catchment Behavior." *A.S.C.E. J. Hydraulics.* 91(HY4):123–37.

Edwards, A. W. F. [1972]. *Likelihood*. London: Cambridge University Press.

Gelb, A., ed. [1974]. *Applied Optimal Estimation*. Cambridge, Mass.: M.I.T. Press.

Gelb, A., and W. E. Vander Velde [1968]. *Multiple-Input Describing Functions and Nonlinear Systems Design*. New York: McGraw-Hill.

Georgakakos, K. P. and R. L. Bras [1980]. "A Statistical Linearization Approach to Real Time Nonlinear Flood Routing." Cambridge, Mass.: Ralph M. Parsons Laboratory for Water Resources and Hydrodynamics, M.I.T. Technical Report 256.

Idem [1982a]. "A Precipitation Model and Its Use in Real-Time River Flow Forecasting." Cambridge, Mass.: Ralph M. Parsons Laboratory for Water Resources and Hydrodynamics, Massachusetts Institute of Technology, technical report no. 286.

Idem [1982b]. "Real-Time Statistically Linearized, Adaptive Flood Routing." *Water Resources Research*. 18(3):513–24.

Ho, Y. C., and R. C. K. Lee [1965]. "Identification of Linear Dynamic Systems." *Information and Control*. 8:93–110.

Ibbitt, R. P. [1974]. "Effects of Random Data Errors on the Parameter Values for a Conceptual Model." *Water Resources Res*. 8(1):70–8.

Ibbitt, R. P., and T. O'Donnell [1971]. "Fitting Methods for Conceptual Catchment Models." *A.S.C.E. J. Hydraulics*. 97(HY9):1331–42.

Jazwinski, A. H. [1969]. "Adaptive Filtering." *Automatica*. 5:475–85.

Johnston, P. R., and D. H. Pilgrim [1976]. "Parameter Optimization for Watershed Models." *Water Resources Res*. 12(3):477–86.

Kailath, T. [1968]. "An Innovations Approach to Least-Squares Estimation, Part I: Linear Filtering in Additive White Noise." *I.E.E.E. Trans. Automatic Control*. AC-13(6):646–55.

Kalman, E. R., and S. R. Bucy [1961]. "New Results in Linear Filtering and Prediction Theory." *A.S.M.E. J. Basic Eng*. 83D:95–108.

Kitanidis, P. K., and R. L. Bras [1978]. "Real-Time Forecasting of River Flows." Cambridge, Mass.: Ralph M. Parsons Laboratory for Water Resources and Hydrodynamics, M.I.T. Technical Report 235.

Idem [1979]. "Error Identification in Conceptual Hydrologic Models." In: *Applications of Kalman Filter to Hydrology, Hydraulics, and Water Resources*. C. L. Chiu, ed. Pittsburgh: Department of Civil Engineering, University of Pittsburgh, pp. 325–53.

Idem [1980a]. "Adaptive Filtering Through Detection of Isolated Transient Errors in Rainfall-Runoff Models." *Water Resources Res*. 16(4):740–8.

Idem [1980b]. "Real-Time Forecasting with a Conceptual Hydrologic Model, 1: Analysis of Uncertainty." *Water Resources Res.*. 16(6):1025–33.

Lettenmaier, D. P., and S. J. Burges [1976]. "Use of State Estimation Techniques in Water Resource System Modeling." *Water Resources Bull*. 12(1):83–99.

Linsley, R. K., Jr., M. A. Kohler, and J. L. H. Paulhus [1982]. *Hydrology for Engineers*, 3rd ed. New York: McGraw-Hill.

Luenberger, D. G. [1973]. *Introduction to Linear and Non-Linear programming*. Reading, Mass.: Addison-Wesley.

Martin, D. C., and A. R. Stubberud [1976]. "Innovations Process with Applications to Identifications." In: *Advances in Theory and Applications of Control and Dynamics Systems*. C. T. Leondes, ed. New York: Academic Press, pp. 173-257.

Mehra, R. K. [1970]. "On the Identification of Variances and Adaptive Kalman Filtering." *I.E.E.E. Trans. Automatic Control*. AC-15(2):175–84.

Mein, R. G., E. M. Laurenson, and T. A. McMahon [1974]. "Simple Non-Linear Model for Flood Estimation." *A.S.C.E. J. Hydraulic*. 100(HY11):1507–18.

Monro, J. C. [1971]. "Direct Search optimization in Mathematical Modeling and a Watershed Model Application." NOAA Technical Memorandum NWS HYDRO-12. Silver Springs, Md: U.S. Department of Commerce.

Nahi, N. E., and B. M. Schaeffer [1972]. "Decision-Directed Adaptive Recursive Estimators: Divergence Prevention." *I.E.E.E. Trans. Automatic Control.* AC-17(1):61-8.

Nevelśon, M. K., and R. Z. Hasmïnskii [1972]. *Stochastic Approximation and Recursive Estimation.* Providence, R.I.: American Mathematical Society. (Translation [from Russian] of Mathematical Monographs, vol. 47.)

O'Connell, P. E., ed. [1980]. *Real Time Hydrological Forecasting and Control.* Proceedings of First International Workshop held at the Institute of Hydrology, July 4-29, 1977, Wallingford, England.

Ohap, R. F., and A. R. Stubberud [1976]. "Adaptive Minimum Variance Estimation in Discrete Time Linear Systems." In: *Advances in Theory and Applications of Control and Dynamic Systems*, vol. 12. C. T. Leondes, ed. New York: Academic Press. pp. 583-625.

Peck, E. L. [1976]. "Catchment Modeling and Initial Parameter Estimation for the National Weather Service River Forecast System." NOAA Technical Memorandum NWS HYDRO-31. Silver Springs, Md.: U.S. Department of Commerce.

Peterson, D. W. [1975]. "Hypothesis, Estimation and Validation of Dynamic Social Models." Cambridge, Mass.: Ph.D. Thesis presented to the Department of Electrical Engineering, Massachusetts Institute of Technology.

Idem [1976]. "GPSIE—General Purpose System Identifier and Evaluation." User's Manual. Cambridge, Mass.: Massachusetts Institute of Technology. (Draft provided by F. Schweppe.)

Restrepo-Posada, P. J., and R. L. Bras [1982]. "Automatic Parameter Estimation of a Large Conceptual Rainfall-Runoff Model: A Maximum Likelihood Approach." Cambridge, Mass.: Ralph M. Parsons Laboratory for Water Resources and Hydrodynamics, M.I.T. Technical Report 267.

Robins, H., and S. Monro [1951]. "A Stochastic Approximation Method." *Ann. Math. Stat.* 22:400-7.

Sage, A. P., and G. W. Husa [1969]. "Adaptive Filtering with Unknown Prior Statistics." Proceedings of the Joint Automatic Control Conference of the American Automatic Control Council, August 1969, Boulder, Colo. pp. 760-9.

Sakrison, D. J. [1966]. "Stochastic Approximation: A Recursive Method for Solving Regression Problems." In: *Advances in Communication Systems: Theory and Applications.* A. V. Balakrishnan, ed. New York: Academic Press, pp. 51-106.

Saridis, G. N., and G. Stein [1968]. "Stochastic Approximation Algorithms for Linear Discrete-Time System Identification." *I.E.E.E. Trans. Automatic Control.* AC-13(5):515-23.

Schweppe, F. C. [1973]. *Uncertain Dynamic Systems.* New York: Prentice-Hall.

Todini, E. [1979]. "Mutually Interactive State-Parameter (MISP) Estimation." In: *Applications of Kalman Filter to Hydrology, Hydraulics and Water Resources.* C. L. Chiu, ed. Pittsburgh: Department of Civil Engineering, University of Pittsburg, pp. 135-51.

Todini, E., A. Szöllösi-Nagy, and E. F. Wood [1976]. "Adaptive State/Parameter

Estimation Algorithms for Real-Time Hydrologic Forecasting: A Case Study." Presented at the Workshop on the Recent Developments in Real-Time Forecasting/Control of Water Resource Systems, October 18–20, 1976, Laxenburg, Austria.

Willsky, A. S., and H. L. Jones [1976]. "A Generalized Likelihood Ratio Approach to the Detection and Estimation of Jumps in Linear Systems." *I.E.E.E. Trans. Automatic Control.* 21(1):108–12.

Index